Mike Ellis

INDOOR AIR QUALITY ENGINEERING

Yuanhui Zhang

CRC PRESS

Boca Raton   London   New York   Washington, D.C.

Library of Congress Cataloging-in-Publication Data

Zhang, Yuanhui.
    Indoor air quality engineering / Yuanhui Zhang.
        p. cm.
    Includes bibliographical references and index.
    ISBN 1-56670-674-2
    1. Indoor air pollution. 2. Air quality—Measurement. 3. Air quality management. I.
Title.

TD883.17.Z43 2004
628.5′3—dc22   2004049667

Visit the CRC Press Web site at www.crcpress.com

No claim to original U.S. Government works
International Standard Book Number 1-56670-674-2
Library of Congress Card Number 2004049667
Printed in the United States of America 1 2 3 4 5 6 7 8 9 0
Printed on acid-free paper

Preface

Airborne pollutants are present all the time and everywhere in the form of gases, odors, and particulate matter. These pollutants have an increasing influence on our daily life. In developed countries, an average working person spends over 90% of his or her lifetime indoors. Growing concerns about indoor air quality (IAQ) and apparently increasing incidences of IAQ-related illness in people has inspired many efforts in this area from industry, academia, the general public, and government agencies.

During my 10 years of teaching about indoor air pollutant measurement and control for a senior and graduate-level engineering class, one of the difficulties that I encountered was finding a suitable textbook. Although there is a considerable amount of literature on the topic and several books cover related topics (such as aerosol science and technology, industrial ventilation, and pollutant measurement instrumentation), there has been a need for a comprehensive textbook to cover the principles and applications of indoor air quality engineering. This book is intended to fill that need. Although this book is primarily written for engineering students and professionals, its principles and approach can be useful to individuals in industrial hygiene, environmental sciences, and public health who are engaged in the measurement and abatement of indoor air quality problems.

This book is an accumulation of my lecture notes. Its aim is to explain the principles in a direct way so that a reader can study them independently. An instructor using this book as a textbook may tailor the contents according to the designated credit hours. The 15 chapters can be divided into three modules:

- Airborne pollutant properties and behavior
- Measurement and sampling efficiency
- Air-cleaning engineering

The first module includes Chapter 1 through Chapter 6 and discusses the properties and mechanics of airborne pollutants, including gases and particulate matter. Properties include physical, chemical, and biological aspects of particulate and gaseous pollutants. Behavior includes particle mechanics, gas kinetics, diffusion, and transportation of pollutants. This module provides the science and is fundamental to identifying and analyzing problems of indoor air quality.

The second module, on measurement and sampling efficiency, includes Chapter 7 through Chapter 9. This module discusses airborne pollutant measurements, including particle impaction, gravitational settling, collection of pollutants by means of diffusion, particle sampling efficiency, and particle production rate in a ventilated airspace. It provides the tools for readers to collect air quality data, analyze a problem, and get ready for engineering solutions.

The third module consists of Chapter 10 through Chapter 15, which discuss IAQ control technologies. Air-cleaning technologies covered in this module include filtration, aerodynamic air cleaning, electrostatic precipitation, wet scrubbing, adsorption, and ventilation effectiveness. This module provides typical methods and technologies to solve IAQ problems.

For those who may choose this book as a textbook, I would like to share my own experience in using this material in my class. On average, one week's course load (typically 3 lecture hours and a 1- to 2-hour laboratory exercise) corresponds to one chapter. Homework is given on a weekly basis. Depending on the availability of equipment, laboratory exercises can be based on chapters or can be arranged according to modules. Four laboratory topics and reports are required in my class:

- Airborne pollutant size distribution and particle statistics
- Concentration measurement and particle mechanics
- Sampling efficiency
- Air-cleaning technologies

In addition, the class is divided into several teams to work on different class projects. Many students have brought their own real-world projects into the class. For example, one team evaluated several types of commercial air cleaners from superstores and made suggestions for improvement.

Although many people have contributed to this book, I am particularly eager to acknowledge the following individuals. Professor Ernest M. Barber, my mentor and friend, inspired me to get into air-quality research and encouraged me to continue. I am especially thankful to Dr. Zhongchao Tan for his tireless efforts on the solutions for this book. He solved and edited all working problems while he was a graduate teaching assistant for this class at the University of Illinois. Dr. Yigang Sun contributed two sections and provided invaluable comments. I am grateful to my former graduate TAs (now professors) Xinlei Wang, Brian He, and Joshua W. McClure, and to the students in my classes over the years, for their comments, critiques, and editorials on the manuscript. There are still inevitable errors, and I am sure that this debugging process will continue after publication. I thank Megan Teague, Malia Appleford and Simon Appleford for their comments and editorial assistance. Finally, I completed this book without spending a sabbatical leave, but the writing occupied many family hours. Thus, I am indebted to my wife, Yanhui Mao, who is a mechanical engineer — enjoying her own work and understanding mine — and to our four-year old son, Matthew, who visited my office so often that he eventually referred to my office as his office and to my office building as "the coolest building in the world." It was nice that I did not really consider this project as work, because I enjoyed it so much. It was a wonderful learning experience.

And I hope that you will enjoy reading it.

Yuanhui Zhang

The Author

Dr. Yuanhui Zhang is a professor of bioenvironmental engineering in the Department of Agricultural and Biological Engineering, and a professor of bioengineering in the Department of Bioengineering, at the University of Illinois at Urbana-Champaign. He is a registered professional engineer. His teaching includes both undergraduate- and graduate-level courses in biosystems engineering, indoor air quality, and waste treatment technologies. His research interests include measurement, modeling, and control of airborne pollutants in indoor environments; characterization of room airflows; effects of indoor air quality on the health and well-being of occupants; bioenergy and biofuel conversion from organic waste; and heating, ventilation, and air conditioning for biological housing systems and environment control strategies.

Nomenclature

Symbol	Description	Dimension
A	Area	m^2
A_c	Collision area of particles	m^2
A_w	Wetted area	m^2
B	Mechanical mobility	$m \cdot N^{-1} s^{-1}$
B_E	Electrical mobility	$m^2 \cdot V^{-1} s^{-1}$
B_{Ei}	Electrical mobility of ions; for air ions, $B_{Ei} = 1.5 \times 10^{-4}$ $m^2/V \cdot s$	$m^2 \cdot V^{-1} s^{-1}$
B_f	Solidity factor	
B_h	Hydrodynamic factor for particle diffusion	
B_i	Particle interception factor	
C	Concentration	
C_c	Slip correction factor for particles	
C_{ca}	Slip correction factor for aerodynamic diameter	
C_{ce}	Slip correction factor for equivalent volume diameter	
C_D	Drag coefficient	
C_e	Equilibrium concentration on the surface of the solid phase	$mg \cdot m^{-3}$
C_{gm}	Concentration of gas in mass	$ppmm, mg \cdot kg^{-1}$
C_{gn}	Concentration of gas in molecules per unit volume	$molecule \cdot m^{-3}$
C_{gn0}	Initial concentration of gas at the diffusion interface	$molecule \cdot m^{-3}$
C_{gv}	Concentration of gas in volume	$ppmv, ml \cdot m^{-3}$
C_i	Concentration of ions	$ion \cdot m^{-3}$
C_0	Undisturbed ion concentration	$ion \cdot m^{-3}$
C_{NC}	Normalized air contaminant concentration	
C_p	Concentration of particles	
C_{pm}	Concentration of particles in mass	$mg \cdot m^{-3}$
C_{pn}	Concentration of particles in number	$particles \cdot m^{-3}$
C_{pn0}	Initial concentration of particles in number	$particles \cdot m^{-3}$
C_r	Pollutant concentration at the exit of a regeneration bed	
C_t	Total concentration of particles and droplets	$particles \cdot m^{-3}$
C_x	Solid-phase pollutant concentration	$g \cdot g^{-1}$
C_{xs}	Saturated solid-phase pollutant concentration, adsorption capacity	$g \cdot g^{-1}$
CMD	Count median diameter	m
c	Specific heat	$kJ \cdot kg^{-1} \cdot K^{-1}$
c_a	Discharge coefficient	
D	Diffusion coefficient; diameter	$m^2 \cdot s^{-1}; m$
D_e	Equivalent diameter	m
D_h	Hydraulic diameter	m
D_p	Diffusion coefficient of particles	$m^2 \cdot s^{-1}$
D_{si}	i^{th} sampling point on a duct diameter in duct sampling	m
D_s	Diameter of sampling head	m
D_{smax}	Maximum diameter of sampling head	m
D_{smin}	Minimum diameter of sampling head	m
D_t	Diameter of tube	m
D_w	Diameter of wire	m
d	Diameter of particles	m
d_{100}	Diameter at which impaction/collection efficiency is 100%	m
d_{50}	Cutsize, diameter at which collection efficiency is 50%	m

Symbol	Description	Dimension
d_c	Colliding diameter	m
d_{cm}	Count mean diameter	m
d_d	Diameter of droplet (solid or liquid)	m
d_e	Equivalent volume diameter in cm	m
d_f	Diameter of a filter fiber	m
d_g	Geometric mean diameter, count median diameter	m
d_i	Midpoint diameter of ith group	m
$d_{\bar{m}}$	Diameter of average mass	m
d_{mm}	Mass mean diameter	m
d_p	Diameter of particles	m
d_{pc}	Critical particle diameter that can be deposited on a ceiling surface	m
d_t	Diameter of tubing	m
$d_{\bar{s}}$	Diameter of average surface	m
d_{sm}	Surface mean diameter	m
d_x	A type of average diameter of particles	m
E	Energy	J
	Electrostatic field intensity, activation energy	$N \cdot C^{-1} (V \cdot m^{-1})$
E_g	Kinetic energy of gases	$N \cdot m$, $kg \cdot m^2 \cdot s^{-2}$
E_p	Kinetic energy of airborne particles	$N \cdot m$, $kg \cdot m^2 \cdot s^{-2}$
E_v	Effectiveness coefficient of mixing	
e	Charge of an electron	C
F	Force	N
F_{ah}	Adhesion force of particles	N
F_D	Drag force	N
F_I	Force attributed to inertia	N
F_i	Frequency of particles falling in i^{th} size group	N
F_τ	Force attributed to shear or friction	N
F_E	Electrostatic force	N
F_e	External force	N
F_c	Centrifugal force	N
F_{om}	Force caused by osmotic pressure	N
F_{omi}	Force caused by osmotic pressure on one (or i^{th}) particle	N
f	Fractional function	
f_i	Fraction of concentration for i^{th} size group	
f_{im}	Mass fraction of ith size group	
f_j	Fraction of concentration for j^{th} size group	
G	Gravitational constant Gas flow rate	$m^3 \cdot s \cdot kg$; $kg \cdot s^{-1}$, $mol \cdot s^{-1}$
G'	Flow rate of stagnant carrying gas	$kg \cdot s^{-1}$, $mol \cdot s^{-1}$
G_f	Factor of filtration efficiency due to gravitational settling	
g	Gravitational acceleration	$m \cdot s^{-2}$
H	Height	m
	Henry's law constant	
H_a	Annular gap of cyclones	m
H_e	Distance of exhaust air outlet (center line) to floor	m
H_f	Thickness of a filter	m
H_s	Distance of supply air inlet (center line) to floor	m
H_v	Number of velocity heads	
h	Enthalpy	$kJ \cdot kg^{-1}$
	Convective heat transfer coefficient	$kJ \cdot m^{-2} \cdot C^{-1} \cdot s^{-1}$
h_s	Enthalpy of supply air	$kJ \cdot kg^{-1}$
h_e	Enthalpy of exhaust air	$kJ \cdot kg^{-1}$

Symbol	Description	Dimension
I	Current	A
	Current of ion toward a particle	ion·s^{-1}
i	i^{th} group of particles, or ith in a group of n variables	
J	Diffusion flux of gases	molecules·m^{-2}·s^{-1}
J_E	Ion flux	ions·m^{-2}·s^{-1}
J_p	Diffusion flux of particles	particles·m^{-2}·s^{-1}
j	j^{th} group of particles	
K	Corrected coagulation coefficient for all particle sizes	
K_0	Coagulation coefficient for $d_p > 0.4$ μm	
$K_{1\text{-}2}$	Coagulation coefficient for two particle sizes	
K_c	Particle capture efficiency of droplet in kinematic coagulation	
K_n	Average coagulation coefficient for particles with n sizes	
Ku	Kuwabara hydrodynamic factor	
K_x	Mass transfer coefficient between gaseous and solid states	s^{-1}
k	Boltzmann's constant, $R/N_a = 1.38 \times 10^{-23}$ (N·m·K^{-1})	N·m·K^{-1}
	Reaction rate constant	s^{-1}
	Turbulence kinetic energy	J
k_{rp}	Particle reentrainment factor at collecting surface	
L	Length, characteristic length	m
	Liquid flow rate	kg·s^{-1}, mol·s^{-1}
L'	Flow rate of stagnant absorbing fluid	kg·s^{-1}, mol·s^{-1}
L_f	Length of flow trajectory	m
LMD	Length median diameter	m
L_w	Length of wetted perimeter	m
l	Liter	10^{-3} m^3
M	Mass per mole of gases	kg.mol^{-1}
	Molar weight	g
MMD	Mass median diameter	m
M_g	Mass of gases	kg
M_{di}	Deposited mass for i^{th} particle size range	kg
$\dot{M}_{di}$	Mass deposition rate for i^{th} particle size range	kg.s^{-1}
M_p	Total mass of particles	kg
$\dot{M}_p$	Mass production rate of particles	kg.s^{-1}
$\dot{M}_{pn}$	Net mass production rate of particles	kg.s^{-1}
M_c	Mass of contaminants	kg
M_{xs}	Adsorption capacity, in g of adsorbate per 100 g of adsorbent	
$\dot{m}_c$	Mass production rate of contaminant	kg·s^{-1}
$\dot{m}$	Mass flow rate	kg·s^{-1}
$\dot{m}_p$	Mass production rate of particle mass	kg·s^{-1}
$\dot{m}_d$	Mass deposition rate of particle mass	kg·s^{-1}
m	Mass of a particle or a molecule	kg
$\overline{m}$	Average mass of a particle population	kg·m^{-3}
m_g	Mass of a gas molecule	kg
m_i	Midsize mass of ith group particles	kg
m_p	Mass of a particle	kg
N	Total number of particles or molecules, or ions	
N_a	Avogadro's number, number of molecules per mole	
N_e	Number of complete turns in a cyclone	
N_p	Total number of particles	
N_{pi}	Total number of particles in i^{th} size range	
N_{pd}	Particle number deposited on unit surface area	particles·m^2
n	Number of particle size groups, number of moles, number of ions	
$\overline{n}$	Average number of particle charges	ion·particle^{-1}

Symbol	Description	Dimension
n_c	Number of collisions among molecules or particles	
n_f	Number of filters	
n_s	Number of collisions between molecules and a surface, number of saturation charges	
n_t	Total number of charges on a particle	
P	Pressure	Pa, N·m⁻²
P_0	Atmospheric pressure	Pa, N·m⁻²
P_A	Partial pressure of pollutant	Pa, N·m⁻²
P_c	Critical pressure	Pa, N·m⁻²
Pe	Peclet number for particle diffusion	
P_e	Pressure loss due to equipment fittings such as elbows and nozzles	Pa, N·m⁻²
P_f	Friction pressure losses	Pa, N·m⁻²
P_n	Penetration rate, penetration for denuders	
P_{om}	Osmotic pressure	Pa, N·m⁻²
P_v	Dynamic pressure losses	Pa, N·m⁻²
P_w	Wetted perimeter	m
P_t	Pressure at throat	Pa, N·m⁻²
pCi	Picocuries, 1 pCi = 0.037 disintegration per second for a radioactive material	s⁻¹
Q	Volumetric airflow rate	m³·s⁻¹
Q_{cp}	Volumetric rate of pollutant production	m³·s⁻¹
Q_e	Volumetric ventilation rate for exhaust air	m³·s⁻¹
Q_g	Flow rate of gas	m³·s⁻¹
Q_j	Flow rate of an air jet	m³·s⁻¹
Q_l	Flow rate of liquid	m³·s⁻¹
Q_{oc}	Inspiration rate of occupants	m³·s⁻¹
Q_r	Airflow rate passing an air cleaning device or a regeneration bed	m³·s⁻¹
Q_s	Sampling rate; volumetric ventilation rate for supply air	m³·s⁻¹
q	Electric charge in coulombs	C
q_c	Particle collector (e.g., cyclone) quality	
q_f	Filter quality	
q_t	Total heat transfer rate	W
R	Radius	m
	Gas constant $R = 8.31$ J·K⁻¹·mol for P in Pa and V in m³	J·K⁻¹·mol
Re	Reynolds number for fluids	
Re_p	Reynolds number for particles	
Re_{p0}	Reynolds number for particles at initial velocity	
RH	Relative humidity	
R_L	Center-to-center distance between two particles	m
r	Radius	m
r_c	Reaction rate of reactants	mole·m⁻³·s⁻¹
r_{sp}	Radius of a cyclone at which particles are separated	m
S	Stopping distance	m
	Absorption factor	
SMD	Surface median diameter	m
Stk	Stokes's number	
Stk_c	Stokes's number for particle capture efficiency	
Stk_I	Stokes's number for particle impactors	
T	Absolute temperature, temperature	K, °C
THP	Total heat production of occupants	kJ·s⁻¹
t	Time	s
t_d	A constant unit time (1 s) for particle deposition	s

Symbol	Description	Dimension
t_{min}	Minimum time required	s
t_x	Breakthrough time for an adsorption bed	s
U	Velocity of air or other carrying fluids	$m \cdot s^{-1}$
$\bar{U}$	Velocity vector of fluid	$m \cdot s^{-1}$
U_0	Initial air velocity, free flow velocity, face flow velocity	$m \cdot s^{-1}$
$\bar{U}_d$	Velocity vector of a particle	$m \cdot s^{-1}$
U_c	Critical velocity of bulk air for particle detachment	$m \cdot s^{-1}$
U_s	Sampling velocity at the face of a sampler	$m \cdot s^{-1}$
U_x	Mean air velocity of an air jet at distance x	$m \cdot s^{-1}$
u	Molecular speed	$m \cdot s^{-1}$
$\bar{u}_i$	Arithmetic mean speed of ions	$m \cdot s^{-1}$
$\bar{u}$	Arithmetic mean speed of molecules	$m \cdot s^{-1}$
$\hat{u}$	Most probable molecular speed	$m \cdot s^{-1}$
$\bar{u}_p$	Arithmetic mean thermal velocity of particles	$m \cdot s^{-1}$
$\hat{u}_p$	Most probable thermal velocity of particles	$m \cdot s^{-1}$
u_c	Colliding speed of molecules	$m \cdot s^{-1}$
u_{rms}	Root-mean-square speed of molecules	$m \cdot s^{-1}$
u_{rmsp}	Root-mean-square thermal velocity of particles	$m \cdot s^{-1}$
u_t	Gas velocity at the throat	$m \cdot s^{-1}$
V	Volume	m^3
	Velocity of particles	$m \cdot s^{-1}$
V_0	Initial velocity of particles	$m \cdot s^{-1}$
V_{az}	Velocity of adsorption zone	$m \cdot s^{-1}$
V_c	Colliding volume swept by a colliding molecule	m^3
V_d	Particle deposition velocity	$m \cdot s^{-1}$
V_{dd}	Particle deposition velocity due to diffusion	$m \cdot s^{-1}$
V_{di}	Deposition velocity for particles in i^{th} size range	
V_M	Molar volume	$m^3 \cdot mol$
V_p	Velocity of particles	$m \cdot s^{-1}$
V_r	Relative velocity of a particle to its carrying fluid	$m \cdot s^{-1}$
V_{rz}	Velocity of desorption (regeneration) zone	$m \cdot s^{-1}$
V_t	Total velocity of a fluid element	$m \cdot s^{-1}$
$\bar{V}_{TE}$	Terminal velocity of particle in an electrical field	$m \cdot s^{-1}$
V_{TE}	Average terminal velocity of particle in an electrical field	$m \cdot s^{-1}$
V_{TS}	Terminal settling velocity of particles	$m \cdot s^{-1}$
V_{TSi}	Terminal settling velocity of ith size particles	$m \cdot s^{-1}$
V_{TF}	Terminal velocity with constant external force	$m \cdot s^{-1}$
v	Molecular kinematic viscosity	$m^2 \cdot s^{-1}$
	Specific volume	$m^3 \cdot kg^{-1}$
v_e	Specific volume of exhaust air	$m^3 \cdot kg^{-1}$
v_s	Specific volume of supply air	$m^3 \cdot kg^{-1}$
$\dot{W}_p$	Moisture production rate	$kg \cdot s^{-1}$
w_s	Moisture content of the supply air	
w_e	Moisture content of the exhaust air	
X_A	Mass ratio of pollutant A in liquid phase	
x, y, z	Cardinal coordinate	m
x_A	Mass fraction (or mole fraction) of pollutant A in liquid phase	
Y_A	Mass ratio of pollutant A in gas phase	
y_A	Mass fraction (or mole fraction) of pollutant A in gas phase	
α	Angle between the sampler axis and flow direction	degree
α_j	Divergent angle of an air jet	degree
α_s	Solidity, solid volume fraction of a screen or a filter	
β	Coagulation coefficient correction factor	

Symbol	Description	Dimension
β_d	Angle between a settling velocity and the deposition surface	degree
β_s	Solubility of the gas to the scrubbing liquid	
γ	Surface tension; for water, $\gamma = 0.073$ N/m	$N \cdot m^{-1}$
	Specific heat ratio	
ΔP	Pressure drop	Pa
ΔW	Potential difference	V
δ	Thickness; thickness of a sampler wall	m
ε	Error in percentage, surface absolute roughness	
	Fluid flow dissipation rate	
ε_0	Permittivity (dielectric constant) of a vacuum	$C^2 \cdot N^{-1} \cdot m^{-2}$.
	$= 8.85 \times 10^{-12}$ $C^2 \cdot N^{-1} \cdot m^{-2}$	
ε_f	Permittivity (dielectric constant) of filter fiber	$C^2 \cdot N^{-1} \cdot m^{-2}$.
ε_p	Permittivity (dielectric constant) of particles	$C^2 \cdot N^{-1} \cdot m^{-2}$.
Γ_k	Effective diffusivity of k; k is the kinetic energy	$m^2 \cdot s^{-1}$
Γ_ω	Effective diffusivity of ω; ω is the specific dispassion rate	$m^2 \cdot s^{-1}$
κ	Concentration factor in the Freundlich equation (Equation 13.4) for adsorption	$kg \cdot m^{-3}$
κ_r	Concentration factor for regeneration	$kg \cdot m^{-3}$
ϕ	Angular variable in spherical coordinate	degree
η	Viscosity of fluid	$Pa \cdot s$
λ	Mean free path of gases	m
ξ	Sampling efficiency, collection efficiency	
ξ_c	Particle collection efficiency by kinematic coagulation	
ξ_d	Particle removal efficiency by deposition	
ξ_f	Overall filtration efficiency of a filter	
ξ_{fp}	Filtration efficiency of a filter due to interception	
ξ_{fI}	Filtration efficiency of a filter due to inertial impaction	
ξ_{fD}	Filtration efficiency of a filter due to diffusion	
ξ_{fS}	Filtration efficiency of a filter due to gravitational settling	
ξ_{fE}	Filtration efficiency of a filter due to electrostatic deposition	
ξ_k	Particle capture efficiency by coagulation	
ξ_r	Efficiency of room air cleaning resulting from an indoor air-cleaning system	
ξ_{oc}	Air-cleaning efficiency of occupants	
ξ_x	Collection efficiency for any given particle size	
θ	Angular variable in spherical coordinate	degree
ρ	Density	$kg \cdot m^{-3}$
ρ_a	Density of air	$kg \cdot m^{-3}$
ρ_0	Standard (aerodynamic) density of particles = 1000 kg/m³	$kg \cdot m^{-3}$
ρ_g	Density of gas	$kg \cdot m^{-3}$
ρ_p	Density of particles	$kg \cdot m^{-3}$
σ	Standard deviation	
σ_g	Geometric standard deviation (also GSD)	
τ	Relaxation time, time constant	s
	Average residence time of air in the room	s
τ_c	Time constant for particle concentration in a ventilated airspace	s
τ_E	Time constant for particle charging	s
τ_i	Initial age of air in the room	s
τ_r	Residual lifetime of air in the room	s
$\tilde{\tau}$	Time period that ventilation rate varies	s
$\bar{\tau}_i$	Local mean air age at an arbitrary point in the room	s
$\bar{\tau}_e$	Mean air age of the room or at the exhaust	s

Symbol	Description	Dimension
ζ	A pure number in the Freundlich equation (Equation 13.4) for adsorption	
ζ_r	A pure number in for regeneration of an adsorption bed	
ω	Specific dissipation rate	
χ	Shape factor	

List of Acronyms

Acronym	Description
ACGIH	American Conference of Governmental Industrial Hygienists
ACH	Air change per hour
APS	Aerodynamic particle sizer
ARI	Air-Conditioning and Refrigeration Institute
ASHRAE	American Society of Heating, Refrigeration and Air-conditioning Engineers
ASAE	American Society of Agricultural Engineers
ASME	American Society of Mechanical Engineers
BMRC	British Medical Research Council
BRI	Building-related illness
CAA	Clean Air Act
CAAA	Clean Air Act Amendments
CCD	Charged coupled device
CFD	Computational fluid dynamics
CHSes	Collimated honeycomb structures
CMD	Count median diameter
CNC	Condensation nucleus counter
DOE	Department of Energy (United States)
DOP	Dioctyl phthalate or bio (2-ethylhexyl) phathalate
EPA	Environmental Protection Agency (United States)
ESP	Electrostatic precipitation
FEV	Fetal exposure value
HEPA	High-efficiency particle attenuation
ISO	International Organization for Standardization
LDV	Laser Doppler velocimetry
LES	Large eddy simulation
LSV	Laser speckle velocimetry
MMD	Mass median diameter
MP	Moisture production
MPPS	Most penetrating particle size
NIOSH	National Institute of Occupational Safety and Health
NRC	National Research Council (United States)
OSHA	Occupational Safety and Health Agency (United States)
PEL	Permissible exposure limit
PIV	Particle image velocimetry
PM	Particulate matter
PSM	Particle streak mode
PTV	Particle tracking velocimetry
RNG	Renormalization group
RSM	Reynolds stress model
SBS	Sick building syndrome
SMD	Surface median diameter
SHP	Sensible heat production
SPIV	Stereoscopic particle image velocimetry
STEL	Short-term exposure limit
THP	Total heat production
TLV	Threshold limit values
TSP	Total suspended particulate matter
TWA	Time weighted average

Acronym	Description
UL	Underwriters Laboratories
ULPA	Ultralow-penetration air
VEF	Ventilation effectiveness factor
VEM	Ventilation effectiveness map
VOC	Volatile organic compound

Contents

CHAPTER 1

Air Quality and You

Air is a critical element to many living things. People, for example, may survive for two weeks without food and for two days without water. But without air, a person may only survive for two minutes. In daily life, an average person consumes approximately 1 kg of food, 2 kg of water, and 20 kg of air. Further, in developed countries, an average working person spends over 90% of his or her lifetime indoors.

1.1 BACKGROUND

In order to define air quality, a definition of *clean air* or *standard air* should first be established. Typically, clean air is the dry atmosphere air found in rural areas or over the ocean far from air pollution sources. The chemical composition of such clean, dry atmospheric air is listed in Table 1.1. The atmospheric air also contains from 0.1 to 3% water vapor by volume, depending on temperature. The clean air defined in Table 1.1 is typical because, in many instances, traces of other components are also found in the atmosphere that are considered clean. These other trace components include ammonia, sulfur dioxide, formaldehyde, carbon monoxide, iodine, sodium chloride, and particulate matter such as dust and pollen.

Based on the definition of clean air, *air quality* refers to the degree of pollution of the clean air. In general, the lower the concentration of airborne pollutants, the better the air quality. *Airborne pollutant* is defined as any substance in the air that can harm the health and comfort of humans and animals, reduce performance and production of plants, or accelerate the damage of equipment. Airborne pollutants can be in the forms of solid (e.g., particulate matter), liquid (e.g., mists), and gaseous substances. Excessively high concentrations or depletion of substances listed in Table 1.1 can impose serious air quality problems. For example, excessive emission of carbon dioxide and methane is suspected to be responsible for the greenhouse effect and global warming.

Table 1.1   Chemical Composition and Volumetric Content
of Typical Dry Atmospheric Air

Substance	Content (%[a])	Concentration (ppm[a])
Nitrogen	78.084 ± 0.004	780,840
Oxygen	20.946 ± 0.002	209,460
Argon	0.934 ± 0.001	9,340
Carbon dioxide	0.033 ± 0.001	330
Neon		18
Helium		5.2
Methane		1.2
Krypton		0.5
Hydrogen		0.5
Xenon		0.08
Nitrogen dioxide		0.02
Ozone		0.01–0.04

[a] In volume.

Source: The Handbook of Air Pollution, PHS Publication, 1968.

1.1.1   Outdoor Air Quality

Although food and water quality was a far more prevalent human concern in the past, air quality concerns can be traced back more than 700 years. In 1272, King Edward I of England banned the use of "sea coal" in an attempt to clean the smoky sky in London.[1] One of the first recorded publications trying to control smoke in the air is a pamphlet published in 1661 by the royal command of Charles II.[2] Yet outdoor air pollution became a serious concern, and many pollution episodes were recorded, only since the late 19th century. The very first air pollutant was referred to as "smog," a combination of smoke and fog, creating a dense haze and foglike condition. In 1873, 268 deaths were reported as the result of a heavy smog in London. In 1930, 60 people died and hundreds became ill during a three-day smog in Meuse Valley, Belgium, a heavily industrialized area. During a nine-day smog in January, 1931, 592 people died in the Manchester, England, area. In 1948, a 4-day smog in Donora, Pennsylvania, caused 20 people to die and nearly 7,000 residents to become ill. Many of these residents eventually died at an earlier age than the average of the townspeople. The most tragic episode of air pollution in human history is the London smog of December, 1952, during which more than 4,000 human lives were claimed and many more residents suffered illness. Public and political reactions resulted in the Clean Air Act of 1956 in the U.K., which is viewed as a turning point in urban pollution legislation. Many notable changes have been made, including how homes are heated and industries are operated, focusing on the reduction of burning soft coal. The United States passed the Clean Air Act (CAA) in 1972, which marked another milestone in air quality control. Since then, there has been no similar smog episode reported, and the atmospheric air quality has improved considerably.

Another air pollutant is acid deposition. Acid deposition is not a recent phenomenon. In the 17th century, scientists noted the ill effects that industry and acidic pollution were having on vegetation and people. However, the term *acid rain* was not coined until two centuries later, when Angus Smith published a book called *Acid*

Rain in 1872. In the 1960s, the problems associated with acid deposition became an international issue when fishermen noticed declines in fish numbers and diversity in many lakes throughout North America and Europe.

Acidic pollutants can be deposited from the atmosphere to the earth's surface in dry (sulfur dioxide, SO_2, and nitrogen oxide, NO_x) and wet forms (nitric acid, HNO_3, sulfuric acid, H_2SO_4, and ammonium, NH_4). The term *acid precipitation* is used specifically to describe wet forms of acid pollution that can be found in rain, sleet, snow, fog, and cloud vapor. An acid can be defined as any substance that, when dissolved in water, dissociates to yield corrosive hydrogen ions. The acidity of substances dissolved in water is commonly measured in terms of pH (defined as the negative logarithm of the concentration of hydrogen ions). According to this measurement scale, solutions with a pH of less than 7.0 are described as being acidic, while a pH greater than 7.0 is considered alkaline. Precipitation normally has a pH between 5.0 and 5.6, because of natural atmospheric reactions involving carbon dioxide. Precipitation is considered acidic when its pH falls below 5.6 (which is 25 times more acidic than pure water).

Acid rain has caused the death of thousands of acres of forest and water pollution in the northeastern United States and Canada. Some sites in eastern North America have precipitation with pHs as low as 2.3 — about 1000 times more acidic than is natural. As a result, the United States and Canada entered an Air Quality Agreement in 1991 to address transboundary air pollution.[3] Under this agreement, the first effort was focused on acid rain issues. The two governments have made significant reductions in emissions of the two major acid rain pollutants — sulfur dioxide and nitrogen oxide. Caps for total permanent national emission of sulfur dioxide were established.[3] Canadian commitment was to achieve a permanent national cap for SO_2 emissions of 3.2 tones by 2000. U.S. commitment is to achieve a permanent national cap for SO_2 emissions of 8.95 tones by 2010. By 1998, the total national SO_2 emissions of both countries were substantially reduced. The U.S. was 30% and Canada was 18% below the national allowance for SO_2 emissions. Nitrogen oxide, on the other hand, is more difficult than sulfur dioxide to control because the emissions involve both stationary and mobile sources. Canadian commitments included reducing NO_x stationary source emissions 100,000 tones below the forecast level of 970,000 tones by 2000, and implementing a NO_x control program for mobile sources. Because the largest NO_x contributor in Canada is the transportation sector, which accounts for 60% of all emissions, major improvements are expected by 2010, with an expected decline in NO_x emissions of 10% from 1990 levels, primarily as a result of improved standards for on-road vehicles. The United States continues to address NO_x emissions from stationary and mobile sources under the 1990 Clean Air Act Amendments (CAAA), which mandated a 2-million ton reduction in NO_x emissions by the year 2000,[4] and a further reduction of 90% in the subsequent 10 years (Tier 4). The Acid Rain Program and the motor vehicle source control program together are expected to achieve this goal, and further annual national emission reduction requirements are under development.

While the worldwide efforts of reducing acid rain have been focused on the stationary (factories) and mobile (on-road vehicles) sources, another acid rain source — animal agriculture — has been receiving attention. In the Netherlands, for example, approximately two-thirds of the acid rain has been attributed to the gas (ammonia and sulfuric gases) emissions from livestock production. The Dutch government called for a 70% reduction of 1993's ammonia emission by the year 2000[5] but had not achieved that goal by the year 2003. Taiwan has already started to phase out pork production on the island. In the United States and Canada, where land bases are large, environmental regulations on the livestock industry are becoming more and more stringent.[6]

The continued awareness of air quality and efforts to protect the natural environment lead to other concerns regarding air pollutants, including greenhouse gases (carbon dioxide and methane), ozone depletion in the stratosphere, synthetic chemical gases emitted from manufacturing and other processes, ground-level ozone, and particulate matter (PM) concentrations.

1.1.2   Indoor Air Quality

Most people are aware that outdoor air pollution can damage the environment and their health, but they may not know that indoor air pollution can also have significant effects. Until the late 1960s, attention to air quality was primarily focused on the outdoors because, by that time, outdoor air pollution was considered responsible for many adverse health effects. In the early 1970s, scientists started to investigate the cause of complaints in indoor working environments. U.S. Environmental Protection Agency (EPA) studies of human exposure to air pollutants indicated that indoor air levels of many pollutants may be two to five times — and occasionally, more than 100 times — higher than outdoor levels.[7] Over the past several decades, our exposure to indoor air pollutants is believed to have increased due to a variety of factors, including the construction of more airtight buildings, reduced ventilation rates to conserve energy, the use of synthetic building materials and furnishings, and the use of chemically formulated personal care products, pesticides, printing inks, and household cleaners. In recent years, comparative risk studies performed by the EPA and its Science Advisory Board (SAB) have consistently ranked indoor air pollution among the top five environmental risks to public health. As a result of these studies, people began to realize that the indoor air quality is important to their comfort and health. Additionally, improvement in the quality of outdoor air since the 1972 Clean Air Act enhances the relative importance of indoor air quality.

The importance of indoor air quality is also due to the sheer amount of time that people spend indoors. Traditionally, people spent more time outdoors than indoors. Today, people in industrialized countries spend more than 90% of their lifetimes indoors.[8] In the United States, for example, every day an average working person spends 22 hours and 15 minutes indoors and one hour in cars or in other modes of transportation — another type of indoor environment.[9] Thus, the total time staying indoors for an average working person is 23 hours and 15 minutes per day, or 97.7% of his or her lifetime! Other people, such as young children and seniors, may spend more time indoors than employed people. Clearly, the quality of indoor air should be as high as possible.

Increasing attention to indoor air quality has contributed to the awareness of poor health associated with a poor indoor environment. Two types of illnesses related to poor indoor air quality have been identified: sick building syndrome (SBS) and building-related illness (BRI). While the definition of SBS varies slightly in the literature, SBS can be defined as the discomfort or sickness associated with poor indoor environments with no clear identification of the source substances.[10, 11] Symptoms of SBS include irritation of sensory organs (eyes, nose, throat, ears, and skin), fatigue, headache, respiratory disorders, and nausea.

Building-related illness (BRI) is defined as a specific, recognized disease entity caused by some known agents that can be identified clinically. Symptoms of BRI include hypersensitivity pneumonities, humidifier fever, asthma, and legionella. The distinction between SBS and BRI is whether the causes of the sickness can be diagnosed clinically. Once the cause of an SBS symptom is identified, the SBS becomes a BRI. On many occasions, people do not distinguish between the two, and sick building syndrome refers to all illnesses associated with poor indoor environments, which are often referred to as *sick buildings*.

Indoor air quality and associated SBS have a profound impact on our quality of life and economy. Approximately 1 million buildings in the United States are sick buildings, within which 70 million people reside or work.[12] These sick buildings do not include agricultural buildings such as animal facilities and grain elevators. Most of these agricultural structures have unique, often more serious, air quality concerns.

The National Institute for Occupational Safety and Health (NIOSH) ranked the top 10 occupational diseases and injuries in 1982 in the following descending order of importance:

1. Occupational lung dysfunctions, including lung cancer, pneumonoconioses, and occupational asthma
2. Musculoskeletal injuries, including back injury, carpal tunnel syndrome, arthritis, and vibration white finger disease
3. Occupational cancers other than lung cancer
4. Traumatic death, amputations, fractures, and eye losses
5. Cardiovascular diseases, including myocardial infarction, stroke, and hypertension
6. Reproductive problems
7. Neurotoxic illness
8. Noise-induced hearing loss
9. Dermatologic problems, including dermatoses, burns, contusions, and lacerations
10. Psychological disorders

The top occupational disease — lung dysfunctions — is undoubtedly related to the air quality to which people are exposed.

A large number of studies have identified confinement animal facilities as particularly hazardous workplaces.[13–18] Particulate matter (dust, endotoxins, living or dead microorganisms and fungi) and gases (ammonia, hydrogen sulfide, and carbon dioxide) in animal buildings[16, 19] have been implicated as contributors to the increased incidence of respiratory disorders among livestock workers compared to grain farmers and nonfarm workers.[13, 20] Young farmers may be at particular risk of developing chronic bronchitis, coughing, wheezing, toxic organic dust syndrome, and/or occupational chronic pulmonary disease.[17]

Particulate matter (PM) in animal buildings is different from other types of building dust in at least three aspects:

First, animal building dust is biologically active in that it contains a variety of bacteria, microorganisms, and fungi.[21, 22]

Second, its concentration is high, typically more than 10 or even 100 times higher than in office and residential buildings.

Third, it is an odor carrier.[23]

Some odorous compounds as low as at one part per trillion may still have a nuisance effect on the building occupants and the neighborhood. Some bacteria carried by PM in animal buildings, such as *Listenia monocytogenes* and *Streptococcus suis,* are zoonotic agents that have caused fatal diseases in people.[24, 25] In Europe, 33% of pork producers suffer chronic respiratory symptoms related to poor indoor air quality.[26] In the Netherlands, 10% of swine producers had to change jobs due to severe respiratory problems caused by poor air quality.[27] In cold climates, such as in the midwestern United States and Canada, the problems appear to be more serious because of (1) larger building size and longer working hours and (2) lower air exchange rates due to cold climates and energy conservation concerns. A study in Saskatchewan, Canada, indicated that nearly 90% of swine producers suffer one or more acute respiratory symptoms such as coughing, wheezing, nasal and throat irritation, and chest tightness, and 50% suffer chronic lung dysfunction, allegedly due to the poor indoor air quality.[28]

Because intensive confinement in animal housing has only become common practice since the late 1960s and early 1970s, the full effect of air quality in animal buildings on human health has not yet been fully revealed. Improving the work environment for livestock producers has become an increasingly important issue for health care officials, governments, and, especially, producers and animal facility designers. Air in animal facilities must be the best possible quality; technically effective and economically affordable air quality control technologies must be developed. It has been proved that improving air quality or reducing dust inhalation resulted in improved human respiratory responses.[18, 19] Thus, intervention for contaminant control has great potential in research, product development, and management practices. Researchers are constantly striving to improve existing air quality control technologies and probing into new areas. Current design of animal facilities focuses on thermal comfort and animal performance. New generations of animal facilities must be designed to be more environmentally friendly, safer for humans and animals, and acceptable to the local communities. To meet these requirements, air quality control in livestock buildings plays a pivotal role. Significant progress has been made in reducing contaminant concentrations and improving the indoor environment.

1.2 TERMINOLOGY AND DEFINITIONS

Rather than using chemical composition to characterize air contaminants, this book defines the following terminologies commonly used in the literature to categorize airborne contaminants.

Acid deposition — Acidic pollutants deposited from the atmosphere to the earth's surface in wet and dry forms. Typical acidic pollutants include sulfur dioxide (SO_2), nitrogen oxide (NO_x), and ammonium (NH_4).

Acid rain — Wet forms of acidic pollutants found in rain, snow, fog, or cloud vapor. Most acid rain forms when nitrogen oxide (NO_x) and sulfur dioxide (SO_2) are converted to nitric acid (HNO_3) and sulfuric acid (H_2SO_4) through oxidation and dissolution, and when ammonia gas (NH_3) from natural sources is converted into ammonium (NH_4).

Aerosol — A general term for airborne particles, in either solid or liquid state, that are usually stable in a gas for at least a few seconds. Some aerosols can be suspended in the air for hours, or even years, depending on the particle size and fluid conditions. Aerosol particle size can range from 0.001 to larger than 100 μm in diameter.

Bioaerosols — A special category of aerosols of biological origin. Bioaerosols include living organisms, such as bacteria and fungi, and parts of products of organisms, such as fungal spores, pollen, and allergens from animals and insects.

Dust — Solid particles formed by mechanical disintegration of a material, such as grounding and crushing, or other decomposition processes, such as decaying.

Fume — Solid particles formed by condensation of vapors or gaseous combustion products. Fume usually consists of a cluster of primary particles, such as smoke particles. Fume particle sizes can be from submicrons to a few microns.

Smog — A combination of smoke and fog and other airborne pollutants that forms a dense haze or foglike atmosphere. The word *smog* is derived from the combination of the words *smoke* and *fog*.

Smoke — A visible aerosol usually generated from incomplete combustion. Smoke particle size is usually smaller than 1 μm.

Mist and fog — Liquid particles formed by condensation or atomization. Mist particles are spheres with a size range from submicrons to 200 μm.

Particulate matter (PM) — The scientific definition of PM is the same as for aerosol. The word *aerosol* is often used in atmospheric science, while the term *particulate matter* is more common in indoor air quality and occupational health science and applications. However, in this text PM primarily refers to solid, airborne particles, as many PM measurement methods involve drying and gravimetric processes. PM is also interchangeable with the terms *particulate contaminant, particle*, and *dust* in this text.

Volatile organic compound (VOC) — An organic compound, generally containing carbon and/or hydrogen, that evaporates easily. VOCs are found in everything from paints and coatings to underarm deodorants and cleaning fluids, from fecal synthetic processes to the decomposition of organic waste materials. VOCs have been found to be a major contributing factor to ground-level ozone, a common air pollutant and proven public health hazard.

1.3 UNITS

Throughout the text, the SI (International Systems) units meter, kilogram and second (m-kg-s) will be used as the primary dimension system. However, due to the nature of particle size, which is usually in the range of microns or submicrons, centimeter, gram, and second (c-g-s) are used as a secondary dimension system. Liters and milliliters are also used in very small volumes or low flow rates. Those secondary units are easily derived from the m-kg-s system. Readers are advised to pay special attention to the consistency of units before applying an equation. Related unit conversions and coefficients are listed in Appendix 1.

DISCUSSION TOPICS

1. What constitutes an air contaminant?
2. Discuss the definition of air quality. "Too much of a good thing can be a bad thing": List examples of this statement in the context of air quality.
3. What are the differences between smoke and fume?
4. Why does particulate matter usually refer to solid particles only, whereas aerosols may include liquid particles?
5. How do you estimate the total ammonia emission from an animal facility? Is such estimation feasible?
6. What are the key factors that make indoor air quality an increasing concern of scientific communities, governments, and the general public?
7. Why may the PM in animal facilities be a more serious concern than in other types of indoor environments?
8. Human history contains fewer episodes of indoor air quality problems than outdoor air pollution tragedies. Does this suggest that indoor air quality is a much lesser concern? Explain.
9. How is dry and wet acid deposition formed? What are the potential sources of acid rain, and how can you reduce these sources?
10. How does outdoor air pollution affect indoor air quality?
11. Do you prefer that outdoor air be filtered before entering your home, or that outdoor air be let directly into your home? Discuss the pros and cons of each option.
12. Name some major adverse effects of outdoor air pollution and of poor indoor air quality. Discuss the differences between these two environmental problems.
13. Visit Web sites to keep up with recent indoor air quality issues. Such Web sites may include the U.S. EPA, Environmental Canada, the National Institute of Occupational Safety and Health (NIOSH), and the Occupational Safety and Health Agency (OSHA).

REFERENCES

1. Stern, A.C., *Air Pollution*, 2nd ed., Academic Press, New York, NY, 1976.
2. Wark, K., Warner, C.F., and Davis, W.T., *Air Pollution — Its Origin and Control*, 3rd ed., Addison-Wesley, Menlo Park, CA, 1998.
3. U.S. EPA, *Air Quality Agreement — 1998 Progress Report*, Acid Rain Division, U.S. Environmental Protection Agency, Washington, DC, 1998.
4. U.S. EPA, *Clean Air Act Amendment*, U.S. Environmental Protection Agency, Washington, DC, 1990.
5. C.A.B. International Information Services, *General News*, Wallingford, Oxon, U.K., 13(4):139, 1992.
6. IDA, Livestock Management Facilities Act, State of Illinois Department of Agriculture, Springfield, IL, 1997.
7. U.S. EPA, *Why Is the Environment Indoors Important to Us?* Indoor Environment Division, U.S. Environmental Protection Agency, Washington, DC, 2000.
8. NRC, *Committee on Indoor Air Pollutants: Indoor Pollutants*, National Research Council, National Academy Press, Washington, DC, 1981.
9. Meyer, B., *Indoor Air Quality*, Addison-Wesley, Reading, MA, 1983.

10. Molhave, L., The sick buildings — a sub-population among the problem buildings, in *Indoor Air '87*, Seifert B., Esdorn, H., and Fischer, M., eds., Proceedings of the IV International Conference on Indoor Air Quality and Climate, Vol. 2, Berlin, Institute for Water, Soil, and Air Hygiene, 469–473, 1987.

11. Burge, S., Hedge, A., and Wilson, S., Sick building syndrome: a study of 4373 office workers, *Am. Occup. Hyg.*, 31:493–504, 1987.

12. Cone, J.E. and Hodgson, M.J., Problem buildings: building associated illness and the sick building syndrome, in *State of the Art Reviews — Occupational Medicine*, Hanley & Belfus, Philadelphia, PA., 1989.

13. Donham, J.K. et al., Environmental and health studies of workers in Swedish swine confinement buildings, *Br. J. of Ind. Med.*, 40:31–37, 1989.

14. Barber, E.M., Rhodes, C.S., and Dosman, J.A., A survey of air quality in Saskatchewan pig buildings, Can. Soc. Agr. Eng., Paper No. 91-216, CSAE, Ottawa, ON, Canada, 1991.

15. Atwood, P. et al., A study of the relationship between airborne contaminants and environmental factors in Dutch swine confinement buildings, *Am. Ind. Hygin. Assoc. J.*, 48:745–751, 1987.

16. Zejda, J.E. et al., Respiratory health of swine producers, focus on young workers, *CHEST*, 103:702–709, 1993.

17. Zejda, J.E. et al., Respiratory health status in swine producers relates to endotoxin exposure in the presence of low dust levels, *J. Occup. Med.*, 36(1):49–56, 1994.

18. Zhang, Y. et al., Acute respiratory responses of human subjects to air quality in a swine building, *J. Agr. Eng. Res.*, 70:367–373, 1998.

19. Senthilselvan A. et al., Positive human health effects of dust suppression with canola oil in swine barns, *Am. J. Respir. and Crit. Care Med.*, 156:410–417, 1997.

20. Dosman, J.A. et al., Respiratory symptoms and alterations in pulmonary function tests in swine producers in Saskatchewan: results of a survey of farmers, *J. Occup. Med.*, 30:715–720, 1988.

21. Martin W.T. et al., Bacterial and fungal flora of dust deposits in a swine building, *Br. J. Occup. Env.l Med.*, 53:484–487, 1996.

22. Wiegard, B. and Hartung, J. Bacterial contaminant of air and floor surfaces in an animal house of a cattle clinic, *Proc. of 4th Intern. Livestock Environ. Symp.*, ASAE, St. Joseph, MI, 643–649, 1993.

23. Wang, X. et al., Odor carrying characteristics of dust from swine facilities, ASAE Paper 98, ASAE, St. Joseph, MI, 1998.

24. Fachlam, R.R. and Carey, R.B., Streptococci and aerococci, in *Manual of Clinical Microbiology*, Lennette, E.H., ed., Amer. Soc. Microbiology, Washington, DC, 1985.

25. Bortolussi, R., Schlech, W.F., and Albritton, W.L., Listeria, in *Manual of Clinical Microbiology*, Lennette, E.H., ed., Amer. Soc. Microbiology, Washington, DC, 1985.

26. Van't Klooster, C.E. and Voermans, J.A.M., European perspectives — how are they solving their problems? Symposium on Meeting the Environmental Challenge, National Pork Producers Council, November 17–18, 1993, Minneapolis, MN, 1993.

28. Zhou, C. et al., Increased airways responsiveness in swine farmers, *CHEST*, 99:941–944, 1991.

27. Preller, L. et al., *Proc. of 3rd Inter. Symp: Issues in Health, Safety and Agriculture*, Centre for Agr. Medicine, University of Saskatchewan, Saskatoon, SK, Canada, 23–24, 1992.

CHAPTER **2**

Properties of Indoor Air Contaminants

This chapter discusses basic properties of airborne contaminants, including shape, density, size, concentration, source, and microbiological composition. In this book, particulate contaminants primarily refer to solid particles larger than 0.01 μm in diameter, and thus include smoke and fumes. Gaseous molecules, though very small particles, have very different characteristics and behavior from particulate matter. These characteristics will be discussed in more detail in Chapter 5. In the last section of this chapter, threshold limit values (TLVs) for typical indoor air contaminants are discussed. These TLVs will serve as the criteria for many indoor environment qualities and therefore are presented early in this book.

By completing this chapter, the reader will be able to

- Specify particle densities — including actual density, standard density, and density relative to air— and interpret particle shape
- Define particle size using three diameters: equivalent volume diameter, Stokes diameter, and aerodynamic diameter
- Classify particle sizes based on respiratory characteristics or EPA criteria
- Describe and convert pollutant concentrations in different units, such as ppmm, ppmv, mg/m^3, particle/m^3 (count/m^3, molecule/ m^3)
- Determine TLVs, including TWA, STEL, and FEV, from existing literature
- Calculate the normalized TLVs for an environment containing more than one type of pollutant

2.1 PARTICLE DENSITY

The *density* of particles refers to the mass per unit volume of the particle. Unlike the density of other commercial materials, the density of airborne particles varies widely because of its complex composition. Particles collected indoors may consist of various materials, including carpet, furniture, dander from the occupants, insects, fecal materials, smoke, cleaning agents, and outdoor pollutants. For example, particles from a bedroom may consist of a large amount of cloth fiber, hair, and dander, and the density may be low. Particles from an animal building may consist of a considerable amount of feedstuff, and the density may be close to that of the animal feed. In a welding shop, airborne particles may contain a large amount of sanding powder and metals and thus be considerably heavier than residential building particles.

When the actual density of a particulate contaminant is needed, measurements must be taken for that particulate contaminant sample. In this book, unless specified, we assume the density of particles of concern to be standard; that is, the density of particles is 1,000 kg/m³ (1 g/cm³), the same as the density of water under standard conditions. This standard density, combined with aerodynamic diameter, simplifies measurement and analysis of complicated airborne particles.

2.2 PARTICLE SHAPE

Particles in indoor environments are found in all kinds of shapes: sphere, rectangle, flake, fiber, and many other, irregular shapes. The shapes of most solid particles are nonspherical and highly irregular. Gaseous molecules, which are very small particles, are considered spheres in the context of this book. Airborne liquid particles, such as water droplets in a spray, are usually considered spheres. Figure 2.1 shows a microscopic view of a dust sample collected in an animal facility. From a typical indoor environment with occupants, this dust sample contains some fine hair (near the upper-right corner), dander flakes (lower-left corner), feedstuffs, and dead microorganisms (most are very small in Figure 2.1).

Particle shape is an important variable, affecting particle behavior and transportation, air-cleaning technology, and the effect on a respiratory system. For example, fiber-shaped particles may be particularly harmful to the lungs when they are inhaled, because they are more difficult to remove once they are settled in or have clung to

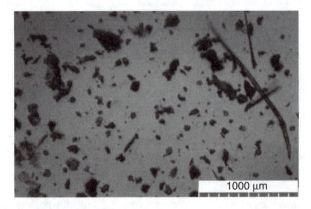

1000 μm

Figure 2.1   A dust sample collected from an animal facility. The particles are found in many different irregular shapes. Particles include feed, serum, dander, and hair from animals and human workers.

airways. On the other hand, the same nature of the fiber shape makes these particles easier to remove from an air stream by filtration. Most importantly, the shape of a particle determines its aerodynamic characteristics, the basis for developing many particle instrumentations. Consider two pieces of identical paper: One is squeezed into a paper ball and the other remains flat, and they are tossed into the air. The paper ball will drop quickly, while the flat piece may float in the air for a long period of time. Detailed particle shape and its effect on the particles will be described in Chapter 4.

2.3 PARTICLE SIZE

Particle size is the most important property of particles because it affects the behavior, transport, and control technologies of the particles. Sizes of particles can range widely. It is common for particle size to vary several 100- or several 1000-fold. The relative size of a large particle and a small particle can be illustrated in the astronomical image (Figure 2.2) of the sun and the earth. The little dot at the upper right corner represents the relative size of the earth to the sun. The relative sizes of the largest to the smallest particles in Figure 2.2 are analogous to the relative sizes of the moon to a space station, a space station to a grain of sand, a grain of sand to a cigarette smoke particle, and a cigarette smoke particle to a gas molecule. The range of particle sizes in Figure 2.2 can be found in the same type of particles. For example, dust in an animal production building may contain particles as large as 100 microns and as small as 0.01 micron. A dust fall following a volcano eruption contains large particles (in the range of millimeters) that can settle down in a few hours, and small particles (in the range of submicrons) that can stay airborne for years.

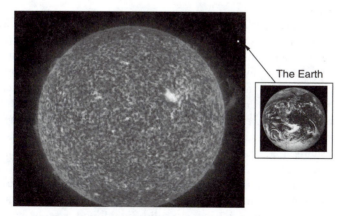

The Earth

Figure 2.2 Commonly, the sizes of concerned particles vary several 100- or 1000-fold. Relative sizes of a large particle to a small particle are illustrated in this astronomical image of the sun and the earth. (Courtesy of SOHO/EIT consortium, a project of international cooperation between ESA and NASA.)

Table 2.1 Dimensions of Length, Symbols, and Typical Particles at Respective Size Ranges

Terminology	Symbol	Dimension (m)	Typical Particles in the Size Range
Kilometer	km	10^3	A small moon
Meter	m	1	A concrete road block
Decimeter	dm	10^{-1}	A soft ball
Centimeter	cm	10^{-2}	A mint candy
Millimeter	mm	10^{-3}	A grain of sand
Micrometer	μm	10^{-6}	A cigarette smoke particle
Nanometer	nm	10^{-9}	A large gas molecule
Angstrom	Å	10^{-10}	A small gas molecule

An appreciation of particle size ranges is fundamental to understanding particle properties. Particle size not only changes the particle properties but also the laws governing these properties, such as interaction and transportation. For example, gravitational force is the predominant bonding force among the very large particles, such as the sun and the earth. Drag force is the predominant force defining the motion of airborne particles ranging from submicrons to millimeters. Diffusion force, on the other hand, determines the mass transfer of gaseous molecules and the behavior of very small particles (typically smaller than 0.01 μm).

Airborne particles are one of the major concerns in indoor environments. Airborne particles are defined as particles that can be suspended in the air for an extended period of time under normal conditions. Thus, the particle size of concern for indoor air quality is no larger than a fine grain of sand (in a range of 100 μm in diameter), because such large particles will quickly settle down to ground surfaces. The smallest particles of concern are usually larger than 0.01 μm. Particles smaller than 0.01 μm can be treated as gas. Gas molecules, which are considered spherical particles ranging in size from 1 to 10 Å, will not be treated in the context of particulate matter.

To determine the size of a particle, it is important to have a yardstick for measurement. A commonly used yardstick is micrometers, or microns (μm). In this book, particle size refers to a particle's diameter d in microns. When particle type must be distinguished to avoid confusion, subscripts are used to refer to specific particles. For example, d_p is used for solid particles, d_l for liquid particles, and d_g for gaseous molecule particles. Table 2.1 lists the dimensions commonly used in daily life and in particle sciences. Examples of typical particles within the size range gives a perspective on the different dimensions.

In bright sunlight, a naked human eye can visualize an airborne particle approximately 50 μm in diameter, which is about the diameter of a human hair. Some typical particles and their size ranges are shown in Figure 2.3. Particles in Figure 2.3 range from 1 Å to 3 cm, or from about the size of a gas molecule to that of a golf ball. Particles in Figure 2.3 include solid, liquid, and gaseous particles. Very small particles, such as gaseous molecules, have very different properties and behave differently from particles larger than 1 μm. This fact is the primary reason for discussing gaseous contaminants and particulate contaminants separately.

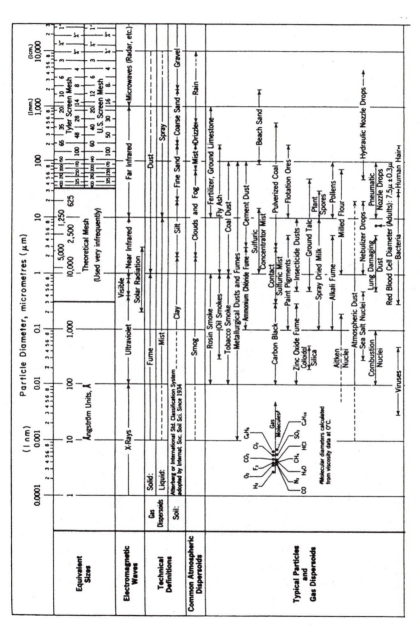

Figure 2.3 Typical particles and their size ranges in indoor environments. *Source:* ASHRAE Fundamentals, 1997. (Courtesy of Stanford Research Institute.) *Continued.*

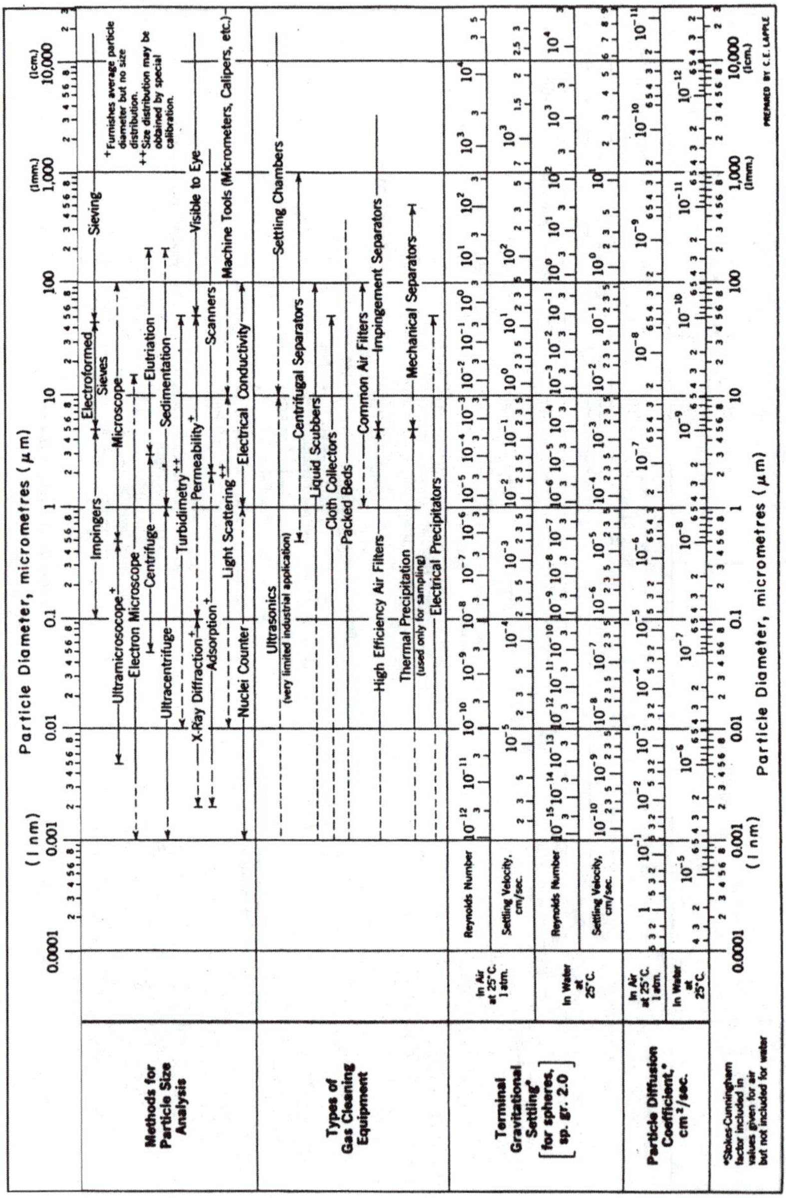

Particle Diameter, micrometres (μm)

(1 nm) 0.001   0.01   0.1   1   10   100   1,000 (1mm)   10,000 (1cm)

Methods for Particle Size Analysis

Sieving
+Furnishes average particle diameter but no size distribution.
++Size distribution may be obtained by special calibration.
Electroformed Sieves
Microscope
Elutriation
Centrifuge
Sedimentation
Ultramicroscope+
Electron Microscope
Ultracentrifuge
Turbidimetry++
Scanners
Visible to Eye
Machine Tools (Micrometers, Calipers, etc.)
X-Ray Diffraction+
Adsorption+
Permeability++
Nuclei Counter
Light Scattering++
Electrical Conductivity
Impingers

Types of Gas Cleaning Equipment

Settling Chambers
Centrifugal Separators
Liquid Scrubbers
Cloth Collectors
Packed Beds
Common Air Filters
Impingement Separators
Mechanical Separators
Ultrasonics (very limited industrial application)
High Efficiency Air Filters
Thermal Precipitation (used only for sampling)
Electrical Precipitators

Terminal Gravitational Settling* [for spheres, sp. gr. 2.0]

Reynolds Number (In Air at 25°C, 1 atm.)
Settling Velocity, cm/sec. (In Air at 25°C, 1 atm.)
Reynolds Number (In Water at 25°C)
Settling Velocity, cm/sec. (In Water at 25°C)

Particle Diffusion Coefficient,* cm²/sec.

In Air at 25°C, 1 atm.
In Water at 25°C

*Stokes-Cunningham factor included in values given for air but not included for water.

Particle Diameter, micrometres (μm)

PREPARED BY C.E. LAPPLE

Figure 2.3 Continued.

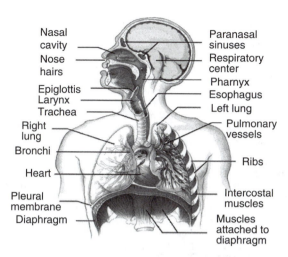

Nasal cavity
Nose hairs
Epiglottis
Larynx
Trachea
Right lung
Bronchi
Heart
Pleural membrane
Diaphragm

Paranasal sinuses
Respiratory center
Pharnyx
Esophagus
Left lung
Pulmonary vessels
Ribs
Intercostal muscles
Muscles attached to diaphragm

Figure 2.4   An overview of the human respiratory system. (From *The Atlas of the Human Body*, American Medical Association Web site http://www.ama-assn.org, 1999. With permission.)

2.3.1   Human Respiratory System

One common characterization of particle size derives from occupational health and safety, in which the primary concern is particle size and deposition in a human respiratory system. Humans have developed effective defense mechanisms to prevent airborne particles from entering the respiratory system, especially some sensitive regions. Figure 2.4 gives an overview of a human respiratory system.

The human respiratory system can be divided into three major regions:

- Head airways region
- Tracheobronchial region
- Pulmonary or alveolar region

The head airways region includes the nasal cavity, pharynx, epiglottis, and larynx. This region is also called the *extrathoracic* or *nasopharyngeal* region. Inhaled air is pretreated (heating and humidification) and large particles are intercepted in this region. The tracheobronchial region, also called *thoracic* or *lung airways* region, includes the airways from trachea to bronchi at both sides of the lung. The pulmonary region includes the pulmonary vessels or alveolar ducts, shown as the inverted tree area in Figure 2.4. This is the working area where gas exchanges take place. For an adult person, the total area of gas exchange is approximately 75 m², and the total length of the pulmonary vessels is about 2,000 km. An adult person breathes about 12 to 20 times per minute and inhales about 10 to 20 liters of air per minute at a moderate workload.

2.3.2   Particle Size Categories

According to the behavior of particles in the human respiratory system, particle size is divided into three major categories: inhalable particles, thoracic particles, and

Table 2.2   Commonly Used Terms and Size Ranges of Particles

Particles	Size Range
Total[a]	All sizes of particles of concern in the air
Inhalable (inspirable)[a]	≤ 100 μm
Thoracic[b]	≤ 10 μm
Respirable[b]	≤ 4 μm
Diminutive[c]	≤ 0.5 μm
PM10 (coarse particles)[d]	≤ 10 μm
PM2.5 (fine particles) [d]	≤ 2.5 μm

[a] *Source:* Vincent, J. H., *Aerosol Sampling — Science and Practice*, John Wiley & Sons, New York, NY, 1989.
[b] *Source:* ACGIH, *Threshold Limit Values for Chemical Substances and Physical Agents and Biological Exposure Indices*, American Conference of Governmental Industrial Hygienists, Cincinnati, OH, 2000.
[c] *Source:* Zhang et al., *Am. Soc. Heat. Refrig. Air Cond. Eng.* 100(1): 906–912, 1994.
[d] *Source:* EPA, *Terms of Environment*, U.S. Environmental Protection Agency, Washington, DC, 1998.

respirable particles.[1] These three particle size categories correspond to the three regions of the human respiratory system where the particles can deposit: head airways region, tracheobronchial region, and pulmonary or alveolar region. A particle category can more simply (but less accurately) be defined as the particle sizes at which the sampling efficiency is 50% in mass with the respective sampler. The particle size at 50% sampling efficiency is also called *particle cutsize* or *particle cutpoint*. Obviously, such a definition and the median point vary with the type of sampler and its sampling (also called *aspiration*) efficiency. Sampling efficiencies will be discussed in more detail in later chapters. Other particle categories include total particle,[2] diminutive particle,[3] PM2.5, and PM10,[4] which have different definitions from the sampling efficiency-based definition. The terminologies and size ranges of particles are summarized in Table 2.2.

More accurate definitions and characteristics of particle size categories are given in the following paragraphs and illustrated in Figure 2.5. In a practical respiratory system, the particle mass fraction collected in a region is a gradual curve, rather than a sharp cut, as given in Table 2.2. The cutsize only indicates the particle size at which the mass collection efficiency reaches 50%. It is obvious that, in terms of particle mass concentrations, the total particle is greater than the inhalable, the inhalable is greater than the thoracic, the thoracic is greater than the respirable, and the respirable is greater than the diminutive.

The term *inhalable particle* refers to the particles that can be inhaled by an air sampler during a measurement process. The air sampler could be a sampling instrument or a human respiratory tract. One of the widely used cutsizes for inhalable particle mass is defined as 100 μm.[1] The inhalable particle size may be different from the actual particle size in the air sample, depending on the efficiency of inhalation (or aspiration) of the sampler. Although the definition of inhalable particles includes all sizes, it is based on the study of particles smaller than 100 μm.[5] One reason for this definition is that particles larger than 100 μm will settle down quickly under normal conditions. Some early studies defined inhalable particles as those

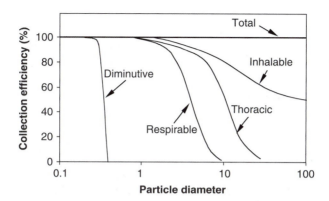

Figure 2.5   Particle categories (in mass fraction) defined by their collection efficiencies. Note that inhalable, thoracic, and respirable particles refer to the particles at which the collection efficiencies are 50%.

particles smaller than 30 μm,[6] because there are rarely particles larger than 30 μm in indoor environments. As a practical matter, larger particles are more difficult to inhale into a sampler. Therefore, some particle measurement instruments have particle upper size limits smaller than 30 μm. An alternative term for inhalable particles is *inspirable particles*.[7] The International Organization for Standardization (ISO) defines inspirability as the orientation-averaged aspiration efficiency for a human head. This definition was largely based on work done by Ogden and Birkett[6] for particles smaller than 30 μm. Sampling/aspiration efficiency for different types of samplers is discussed in later chapters.

Thoracic particles are defined as particles that can be deposited in the lung airways and gas exchange regions of a human respiratory system. The cutsize of thoracic particle mass is 10 μm.[1] This definition is the same as that of the ISO.[7]

The term *respirable particles* refers to the particles that can travel to and deposit in the gas exchange region of a human respiratory system. There are differences among the size ranges of respirable particles as defined by different agencies or scientific societies. For example, the American Conference of Governmental and Industrial Hygienists (ACGIH) defines respirable particles as particles smaller than 4 μm.[1] This definition is in accord with the ISO[7] and the European Standardization Committee.[8] The British Medical Research Council (BMRC) considers respirable particles to be smaller than 7 μm.[2] In occupational health studies regarding working conditions in biological structures such as animal buildings, the respirable particle is suggested to be a particle smaller than 5 μm.[9] In this book, the term *respirable particle* refers to the particles smaller than 4 μm.

The concept of total particles was introduced in the early days of environmental and occupational health studies. The total particle refers all airborne particles in an air sample. If an air sampler has a 100% sampling efficiency for all sizes of particles, particles sampled by the sampler would be the *true* total particle in the air. In practice, such an air sampler with a particle sampling efficiency of 100% for all particle sizes at all sampling conditions does not exist. Thus, the total particle was effectively defined entirely by the particular instrumentation under given sampling conditions.

By carefully designing a sampler and choosing sampling conditions, sampling efficiency close to 100% can be achieved; thus, the total particle can be measured, at
least for the particles with aerodynamic diameters smaller than 30 μm.

In other practical scenarios, such as evaluation of particle control technologies,
some extremely small particles must be excluded, especially when the particle
concentration is measured in numbers rather than mass. *Diminutive particles* were
introduced for such a purpose. Diminutive particles are defined as particles smaller
than 0.5 μm.[3] The reason for introducing the diminutive particles into the study was
to isolate the effect of background particles contained in the supply air from atmosphere on indoor air particle number concentrations. In many indoor air quality
problems, the supply air to the indoor environment is not filtered. The supply air
(often the atmospheric air outdoors) contains a large number of small particles,
typically smaller than 0.5 μm. When particle count concentration is used instead of
particle mass concentration, very small particles may have a predominant majority
of particle count; even the mass of these large numbers of small particles is very
small compared to the total mass of particles. Diminutive particles are useful in
identifying or analyzing the particles generated within a concerned airspace. Some
particle statistics exclude particles smaller than 0.3 μm in diameter, because of the
optical cut-off phenomenon.[10] Particles smaller than 0.3 μm in diameter cannot be
measured using the light scattering principle.

Concerning emissions and outdoor air pollution, the particle size is often characterized using the terms PM2.5 and PM10. PM2.5 refers to particulate matter
smaller than 2.5 μm in diameter and is also called *fine particles*. PM10 refers to
particulate matter nominally smaller than 10 μm and also is called *coarse particles*.[4]

2.3.3 Particle Diameters

The highly irregular shape of particles, together with variations in density, calls
for a term to describe the size of a particle. A particle's size refers to its *diameter*.
However, a diameter can have different meanings. Understanding the meaning of
different types of diameters is important to the study of particle properties and
behavior. The following diameters are relevant to airborne particle behavior and are
frequently encountered in the literature:

Equivalent volume diameter (d_e) of an irregularly shaped particle is defined as the
   diameter of the sphere that would have the same volume and density as the particle.
   As an example of determining the equivalent diameter, consider a piece of irregularly shaped glass. If the glass piece were melted and filled up a sphere (the sphere
   must be full) with a diameter d, then the d is the equivalent diameter (d_e) of that
   piece of glass.
Stokes diameter (d_s) is defined as the diameter of the sphere that would have the same
   density and settling velocity as the particle. To determine the Stokes diameter,
   consider the same piece of irregularly shaped glass as an example. If a ball made
   of the same glass had the same settling velocity as that of the irregularly shaped
   piece, the diameter of the glass ball would be the Stokes diameter of the irregularly
   shaped glass piece. Settling velocity can be determined by releasing the glass ball
   at a sufficient height H and recording the traveling time t. The settling velocity

will be H/t. This method is usually sufficiently accurate for particles smaller than 200 μm.

Aerodynamic diameter (d_a) of an irregularly shaped particle is defined as the diameter of the sphere with a standard density (1,000 kg/m³) that would have the same settling velocity as the particle. If we use the irregularly shaped piece of glass as an example, the aerodynamic diameter can be determined by comparing the settling velocities of the glass piece with a water droplet (an approximate sphere with a standard density). If the water droplet had the same settling velocity as the glass piece, the diameter of the water droplet would be the aerodynamic diameter of the glass piece.

The equivalent volume diameter only standardizes the shape of the particle by its equivalent spherical volume. The Stokes diameter standardizes the settling velocity of the particle but not the density. The aerodynamic diameter standardizes both the settling velocity and the density of the particles; thus the aerodynamic diameter is a convenient variable to use when analyzing particle behavior and design of particle sampling or control equipment. Aerodynamic diameter is a critical property of particles. More analysis of aerodynamic diameter and its applications is found in later chapters. In this book, particle size refers to its aerodynamic diameter unless otherwise specified.

Other statistically averaged diameters, such as median diameter and mean diameters for mass, volume, surface, length, and count, will be discussed in Chapter 3.

2.4 CONCENTRATION OF GASES AND PARTICLES

A concentration of an airborne contaminant is a quantitative measurement of pollution. The concentration can be expressed in terms of mass, volume, or particle number of the concerned contaminant per unit volume of air.

Particle mass concentration C_{pm} is the mass of the particles per unit volume of air in which the particles are airborne, usually in mg/m³ (milligram of particles per cubic meter of air). Particle number concentration C_{pn} is the number of particles per unit volume of air in which the particles are airborne, in particles/m³ or count/m³. In practice, often the C_{pn} is in particles/cm³ (number of particles per milliliter) for convenience, because the C_{pn} usually exceeds 10^4 in particles/m³. Particle mass and number concentration can be converted back and forth if the density and size of the particles or the size distribution of the particles is known.

Assume that the particles in an air sample are composed of n sizes, each with an equivalent volume diameter d_{ei}, a density ρ_{pi}, and a fractional mass concentration, C_{pmi}. The mass of a particle at i^{th} size, m_{pi}, is

$$m_{pi} = \frac{\pi d_{ei}^3 \rho_{pi}}{6} \tag{2.1}$$

The particle number concentration at i^{th} size can be calculated as

$$C_{pni} = \frac{C_{pmi}}{m_{pi}} = \frac{6\,C_{pmi}}{\pi\,d_{ei}^{3}\,\rho_{pi}} \qquad (2.2)$$

The total particle number concentration C_{pn} can be obtained by adding particle number concentrations over the n size ranges:

$$C_{pn} = \sum_{i=1}^{n} \frac{C_{pmi}}{m_{pi}} = \frac{6}{\pi} \sum_{i=1}^{n} \frac{C_{pmi}}{d_{ei}^{3}\,\rho_{pi}} \qquad (2.3)$$

where C_{pn} is in particles/m^3, C_{pm} is in kg/m^3, d_e is the equivalent volume diameter in m, and ρ_p is the density of the particles in kg/m^3. Care should be exercised in applying consistent units in Equation 2.1 through Equation 2.3. Often, conversions are needed before values can be substituted into the equations.

Assuming that the sampling rate is Q_s in m^3/s and the sampling time is t in seconds, the total mass m_s or total number of particles N_s for a given size of particles can be calculated as follows:

$$m_s = Q_s\,C_{pm}\,t \qquad (2.4)$$

$$N_s = Q_s\,C_{pn}\,t \qquad (2.5)$$

Example 2.1: *An employee works in a building for eight hours per day. Total dust particle concentration in that building is measured at 1.2 mg/m^3, of which approximately 10% consists of thoracic particles (< 10 μm). An average person's breathing rate is approximately 20 L/min. Assume the thoracic particles have an average equivalent volume diameter of 1.5 μm, other particles have an average equivalent volume diameter of 12 μm; the density for all particles is 2,000 kg/m^3. Assume only thoracic particles can be inhaled. Find the following:*

    a. The total particle number concentration of the building air (in particles/ml)
    b. How many milligrams of particles the worker inhales during a 2-hour work period
    c. How many particles the worker inhales during the 2-hour work period

Solution:

    a. Because the particles are divided into two size ranges with average equivalent volume diameters of 1.5 μm and 12 μm

$$C_{pm1} = 1.2 \text{ mg/m}^3 \times 10\% = 0.12 \text{ mg/m}^3 = 1.2 \times 10^{-7} \text{ kg/m}^3$$

$$C_{pm2} = 1.2 \text{ mg/m}^3 \times 90\% = 1.08 \text{ mg/m}^3 = 1.08 \times 10^{-6} \text{ kg/m}^3$$

Substituting the preceding values into Equation 2.3 gives the total particle number concentration:

$$C_{pn} = \frac{6}{\pi} \sum_{i=1}^{n} \frac{C_{pmi}}{d_{ei}^3 \rho_{pi}}$$

$$= \frac{6}{\pi} \left(\frac{1.2 \times 10^{-7}\,(kg/m^3)}{(1.5 \times 10^{-6}\,m)^3 \times 2000(kg/m^3)} + \frac{1.08 \times 10^{-6}\,(kg/m^3)}{(12 \times 10^{-6}\,m)^3 \times 2000(kg/m^3)} \right)$$

$$= 3.51 \times 10^7 \ (particle/m^3)$$

b. Because $Q_s = 20$ l/min $= 1.2 \times$ m³/h and $t = 2$ h, using Equation 2.4 gives

$$m_s = Q_s\,C_{pm1}\,t = 1.2 \left(\frac{m^3}{h} \right) \times 1.2 \times 10^{-7} \left(\frac{kg}{m^3} \right) \times 2\,(h)$$

$$= 2.88 \times 10^{-7}\,(kg) = 0.288\,(mg)$$

$$= 2.88 \times 10^{-4}\,(g) = 0.288\,(mg)$$

c. Using the mass concentration, $C_{pml} = 0.12$ mg/cm³, and the average diameter 1.5 μm for thoracic particle, yields

$$N_s = Q_s\,C_{pm1}\,t = 1.2 \left(\frac{m^3}{h} \right) \times \frac{6 \times 1.2 \times 10^{-7}\,(kg/m^3)}{\pi(1.5 \times 10^{-6}\,m)^3 \times 2000\,(kg/m^3)} \times 2\,(h)$$

$$= 8.15 \times 10^7 \ (particles)$$

Concentration of gaseous contaminant C_g is usually expressed in terms of part of gas of concern per million parts of air (ppm):

$$1\,ppm = 10^{-6}\,\frac{amount\ of\ gas\ of\ concern}{amount\ of\ carring\ fluid} \tag{2.6}$$

The ppm could be in either mass concentration C_{gm} or volume concentration C_{gv}. To distinguish the difference between mass and volume, ppmm refers to the gas concentration in mass, and ppmv refers to the gas concentration in volume. Unless specified, ppm is in volumetric concentration in this book. When a gas is expressed in mass concentration, the value of ppmm is in mg/kg (milligram of gas of concern per kilogram of air). For example, 300 ppmm is equal to 300 mg of gas per kg of air, or 0.0003 kg of gas per kg of air. When a gas is expressed in volume concentration, the value of the ppm is in ml/m^3 (milliliter of gas of concern per cubic meter of air). For example, 100 ppm is equal to 100 ml of gas of concern per m^3 of air, or 0.0001 m^3 of gas per m^3 of air. The mass and volume concentration of a gas can be converted back and forth if the densities of the gas and the air are known at the given conditions (temperature and pressure). Unless specified, the temperature and atmospheric pressure in this book are assumed to be at standard conditions, 20°C and 101.315 kPa, respectively.

$$C_{gv} = C_{gm} \frac{\rho_a}{\rho_g} \tag{2.7}$$

where C_{gv} is in ppmv, C_{gm} is in ppmm, ρ_a and ρ_g are densities of air and gas, in kg/m^3, respectively.

In some instances, the concentration of a gaseous contaminant is expressed in mg of the contaminant per cubic meter of air: mg/m^3. To convert the concentration of gases or vapor between mg/m^3 and ppmv, the following equation can be used:

$$C\left(in \ \frac{mg}{m^3 \ air}\right) = C_{gv} \ (in \ ppmv) \times \rho_g \tag{2.8}$$

When the molar weight of the gas is known, an alternative conversion can be expressed as

$$C\left(in \ \frac{mg}{m^3 \ air}\right) = \frac{C_{gv} \ (in \ ppmv) \times M}{24.5} \tag{2.9}$$

where M is the molar weight of the gas, and 24.5 is the molar volume of air, in liter per mole, under standard atmospheric conditions (20°C and 101.325 kP). It should be noted that Equation 2.7 and Equation 2.8 are the same except for the units.

Example 2.2: *The carbon dioxide concentration in the ground-level atmosphere is 350 ppmv. The densities of air and carbon dioxide are 1.2 and 1.82 kg/m³, respectively. What is the carbon dioxide concentration in ppmm and in mg/m³ of air?*

Solution: Using Equation 2.7 and Equation 2.8

$$C\left(in \ \frac{mg}{m^3 \ air}\right) = C_{gv} \ (in \ ppmv) \times 1.82 = 350 \times 1.82 = 637 \ (mg \ / \ m^3)$$

Table 2.3   Particulate Contaminant Sources in Typical Residential (Including Office and Commercial) and Animal Buildings

Residential Buildings		Animal Buildings	
Contaminant	Source	Contaminant	Source
Asbestos	Insulation, siding shingles, roofing materials, textured paints, adhesives, paper, millboard	Feed particles	Grain dust, antibiotics, growth promotants
Dust and molds	Foodstuff, furniture, household products, office machinery, pollen	Fecal material	Dry feces, dead organisms, endotoxins
Dust mites	Carpet, foodstuff, dander, soil	Animal protein	Urine, dander, serum, hair or feathers, mold, pollen, grain mites, insects
Allergens	Pollen, molds, dust mite feces, organisms, dander and animal hairs, endotoxins, household chemicals	Smoke and fumes	Heating systems, sterilizing fumigation, tattooing
Smoke and fumes	Combustion, cooking, smoking, evaporation	Gases and odor	Fecal material, urine, animal bodies
VOC	Formaldehyde, fragrance products, pesticides, solvents, cleaning agents	Other agents	Mineral ash, gram-negative bacteria, microbial proteases, infectious agents
Radon	Soils, well water, building materials		

2.5 AIR CONTAMINANTS AND SOURCES

Identification of sources of particulate contaminants in an indoor environment is important for implementation of appropriate air quality control strategies. Due to the differences in functions and structures of buildings, it is useful but very difficult to have a complete list of sources of particulate contaminants. Most of the airborne contaminants in residential buildings are also present in animal buildings. Additionally, more contaminants, usually in much higher concentrations, are found in animal buildings. Common indoor airborne contaminants in residential and animal buildings and their sources are listed in Table 2.3. When considering typical airborne contaminants with sources found indoors, some deserve more detailed descriptions. The composition of typical contaminants is listed in Table 2.4.

2.5.1   Asbestos

Asbestos is a generic term for a group of naturally occurring mineral silicates in various forms that have been used extensively for building materials including insulation, siding shingles, roofing materials, textured paints, and adhesives. Asbestos, like many minerals, may be present in fibrous (asbestiform) or crystalline (nonasbestiform) forms. There are two basic types of asbestiform asbestos:

- Serpentine, of which chrysotile $\{Mg_3[Si_2O_5](OH)_4\}$ is the sole member
- The amphiboles, of which there are several types, including anthophyllite $(Mg,Fe^{2+})_7[Si_8O_{22}](OH,F)_2\}$, amosite $\{(Fe^{2+})_4(Fe^{2+}, Mg)_3[Si_8O_{22}](OH)_2\}$, and crocidolite $\{Na_2Fe_3^{2+}Fe_2^{3+}[Si_8O_{22}](OH,F)_2\}$

Asbestos fibers are characterized by their large aspect (or length-to-width) ratio and small diameter. Breathing of high levels of asbestos fibers can lead to an increased risk of lung cancer and asbestosis, in which the lungs become scarred with fibrous tissue. The symptoms of these diseases do not usually appear until about 20 to 30 years after the first exposure to asbestos.

There are thousands of commercial products that use asbestos because of its high tensile strength, flexibility, and resistance to temperature, acids, and alkali. Asbestos fibers have been used widely as a building material for fire prevention, thermal and acoustical insulation, and roofing; as a friction material in brake pads; and as a reinforcing material in cement. Because of the widespread use of asbestos, its fibers are ubiquitous in indoor environments. Although there has been declining production due to bans and voluntary phase-outs, it is still common to find asbestos products in older buildings, where they were often the materials of choice for insulation, tiles, and other building products prior to 1975. In a survey of public buildings, the EPA[11] found that friable asbestos-containing materials (ACMs) were present in 16% of the buildings surveyed (511,000 of 3.2 million).

The permissible exposure limit (PEL) for asbestos is 0.1 fiber/cm^3 for occupational environments, regardless of the condition of the asbestos. There have been many determinations of asbestos fiber levels in outdoor and indoor air in the U.S. and Europe. The National Research Council[12] summarizes that the range of fibers in urban air is between 0.0002 and 0.00075 fibers per cubic centimeter of air. Indoor air in public buildings has an average concentration of 0.0000770 fiber/cm^3 in buildings with intact asbestos levels, which is at least 1000 times below the PEL.

Although asbestos is present in almost every building, it is usually not a serious problem, because the mere presence of asbestos in a building is not considered air pollution or hazardous. The danger is that asbestos materials may become damaged over time. The detached asbestos is then airborne and becomes a health hazard. Burdette et al.[13] have shown that asbestos removal can lead to the increase of indoor airborne asbestos fiber levels (< 0.0002 to 0.004 fiber/cm^3 18 weeks after removal) that can remain for a year or longer. Thus, caution should be exercised in removing asbestos: Improper removal can be more hazardous than leaving it in place. The EPA[14] recommends the removal of badly damaged friable asbestos, leaving intact, well-maintained asbestos in place.

2.5.2   Formaldehyde

Formaldehyde, a colorless, pungent-smelling gas, is a major allergy agent in indoor environments. It can cause watery eyes, burning sensations in the eyes and throat, nausea, and difficulty breathing in some humans who are exposed at elevated levels (above 0.1 ppm). High concentrations may trigger attacks in people with asthma. There is evidence that some people can develop a sensitivity to

Table 2.4   Chemical and Biological Composition of Typical Airborne Contaminants in Indoor Environments

Residential Buildings		Animal Buildings[d]
Toxic substances:[a]	1,2,4-trimethyl benzene	*Gram positive cocci:*
Benzene	1,2,3-trimethyl benzene	Staphylococcus species
Dichloromethane	Vinyl toluene	(coagulase negative)
Tetrachloroethylene	Propyl toluene	Staphylococcus
Trichloroethylene	Σ ethyl dimethyl benzenes	haemolyticus
1,2-dichloraethane	1,1-dimethyl butyl benzene	Staphylococcus hominis
	1,1-diethylpropyl benzene	Staphylococcus simulans
Molds and fungi:[b]	1-propylheptyl benzene	Staphylococcus sciuri
Alternaria	1-ethyloctyl benzene	Staphylococcus warneri
Cladosporium	1,1,4,6,6-pentamethyl heptyl	Micrococcus species
Aureobasidium pullulans	benzene	Aerococcus species
Aspergilus	1,1-dimethyldecyl benzene	Streptococcus equinus
Ulocladium	2-methyl 2-phenyl tridecane	Streptococcus suis
Epiciccum	1,1-dimethylnonyl benzene	(presumptive)
Geotrichum	Chloride hydrocarbons	Enterococcus durans
Aspergillus	1,1,2-trichloro	
Versicolor	1,2,2-triflouro ethane	*Gram positive bacilli:*
Penicillium species	Ketones	Corynebacterium species
Paecilomuces species	2-butanone	Corynebacterium xerosis
Helminthosporium	Cyclohexanone	Bacillus species
Mucor	1-phenyl ethanone	
Rhizopus	5-Methyl 5-phenyl 2-	*Gram negative bacilli:*
Verucosum	hexanone	Acinetobacter calcoaceticus
Fusarium poae	Alcohols	Nonfermentative Gram
Sporotrichioides	1-methoxy 2-propanol	negative bacillus
Yeasts	1-ethoxy 2-ethanol	Enterobacter agglomerans
Algae	1-ethoxy 2-propanol	Pasteurella species
	2-ethyl hexanol	Vibrio species
VOCs[c]	2-(butoxyethoxy) ethanol	
Alkane	Phenol/phenolic derivatives	*Fungi:*
Octane	2-i-propyl phenol	Alternaria species
Tridecane	3-i-propyl phenol	Cladosporium species
Trimethyl decane	2,4-di-i-propyl phenol	Penicillium species
Tetradecane	Acids/acid derivatives	
Pentadecane	Acetic acid	
Aromatic hydrocarbons	2,6-di-i-propyl	
Toluene	4-methyl phenomethylamide	
m,p-xylene	2-ethyl hexanoic acid	
Propyl benzene	2-methyl propanoic acid, 1-	
2-ethyl toluene	(1,1-dimethylethyl)-2-methyl	
3-ethyl toluene	propyl diester	
4-ethyl toluene	Others	
Cumene	Hexanal	
1,3,5-trimethyl benzene	Nonanal	
	2-vinyl pyrazine	

[a] *Source:* Otson, R., Meth, F., and Fellin, F., Identification of important sources of five priority substances in Canadian indoor air, *Indoor Air*, Proc. of 7th Int. Conf. on Indoor Air Quality and Climate, Nagoya, 1:565–573, 1996.

[b] *Source:* Byrd, R.R., Prevalence of microbial growth in cooling coils of commercial air conditioning systems, *Indoor Air*, Proc. of 7th Int. Conf. on Indoor Air Quality and Climate, Nagoya, 3:203–207, 1996.

[c] *Source:* Kichner, S. and Karpe, P., Contribution of volatile organic compounds to the odor of building materials, *Indoor Air*, Proc. of 7th Int. Conf. on Indoor Air Quality and Climate, Nagoya, 2:663–668, 1996.

[d] *Source:* Martin W.T., Zhang, Y., Willson, P.J., Archer, T.P., Kinahan, C., and Barber, E.M., Bacterial and fungal flora of dust deposits, in a Wiegard, B. and J. Hartung, Bacterial contaminant of air and floor surfaces in an animal house of a cattle clinic, Proceedings of 4th Intern. Livestock Environ, Symp, 643–649,1993. swine building, *Br. J. Occup. and Env. Med.*, 53:484–487, 1996.

formaldehyde. It has also been shown to cause cancer in animals and may cause cancer in humans.

Formaldehyde is an important chemical used widely by industries to manufacture building materials and numerous household products. It is also a by-product of combustion and certain other natural processes. Thus, it may be present in substantial concentrations both indoors and outdoors.[15] Sources of formaldehyde in homes include building materials, smoking, household products, and unvented, fuel-burning appliances such as gas stoves or kerosene space heaters. Formaldehyde, by itself or in combination with other chemicals, serves a number of purposes in manufactured products. For example, it is used to add permanent-press qualities to clothing and draperies, as a component of glues and adhesives, and as a preservative in some paints and coating products. During the 1970s, many homeowners had urea-form-aldehyde foam insulation (UFFI) installed in the wall cavities of their homes as an energy conservation measure. However, many of these homes were found to have relatively high indoor concentrations of formaldehyde soon after the UFFI installa-tion. Average concentrations in older buildings (before 1980) without UFFI are generally well below 0.1 (ppm). In homes with significant amounts of new pressed wood products, levels can be greater than 0.3 ppm.[15] Since 1985, the U.S. Department of Housing and Urban Development (HUD) has permitted only the use of plywood and particleboard that conform to specified formaldehyde emission limits in the construction of prefabricated and mobile homes.

In buildings, the most significant sources of formaldehyde are likely to be pressed wood products made using adhesives that contain urea-formaldehyde (UF) resins. Pressed wood products made for indoor use include particleboard (used as subflooring and shelving and in cabinetry and furniture), hardwood plywood pan-eling (used for decorative wall covering and in cabinets and furniture), and medium-density fiberboard (used for drawer fronts, cabinets, and furniture tops). Medium-density fiberboard contains a higher resin-to-wood ratio than any other UF pressed wood product and is generally recognized as being the pressed wood product that emits the highest levels of formaldehyde. Other pressed wood products, such as soft plywood and flake or oriented strand board, are produced for exterior construc-tion use and contain the dark or red/black phenol-formaldehyde (PF) resin. Although formaldehyde is present in both types of resins, pressed woods that contain PF resin generally emit formaldehyde at considerably lower rates than those containing UF resin.

2.5.3   Molds

Molds are almost ubiquitous in indoor environments and play an important part in life cycles: They are the primary forces in assisting with decomposition of organic materials. Some molds cause illness and some, such as penicillin, cure illness. Certain molds help to develop the flavor of wines and cheeses, whereas others can cause them to spoil. There are thousands of types of molds and yeasts, the two groups of plants in the fungus family. Yeasts are single cells that divide to form clusters. Molds consist of many cells that grow as branching threads called *hyphae*. Although both groups can probably cause allergic reactions, only a few dozen types of molds are significant allergens. In general, *Alternaria* and *Cladosporium (Hormodendrum)* are the molds most commonly found both indoors and outdoors. *Aspergillus, Penicillium, Helminthosporium, Epicoccum, Fusarium, Mucor, Rhizopus,* and *Aureobasidium (Pullularia)* are also common. The seeds or reproductive particles of fungi are called *spores*. They differ in size, shape, and color among species. Each spore that germinates can give rise to new mold growth, which in turn can produce millions of spores.

Along with pollens from trees, grasses, and weeds, molds are an important cause of seasonal allergic rhinitis. Molds can live for years in a dormant state. To grow, the spores need moisture, warm temperature, low air movement, and a food source. Molds can become airborne when their habitats are disturbed by shaking or sweeping. The airborne molds can travel with air current, land on surfaces, and settle into the tiniest cracks and crevices of carpets, furniture, draperies, insulation, rough textures, and smooth surfaces. Heating and cooling ducts, wet carpets, damp upholstery, and air filters on air conditioners and furnaces become common habituating places for molds. People allergic to molds may have symptoms from spring to late fall. The mold season often peaks from July to late summer. Unlike pollens, molds may persist after the first killing frost. Some can grow at subfreezing temperatures, but most become dormant. Snow cover lowers the outdoor mold count dramatically but does not kill molds. After the spring thaw, molds thrive on the vegetation that has been killed by the winter cold. In temperature-controlled indoor environments, molds can thrive at all times, especially when the relative humidity is above 70%. When relative humidity is below 70%, the growth rate of molds quickly decreases. When the relative humidity is below 30%, the molds are primarily in the dormant stage.

Many airborne molds in indoor environments are responsible for human respiratory allergies. Reactions to molds can result in sneezing, itching, coughing, wheezing, shortness of breath, and chest pain. Molds are measured in spores/m^3. Specific threshold limit values for molds have not been established to prevent hypersensitivity, irritant, or toxic responses. Like pollen counts, mold counts may suggest the types and relative quantities of fungi present at a certain time and place. For several reasons, however, these counts probably cannot be used as a constant guide for daily activities. One reason is that the number and types of spores actually present in the mold count may have changed considerably in 24 hours because weather and spore dispersal are directly related. Currently, information relating to molds and other bioaerosols consists largely of case studies and qualitative exposure assessment.[1] One rule of thumb for judgment is that mold concentration below 500 spores/m^3 is considered low, 500 to 1500 spores/m^3 is moderate, and above 1500 spores/m^3 is high.

2.5.4 Dust Mites

Dust mites (*Dermatophagoides Farinae*) are microscopic, spiderlike insects found everywhere. An adult dust mite is approximately 200 μm long and is usually invisible to the naked human eye (Figure 2.6). Dust mites are in the arachnid family, which includes spiders, scorpions, and ticks. Dust mites feed on dead skin that sloughs from our bodies (and probably on potato chips and cookie crumbs). They live their whole lives in dark-corner dust bunnies: hatching, growing, eating, defe-

Figure 2.6 A microscopic dust mite image. An adult dust mite is approximately 200 μm long and is usually invisible to the naked human eye.

cating, mating, and laying eggs. During warm weather, when the humidity is above 50%, they thrive and produce waste pellets. They live about 30 days, and the female lays approximately one egg each day. In less-than-ideal conditions, they can go into dormancy. When they die, their bodies disintegrate into small fragments, which can be stirred into the air and inhaled by people in the environment.

When they are present in indoor environments and inhaled by sensitive people, dust mites are responsible for many sick building symptoms. It is their "bathroom habits" that make us itch and wheeze. Many people develop severe allergies to dust mite droppings. Lie on a rug where they live and you may find itchy red bumps on your skin. Breathe in dust and you may have more serious symptoms, such as difficulty breathing or even a severe asthma attack. Dust mite allergy is an allergy to a microscopic organism that lives in dust that is found in all dwellings and workplaces. Dust mites are perhaps the most common cause of perennial allergic rhinitis. Dust mite allergy usually produces symptoms similar to pollen allergy and also can produce symptoms of asthma.

2.5.5 Smoke and Fumes

Environmental tobacco smoke (ETS) is a mixture of particles that are emitted from the burning end of a cigarette, pipe, or cigar, and from smoke exhaled by the smoker. Smoke can contain any of more than 4000 compounds, including carbon monoxide and formaldehyde. More than 40 of the compounds are known. Exposure to ETS is often called *passive smoking*. Fumes are a combination of particles and gases that are the result of combustion, high-temperature operations (such as cooking, welding, grounding), and adhesive applications. The chemical composition can be more complicated than ETS because fumes can come from a variety of sources.

ETS has been classified as a Group A carcinogen by the U.S. Environmental Protection Agency (EPA), a rating used only for substances proven to cause cancer in humans. The EPA[16] concluded that each year approximately 3,000 lung cancer deaths in nonsmoking adults are attributable to ETS. Exposure to secondhand smoke also causes eye, nose, and throat irritation. It may affect the cardiovascular system, and some studies have linked exposure to secondhand smoke with the onset of chest pain. ETS is an even greater health threat to people who already have heart and lung illnesses. Infants and young children whose parents smoke in their presence are at increased risk of lower respiratory tract infections (pneumonia and bronchitis) and are more likely to have symptoms of respiratory irritation, such as coughing, wheezing, and excess phlegm. In children under 18 months of age, passive smoking causes between 150,000 and 300,000 lower respiratory tract infections, resulting in 7,500 to 15,000 hospitalizations each year, according to EPA estimates.[16] These children may also have a buildup of fluid in the middle ear, which can lead to ear infections. Slightly reduced lung function may occur in older children who have been exposed to secondhand smoke. Children with asthma are especially at risk from ETS. The EPA estimates that exposure to ETS increases the number of asthma episodes and the severity of symptoms in 200,000 to 1 million children annually.[16] Secondhand smoke may also cause thousands of nonasthmatic children to develop the disease each year.

2.5.6 Radon (Rn)

Radon is a gaseous radioactive element with the symbol Rn, a molar weight of 222 g, a melting point of −71°C, a boiling point of −62°C, and 18 radioactive isotopes. It is an extremely toxic, colorless, and odorless gas; it can be condensed to a transparent liquid and to an opaque, glowing solid. It is used in cancer treatment, as a tracer in leak detection, and in radiography. Radon is a natural product of most soils and rocks as a result of radioactive decay of radium. An unventilated basement is the most likely place to have high radon concentrations.

Radon concentration indoors is in picocuries per liter of air (pCi/l). One curie (Ci) is 3.7×10^{10} radioactive disintegration per second. Thus, one pCi/l = $10^{-12} \times$ Ci/l = 0.037 radioactive disintegration per second per liter of air. The average radon concentrations for outside and indoor air in the United States are 0.4 and 1.3 pCi/l, respectively.[17]

There is no scientific doubt that radon gas is a known human lung carcinogen. Prolonged exposure to high levels of radon gas can cause lung cancer. Millions of homes and buildings contain high levels of radon gas. As a means of prevention, the EPA and the Office of the Surgeon General recommend that all homes below the third floor be tested for radon. Because radon is invisible and odorless, a simple test is the only way to determine whether a home has high radon levels. Most homes do not have a radon problem. If a radon problem does occur, it is relatively easy to fix. Ventilation is the most effective method for diluting the radon concentration in an airspace.

2.5.7 Volatile Organic Compounds (VOCs)

Generally speaking, an organic compound is a compound that contains carbon and hydrogen. Some examples of organic compounds are octane, butane, and sugar. Organic compounds that evaporate easily are called *volatile organic compounds* (VOCs). VOCs in the atmosphere come from all combustible engines, industry, fuel spills, animal and plant production facilities, etc. Certain other fumes, such as those released from industrial plants and print shops, also can contain significant amounts of VOCs. When released into the atmosphere, VOCs contribute to ozone and smog formation. In an indoor environment, VOCs can have direct adverse effects on human health and comfort. Many VOCs have been classified as toxic and carcinogenic, and it is therefore unsafe to be exposed to these compounds in large quantities or over extended periods of time. Some health effects from overexposure to VOCs are dizziness, headaches, and nausea. Long-term exposure to certain VOCs, such as benzene, has also been shown to cause cancer and, eventually, death.

2.5.8   Airborne Contaminants in Animal Environments

The most important contributor to airborne contamination in animal buildings may be dust. Animal building dust sources include feed grains, fecal material, animal skin and hair, insects, and dead microorganisms (Figure 2.6). Previously, dust in animal buildings was considered to originate primarily from feed,[9] but a number of studies have shown that animal building dust is very complicated. Feddes et al.[18] characterized dust in a turkey building and showed that most dusts may not be from feed. Fecal material was found to be the main constituent of airborne dust in turkey barns. In animal production buildings, especially when oil is added to the feed, it has been suspected that fecal material and animals are major dust sources. Adding oil to animal feed is one of the methods to control dust sources. In one study,[19] it was proved that adding 2% oil to a swine feed formula reduced inhalable mass concentration by 31% and increased the respirable particle count by 45%, compared with the treatment where oil was not added.

Dust in animal buildings is biologically active and different from ordinary dust, such as field dust. Dust particles from animal buildings may combine components of viable organic compounds, mites, fungi, endotoxins, toxic gases, and other hazardous agents.[3,20] Although the pathology of dusts in animal facilities remains unclear, there is little argument about their adverse effect on animal and human health. Some of the components, such as *Streptococcus suis,* are considered particularly hazardous to animals; others, such as *Alternaria* species, are suspected to be responsible for allergic symptoms in humans. In many animal production facilities, dusts tend to be accepted as a normal part of the job and may be ignored until permanent damage occurs. Toxic gases cause discomfort (e.g., nuisance odor or more dramatic symptoms), which attracts immediate attention and quick action in solving the problem. Dust control is one of the main tasks that will improve air quality in production animal facilities.

Household pets are a common source of allergic reactions to animals. Many people think that a pet allergy is provoked by the fur of cats and dogs. However, researchers have found that the major allergens are proteins secreted by oil glands in the animals' skin and shed in dander, as well as proteins in the saliva, which sticks to the fur when the animal licks itself. Urine is also a source of allergy-causing proteins. When the substance carrying the proteins dries, the proteins can then become airborne. Cats may be more likely than dogs to cause allergic reactions because they lick themselves more frequently, may be held more, and spend more time indoors and close to humans. Some rodents, such as guinea pigs and gerbils, have become increasingly popular as household pets. They, too, can cause allergic reactions in some people, as can mice and rats. Urine is the major source of allergens from these animals.

Allergies to animals can take two years or longer to develop and may not subside for six months or more after contact with the animal has ended. Carpets and furniture are a reservoir for pet allergens, and the allergens can remain in them for four to six weeks. In addition, these allergens can stay in household air for months after the animal has been removed. Therefore, it is wise for people with an animal allergy to check with the landlord or previous owner to find out whether furry pets have lived previously on the premises.

2.6 THRESHOLD LIMIT VALUES FOR TYPICAL INDOOR AIR CONTAMINANTS

Threshold limit values (TLVs) refer to concentrations of airborne contaminants and represent conditions under which, it is believed, nearly all workers may be repeatedly exposed day after day without adverse health effects.[1] TLVs are based on available information from industrial experience, from experimental human and animal studies, and, when possible, from a combination of the three. The basis on which the values are established may differ from substance to substance; protection against impairment of health may be a guiding factor for some, whereas reasonable freedom from irritation, narcosis, nuisance, or other forms of stress may form the basis for others. Health impairments considered include those that shorten life expectancy, compromise physiological function, impair the capability for resisting other toxic substances or disease processes, or adversely affect reproductive or developmental processes.

Because of wide variation in individual susceptibility, however, a small percentage of workers may be affected more seriously by aggravation of a preexisting condition or by development of an occupational illness. For example, smoking may act to enhance the biological effects of chemicals encountered in the workplace and may reduce the body's defense mechanisms against toxic substances. Individuals may be hypersusceptible or unusually responsive to some airborne contaminants because of genetic factors, age, personal habits (e.g., smoking, alcohol consumption, or the use of other drugs), medication, or previous exposure. Such workers may not be adequately protected from adverse effects of certain chemicals at concentrations at or below the threshold limits.

There are two categories of TLVs, differentiated by their legal status:

- Mandatory and issued by governmental regulatory agencies, such as the EPA and the Occupational Safety and Health Agency (OSHA) in the United States
- Recommended and published by scientific communities, such as the American Conference of Governmental and Industrial Hygienists (ACGIH) or individual researchers

Mandatory TLVs are enforced by law. Recommended TLVs are intended for use in the practice of industrial hygiene as guidelines or recommendations in the control of potential health hazards in the workplace. These limits are not fine lines between safe and dangerous concentrations; neither are they a relative index of toxicity. In spite of the fact that serious adverse health effects are not believed to occur as a result of exposure to TLVs, the best practice is to maintain concentrations of all atmospheric contaminants as low as is practical.

Three categories of TLVs, according to the exposure, are the following:[1]

Time-Weighted Average (TLV-TWA) — The time-weighted average concentration for a conventional eight-hour workday and a 40-hour work week, to which it is believed that nearly all workers may be repeatedly exposed, day after day, without adverse effect

Short-Term Exposure Limit (TLV-STEL) — The concentration to which it is believed that workers can be exposed continuously for a short period of time without suffering from 1) irritation, 2) chronic or irreversible tissue damage, or 3) narcosis of sufficient degree to increase the likelihood of accidental injury, impair self-rescue, or materially reduce work efficiency, provided that the daily TLV–TWA is not exceeded. STEL is not a separate, independent exposure limit; rather, it supplements the TWA limit where there are recognized acute effects from a substance whose toxic effects are primarily of a chronic nature. STELs are recommended only where toxic effects have been reported from high, short-term exposures in either humans or animals. A STEL is defined as a 15-minute TWA exposure that should not be exceeded at any time during a workday, even if the eight-hour TWA is within the TLV-TWA. Exposures above the TLV-TWA up to the STEL should not be longer than 15 minutes and should not occur more than four times per day. There should be at least 60 minutes between successive exposures in this range. An averaging period other than 15 minutes may be recommended when this is warranted by observed biological effects.

Ceiling (TLV-C) — The concentration that should not be exceeded during any part of the working exposure. An alternative term for TLV-C is *fatal exposure value* (FEV).

Not all three categories of TLVs are available for every airborne contaminant. Many airborne contaminants have only one or two of the TLVs. It is important to observe that if any one of these types of TLVs is exceeded, a potential hazard from that contaminant is presumed to exist. Typical indoor air contaminants are listed in Table 2.5. More complete lists of TLVs can be obtained from ACGIH.[1]

Whereas the ceiling limit places a definite boundary that concentrations should not be permitted to exceed, the TWA requires an explicit limit to the excursions that are permissible above the TLV-TWAs. TWAs permit excursions above the TLV, provided they are compensated by equivalent excursions below the TLV-TWA during the workday. In some instances, it may be permissible to calculate the average concentration for a work week rather than for a workday. The relationship between the TLV and permissible excursion is a rule of thumb and, in certain cases, may not apply. The amount by which the TLVs may be exceeded for short periods without injury to health depends on a number of factors, such as the nature of the contaminant, whether very high concentrations — even for short periods — produce acute

Table 2.5 Typical Indoor Air Contaminant Threshold Values for Work Places

Contaminant	Symbol	Density (relative to air at room temperature)	Molar Weight (g)	Odor	TWA	STEV	FEV
					(in parts per million [ppm] for gases and mg/m³ for dust)		
				Gaseous (ppmv)			
Ammonia	NH_3	0.61	17	Sharp, pungent	25	35	300[c]
Carbon dioxide	CO_2	1.57	44	None	5,000	30,000	
Carbon monoxide	CO	1.00	28	None	25	400[c]	
Formaldehyde	HCHO	1.06	30	Pungent		0.3	
Hydrogen sulfide	H_2S	1.20	34	Rotten eggs, nauseating	5	50	150
Methane	CH_4	0.57	16	None	1,000		50,000
Ozone	O_3	1.70	48	Odorous	0.1		
Radon	Rn	7.84	222	None		4pCi/l[b]	
				Particulate matter (mg/m³)			
Animal building dust, inhalable					2.3[a]		
Animal building dust, respirable					0.23[a]		
Asbestos					0.1 fiber/ml		
Coal dust, respirable					0.4		
Flour dust, inhalable					0.5		
Grain dust, inhalable					4.0		
Iron oxide dust and fume, inhalable					5.0		
Metal dust, inhalable					10.0		
Welding fumes					5.0		
Particulates not otherwise classified (PNOC), inhalable					10.0		
Particulates not otherwise classified (PNOC), respirable					3.0		

[a] From Canada Plan Service, M-8710, 1985; M-9707, Ottawa, Canada, 1992.

[b] EPA, Radon: Is Your Family at Risk? Environmental Protection Agency, Washington, DC, 1996.

[c] Donham, J.K., Haglind, P., Peterson, Y., Rylander, R., and Belin, L., Environmental and health studies of workers in Swedish swine confinement buildings, Br. J. Industr. Med., 40:31–37, 1989.

Source: If not specified, data are from ACGIH, Threshold Limit Values for Chemical Substances and Physical Agents and Biological Exposure Indices, American Conference of Governmental Industrial Hygienists, Cincinnati, OH, 2000.

poisoning, whether the effects are cumulative, the frequency with which high concentrations occur, and the duration of such periods. All factors must be taken into consideration in arriving at a decision about whether a hazardous condition exists.

For the vast majority of substances with a TLV-TWA, there is not enough toxicological data available to warrant a STEL. Nevertheless, excursions above the TLV-TWA should be controlled even where the eight-hour TLV-TWA is within recommended limits. Although no rigorous rationale was provided for these particular values, the basic concept was intuitive: In a well-controlled process exposure, excursions should be held within some reasonable limits. Unfortunately, neither toxicology nor collective industrial hygiene experience provides a solid basis for quantifying what those limits should be. For those air contaminants that have no TLVs, two simplified STEL criteria, or excursion limit values, are recommended by ACGIH 1:

- Excursions in worker exposure levels may reach three times the TLV-TWA for no more than a total of 30 minutes during a workday.
- Under no circumstances should they exceed five times the TLV-TWA.

This simplification is based on the maximum recommended excursion observed in actual industrial processes. When the toxicological data for a specific substance are available to establish a STEL, this value takes precedence over the excursion limit, regardless of whether it is more or less stringent.

2.6.1   Normalized Air Contaminant Concentration

In many practical indoor air quality problems, several air contaminants exist simultaneously. For example, ammonia and airborne dust are almost always present in an animal facility. The combined effect of these air contaminants on the health and comfort of the occupants is generally considered greater than any one of the existing contaminants in the airspace. When two or more air contaminants that act upon the same organ system are present, their combined effect, rather than that of either individually, should be given primary consideration. The effect on human health of some contaminants may be independent of other contaminants. In this case, TLVs for individual contaminants should be used. In other instances, the effect on health of some contaminants may be compounding and greater than the sum of individual effects.

The synergy of the mixture of air contaminants and their combined effect on human health and comfort is extremely complex. In the absence of information to the contrary, the effects of the different hazards should be considered additive.[1] In order to quantify the combined effect, the concentration of the concerned airborne contaminants is normalized. The normalized air contaminants concentration C_{NC} is defined in Equation 2.10:

$$C_{NC} = \sum \frac{C_i}{TLV_i} \qquad (2.10)$$

where C_i and TLV_i are the concentration and threshold limit value, respectively, for the ith concerned air contaminant. When C_{NC} is smaller than unity, the air quality is acceptable. When C_{NC} is greater than unity, the air quality is not acceptable, and one or several air contaminant concentrations should be reduced. The normalized contaminant concentration is dimensionless and can operate with different contaminants with different concentrations and TLVs.

Example 2.3: In a newly built building, the measured concentrations of formaldehyde and radon are 0.2 ppm and 3 pCi/l, respectively. The recommended TLVs for formaldehyde and radon are 0.4 ppm and 4 pCi/l, respectively. What is the normalized air contaminant concentration?

Solution: For formaldehyde, $C_1 = 0.2$ ppm, $TLV_1 = 0.4$ ppm; for radon, $C_2 = 3$ pCi/l, $TLV_2 = 4$ pCi/l:

$$C_{NC} = \frac{C_1}{TLV_1} + \frac{C_2}{TLV_2} = \frac{0.2\,ppm}{0.4\,ppm} + \frac{3\,pCi\,/\,l}{4\,pCi\,/\,l} = 0.5 + 0.75 = 1.25$$

Because the $C_{NC} > 1$, the air quality is not acceptable.

In a practical indoor air quality scenario, the concerned air contaminants should first be identified for a given airspace containing a number of harmful dusts, fumes, vapors, or gases. Frequently, it will only be feasible to attempt to evaluate the hazard by measuring a single contaminant at one time. Then, Equation 2.10 should be applied to establish acceptable TLVs for each contaminant concerned.

DISCUSSION TOPICS

1. Why is the threshold limit value for particulates not otherwise classified (PNOC) 10 mg/m³, whereas for inhalable flour dust it is only 0.5 mg/m³?
2. The size of particles is important because it affects behavior and deposition in airways. What do you think about particle shape? Is the particle shape an important factor? If so, can you think of an example to justify?
3. Why do particles in different size ranges behave very differently? Can a piece of airborne particle cycle around a basketball just as the earth cycles around the sun?
4. What is the rationale for the distinction between total and inhalable particles? Are they the same thing?
5. Why is the diminutive particle a separate particle size range? Are respirable particles good enough for all health and engineering purposes?
6. It seems that the threshold limit value for asbestos (0.1 fiber/ml) is quite low compared with other particulate matter TLVs. Why?

PROBLEMS

1. An employee works in an animal building 8 hours per day. The average ammonia (NH_3) concentration in the building is 10 ppmv. The density of building air is 1.2 kg/m³, and the density of ammonia is 0.72 kg/m³.
   a. What is the mass concentration of ammonia in ppmm?

   b. What is the ammonia concentration in mg/m^3 of air?

2. Dust mass concentration in a confinement building is measured at 2.8 mg/m^3 using a filter sampler. Assume the particles are spherical and have an average diameter of 2.5 μm with a density of 2,000 kg/m^3. What is the concentration in particles/ft^3, particles/ml, and particles/l?

3. The dust concentration was measured in a classroom with a laser particle counter. The measured data show that the dust concentration is 32,800 particles/ft^3. If these particles are standard-density (1,000 kg/m^3) spheres with an average diameter of 0.5 μm, what is the particle concentration in mg/m^3?

4. Assume that the material is divided into spheres each having a diameter d and a density of $\rho = 1000$ kg/m^3.
   a. Derive an expression for the surface area per gram of material as a function of particle size.
   b. What is the total surface area of 1g of 0.1 μm diameter particles?

5. The time-weighted average (TWA) threshold limit for a particular contaminant is the value to or below which a human worker can be exposed 40 hours per week without health problems resulting. One suggested TWA value of respirable particles for biological systems is less than 0.23 mg/m^3. If the particles are spherical and have an equivalent volume diameter of 1.2 μm, and if the particle density is 1500 kg/m^3, what is the TWA value in particle number concentration?

6. One cigarette contains approximately 20 mg of smoking particles. The average diameter of the particles is 0.3 μm and density is 1000 kg/m^3. The smoker consumes 20 cigarettes per day. Assuming that 99.9% of the particles are exhaled during the smoking process, how much mass was deposited in the smoker's lung per day, per year? How many particles were deposited in the lung per day, per year?

7. In question 6, the smoker consumes all 20 cigarettes within 8 hours, and all smoke is inhaled. An employee works 8 hours per day in a building with a respirable dust concentration of 50 particles/ml. Assume that the average breathing rate of a person is 20 l/min.
   a. Compare the total particles inhaled per day by the smoker and by the employee. Who inhales more particles?
   b. If the average particle diameter is 1 μm and density is 1000 kg/m^3 for the building dust, how much dust was inhaled by the worker?

8. In smoking one pack of cigarettes per day, a smoker inhales 8 L of smoke containing 200 mg of tobacco smoke particles. Assume that these particles are standard density (1000 kg/m^3) spheres.
   a. Find the ratio of the smoke to the TLV of nuisance dusts (10 mg/m^3).
   b. Find the ratio of the smoke to the U.S. national ambient air quality standard for particulate matter (0.08 mg/m^3).

9. If airborne particles were considered extremely large gas molecules, what would the molecular weight be of a "gas" of 1.0 μm particles having a density of 1000 kg/m^3? Hint: One mole of any matter has the same number (Avogadro's number) of molecules.

10. In a heated building, concentrations of ammonia, hydrogen sulfide, and carbon dioxide are measured at 18, 0.5, and 2000 ppm, respectively. The TWA threshold limit for ammonia, hydrogen sulfide, and carbon monoxide are 25, 5, and 5000 ppm, respectively. Is the building air quality acceptable, i.e., is the normalized air contaminant concentration smaller than unity?

11. An animal building contains 20 ppmv ammonia and 1.5 mg/m^3 of inhalable dust in the air. How much dust must be removed from the air to bring the combined

TWA below the recommended threshold limit value? TLV-TWA values for ammonia and dust in animal buildings are 25 ppm and 2.3 mg/m^3, respectively.

12. The concentrations of welding fumes and PNOC (particluates not otherwise classified) are 2 mg/m^3 and 4 mg/m^3, respectively, in a machine shop. TLV-TWA values for welding fumes, PNOC, and carbon monoxide are 5 mg/m^3, 10 mg/m^3, and 25 ppm, respectively. In order to maintain an acceptable air quality, what can the maximum concentration of carbon monoxide be in the building?

REFERENCES

1. ACGIH, *Threshold Limit Values for Chemical Substances and Physical Agents and Biological Exposure Indices.* American Conference of Governmental Industrial Hygienists, Cincinnati, OH, 2000.
2. Vincent, J.H., *Aerosol Sampling — Science and Practice*, John Wiley & Sons, New York, 1989.
3. Zhang, Y, Barber, E.M., Martin, W.T., and Wilson, P., Dynamics of aerosol deposition rates in a commercial swine building: a pilot study. *Amer. Soc. Heat. Refrig. Air Cond. Eng.,* 100(1): 906–912, 1994.
4. U.S. EPA, *Terms of Environment,* U.S. Environmental Protection Agency, Washington, DC, 1998.
5. U.S. EPA, *Source of Information on Indoor Air Quality*, Environmental Protection Agency, Washington, DC, 1999.
6. Ogden, T.L. and Birkett, J.L., An inhalable dust sampler for measuring the hazard from total airborne particulate, *Ann. Occup. Hyg.* 21:41–50, 1978.
7. ISO, Air quality — particle size fraction definitions for health related sampling. *Technical Report ISO/TR/7708*, International Organization for Standardization, Geneva, Switzerland, 1991.
8. CEN, *Size Fraction Definitions for Measurement of Airborne Particles in the Workplace*, approved for publication as EN 481, European Standardization Committee (CEN), Luxembourg, 1991.
9. Donham, J.K., Haglind, P., Peterson, Y., Rylander, R., and Belin, L., Environmental and health studies of workers in Swedish swine confinement buildings, *Br. J. Industr.Med.,* 40:31–37, 1989.
10. Hinds, C.W., *Aerosol Technology*, John Wiley & Sons, New York, 1999.
11. U.S. EPA. *Asbestos in Buildings: A National Survey of Asbestos-Containing Friable Materials* Office of Toxic Substances, EPA 560/5-84-006. Environmental Protection Agency., Washington, DC, 1984.
12. NRC, *Asbestiform Fibers — Nonoccupational Health Risks,* National Research Council, National Academy Press, Washington, DC, 1984.
13. Burdette, G.F., Jaffrey, S.A.M.T., and Rood, A.P., Airborne asbestos fiber levels in buildings: a summary of UK measurements. In Bignon J., Peto, J., Saracci, R., eds. *Non-Occupational Exposure to Mineral Fibres,* IARC Scientific Publications, Lyon, France, 1989.
14. U.S. EPA, *A Building Owner's Guide to Operations and Maintenance Programs for Asbestos-Containing Materials*, Office of Pesticides and Toxic Substances, Environmental Protection Agency, Washington DC, 1990.
15. U.S. EPA. *Source of Information on Indoor Air Quality*, Environmental Protection Agency, Washington, DC, 1999.

16. U.S. EPA, *Respiratory Health Effects of Passive Smoking: Lung Cancer and Other Disorders*, EPA/600/6-90/006F, Environmental Protection Agency, Washington, DC, 1993.

17. U.S. EPA, *Radon: Is Your Family at Risk?* Environmental Protection Agency, Washington, DC, 1996.

18. Feddes, J.J.R., Cook, H., and Zuidhof, M., Characterization of airborne dust particles in turkey housing, ASAE Paper 89-4021, *Am. Soc. of Agri. Eng.*, St. Joseph, MI, 1989.

19. Welford, R.A., Feddes, J.J.R., and Barber, E.M., Pig building dustiness as affected by canola oil in the feed, *Can. Agr. Eng.*, 34(2):365–373, 1992.

20. Martin W.T., Zhang, Y., Wilson, P.J., Archer, T.P., Kinahan, C., and Barber, E.M., Bacterial and fungal flora of dust deposits in a swine building. *Br. J. Occup. and Environ. Med.*, 53:484–487, 1996.

CHAPTER 3

Particle Size Statistics and Distribution

Statistics and distribution of particle sizes are important means by which we can identify and solve air quality problems. For example, in a dusty airspace, how much of the total dust falls into the respirable particle size range? If the total particle mass in an air stream needs to be reduced by 90% by passing through a filter, what is the maximum porous size of the filter? In other cases, one needs to know other particle statistics based on measured data or a known distribution function. In this chapter, we will begin to characterize the particle number and mass distribution with respect to particle size. Different particle size averages can be calculated statistically with measurement data or calculated with a known size distribution function. Lognormal distribution, one of the most commonly used particle size distributions, is discussed in detail.

By completing this chapter, the reader will be able to

- Determine particle distributions in count or in mass, both mathematically and graphically
- Determine other moment averages of particle diameters based on a set of measured data, such as particle size and its corresponding concentration (Table 3.7)
- Understand and apply the lognormal size distribution function to real-life particles
- Determine the fraction of particles that are smaller than a particular diameter (or size classification) for particles with a lognormal size distribution
- Apply the log-probability graph to determine the CMD (or MMD) and the σ_g
- Apply the Hatch–Choate equation to find other moment averages of diameter for particles with a lognormal size distribution
- Be aware of other particle size distributions that can be applied to particular indoor pollutants

3.1 NUMBER DISTRIBUTION

As far as number distribution is concerned, particles are considered to be spherical, regardless of their shapes and densities. This assumption is necessary for the

Table 3.1   Illustration of Statistical Variables of Particle Count Distribution for
Grouped Data

Size Range,[a] $d_{i-1} - d_i$ (µm),	Δd_i (µm)	Midsize Diameter, d_i (µm)	Frequency, F_i	Fraction, Δf_i, (µm^{-1})	Accumulative Fraction, f
0–3	3	1.5	90	0.041667	0.125
3–5	2	4	125	0.086806	0.299
5–7	2	6	165	0.114583	0.528
7–10	3	8.5	165	0.076389	0.757
10–20	10	15	115	0.015972	0.917
20–50	30	35	60	0.002778	1.000
>50			0		
Total:			$N = 720$		

[a] Upper size range is smaller than, and lower size range is equal to, the size indicated.

discussion of particle statistics. There are two categories of spherical particles: monodisperse and polydisperse. The "size" of monodisperse particles is completely defined by a single parameter, the particle diameter d. Most particles, however, are polydisperse and have particle sizes that range over two or more orders of magnitude. This wide range of particle sizes even applies to the same particle source. For example, cigarette smoke contains tobacco particles ranging from less than 0.01 µm to larger than 1 µm. Because of the wide range and the fact that the physical properties of particles are strongly dependent on particle size, it is necessary to characterize these size distributions accurately by statistical means.

In measuring particles, we often encounter grouped particle data. From a statistical point of view, we consider a total population of N polydisperse particles, each particle with a diameter d_p. This N number of particles can be divided further into n groups of particles. Each group has F_i particles, or an F_i frequency of particle occurrences, with a size range of Δd_i and a midpoint diameter d_i.

From the data example in Table 3.1, the total particle population N is the sum of frequencies for all groups:

$$N = \sum_{i=1}^{n} F_i \qquad (3.1)$$

where F_i is the number of particles, or the frequency, in the ith group. The fraction for the ith group of particles, f_i, with respect to the total particle population is

$$f_i = \frac{F_i}{N} \qquad (3.2)$$

Therefore, the sum of the fractions for the n-groups of particles, or the total population of the particles, is

$$\sum_{i}^{n} f_i = \sum_{i}^{n} \frac{F_i}{N} = \frac{1}{N} \sum_{i}^{n} F_i = 1 \qquad (3.3)$$

The fraction for the ith group particles includes the variable of size range Δd_i, which can be misleading in some instances because Δd_i can vary by many magnitudes. A clearer representation of fraction for particle size d_p is standardized as the fraction per micrometer of the particle size:

$$\Delta f_i = \frac{F_i / \Delta d_i}{N} \qquad (3.4)$$

Obviously, the nominator on the right side of Equation 3.4 is the standardized frequency of the total particle population:

$$F = \frac{F_i}{\Delta d_i} \qquad (3.5)$$

Unless otherwise specified, the frequency and fraction refer to the standardized frequency and standardized fraction, respectively. They are both in units of μm^{-1}. When the particle fraction is a continuous function, $df(d_p)$, the standardized fraction for the size range of $d(d_p)$ can be expressed as

$$df = df(d_p) d(d_p) \qquad (3.6)$$

The accumulative fraction of the total particle population f is

$$f = \sum f_i = \sum \frac{F_i}{N} = \int df(d_p) d(d_p) = 1 \qquad (3.7)$$

Figure 3.1 shows a graphical representation of data in Table 3.1. The height and width of each bar in Figure 3.1 are the frequency F_i and size range Δd_i, respectively, of each particle group. The curve is the standardized fraction function $f(d_p)$. The total area of the bars represents the total particle population N. The total area under the curve $f(d_p)$ is unity, which represents the accumulated fraction, as shown in Figure 3.2.

Many quantities can be used to define a size distribution. The most commonly used quantities for defining the location of a distribution are the mean, mode, median, and geometric mean. These quantities are described by the general term *averages*, that is, they are different types of averages of a given population. The mean diameter, also called *arithmetic average* $\overline{d}_p$, is the sum of all the particle sizes divided by the number of particles. The mean diameter for listed data, for grouped data, and for the frequency function is given by

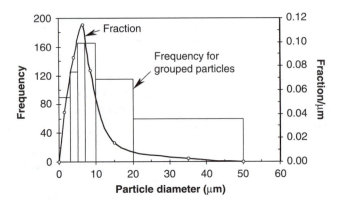

Figure 3.1   Particle size distribution in terms of frequency in a size range and fraction function.

$$\overline{d}_p = \frac{\sum F_i d_i}{N} = \frac{\sum_i F_i d_i}{\sum_i F_i} = \int_0^\infty d_p f(d_p) d(d_p) \qquad (3.8)$$

where $f(d_p)$ is the fractional function of particles with diameter d_p (Figure 3.1).

The *count median diameter* (CMD) is defined as the diameter for which one-half of the particles are smaller and one-half are larger. It is the diameter corresponding to the 0.5 accumulative fraction in Figure 3.2. The *mode* is the most frequent size, or the diameter associated with the highest point on the fraction function curve. It can be determined by setting the derivative of the fraction function equal to zero and solving for d_p. For symmetrical distributions, such as the normal distribution, the mean, median, and mode will have the same value, which is the diameter of the axis of symmetry. For an asymmetrical or skewed distribution, these quantities will have different values. The median is commonly used with skewed distributions because extreme values in the tail have less effect on the median than

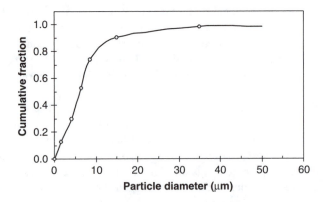

Figure 3.2   Cumulative fraction of the particle population.

on the mean. Most particle size distributions for indoor environments are skewed, with a long tail to the right:

$$\text{mode} < \text{median} < \text{mean} \tag{3.9}$$

For such a skewed distribution, the geometric mean can be directly read from an accumulative fraction diagram.[1] However, when such a cumulative graph is not available, the geometric mean must be calculated from the particle population or the grouped data. The geometric mean d_g is mathematically defined as the Nth root of the product of N particle diameters

$$d_g = \left(d_1 d_2 d_3 ... d_N\right)^{1/N} \tag{3.10}$$

Or, for grouped data with n groups, each group has F_i particles:

$$d_g = \left(d_1^{F_1} d_2^{F_2} d_3^{F_3} ... d_n^{F_n}\right)^{1/N} \tag{3.11}$$

The geometric mean can also be expressed in terms of $\ln(d)$ by converting Equation 3.10 to natural logarithms:

$$\ln d_g = \frac{\Sigma F_i \ln d_i}{N} \tag{3.12}$$

or

$$d_g = \exp\left[\frac{\Sigma F_i \ln d_i}{N}\right] \tag{3.13}$$

For monodisperse particles, the mean is equal to the geometric mean $d_{cm} = d_g$, where d_{cm} refers to count mean; otherwise, $d_g < d_{cm}$. The geometric diameter is widely used in lognormal distribution.

The *count mean diameter*, or the arithmetic mean of the number distribution d_{cm} for a particle population N or for grouped particles, is defined as

$$d_{cm} = \left(\frac{\Sigma d_p}{N}\right) = \frac{1}{N}\sum_{i=1}^{n} F_i d_i \tag{3.14}$$

Example 3.1: *The data in Table 3.2 were obtained using a particle counter within a building airspace. What is the most frequent particle diameter? What is the count median and count mean diameter?*

Table 3.2   Measured Particle Size Distribution

Particle Diameter Range (μm)	Count/ml
0.3–0.5	28
0.5–0.7	24
0.7–1.0	26
1.0–5.0	32
5.0–10	12
10–50	3
Total	N = 125

Solution: First, the midpoint diameter for each size range is determined. The standardized frequency for each size group can be calculated using Equation 3.5. Other terms related to the CMD and count mean diameter are $\Sigma F_i ln(d_i)$ and $\Sigma F_i d_i$. All calculated variables based on the given data are summarized in Table 3.3.

Table 3.3   Calculated Particle Statistics

Δd_i	d_i	F_i	$F = Fi/\Delta d_i$	$F_i lnd_i$	$F_i d_i$
0.2	0.4	28	140	−25.656	11.2
0.2	0.6	24	120	−12.260	14.4
0.3	0.85	26	86.7	−4.225	22.1
4	3	32	8	35.156	96
5	7.5	12	2.4	24.179	90
40	30	3	0.075	10.204	90
Total		$N = \Sigma F_i = 125$		$\Sigma F_i ln(d_i) = 27.4$	$\Sigma F_i d_i = 323.7$

According to the definition of the count mode, which is the most frequent particle size, the count mode is 0.4μm.

Using Equation 3.13, the count median, or geometric mean diameter of the particles, is

$$d_g = \exp\left[\frac{\Sigma F_i lnd_i}{N}\right] = \exp\left[\frac{27.4}{125}\right] = 1.255 \ (\mu m)$$

Using Equation 3.14, the count mean diameter, or the arithmetic mean of the number distribution, is

$$d_{cm} = \frac{1}{N}\sum_{i=1}^{n} F_i d_i = \frac{323.7}{125} = 2.59 \ (\mu m)$$

3.2 MASS DISTRIBUTION

Mass distribution, or distribution of mass with respect to particle size, is the most important particle distribution as far as indoor air quality is concerned. Most threshold limit values of particulate matter for human exposure are based on mass concentrations, as described in Chapter 2. Using the same example data set in Table 3.1 and assuming the particle population has a density ρ_p, the midsize particle mass and the total mass M of the particle population can be calculated.

$$m_i = \frac{\pi d_i^3}{6} \rho_p \qquad (3.15)$$

$$M = \sum m_i F_i \qquad (3.16)$$

where m_i is in kg, d_i is in meters, and ρ_p is in kg/m³. The mass fraction, or, to be more exact, the standardized mass fraction Δf_{im}, is in µm⁻¹, and the accumulative mass fraction f_m can be calculated:

$$\Delta f_{im} = \frac{m_i F_i / \Delta d_i}{M} \qquad (3.17)$$

$$f_m = \sum f_{im} = \sum \frac{m_i F_i}{M} = 1 \qquad (3.18)$$

Figure 3.3 and Figure 3.4 show the standardized mass fraction and accumulative mass fraction, respectively, with respect to particle size. In practice, many particle

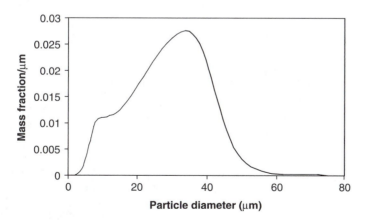

Figure 3.3   Mass fraction distribution of a particle population in Table 3.1.

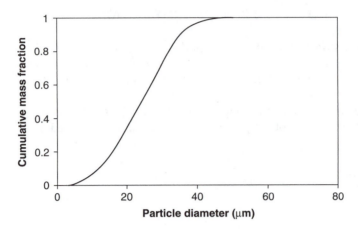

Figure 3.4   Cumulative mass fraction of a particle population in Table 3.1.

populations do not have a smooth normal distribution. When a distribution function is not available or does not fit a given function, statistical parameters should be calculated using their definitions. For example, the count mean diameter should be calculated using Equation 3.13.

The commonly used averages for mass distribution include the diameter of average mass, mass mean diameter, and mass median diameter. The averages of mass distribution are perhaps most confusing and deserve careful clarification. Let us start with the example data set in Table 3.4. The average mass of the particle population $\overline{m}$, can be obtained:

$$\overline{m} = \frac{\sum m_i F_i}{N} = \frac{M}{N} \qquad (3.19)$$

Experimentally, $\overline{m}$ can be obtained either by weighing each size group or by weighing the entire particle population and dividing by the total particle number N.

Table 3.4   Illustration of Statistical Variables for Mass Distribution for Grouped Data, Assuming that $\rho_p = 1000$ kg/m³ and that All Particles Are Spheres

Size Range,[a] (μm)	Δd_i (μm)	Midsize Diameter, d_i (μm)	Frequency F_i	Midsize Mass, m_i (mg)	Mass Fraction, Δf_{im} (μm⁻¹)	Accumulative Mass Fraction, f_m
0–3	3	1.5	90	1.7672E-09	3.2498E-05	9.75E-05
3–5	2	4	125	3.351E-08	0.00128388	0.002665
5–7	2	6	165	1.4379E-07	0.00727206	0.017209
7–10	3	8.5	165	3.2156E-07	0.01084133	0.049733
10–20	10	15	115	1.7672E-06	0.01245761	0.174309
20–50	30	35	60	2.2449E-05	0.02752309	1
> 50			0		0	1
Total:			$N = 720$	$M = 0.00163$		1

[a] Upper size range is smaller than, and lower size range is equal to, the size indicated.

Assuming that all particles are spheres and have a density ρ_p, the diameter related to this average mass, called the *diameter of the average mass* $d_{\overline{m}}$, has the following relationship with the average mass:

$$\overline{m} = \frac{\pi d_{\overline{m}}^3}{6} \rho_p \tag{3.20}$$

Note that the total mass M is

$$M = \sum \frac{\pi d_p^3}{6} \rho_p = \frac{\pi \rho_p}{6} \sum d_p^3 = N \frac{\pi \rho_p}{6} d_{\overline{m}}^3 \tag{3.21}$$

$$d_{\overline{m}} = \left(\frac{6\overline{m}}{\pi \rho_p}\right)^{1/3} = \left(\frac{6M}{\pi \rho_p N}\right)^{1/3}$$

$$= \left(\frac{6}{\pi \rho_p N} \frac{\pi \rho_p}{6} \sum d_p^3\right)^{1/3} = \left(\frac{\sum F_i d_p^3}{N}\right)^{1/3} \tag{3.22}$$

The *mass mean diameter* d_{mm} for the particle population N with a total mass M, is defined as

$$d_{mm} = \left(\frac{m_1}{M} d_1 + \frac{m_2}{M} d_2 + \cdots + \frac{m_N}{M} d_N\right) \tag{3.23}$$

Or, for grouped particles with midpoint diameter d_i and midpoint mass, m_i:

$$d_{mm} = \frac{1}{M} \sum_{i=1}^{n} F_i m_i d_i$$

$$= \frac{6}{\pi \rho_p \sum F_i d_i^3} \frac{\sum F_i \pi \rho_p d_i^3 d_i}{6} = \frac{\sum F_i d_i^4}{\sum F_i d_i^3} \tag{3.24}$$

Although it is analogous to the count mean diameter and has an explicit mathematical expression, the mass mean diameter has no physical meaning because it averages the diameter (first order of length), yet weighs according to its mass (third order of length), which contributes to the total mass of the particle population.

The mass median diameter (MMD), also called the volume median diameter, is the diameter at which half of the mass has smaller diameters and half of the mass

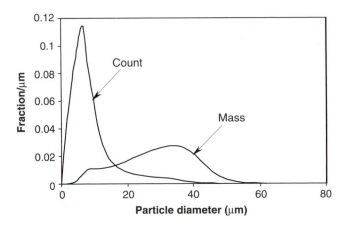

Figure 3.5   Comparison of count and mass fraction distribution for the same particle population as shown in Table 3.1.

has larger diameters for a particle population. For grouped particles, MMD can be
calculated using the following equation:

$$MMD = \exp\left(\frac{\sum F_i d_i^3 \ln d_i}{\sum F_i d_i^3}\right)$$ (3.25)

Comparing the fraction distributions of particle count and particle mass for the
same particle population as shown in Table 3.1, the mass fraction distribution is
shifted toward the larger particle size. This is because the few large particles make
up a large portion of the total mass. Figure 3.5 reveals the following:

- The averages of mass are larger than their counterpart averages of count. For
  example, the mass mean diameter is larger than the count mean diameter.
- The two fraction distributions are not necessarily in the same pattern, which
  indicates that the distributions may not be the same.

For these reasons, care should be taken in using distribution functions.

Example 3.2: *Using the same data as in Example 3.1, find the mass median diameter
(MMD) and the mass mean diameter* (d_{mm})*.*

Solution: Using the midpoint diameters d_i and frequencies F_i for each size group,
sums of terms are calculated and summarized in Table 3.5.
    Using Equation 3.25, the mass median diameter of the particles is

$$MMD = \exp\left[\frac{\sum F_i d_i^3 \ln d_i}{\sum F_i d_i^3}\right] = \exp\left[\frac{286,639.34}{86,949.44}\right] = 27.02 \ (\mu m)$$

Table 3.5   Calculated Particle Fractions and Statistics

d_i	F_i	$F_i d_i^3 \ln d_i$	$F_i d_i^3$	$F_i d_i^4$
0.4	28	−1.64	1.79	0.72
0.6	24	−2.65	5.18	3.11
0.85	26	−2.59	15.97	13.57
3	32	949.20	864.00	2592.00
7.5	12	10200.45	5062.50	37968.75
30	3	275496.99	81000.00	2430000.00
	$N = \Sigma F_i = 125$	$\Sigma F_i d_i^3 \ln d_i$ $= 286639.76$	$\Sigma F_i d_i^3$ $= 86949.44$	$\Sigma F_i d_i^4$ $= 2470578.15$

Using Equation 3.24, the mass mean diameter is

$$d_{mm} = \frac{\sum F_i d_i^4}{\sum F_i d_i^3} = \frac{2,470,578.15}{86,949.44} = 28.41 \ (\mu m)$$

3.3 OTHER MOMENT AVERAGES AND DISTRIBUTIONS

The statistics of particles are much more complicated than many other number statistics, such as birth rate, financial gain, or tourist population, because many particle statistics are derived from a certain power (moment) of the particle size. For example, the statistics of particle surface are proportional to the square (second moment) of particle diameter. The statistics of particle volume or mass are proportional to the cubic of particle diameter (third moment). The particle number statistics, which are dimensionless, can be considered the zero power (zero moment) of the particle diameter. These characteristics, related to a certain moment of the particle size, make particle statistics unique. For example, what does $\$^2$ or $baby^3$ mean statistically? For particles, d^0 is proportional to the particle count and d^3 is proportional to the particle mass. In addition to these physical implications, a more practical reason for the particle moment expression is that particle sizes are often measured directly. The particle size then can be calculated using the moment statistics. For example, particle mass distribution and density can be measured directly using weighing methods. The particle diameters can then be calculated using the moment statistics.

The averages of particle count (zero moment) and mass (third moment) are used mostly in particle statistics and indoor air quality, due largely to the available measurement technology and the threshold limit values. There are two other types of averages: particle length or diameter (d^1 or first moment), and particle surface area (d^2 or second moment). Of these two, the averages of particle surface areas deserve special consideration, even though the measurement may be extremely difficult. For example, many air-cleaning devices, such as filtration and electrostatic precipitation, depend on the particle surface area. Therefore, averages of particle surfaces could be particularly useful statistics in the design of air-cleaning technologies.

The averages for particle surface areas can be obtained in a similar way to those of their counterparts for averages of particle mass. Let us consider a particle population as shown in Table 3.1. The average particle surface area $\bar{s}$ can be calculated as

$$\bar{s} = \frac{\sum F_i s_i}{N} = \frac{\pi}{N} \sum F_i d_i^2 \tag{3.26}$$

where F_i is the frequency of ith group, and N is the total number of particles. The *diameter of average surfaces* $d_{\bar{s}}$ can then be calculated as

$$d_{\bar{s}} = \left(\frac{\bar{s}}{\pi} \right)^{1/2} = \left(\frac{\sum F_i d_i^2}{N} \right)^{1/2} \tag{3.27}$$

The *surface mean diameter* d_{sm}, also called Sauter diameter or mean volume–surface diameter, for the particle population N with a total surface area S, is defined as

$$d_{sm} = \left(\frac{S_1}{S} d_1 + \frac{S_2}{S} d_2 + \cdots + \frac{S_N}{S} d_N \right) \tag{3.28}$$

Or, for grouped particles with midpoint diameter d_i and midpoint mass m_i

$$d_{sm} = \frac{1}{S} \sum_{i=1}^{n} F_i s_i d_i = \frac{\pi \sum F_i d_i^2 d_i}{\pi \sum F_i d_i^2} = \frac{\sum F_i d_i^3}{\sum F_i d_i^2} \tag{3.29}$$

Again, the surface mean diameter has no physical meaning because it averages the diameter (first order of length) yet weighs according to its surface area (second order of length), which contributes to the total surface area of the particle population.

The surface median diameter (SMD) is the diameter at which half of the total surface has smaller diameters and half of the total surface area has larger diameters for a particle population. For grouped particles, the surface median diameter can be calculated using the following equation:

$$SMD = \exp \left(\frac{\sum F_i d_i^2 \ln d_i}{\sum F_i d_i^2} \right) \tag{3.30}$$

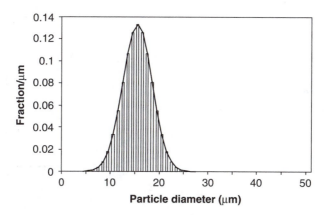

Figure 3.6   A particle population with a normal count distribution. The geometric mean diameter $d_g = 15$ μm, and the standard deviation σ = 4 μm.

3.4 THE LOGNORMAL DISTRIBUTION

This section discusses the best defined and most widely used particle distribution: *lognormal distribution*. Lognormal distribution is a transformation of normal distribution. In nature, many populations are normally distributed. Such normal distributions include the heights and weights of a group of grown people, exam grades of a class, and service life of a batch of light bulbs. The plot of a normal distribution is symmetrical with its arithmetic mean (Figure 3.6). In a normally distributed particle population, the number fraction function *df* can be uniquely defined by the geometric mean d_g and the standard deviation σ.

$$df = \frac{1}{\sigma\sqrt{2\pi}} \exp\left(-\frac{(d_p - d_g)^2}{2\sigma^2}\right) d(d_p) \qquad (3.31)$$

$$\sigma = \left(\frac{\sum F_i (d_i - d_g)^2}{N-1}\right)^{\frac{1}{2}} \qquad (3.32)$$

Normal distribution has many advantages, especially in its ease of analysis and the simple relationships among its properties. However, the size distribution of airborne particles is usually not normally distributed. It has a skewed distribution (with a long tail at the large particle sizes). Figure 3.7 shows such a skewed size distribution with the same CMD (15 μm) as in Figure 3.6. Apparently, the distributions in Figure 3.6 and Figure 3.7 are quite different. Such skewed distribution is very difficult to analyze, and the relationships among the particle properties can only be calculated individually using the equations described previously.

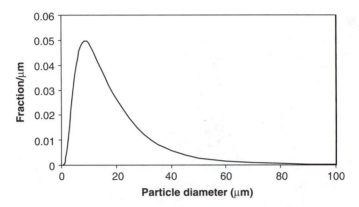

Figure 3.7   A skewed normal distribution with a long tail to the larger side. *CMD* = 15 μm and $\sigma_g = 2$.

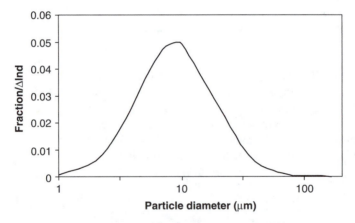

Figure 3.8   When a skewed distribution with a long tail to the larger side is plotted on a log scale, the distribution appears to be normally distributed.

Properties and rules for analyzing the normal distribution can no longer be applied directly to such a skewed distribution. However, when the fraction is plotted against the log scale of the particle size, the fraction distribution is approximately normally distributed. Figure 3.8 shows the same particle population as in Figure 3.7, with the particle size on a logarithmic scale.

Because the *x*-axis is in log scale, the distribution becomes lognormal. In a lognormal distribution, the fraction function can be derived directly from that of the normal distribution by replacing d_g with *lnCMD* and σ with $ln\sigma_g$:

$$df = \frac{1}{\ln \sigma_g \sqrt{2\pi}} \exp\left(-\frac{(\ln d_p - \ln CMD)^2}{2(\ln \sigma_g)^2}\right) d(\ln d_p) \qquad (3.33)$$

$$\ln \sigma_g = \left(\frac{\sum F_i (\ln d_{pi} - \ln CMD)^2}{N-1} \right)^{\frac{1}{2}} \tag{3.34}$$

where CMD is the count median diameter and σ_g is the geometric standard deviation (GSD). In a lognormal distribution, the geometric mean diameter equals the count median diameter. Note that either the natural logarithm or the logarithm to the base 10 can be used. The natural logarithm is used in this book. It also should be pointed out that the terms lnd_p and $lnCMD$ have already been standardized in Equation 3.33 and Equation 3.34, and they really should be $ln(d_p/d_0)$ and $ln(CMD/d_0)$, respectively, where $d_0 = 1$ μm.

Lognormal distribution has been applied in many real-world phenomena. Incomes, populations (of humans, animals, and bacteria), stock processes, and environmental contaminants are some examples. It is particularly widely used in particle size distribution, especially for those populations from single sources.

Equation 3.33 gives the fraction function of particles having a diameter range of (lnd_p) and $[lnd_p + d(lnd_p)]$. Because $d(lnd_p) = d(d_p)/d_p$, Equation 3.33 can be written in terms of particle size d_p rather than $d(lnd_p)$:

$$df = \frac{1}{d_p \ln \sigma_g \sqrt{2\pi}} \exp\left(-\frac{(\ln d_p - \ln CMD)^2}{2(\ln \sigma_g)^2} \right) d(d_p) \tag{3.35}$$

Although Equation 3.35 is written in terms of CMD and σ_g, it is important to understand that the particle surface area and mass distributions are also lognormal and can be written in terms of SMD and MMD, respectively, as

$$df = \frac{1}{d_p \ln \sigma_g \sqrt{2\pi}} \exp\left(-\frac{(\ln d_p - \ln SMD)^2}{2(\ln \sigma_g)^2} \right) d(d_p) \tag{3.35b}$$

$$df = \frac{1}{d_p \ln \sigma_g \sqrt{2\pi}} \exp\left(-\frac{(\ln d_p - \ln MMD)^2}{2(\ln \sigma_g)^2} \right) d(d_p) \tag{3.35c}$$

For normal distribution, 95% of the particles fall within a size range defined by $d_g \pm 2\sigma$. For lognormal distribution, the distribution is normal with respect to lnd_p and has a standard deviation of $ln\sigma_g$. Analogous to normal distribution, the 95% of the particles with a lognormal distribution fall within a size range defined by

$$\exp\left[lnCMD \pm 2ln\sigma_g \right] \tag{3.36}$$

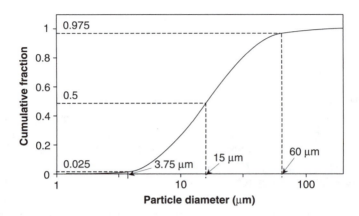

Figure 3.9 Cumulative fraction of a particle population with a lognormal distribution.

This size range is asymmetrical and goes from CMD/σ_g^2 to $CMD\,\sigma_g^2$. For $\sigma_g = 2.0$, 95% of the particles would have sizes between one-fourth and four times the count median diameter. For example, for the particle sample in Figure 3.8 with $CMD = 15$ μm and $\sigma_g = 2$, 95% of the total particles fall in the size range of 3.75 to 60 μm.

The cumulative count fraction for the particle population represented in Figure 3.7 is plotted on the logarithmic scale of d_p and shown in Figure 3.9. The cumulative fraction 0.5 value intersects with the CMD (15 μm). Particles smaller than CMD/σ_g^2 (3.75 μm) and particles larger than $CMD\sigma_g^2$ (60 μm) are less than 5% of the total particle count.

The total fraction between the particle size range of [0, ∞], or the ranges of lnd_p [-∞, ∞] is a known function:

$$f = \int_0^\infty \frac{1}{d_p \ln\sigma_g \sqrt{2\pi}} \exp\left(-\frac{(\ln d_p - \ln CMD)^2}{2(\ln\sigma_g)^2}\right) d(d_p)$$

$$= \int_{-\infty}^\infty \frac{1}{\ln\sigma_g \sqrt{2\pi}} \exp\left(-\frac{(\ln d_p - \ln CMD)^2}{2(\ln\sigma_g)^2}\right) d(\ln d_p)$$

$$= 1 \tag{3.37}$$

In many air quality problems, we need to know the fraction between two particle diameters d_{p1} and d_{p2} ($0 < d_{p1} < d_{p2}$), or the fraction of particles smaller than a given diameter in a particle population with a lognormal distribution. Thus, we encounter problems in solving the following equation:

$$f(d_{p2} - d_{p1}) = \int_{d_{p1}}^{d_{p2}} \frac{1}{d_p \ln \sigma_g \sqrt{2\pi}} \exp\left(-\frac{(\ln d_p - \ln CMD)^2}{2(\ln \sigma_g)^2}\right) d(d_p)$$

(3.38)

$$= \int_{\ln d_{p1}}^{\ln d_{p2}} \frac{1}{\ln \sigma_g \sqrt{2\pi}} \exp\left(-\frac{(\ln d_p - \ln CMD)^2}{2(\ln \sigma_g)^2}\right) d(\ln d_p)$$

Integrating Equation 3.38 has been proved to be extremely difficult. In fact, it is one of the unsolved mathematical problems with a definite range. However, with indefinite ranges of the integrals, Equation 3.38 can be solved. The following procedure is useful in obtaining the value of $f(d_{p2}-d_{p1})$, the fraction of a given particle size range. Let

$$K_p = \frac{\ln d_p - \ln CMD}{\ln \sigma_g}$$

(3.39)

Similar to Equation 3.35b and Equation 3.35c, K_p values for particle surface area fractions and mass fractions can be written in terms of surface median diameter or mass median diameter, respectively:

$$K_p = \frac{\ln d_p - \ln SMD}{\ln \sigma_g}$$

(3.39b)

$$K_p = \frac{\ln d_p - \ln MMD}{\ln \sigma_g}$$

(3.39c)

Taking derivatives of both sides of Equation 3.39 (noting that CMD and σ_g are constants) gives

$$d(\ln d_p) = \ln \sigma_g dK_p$$

(3.40)

Substituting Equation 3.39 and Equation 3.40 into Equation 3.38 gives

$$f(d_{p2} - d_{p1}) = \int_{K_{p1}}^{K_{p2}} \frac{1}{\sqrt{2\pi}} \exp\left(-\frac{K_p^2}{2}\right) dK_p$$

$$= \int_{-\infty}^{K_{p2}} \frac{1}{\sqrt{2\pi}} \exp\left(-\frac{K_p^2}{2}\right) dK_p - \int_{-\infty}^{K_{p1}} \frac{1}{\sqrt{2\pi}} \exp\left(-\frac{K_p^2}{2}\right) dK_p$$

$$= f(d_{p2}) - f(d_{p1}) \tag{3.41}$$

The two terms at the right side of Equation 3.41 represent two integrals of a normal distribution function with respect to K_p: $f(d_{p2})$ is the fraction of particles smaller than d_{p2}, and $f(d_{p1})$ is the fraction of particles smaller than d_{p1}.

For indoor air quality problems, the particles of concern are usually within the size range of 0.01 to 200 μm, and the K_p values within a range of −3 to 3. Typical values for $f(d_p)$ with respect to K_p are listed in Table 3.6. More detailed normal distribution values are listed in Appendix 4. If the K_p falls between the two adjacent values, an interpoint of the two adjacent $f(d_p)$ values can be selected using linear interpolation.

Equation 3.41 and Table 3.6 combined are very useful for finding the fraction of a given particle size range, if the CMD and GSD are given. The fraction of mass can also be obtained by substituting the CMD with the mass median diameter (MMD). A variety of commercial particle instruments measure the CMD and GSD of particles. Thus, the particle count or mass of any specified size range can be obtained.

Example 3.3: *The total particle count in a building is 100 particles/ml. The particle size distribution is lognormal. The particle analyzer also gives the geometric mean diameter as 1.5 μm, and the geometric standard deviation is 4. What is the number concentration of particles between 0.3 and 10 μm?*

Solution: Because $d_{p1} = 0.3$μm and $d_{p2} = 10$ μm, $CMD = 1.5$μm and $\sigma_g = 4$.

$$K_{p1} = \frac{\ln d_{p1} - \ln CMD}{\ln \sigma_g} = \frac{\ln 0.3 - \ln 1.5}{\ln 4} = -1.161$$

$$K_{p2} = \frac{\ln d_{p2} - \ln CMD}{\ln \sigma_g} = \frac{\ln 10 - \ln 1.5}{\ln 4} = 1.368$$

From Table 3.6, we find $f(d_{p1}) = 0.123$, and $f(d_{p2}) = 0.9148$.
The fraction between the size range d_{p1} and d_{p2} is

$$f(d_{p2} - d_{p1}) = f(d_{p2}) - f(d_{p1}) = 0.7918$$

Because the total number concentration $C_N = 100$ particles/ml, the number concentration between the size range d_{p1} and d_{p2} is

$$C_{N2-1} = f(dp2 - dp1) \times CN = 0.7918 \times 100 = 79.2 \text{ (particles/ml)}$$

Table 3.6 Typical Values for $f(d_p)$ with Respect to K_p, where $f(d_p)$ Is the Fraction of Particles Smaller than d_p

K_p	$f(d_p)$	K_p	$f(d_p)$	K_p	$f(d_p)$
-3.00	0.001350	-1.00	0.1587	1.00	0.8413
-2.95	0.001589	-0.95	0.1711	1.05	0.8531
-2.90	0.001866	-0.90	0.1841	1.10	0.8643
-2.85	0.002186	-0.85	0.1977	1.15	0.8749
-2.80	0.002555	-0.80	0.2119	1.20	0.8849
-2.75	0.002980	-0.75	0.2266	1.25	0.8944
-2.70	0.003467	-0.70	0.2420	1.30	0.9032
-2.65	0.004025	-0.65	0.2578	1.35	0.9115
-2.60	0.004661	-0.60	0.2743	1.40	0.9192
-2.55	0.005386	-0.55	0.2912	1.45	0.9265
-2.50	0.006210	-0.50	0.3085	1.50	0.9332
-2.45	0.007143	-0.45	0.3264	1.55	0.9394
-2.40	0.008198	-0.40	0.3446	1.60	0.9452
-2.35	0.009387	-0.35	0.3632	1.65	0.9505
-2.30	0.01072	-0.30	0.3821	1.70	0.9554
-2.25	0.01222	-0.25	0.4013	1.75	0.9599
-2.20	0.01390	-0.20	0.4207	1.80	0.9641
-2.15	0.01578	-0.15	0.4404	1.85	0.9678
-2.10	0.01786	-0.10	0.4602	1.90	0.9713
-2.05	0.02018	-0.05	0.4801	1.95	0.9744
-2.00	0.02275	0	0.5000	2.00	0.9773
-1.95	0.02559	0.05	0.5199	2.05	0.9798
-1.90	0.02872	0.10	0.5398	2.10	0.9821
-1.85	0.03216	0.15	0.5596	2.15	0.9842
-1.80	0.03593	0.20	0.5793	2.20	0.9861
-1.75	0.04006	0.25	0.5987	2.25	0.9878
-1.70	0.04457	0.30	0.6179	2.30	0.9893
-1.65	0.04947	0.35	0.6368	2.35	0.9906
-1.60	0.05480	0.40	0.6554	2.40	0.9918
-1.55	0.06057	0.45	0.6736	2.45	0.9929
-1.50	0.06681	0.50	0.6915	2.50	0.9938
-1.45	0.07353	0.55	0.7088	2.55	0.9946
-1.40	0.08076	0.60	0.7257	2.60	0.9953
-1.35	0.08851	0.65	0.7422	2.65	0.9960
-1.30	0.09680	0.70	0.7580	2.70	0.9965
-1.25	0.1056	0.75	0.7734	2.75	0.9970
-1.20	0.1151	0.80	0.7881	2.80	0.9974
-1.15	0.1251	0.85	0.8023	2.85	0.9978
-1.10	0.1357	0.90	0.8159	2.90	0.9981
-1.05	0.1469	0.95	0.8289	2.95	0.9984

3.5 LOG-PROBABILITY GRAPHS

In the previous sections, our particle analysis has been based on known statistics, such as CMD and GSD. However, many particle instruments measure only the size and the corresponding count or mass concentration. If the particle distribution is lognormal, using log-probability graphs can facilitate much of the practical appli-

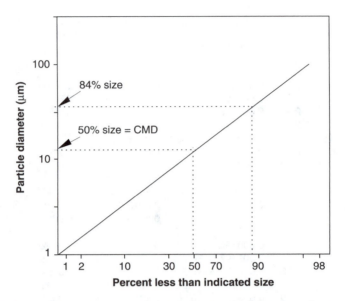

Figure 3.10 The cumulative fraction of a particle population with a lognormal size distribution is a straight line when plotted on a log-probability graph.

cation of lognormal size distribution. In a log-probability graph (Figure 3.10), the *x*-axis is the cumulative particle fraction in probability scale. A *probability scale* is a percentage scale symmetric to the 50% point. It is compressed around the median (50% of the cumulative value) and expands near the both ends of the axis. The *y*-axis is the particle diameter in log scale. Such graphs are available commercially.

The cumulative particle fraction appears to be a straight line when it is plotted in such a log-probability graph (Figure 3.10). The median diameter can be read directly from the intersection point with 50% cumulative fraction. The slope of the line represents the range of particle sizes. A steep slope indicates a large particle size range, and a shallow slope indicates a small particle size range. If the line is horizontal, the particles are monodisperse, that is, all particle sizes coincide with the median size. If the distribution is normal and the *y*-axis is linear, the cumulative fraction is also a straight line and the graph is called *probability graph.*

In using a log-probability graph, one needs two sets of data: particle size (or size range) and cumulative fraction of the particles. An example of such data sets is shown in Table 3.1, where cumulative particle count fractions are calculated against size ranges. When particle size range is given, the midpoint diameter can be used. However, other weighted mid-diameters can be used to plot against the corresponding cumulative fraction. The weighing methods depend on the types of statistics, such as count distribution or mass distribution. For example, for the size range of 5 to 10 μm, the midpoint diameter is 7.5 μm. However, the weighted midpoint diameter may be 7 μm (less than the geometric midpoint) for particle count cumulative fraction, because most of the particles in that size range may have smaller diameters than 7.5 μm. Also, the weighted midpoint diameter may be 8 μm for particle mass cumulative fraction, because most of the particle mass in that size

Probability by 3 Cycle Log (Long-axis)

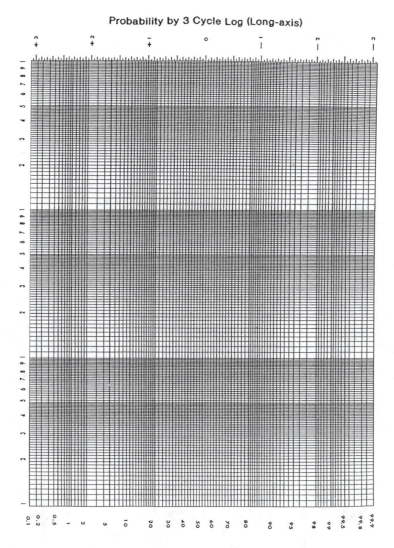

Figure 3.11 A log-probability graph. The horizontal axis should be the cumulative fraction of particles, and the vertical axis should be the particle size.

range is contributed by particles larger than 7.5 µm. Computer programs are available to construct log-probability graphs. These programs can calculate the GSD and medians, but the user must provide the weighted data.

For a normal distribution, the standard deviation represents the difference between d at 84% of its cumulative value ($d_{84\%}$) and d at 50% of its cumulative value ($d_{50\%}$), or the difference between d at 50% of its accumulated value and d at 16% of its accumulated value ($d_{16\%}$).

$$\sigma = d_{84\%} - d_{50\%} = d_{50\%} - d_{16\%} \tag{3.42}$$

For a lognormal distribution that is normal with respect to *lnd*, as shown in Figure 3.8, the differences are in log scale:

$$\ln \sigma_g = \ln d_{84\%} - \ln d_{50\%} = \ln \left(\frac{d_{84\%}}{d_{50\%}} \right)$$

$$= \ln d_{50\%} - \ln d_{16\%} = \ln \left(\frac{d_{50\%}}{d_{16\%}} \right) \tag{3.43}$$

$$\sigma_g = GSD = \left(\frac{d_{84\%}}{d_{50\%}} \right) = \left(\frac{d_{50\%}}{d_{16\%}} \right) = \left(\frac{d_{84\%}}{d_{16\%}} \right)^{\frac{1}{2}} \tag{3.44}$$

In Equation 3.44, the geometric standard deviation of particles with a lognormal distribution is a ratio of two particle diameters; hence, it is dimensionless and must be equal to or greater than 1. When $\sigma_g = 1$, the particles are monodisperse, that is, all particles have the same diameter. The dimensionless feature of GSD for lognormal distribution has one unique advantage in particle analysis: Because the GSD is only related to a particle size ratio, not the particle size, all weighted (moment) distributions of any lognormal distribution will be lognormal and have the same GSD. This means they will have the same shape and be parallel to each other when plotted on a logarithmic scale. Figure 3.12 shows the distribution of count and mass plotted on the same logarithmic diameter scale.

The mass distribution has the same shape as the count distribution, but it is displaced along the axis by a constant amount equal to *MMD/CMD*. The ratio

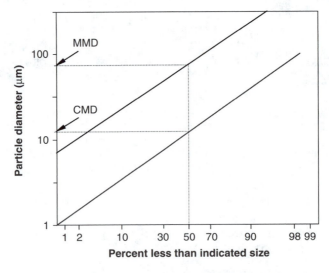

Figure 3.12 Cumulative count and cumulative mass distributions for the same particle population on a log-probability graph.

MMD/CMD can be calculated knowing only the GSD. If the two lines are not parallel, the midpoint diameters are not weighed properly. Using a computer program, one can adjust the weighed diameters to make the two lines parallel. The relationship among the GSD and average diameters of a particle population with a lognormal distribution is described by the Hatch–Choate conversion equation.

3.6 THE HATCH–CHOATE CONVERSION EQUATION

It is often necessary to know other characteristics, such as the mass median diameter or the surface median diameter, based on the measurement of one characteristic of particle size, such as the count median diameter. The statistical parameters of particle size distributions such as the medians, means, and diameters of average properties can be calculated from a detailed data set containing the sizes and the counts or masses of particles using the equations described in Section 3.3. This process is time consuming and may not be accurate if the detailed data are insufficient. However, if the size distribution is lognormal, any other parameter d_x can be converted from the count median diameter *CMD* and the geometric standard deviation σ_g.

$$d_x = CMD \exp(b \ln^2 \sigma_g) \tag{3.45}$$

Equation 3.45 is called the Hatch–Choate equation to commemorate the original study of Hatch and Choate (1929).[2] The b value in the Hatch–Choate equation depends on the types of averages and the distribution momentum. The b values for different average diameters are listed in Table 3.7. Detailed descriptions of the b values and the derivation of the Hatch–Choate equation can be found in Hinds (1999).[1] The corresponding equations to calculate the averages from raw data (total particle population or grouped particles) are also listed in Table 3.7. It is important to remember that those equations can be applied to any type of size distribution, whereas the Hatch–Choate equation is applicable only to the lognormal distribution.

Although Equation 3.45 is written in terms of CMD, it can be applied to any other type of averages; that is, any other parameters can be converted from one type of average and the geometric standard deviation σ_g. If one type of average rather than CMD is known and we need to find another type of average, we use the known value to find CMD first, then use the CMD to find the parameter in question.

Example 3.4*: The particle concentration in a room is measured using two instruments: a particle counter and a mass sampler. The particle size distribution is lognormal with a GSD of 2.5. In two air samples with the same volume, the particle counter recorded 6×10^7 particles, the mass sampler collected 0.8 mg of total mass, and the density of the particles was 1000 kg/m³. What is the mass median diameter (MMD)?*

Table 3.7   The *b* Values in the Hatch–Choate Equation for Different Types of Average Diameters

Type of Average Diameter	Symbol	Statistical Equation	*b* Values
Mode	$\hat{d}$		−1
Count median diameter, geometric mean	CMD, d_g	$\exp\left(\dfrac{\sum F_i \ln d}{N}\right)$	0
Count mean diameter	$\bar{d}$	$\dfrac{\sum F_i d}{N}$	0.5
Length median diameter	LMD	$\exp\left(\dfrac{\sum F_i d \ln d}{\sum F_i d}\right)$	1
Diameter of average surface	$d_{\bar{s}}$	$\left(\dfrac{\sum F_i d^2}{N}\right)^{\frac{1}{2}}$	1
Length mean diameter	$d_{\bar{l}}$	$\dfrac{\sum F_i d^2}{\sum F_i d}$	1.5
Diameter of average mass	$d_{\bar{m}}$	$\left(\dfrac{\sum F_i d^3}{N}\right)^{\frac{1}{3}}$	1.5
Surface median diameter	SMD	$\exp\left(\dfrac{\sum F_i d^2 \ln d}{\sum F_i d^2}\right)$	2
Surface mean diameter, Sauter diameter, mean volume–surface diameter	d_{sm}	$\dfrac{\sum F_i d^3}{\sum F_i d^2}$	2.5
Mass median diameter, volume median diameter	MMD	$\exp\left(\dfrac{\sum F_i d^3 \ln d}{\sum F_i d^3}\right)$	3
Mass mean diameter, volume mean diameter	d_{mm}	$\dfrac{\sum F_i d^4}{\sum F_i d^3}$	3.5

Solution: Because the average mass per particle $\overline{m}$, can be calculated from the total number and the total mass of the particles

$$\overline{m} = \frac{M}{N} = \frac{8 \times 10^{-7}(kg)}{6 \times 10^{7}(particles)} = 1.333 \times 10^{-14}\,(kg)$$

The diameter of average mass $d_{\bar{m}}$ can be obtained:

$$d_{\bar{m}} = \left(\frac{6 \times \bar{m}}{\pi \rho}\right)^{1/3} = \left(\frac{6 \times 1.333 \times 10^{-14}(kg)}{\pi \times 1000 (\frac{kg}{m^3})}\right)^{1/3}$$

$$= 2.94 \times 10^{-6} (m) = 2.94 (\mu m)$$

From Table 3.7, the b value for the diameter of average mass is 1.5. Substituting $b = 1.5$, $d_{\bar{m}} = 2.94$ μm, and $\sigma_g = 2.5$ into the Hatch–Choate equation and rearranging gives

$$CMD = \frac{d_{\bar{m}}}{\exp(b \ln^2 \sigma_g)}$$

$$= \frac{2.94 (\mu m)}{\exp(1.5 \times \ln^2 2.5)} = 0.834 (\mu m)$$

Applying the Hatch–Choate equation again and noting that $b = 3$ for MMD yields

$$MMD = CMD \exp(b \ln^2 \sigma_g)$$

$$= 0.834 \times \exp(3 \times \ln^2 2.5) = 10.4 (\mu m)$$

3.7 OTHER TYPES OF PARTICLE SIZE DISTRIBUTIONS

The most commonly used distribution for characterizing aerosol particle size is the lognormal distribution described previously. Many particle populations, especially particles from a single source, are lognormal in size distribution. Several less common distributions, however, have been found useful for specific types of aerosols.[1] In the following equations, a and b are empirical constants having different values for each distribution.

3.7.1 The Rosin–Rammler Distribution

Rosin and Rammler (1933) developed this distribution, originally based on sizing crushed coal, and it is applicable to coarsely dispersed dusts and sprays.[3] It is particularly useful for size distributions that are more skewed than the lognormal distribution and for sieve analysis. The mass fraction between d_p and $d_p + dd_p$ is defined as

$$df_m = abd_p^{b-1} \exp(-ad_p^b) dd_p \tag{3.46}$$

where a depends on the fineness of the particles and b depends only on the material. This distribution can be used when cutoff points for the smallest and largest diameters are well defined. Expressed in particle count distributions, as in Equation 3.46, the Rosin–Rammler distribution has the same form as the Weibull distribution:

$$df = abd_p^{b-1} \exp(-ad_p^b) \, dd_p \qquad (3.47)$$

3.7.2   The Nukiyama–Tanasawa Distribution

Nukiyama and Tanasawa (1939) developed this distribution, which is used for sprays having extremely broad size ranges.[4] The count distribution of particles with diameters between d_p and $d_p + dd_p$ is given by

$$df = ad_p^2 \exp\left(-\frac{b}{d_p^3}\right) dd_p \qquad (3.48)$$

where the empirical constants a and b are functions of a nozzle constant.

3.7.3   The Power-Law Distribution

This size distribution function has been applied to the size distribution of atmospheric aerosols. The number of particles with sizes between d_p and $d_p + dd_p$ is given by

$$df = ad_p^{-b} \, dd_p \qquad (3.49)$$

3.7.4   The Exponential Distribution

The exponential distribution has been applied to powdered materials. The number of particles with sizes between d_p and $d_p + dd_p$ is given by

$$df = a \exp(-bd_p) \, dd_p \qquad (3.50)$$

The total number of particles is a/b.

3.7.5   The Khrgian–Mazin Distribution

The size distributions of cloud droplets are described by the Khrgian–Mazin distribution.[5] The number of droplets per unit volume with sizes between d_p and $d_p + dd_p$ is given by

$$df = ad_p^2 \exp(-bd_p) \, dd_p \qquad (3.51)$$

Total number concentration is $2a/b^3$, and the mean diameter is $3/b$.

3.7.6 Chen's Empirical Distribution

This size distribution may be used for particle populations with mixed sources. Chen et al. (1995) studied dust particles in animal buildings that contained a variety of dust sources such as grain, fecal material, dander, and hair.[6] The number of droplets per unit volume with sizes between d_p and $d_p + dd_p$ is given by

$$df = \exp(-ad_p)\sin(bd_p) \tag{3.52}$$

All of the previous models are two-parameter models. The empirical constants a and b for each model must be determined before the model can be used. Many methods, such as parameter analysis, can be used to calculate a and b. However, care must be exercised in obtaining representative data for application of any of these statistical methods.

DISCUSSION TOPICS

1. Can you name one practical example of a particle surface distribution that may be useful?
2. Why are particle count distribution and mass distribution with respect to particle size more frequently used?
3. Why are statistics of particle size more complicated than other types of statistics, such as the height of adult people or the valuation of a stock?
4. Intuitively, mass is proportional to volume and density. Why is there no density involved in those particle average diameters related to mass, such as diameter of average mass or mass mean diameter?
5. For a particle population, the standard deviation for diameter, σ, has the same unit as the diameter, while the geometric standard deviation, GSD or σ_g, is dimensionless. Why?
6. If the particle size distribution is unknown, how do you plan to obtain the statistics of a particle population? Is a particle counter that gives particle concentrations at different sizes adequate to do the job?
7. Two methods are used to characterize particle statistics and distributions: calculation and graph (log-probability graph). What are the limitations of each method?

PROBLEMS

1. Use the measured data in Table 3.8 to calculate the arithmetic mean (or count mean), geometric mean (or CMD), mass median (MMD), and diameter of average mass. F_i is the frequency of the ith size particles.

Table 3.8   Particle Data for Problem 1

d (µm)	F_i
1	3
3	5
5	2
8	1

2. A sample of particle count concentrations in an indoor airspace was measured as in Table 3.9. Assume that the particles are spherical with standard density (1000 kg/m³). Calculate and plot the fractions vs. particle size for particle count, surface area, and mass. Find the modes and median diameters for particle count, surface area, and mass.

Table 3.9   Particle Data for Problem 2

d (µm)	F_i
1	30
3	70
5	100
8	80
10	50
12	30
14	15
16	5
18	1

3. Use a log-probability graph to determine the CMD and GSD (σ_g) of the dust sample shown in Table 3.10, assuming that the dust sample has a lognormal distribution.

Table 3.10 Particle Data for Problem 3

d (µm)	F_i
0.3–0.5	7
0.5–1	6
1–2	7
2–3	5
3–5	7
5–10	12
10–20	5
20–30	1
Total	50

4. Repeat problem 3, using the following particle diameters instead of the mid-point diameter:
   a. Lower size of the range
   b. Upper size of the range
5. A dust sample has a lognormal distribution. The count median diameter is 0.5 µm and the GSD is 1.8. The dust number concentration is measured at 50 particles/ml.

What is the mass concentration in mg/m³? The particles are spherical with a density of 2000 kg/m³.

6. A dust has a lognormal particle size distribution. It is known that 95% of the particle count is within the size range of 0.3 to 12 μm. What are the CMD and the GSD of the dust sample?

7. Dust mass concentration in an animal building is measured at 2.8 mg/m³ using a filter sampler. Assume that the dust has a lognormal distribution, with a count median diameter (CMD) of 2.5 μm and a geometric standard deviation (GSD) of 2.5. The dust particles are spherical and have a density of 2000 kg/m³.

    a. Find the dust number concentration.

    b. Find the mass concentration of particles smaller than 4 μm.

8. The particle count distribution is lognormal as follows:

$$df = 0.576 \exp\left(-\frac{(\ln d_p + 0.223)^2}{0.96}\right) d(\ln d_p)$$

    a. What is the particle mass distribution?

    b. What is the particle surface area distribution?

9. Dust concentration in a vehicle-assembling building is measured at 120 parti-cles/ml using a particle counter. The particle counter also gives a CMD of 1.5 μm and GSD of 2.5. The dust particles are spherical and have a standard density. What are the number concentration and the mass concentration for particles smaller than 10 μm in diameter?

10. In specifying a lognormal dust distribution in a problem building, a two-stage cascade particle impactor, it was found that 10% of the total particle mass are particles smaller than 10 μm, and 99% is smaller than 100 μm. Calculate the MMD and GSD of the dust particles.

11. With the same measurement data as in problem 10, use a log-probability graph to determine the MMD and GSD.

12. Particle concentration in a dust storm is 5 mg/m³. The particle distribution is lognormal, with CMD = 5 μm and GSD = 2.5. In order to reduce the dust concentration to 0.3 mg/m³ indoors, what should the filter porous size be? Assume that the particles are spherical with a density of 1500 kg/m³ and that all particles smaller than the porous size will pass through the filter.

REFERENCES

1. Hinds, C.W., *Aerosol Technology*, John Wiley & Sons, New York, 1999.

2. Hatch, T. and Choate, S.P., Statistical description of the size properties of non-uniform particulate substances, *J. Franklin Inst.*, 207:369, 1929.

3. Rosin, O.G. and Rammler, E., Particle mass distribution for coarse crushed coal, *J. Inst. Fuel*, 7:29, 1933.

4. Nukiyama, S. and Tanasawa, Y., *Transactions of Mechanical Engineering (Japan)*, 5:63, 1939.

5. Pruppacher, H.R. and Klett, J.D., *Microphysics of Clouds and Precipitation*, 2nd ed., Kluwer, Dordrecht, the Netherlands, 1997.

6. Chen, Y., Zhang Y., and Barber, E.M. A new mathematical model of particle size distribution for swine building dust, *Trans. Am. Soc. Heat. Refrig. Air Cond. Engineers*, 101(2):1169–1178, 1995.

CHAPTER 4

Mechanics of Particles

Particle mechanics governs many principles of particle measurement instrumentation and air-cleaning technologies. This chapter provides the fundamentals of particle mechanics. Because particle mechanics involves both the particles (which are treated as discrete subjects) and their carrying fluids (which are treated as a continuous medium), particle mechanics is inevitably associated with fluid mechanics. Unlike fluid mechanics, in which the primary concerns are mass and energy transfer carried by the fluid itself, particle mechanics is concerned with the mechanical behaviors of the particles relative to the carrying fluid. These relative behaviors include relative velocity, particle Reynolds number, terminal velocity, relaxation time, and stopping distance. Two unique standardized parameters are introduced and will be widely used in this book: particle Reynolds number and aerodynamic diameter. The former standardizes the fluid conditions around the particle of concern, and the latter standardizes the particle sizes based on their aerodynamic behavior.

By completing this chapter, the reader will be able to

- Determine Reynolds numbers for fluid (Re) and for particles (Re_p), and explain their implications to flow conditions and drags on particles.
- Apply Newton's resistance law to particles at any flow conditions that are a wide range of Re, the Stokes region, and beyond.
- Apply Stokes's law to particle mechanics, and understand that Stokes's law is just a particular case of Newton's resistance law. The majority of indoor air particulate matter transportation falls into the Stokes region.
- Determine the following mechanical properties of particles by means of analytical methods, experimental methods, or a combination of the two, depending on the availability of variables:
  - Cunningham slip correction factor C_c
  - Terminal settling velocity V_{TS}
  - Mechanical mobility B
  - Dynamic shape factor χ
  - Aerodynamic diameter d_a

- Relaxation time τ
- Stopping distance s

Of the properties just listed, aerodynamic diameter and terminal settling velocity are perhaps the most important and commonly used, because d_a is a standardized description of the particle, and the V_{TS} can be measured with relative ease. The other mechanical properties can then be derived.

4.1 REYNOLDS NUMBERS FOR FLUIDS AND PARTICLES

General fluid dynamic equations (continuity, energy conservation, and momentum) include the shear stress, which is a function of the viscosity. These general equations are complicated, nonlinear partial differential equations, and usually there are no general solutions. To understand the dynamic properties and behavior of a fluid, Osborne Reynolds investigated the dynamic similarity of different flows.[1] Two flow cases can be considered dynamically similar when they are geometrically similar and have a similar pattern of streamlines. Geometric similarity refers to the corresponding linear dimensions that have a constant ratio. Streamline similarity refers to the pressures at corresponding points that have a constant ratio. Reynolds found that the dimensionless group, $\rho UL/\eta$, must be the same for two similar flows, where ρ is the density of the fluid, U is a characteristic velocity, L is a characteristic length, and η is the viscosity of the fluid. This dimensionless group is now called the *Reynolds number:*

$$\text{Re} = \frac{\rho U L}{\eta} \tag{4.1}$$

Of the four variables in determining the Reynolds number, the characteristic length L is the most difficult variable to define for a poorly defined flow, such as most indoor air flows. However, the other three variables, ρ, U, and η, can be directly measured. For most flows with physical boundaries, the characteristic length can be estimated using the following equation:

$$L = \frac{4 A_w}{P_w} \tag{4.2}$$

where A_w and P_w are the wetted area and the wetted perimeter, respectively. The ratio of A_w/P_w is called the *hydraulic radius*. For a circular duct, the hydraulic radius is only half of its geometric radius. From Equation 4.2, the characteristic length of a circular duct is the duct diameter D.

Because the initial force F_I of an element is proportional to its mass (which is proportional to the density ρ and its dimensions) and velocity, and because the shear or frictional force F_τ is proportional to the viscosity, the Reynolds number can be

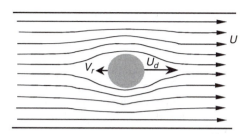

Figure 4.1   Streamlines of a fluid flow between two parallel walls with a sphere in the fluid. The fluid velocity is U, and the sphere has a relative velocity V_r to the fluid flow.

viewed as the ratio of inertial force to shear or frictional force acting on an element of the flow in steady state.

$$\text{Re} \propto \frac{F_I}{F_\tau} \tag{4.3}$$

The inertial force, according to Newton's second law, is equal to the fluid element's change of momentum

$$F_I = \frac{d(mV_t)}{dt} = V_t \frac{dm}{dt} + m \frac{dV_t}{dt} \tag{4.4}$$

where m is the mass of the fluid element, and dV_t/dt is the total acceleration of the fluid element.

The total acceleration dV_t/dt contains two components:

- An acceleration due to the change in the total flow system dU/dt
- An acceleration resulting from the change of positions of that fluid element along the moving direction UdU/dx

The UdU/dx is proportional to the fluid velocity and the change of velocity with position. For example, when the fluid flows around an elbow tube, the acceleration of a fluid element is related to its velocity and the location in the elbow. Thus, the inertial force can be expressed as

$$F_I = V_t \frac{dm}{dt} + m \left(\frac{dU}{dt} + U \frac{dU}{dx} \right) \tag{4.5}$$

For incompressible flow at steady state, $dm/dt=0$ and $dU/dt =0$. The mass of the fluid element is the density ρ times its volume, which is proportional to its L^3, where L is a characteristic length to represent any linear length, such as the diameter of a tube, a distance of travel, or the diameter of a sphere. Thus, the inertial force can be written as

$$F_I \propto \rho\, L^3 U \frac{dU}{dL} \tag{4.6}$$

The frictional force F_τ is proportional to the viscosity η, the surface area A, and the velocity change perpendicular to the moving direction, dU/dy, where y is a direction perpendicular to the moving direction x. Noting that the characteristic length L represents any linear length, the area $A \propto L^2$ and $y \propto L$. Thus, the frictional force can be expressed as

$$F_\tau \propto \eta\, L^2 \frac{dU}{dL} \tag{4.7}$$

Substituting Equations 4.6 and 4.7 into Equation 4.2 yields Equation 4.1.

The nature of a given flow of an incompressible fluid is characterized by its Reynolds number. Reynolds numbers for fluid flow are commonly divided into three regimes:

- $Re < 2000$ for laminar flow
- $2000 < Re < 4000$ for transient between laminar and turbulent flow
- $Re > 4000$ for turbulent flow

For large values of Re, one or all of the terms in the numerator are large compared with the denominator, that is, large volume or expansion of the fluid, high velocity, great density and small viscosity of the fluid, or a combination of all of these. A large Re (> 4000) implies that the inertial force predominates the flow, while the frictional force is often negligible. A large Re indicates a highly turbulent flow with losses proportional to the square of the velocity. The turbulence may be large in scale, such as an air jet that encounters an obstacle and swirls in a room. Such large-scale turbulence creates large eddies, which carry most of the mechanical energy. The large eddies generate many smaller eddies that rapidly convert the mechanical energy into irreversibilities via their viscous action. The turbulent flow may also be small in scale, such as the air flow near the exit of a diffuser. A small-scale turbulent flow has small eddies and small fluctuations in velocity with high frequency. Generally, the turbulence intensity increases as the Reynolds number increases.

For intermediate values of Re ($2000 < Re < 4000$), both inertial and frictional effects on the flow are important. The changes in viscosity result in changes in velocity distribution and the resistance to the flow. Examples of such a transient flow are the room air at some boundaries, such as along a wall or around a human body.

For small values of Re ($Re < 2000$), the flow is said to be laminar. Because gases, including air, usually have very small viscosities (the denominator of the Reynolds number), the Re values are usually large. For indoor air movement, the characteristic length L is large (in the order of meters). Even at very low air velocities, such as 0.2 m/s, the room airflow is still highly turbulent. On the other hand, some indoor air applications can be laminar flow. For example, an air diffuser with an opening

1 cm wide and a discharge air velocity of 2 m/s has an *Re* value of 1326, which is a laminar flow under standard air conditions.

Many Reynolds numbers are in use today, depending on the situations of flow and the problems of concern. For example, the motion of water in a pipe may be characterized by $\rho UD/\mu$, where ρ is the density of water, U is the velocity of water, D is the diameter of the pipe, and η is the viscosity of the water. Another example is the Reynolds number of a sphere moving in a fluid, $\rho UD/\mu$, where ρ is the density of the fluid, U is the velocity of the sphere, D is the diameter of the sphere, and η is the viscosity of the fluid. Although the expressions of the Reynolds numbers are the same in both examples, the physical parameters and the values of *Re* may be quite different. The difference is a result of the flow conditions and the problems of concern. In this book, there are two different Reynolds numbers: *Re* for fluid (usually air) flow, and Re_p for airborne particles. The subscript p distinguishes the Reynolds number of particles from that of fluid.

The Reynolds number for fluid is described in Equation 4.1, where ρ is the density of the fluid, U is the velocity of the fluid, L is a characteristic length of the flow (such as the diameter of an air duct), and η is the viscosity of the fluid.

The Reynolds number for a particle moving in a fluid is expressed as

$$\mathrm{Re}_p = \frac{\rho V_r d_p}{\eta} \tag{4.8}$$

where ρ is the density of the fluid, V_r is the relative velocity of the particle to the fluid, d_p is the diameter of the particle, and η is the viscosity of the fluid. The relative velocity of a particle, V_r, relative to the fluid flow velocity U, can be calculated as

$$V_r = \left| \vec{U}_d - \vec{U} \right| = \left((U_{dx} - U_x)^2 + (U_{dy} - U_y)^2 + (U_{dz} - U_z)^2 \right)^{\frac{1}{2}} \tag{4.9}$$

where $\vec{U}_d$ and $\vec{U}$ are velocity vectors for the particle and the carrying fluid, respectively. Subscripts x, y, and z are components for a cardinal coordinate.

When the motions of the particle and the carrying fluid are on the same line, but not necessarily in the same direction, the relative velocity becomes

$$V_r = U_d - U \tag{4.10}$$

If V_r is a positive value, it is the same direction as the U_d; if V_r is negative, the direction of V_r is opposite to the direction of U_d. There is often some confusion about the calculations of *Re* for the fluid flow and Re_p for a particle in the fluid flow (Figure 4.1, page 73). The following items may be helpful in clarifying this issue:

- The density ρ and viscosity η of the fluid are the same for the fluid flow Reynolds number and the particle Reynolds number.

- For a fluid flow Reynolds number, the characteristic velocity is the fluid velocity U, and the characteristic length L is the characteristic length of the flow field, such as the diameter of the tube containing the fluid flow.
- For a particle Reynolds number, the characteristic velocity is the relative velocity of the particle to the surrounding fluid, V_r, and the characteristic length L is the diameter of the particle, d_p.

The particle Reynolds number characterizes the mechanical properties of a particle in a fluid flow. These particle mechanical properties are discussed in detail in later sections of this chapter. Particle Reynolds numbers have four regions: laminar (more commonly called the Stokes region), transient, turbulent, and Newton. Reynolds numbers and typical characteristics of each region of particle motion are described in the following list:

Stokes region — $Re_p < 1$. Fluid flow around the particle is laminar. Frictional force exerted on the particle is predominant, and the inertial force is negligible.

Transient region — $1 < Re_p < 5$. Turbulence starts to occur around the particle. Both inertial force and frictional force are important to the particle's behavior.

Turbulent region — $5 < Re_p < 1000$. Fluid flow around the particle is turbulent. Drag coefficient of the particle decreases as the Re_p increases.

Newton's region — $Re_p > 1000$. Fluid flow around the particle is highly turbulent. Drag coefficient of the particle remains approximately constant.

In summary, the Reynolds number has the following properties:

It is an index of the flow regime that serves as the benchmark to indicate whether the flow is laminar or turbulent, or how the resistance to a particle changes.

Because it can be viewed as the ratio of inertial force to frictional force, its value is important to determining which parameters are most important to the flow.

It provides a means to study a similar flow using a geometrically similar experimental approach.

Example 4.1: *The cross section of the hatch cabinet at a poultry facility is 1 m wide and 0.5 m high. Air velocity through the cabinet (perpendicular to the cabinet cross section) is 0.2 m/s. A newborn chicken walks downstream (in the same direction as the air flow in the cabinet) along the length of the hatch cabinet at a speed of 0.05 m/s. The chicken can be approximated as a sphere with a diameter of 5 cm. The density of air is 1.2 kg/m³ and the viscosity is 1.81 x 10⁻⁵ N·s/m².*

a. Determine the Reynolds number for the airflow in the hatch cabin.
b. Determine the Reynolds number for the chicken.
c. Determine the Reynolds number if the chicken were a 50 μm particle.

Solution:

a. The characteristic length L is calculated from the width and height of the hatch:

$$L = \frac{4A_w}{P_w} = \frac{4 \times (1 \times 0.5)}{2 \times (1 + 0.5)} = 0.67 \ (m)$$

$$Re = \frac{\rho U L}{\eta} = \frac{1.2 \times 0.2 \times 0.67}{1.81 \times 10^{-5}} = 8,884$$

The airflow in the hatch cabin is turbulent because $Re = 8884 > 2000$.

b. Because the chicken is walking in the same direction as the airflow in the cabinet, the relative velocity of the chicken to the airflow is

$$V_r = 0.05 \, (m/s) - 0.2 \, (m/s) = -0.15 \, (m/s)$$

The minus sign (–) of V_r indicates that the direction of V_r is opposite to the chicken's direction of motion.

$$Re_p = \frac{\rho V_r d_p}{\eta} = \frac{1.2 \times 0.15 \times 0.05}{1.81 \times 10^{-5}} = 497$$

The airflow around the chicken is turbulent because $Re_p = 497 > 1$.

c. If the chicken were a 50 μm particle, the Reynolds number for the particle would be

$$Re_p = \frac{\rho V_r d_p}{\eta} = \frac{1.2 \times 0.15 \times 0.00005}{1.81 \times 10^{-5}} = 0.497$$

The airflow around the particle becomes laminar because $Re_p = 0.497 < 1$.

4.2 NEWTON'S RESISTANCE LAW

One of the most important mechanical properties of particles is the resistance of the carrying gas in which the particle is traveling. In analyzing the resistance to the particle motion, steady-state motion and streamline motion are assumed. This assumption is especially valid for small particles, such as those that are airborne. The rationale will be justified in this chapter by the short relaxation time and stopping distance of airborne particles. For airborne particles typically smaller than 100 μm, the particle motion in the air almost instantaneously becomes steady state, regardless of its initial conditions.

Newton's resistance law applies to particle motion with high particle Reynolds numbers ($Re_p > 1000$). In such a high Re_p region, the predominant force that governs the particle motion is the inertial force. Other forces, such as frictional force and gravity, become negligible. Originally, Newton reasoned that the resistance encountered by a cannonball traveling through air is a result of the acceleration of the air that has to be pushed aside to allow the sphere to pass through (Figure 4.2).

Applying Newton's second law, the resistance force or the drag force is proportional to the change of momentum of the replaced gas:

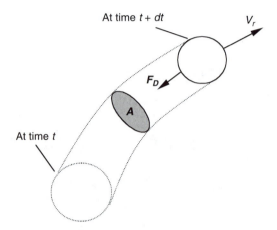

Figure 4.2   The resistance encountered by the sphere is the momentum changes of the gas being pushed aside by the sphere.

$$F_D = K\frac{d(m_g V_r)}{dt} = K\left(V_r \frac{dm_g}{dt} + m_g \frac{dV_r}{dt}\right) \tag{4.11}$$

where K is a constant of proportionality, m_g is the mass of gas that has been displaced by the motion of the sphere, and V_r is the velocity of the sphere. Newton initially thought that K was a constant that did not vary with the velocity. The change of mass with respect to time, dm_g/dt, is equal to the volume of gas that has been pushed aside in one second:

$$\frac{dm_g}{dt} = \rho_g A V_r = \rho_g \frac{\pi d^2}{4} V_r \tag{4.12}$$

where A is the cross-sectional area of the sphere (Figure 4.2). Because the particle motion is steady state, $dV_r/dt = 0$. Substituting Equation 4.12 into 4.11 gives

$$F_D = K\left(V_r \frac{dm_g}{dt}\right) = K\rho_g \frac{\pi d^2}{4} V_r^2 \tag{4.13}$$

Equation 4.13 is the restricted form of Newton's resistance equation, which is only valid for particle motion with very high values of particle Reynolds number ($Re_p > 1000$). This is true because the equation is derived based on the inertial force without considering the viscous effect. At lower values of Re_p, the frictional force becomes more important. Thus, the proportional constant K is a constant only for high Re_p. A modified Newton's resistance equation applicable for entire Re_p ranges can be written as follows:

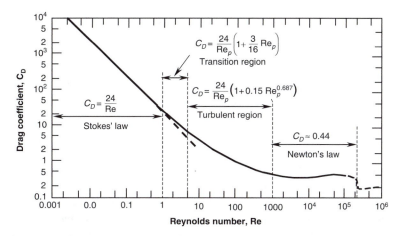

Figure 4.3   Drag coefficient of spheres vs. the particle Reynolds number.

$$F_D = C_D \frac{\pi}{8} d^2 \rho_g V_r^2 \tag{4.14}$$

where C_D is the drag coefficient that is dependent on the particle Reynolds number.

The drag coefficient C_D is the most important property in the mechanics of particle motion and fluid dynamics, because most applications in real life are employed either to overcome or to utilize the drag caused by the fluid. Some examples of these applications are designing airplanes, ships, and air-cleaning devices to separate airborne particles, which will minimize the drag. Parachutes, on the other hand, are designed to increase the drag. Drag force also varies with particle shape. In this book, the particles we deal with are spheres or have been normalized (geometrically or aerodynamically) as equivalent spheres. Figure 4.3 shows the drag coefficient of spheres with respect to the sphere's Reynolds number. The drag coefficient for other shapes can be found from many fluid dynamics references.[2]

Drag coefficients of spheres in Figure 4.3 vary with Reynolds number in a highly nonlinear fashion. For most engineering applications, however, the drag coefficient can be calculated based on four regions of particle Reynolds number:

- Low Reynolds number region ($Re_p < 1$)
- Transient region ($1 < Re_p < 5$)
- Turbulent region ($5 < Re_p < 1000$)
- High turbulent region ($Re_p > 1000$)

Of the four, drag coefficient in the transient region is the most complicated, and several equations have been proposed based on regressions of experimental data.[3, 4] Most of these equations are sufficiently accurate and generally within 4% for $Re_p < 800$ and within 7% for $Re_p < 1000$.

$$C_D = \frac{24}{Re_p} \quad \text{(for } Re_p < 1) \tag{4.15}$$

$$C_D = \frac{24}{Re_p}\left(1 + \frac{3}{16}Re_p\right) \quad \text{(for } 1 < Re_p < 5) \tag{4.16}$$

$$C_D = \frac{24}{Re_p}(1 + 0.15\,Re_p^{0.687}) \quad \text{(for } 5 < Re_p < 1000) \tag{4.17}$$

$$C_D = 0.44 \quad \text{(for } Re_p > 1000) \tag{4.18}$$

For airborne particles, Re_p values rarely exceed 1. In some extreme situations, such as in a dust storm or a volcanic eruption, particles may fall in the lower range of transitional Reynolds numbers. In indoor air quality applications, although the room air is almost always highly turbulent ($Re > 2000$), the particle Reynolds numbers are almost always smaller than 1, which is well within the Stokes region.

4.3 STOKES'S LAW

The general Navier–Stokes equation for incompressible flow is an application of Newton's second law, that is, the change of momentum of a fluid element is equal to the total external forces exerted on the fluid element of concern. For airborne particles, the gravity and the buoyancy are negligible compared with other external forces, such as the frictional force and static pressure (head pressure). Thus, the general Navier–Stokes equation for each of the x, y, and z directions can be simplified as follows:

$$\rho\left(\frac{\partial U}{\partial t} + U\frac{\partial U}{\partial x} + U\frac{\partial V}{\partial y} + U\frac{\partial W}{\partial z}\right) = -\frac{\partial P}{\partial x} + \eta\left(\frac{\partial^2 U}{\partial x^2} + \frac{\partial^2 U}{\partial y^2} + \frac{\partial^2 U}{\partial z^2}\right) \tag{4.19}$$

$$\rho\left(\frac{\partial V}{\partial t} + V\frac{\partial U}{\partial x} + V\frac{\partial V}{\partial y} + V\frac{\partial W}{\partial z}\right) = -\frac{\partial P}{\partial y} + \eta\left(\frac{\partial^2 V}{\partial x^2} + \frac{\partial^2 V}{\partial y^2} + \frac{\partial^2 V}{\partial z^2}\right) \tag{4.20}$$

$$\rho\left(\frac{\partial W}{\partial t} + W\frac{\partial U}{\partial x} + W\frac{\partial V}{\partial y} + W\frac{\partial W}{\partial z}\right) = -\frac{\partial P}{\partial z} + \eta\left(\frac{\partial^2 W}{\partial x^2} + \frac{\partial^2 W}{\partial y^2} + \frac{\partial^2 W}{\partial z^2}\right) \tag{4.21}$$

where U, V, and W are velocity components corresponding to the x, y, and z directions, respectively. P is the pressure. The term $-\partial P/\partial x$ indicates that the pressure force is opposite to the direction of the motion. Equations 4.19 through 4.21 are generally unsolvable, because they are nonlinear partial differential equations. Stokes's law is a solution to Equations 4.19 through 4.21, based on the following assumptions:

- The inertial force is negligible compared to the frictional force. This assumption is true for particle Reynolds numbers smaller than 1. Thus, the term $UdU/dx \approx 0$, or the carrying fluid is incompressible.
- The particle motion is in steady state, that is, $dU/dt = 0$.
- The particle is rigid and there is no other particle nearby.
- The velocity of fluid at the particle surface is zero.

The preceding assumptions are accurate for most indoor air situations.

The incompressibility assumption ($UdU/dx \approx 0$) refers to the fact that the air near the particles does not compress significantly when the particles flow through it; it does not imply that the air is incompressible. A moving particle can significantly compress the air only when it reaches the speed of sound, which is usually not the case in an indoor environment.

The particle motion in the air can achieve the steady state almost instantaneously. This will be discussed in detail in later sections on relax time and stop distance.

Particles with a much larger density than the air can be considered rigid. For example, the settling velocity of a water droplet in air is only 0.6% faster than predicted by Stokes's law, indicating that the deformation of a particle in the air flow and the effect on Stokes's law are negligible.

Only a wall within 10 times of the particle diameter will affect the drag force of the particle. Generally, only a small portion of particles can get near that distance in an indoor environment. Thus, the nearby wall effect can be neglected. For example, considering an extremely high particle concentration of 1000 particles/ml with a mean particle diameter of 10 µm, the mean distance between two particles is 1000 µm, that is, 100 times its mean diameter.

The assumption of zero-velocity at the particle surface can cause significant error for small particles. This error can be corrected by the slip correction factor, which will be discussed in the next section.

With the preceding assumptions, and considering particle motion in the x-direction only, Equation 4.19 can be reduced to only two terms: pressure and shear stress caused by viscosity, both opposite to the direction of the particle motion (or the fluid motion relative to the particle).

$$\eta \left(\frac{\partial^2 U}{\partial x^2} + \frac{\partial^2 U}{\partial y^2} + \frac{\partial^2 U}{\partial z^2} \right) - \frac{\partial P}{\partial x} = 0 \qquad (4.22)$$

Equation 4.22 is linear and can be solved if the boundary conditions are known. The drag force of the fluid exerted on the sphere is the sum of the pressure and frictional forces. Integrating the pressure stress and frictional force over the entire surface of the sphere gives the drag force exerted on the sphere:

$$F_D = \oiint_S \eta \left(\frac{\partial^2 U}{\partial x^2} + \frac{\partial^2 U}{\partial y^2} + \frac{\partial^2 U}{\partial z^2} \right) dS - \oiint_S \frac{\partial P}{\partial x} dS$$

$$= \oiint_S \tau \sin\theta \, dS - \oiint_S P \cos\theta \, dS \tag{4.23}$$

The first term at the right side of Equation 4.23 is caused by friction (or shear or viscous force), and the second term is caused by pressure. If the shear stress and pressure distribution are known, Equation 4.23 can be solved.[5, 6] Stokes found the following shear stress and pressure distribution exerted on a sphere surface with a diameter d_p in a spherical coordinate system, r, θ, and ϕ (Figure 4.4):

$$\tau = \frac{3\eta V_r}{d_p} \left(\frac{d_p}{2r} \right)^4 \sin\theta \tag{4.24}$$

$$P = P_0 - \frac{3\eta V_r}{d_p} \left(\frac{d_p}{2r} \right)^2 \cos\theta \tag{4.25}$$

where P_0 is the ambient pressure. Apparently, when $r >> d_p/2$, $P = P_0$, which means that the pressure caused by fluid is zero. In a spherical coordinate, as shown in Figure 4.4

$$dS = r^2 \sin\theta \, d\theta \, d\phi \tag{4.26}$$

where $0 < r = d_p/2$, $0 \le \theta \le \pi$, $0 \le \phi \le 2\pi$. Substituting Equations 4.24 through 4.26 into Equation 4.23, noting that $r = d_p/2$, and integrating, yields

$$F_D = \oiint_S \frac{3\eta V_r}{d_p} \sin\theta \, dS - \oiint_S \left(P_0 - \frac{3\eta V_r}{d_p} \cos\theta \right) dS$$

$$= \int_0^{2\pi} \int_0^{\pi} \left(\frac{3\eta V_r}{d_p} \sin^3\theta \right) \left(\frac{d_p}{2} \right)^2 d\theta \, d\phi + \int_0^{2\pi} \int_0^{\pi} \left(\frac{3\eta V_r}{d_p} \cos^2\theta - P_0 \right) \left(\frac{d_p}{2} \right)^2 \sin\theta \, d\theta \, d\phi$$

$$= 2\pi\eta V_r d_p + \pi\eta V_r d_p$$

$$= 3\pi\eta V_r d_p \quad \text{(for } Re_p < 1\text{)} \tag{4.27}$$

Equation 4.27 is Stokes's law, stating that the drag force exerted on a particle is proportional to the viscosity of the fluid (η), the relative velocity of the particle to the fluid (V_r), and the diameter of the particle (d_p), for a particle Reynolds number smaller than unity. The first term on the right side is caused by friction, and the second term is caused by pressure. In the Stokes region, the pressure contributes

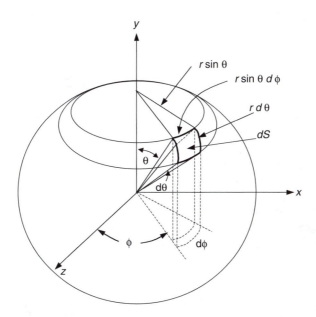

Figure 4.4   A spherical coordinate system where an element surface area defined by the
radius r and two angles θ and ϕ.

one-third of the total drag, and the frictional force contributes two-thirds of the total
drag force.

Now let us revisit Newton's resistance law, expressed by Equation 4.14. Because
Newton's general resistance equation (Equation 4.14) applies to the entire range of
Re_p values, it also holds for $Re_p < 1$. Equating Equation 4.14 and Equation 4.27 gives

$$C_D \frac{\pi}{8} \rho_g V_r^2 d_p^2 = 3\pi\eta V_r d_p \qquad (4.28)$$

Solving for C_D, noting that $Re_p = \rho_g V_r d_p/\eta$, verifies Equation 4.15:

$$C_D = \frac{24}{Re_p} \qquad (4.15)$$

Equation 4.15 represents the straight-line portion in Figure 4.3, where the drag
coefficient is inversely proportional to the particle Reynolds number.

From Newton's law to Stokes's law, there is a change of relationships between
the drag force and its affecting variables, V, d_p, ρ_g, and η. Newton's resistance
equation contains the properties of inertia ρ_g, and the drag force is proportional to
the squares of velocity and the sphere diameter. Although Stokes's law contains the
viscosity η, the drag force is proportional to the velocity and the sphere diameter.
The dependent variable of the drag force gradually changes from V_r^2 to V_r and from

d_p^2 to d_p as the Re_p decreases. This change is represented by the straight line in Figure 4.3.

4.4 SLIP CORRECTION FACTOR

One of the assumptions made during the derivation of Stokes's law was that the velocity of fluid at the surface of the particle was zero. This assumption is true for large particles. For small particles, the actual drag force is smaller than the value predicted in Equation 4.28. It appears that very small particles slip in the fluid. In fact, very small particles do slip with respect to the fluid. When the particle is sufficiently small, especially approaching the size of the mean free path of the gas, the probability of impacting with gas molecules is smaller; hence, the drag force is smaller. To take this slip effect into account, Stokes's law can be modified by a slip correction factor C_c. Thus, Stokes's law becomes

$$F_D = \frac{3\pi\eta V_r d_p}{C_c} \quad \text{(for } Re_p < 1) \tag{4.29}$$

The slip correction factor was first derived by Cunningham in 1910 for particles larger than 0.1 μm. Thus the slip correction factor is often called the *Cunningham correction factor*. Allen and Raabe proposed a slip correction factor equation for oil droplets and for solid particles for all particle sizes within 2.1% of accuracy.[7, 8] In summary, the slip correction factor may be calculated using the following equations:

$$C_c = 1 + \frac{2.52\lambda}{d_p} \quad \text{(for } d_p \geq 0.1 \text{ μm)} \tag{4.30}$$

$$C_c = 1 + \frac{\lambda}{d_p}\left[2.34 + 1.05\exp\left(-0.39\frac{d_p}{\lambda}\right)\right] \quad \text{(for } d_p < 0.1 \text{ μm)} \tag{4.31}$$

where λ is the mean free path of the carrying fluid. For air at standard conditions ($P = 101.325$ kPa and $T = 20°C$), $\lambda = 0.066$ μm. Mean free path for different gases can be calculated and is discussed in Chapter 5. Equation 4.31 can be applied to all sizes of particles. Because of its simplicity, Equation 4.30 is recommended for particles larger than 0.1 μm. For particles smaller than 0.1 μm, Equation 4.31 must be used.

The slip correction factor increases rapidly as the particle diameter decreases. For example, under standard conditions, $C_c = 1.0167$ for a 10 μm particle in air, the slip effect is only 1.67%. The slip effect in air increases to 16.7% for a 1 μm particle and 167% for a 0.1 μm particle. Therefore, the slip correction factor must be used for small particles, typically smaller than 3 μm. The slip correction factor for a 3 μm particle is 1.055, representing a 5.5% error if the slip effect is not considered.

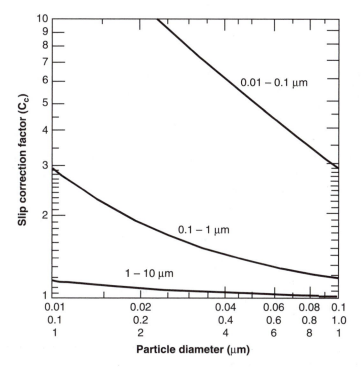

Figure 4.5 Slip correction factor at standard atmospheric conditions, plotted using Equation 4.30 and Equation 4.31.

Substituting $C_c = 1.055$ into Stokes's law, the drag force for the 3 μm particles is 95% of that without a slip effect. For most indoor air problems, slip effect is usually not a concern for particles larger than 3 μm. Typical values for C_c vs. particle diameters carried by air streams are plotted in Figure 4.5.

Generally, indoor air is at or near standard conditions, that is, near the sea level pressure of 101.325 kPa with a temperature of 20°C. Under such standard conditions, Equation 4.30 and Equation 4.31 are sufficiently accurate for most applications without considering the pressure effect. When ambient conditions change substantially, such as when the air is under vacuuming or pressurizing, the effect of the pressure and/or temperature must be considered.

For a given diameter of a particle, the slip correction factor is only dependent on the mean free path λ. From Boyle's ideal gas law and gas kinetics, the mean free path of a gas is inversely proportional to the molecular concentration n. The molecular concentration n is proportional to the absolute pressure and inversely proportional to the absolute temperature. Denoting λ_0, n_0, P_0, and T_0 as the mean free path, gas molecule concentration, absolute pressure, and temperature at standard conditions ($P_0 = 101.325$ kPa and $T_0 = 293$ K), the mean free path λ and the new molecule concentration n at a given absolute pressure P and temperature T have the following relationships with those of standard conditions:

$$\frac{\lambda_0}{\lambda} = \frac{n}{n_0} = \frac{PT_0}{P_0 T} \tag{4.32}$$

The mean free path of the gas molecules from Equation 4.32 is

$$\lambda = \frac{P_0 T}{PT_0}\lambda_0 \tag{4.33}$$

Substituting Equation 4.33 into Equation 4.30 and Equation 4.31, noting that P_0 = 101.325 kPa and T_0 = 293 K, gives the slip correction factor, including the effects of ambient pressure and temperature.

$$C_c = 1 + 0.8715\frac{\lambda_0}{d_p}\frac{T}{P} \quad \text{(for } d_p \geq 0.1 \ \mu m) \tag{4.34}$$

$$C_c = 1 + \frac{\lambda_0}{d_p}\left(\frac{T}{P}\right)\left[0.809 + 0.363\exp\left(-1.128\frac{d_p}{\lambda_0}\frac{P}{T}\right)\right] \quad \text{(for } d_p < 0.1 \ \mu m) \tag{4.35}$$

where λ_0 is the mean free path of the carrying gas molecules at standard conditions, P is in kPa, and T is in K. For air, $\lambda_0 = 0.066 \ \mu m$.

Example 4.2. *At the top of a mountain, the atmospheric pressure is measured as 70 kPa and the temperature as −20°C. What would the error of the slip correction factor be for a 0.3 μm particle if the effects of pressure and temperature were ignored?*

Solution: Without considering the effects of pressure and temperature

$$C_{C0} = 1 + 2.52\frac{\lambda}{d_p} = 1 + 2.52\frac{0.066}{0.3} = 1.55$$

When the effects of pressure and temperature are taken into account

$$C_C = 1 + 0.8715\frac{\lambda_0}{d_p}\frac{T}{P} = 1 + 0.8715\frac{0.066}{0.3}\times\frac{253}{70} = 1.69$$

The error of slip correction factor ε is caused by ignoring the effect of temperature and pressure:

$$\varepsilon = \frac{C_c - C_{c0}}{C_c}\times 100\% = \frac{1.69 - 1.55}{1.69}\times 100\% = 9\%$$

Table 4.1 Terminal Settling Velocity and Mobility for Typical Sizes of Particles with Standard Density under Standard Atmospheric Conditions

d_p (μm)	Slip Correction Coefficient C_c	Mobility B (m/N·s)	Terminal Settling Velocity V_{TS} (m/s)
0.001	224	1.32×10^{15}	6.9×10^{-9}
0.01	23.0	1.35×10^{13}	7.0×10^{-8}
0.1	2.67	1.56×10^{11}	8.0×10^{-7}
1	1.17	6.84×10^{9}	3.5×10^{-5}
10	1.02	5.98×10^{8}	0.0030
100	1.00	5.87×10^{7}	0.25

4.5 SETTLING VELOCITY AND MECHANICAL MOBILITY

The *settling velocity* of a particle is defined as the terminal free-fall velocity of the particle after it is released in still air. The terminal settling velocity is a constant for a given size of particle. When a particle is released into still air, the particle quickly approaches its terminal velocity and settles down at a constant speed. Apparently, in the Stokes region, the drag force exerted on the particle is equal to the relative gravity forces of the particle to the gas:

$$F_D = F_G = m_p g - m_g g \tag{4.36}$$

$$\frac{3\pi \eta V_r d_p}{C_c} = \frac{\pi (\rho_p - \rho_g) d_p^3 g}{6} \tag{4.37}$$

where F_G is the relative gravity of the particle to the gas, m_p is the mass of the particle, m_g is the mass of the gas with the same volume as the particle, and $g = 9.81$ N·m/s². The mass of gas represents the effect of buoyancy, which, in practice, can often be neglected because the ρ_p is much greater than ρ_g. For example, a water droplet has a density 800 times greater than the air. Neglecting the buoyancy effect of the air only gives 0.1% error when calculating the settling velocity. For indoor air quality problems, the buoyancy effect can be neglected. The settling velocity of an airborne particle, V_{TS} (which is the relative velocity of the particle to the air), can be derived from Equation 4.37.

$$V_{TS} = \frac{\rho_p d_p^2 g C_c}{18 \eta} \quad \text{(for } Re_p < 1.0) \tag{4.38}$$

Particle terminal settling velocity increases rapidly with the particle size because it is proportional to the square of the particle diameter (Table 4.1). The density of the gas within which the particle becomes airborne has no significant effect on the settling velocity. For particles 3 μm in diameter, the effect of the slip is approximately

5% on its terminal settling velocity; for particles 1 μm in diameter, the error increases to 17%. In most cases, we will neglect the slip effect for particles larger than 3 μm in diameter and still have 95% or better accuracy. However, C_c should be considered for particles smaller than 3 μm, because the errors become significant.

An important factor affecting the terminal settling velocity is the gravitational acceleration g, which is proportional to the external force $m_p g$, which causes the settling. Similarly, when a particle is subjected to other kinds of external forces, such as centrifugal or electrical force, the acceleration a_e can be calculated as follows:

$$a_e = \frac{F_e}{m_d} \tag{4.39}$$

where F_e is the external force in Newtons and m_d is the mass of the particle in kg. For example, the centrifugal acceleration a_e is V_T^2/R, where V_T is the tangential velocity and R is the radius of the circular motion. The terminal velocity of a particle in a centrifugal force field is

$$V_T = \frac{\rho_p d_p^2 C_c a_e}{18\eta} = \frac{\rho_p d_p^2 C_c V_T^2}{18\eta R} \quad \text{(for } Re_p < 1.0) \tag{4.40}$$

Note that V_T here is the tangential particle velocity, not the relative velocity of the particle to the fluid, V_r. The tangential particle velocity is equal to the fluid velocity of the circular motion in a steady flow. The terminal velocity caused by centrifugal force can be many times higher than the terminal settling velocity caused by gravity. For example, if a particle has a circular motion at a tangential velocity of 8 m/s with an $R = 0.3$ m, the terminal velocity caused by centrifugal force is 21.7 times higher than the terminal settling velocity. In this case, the gravity effect may be neglected. Terminal velocity can be applied to particle separation for particle sampling and air-cleaning technologies and will be discussed in detail in later chapters.

Particle mobility B is defined as the ratio of the particle velocity relative to the fluid (V_r) to the drag force exerted on the particle.

$$B = \frac{V_r}{F_D} = \frac{C_c}{3\pi\eta d_p} \tag{4.41}$$

Particle mobility is also called *mechanical mobility* to distinguish it from other mobilities, such as electrical mobility. The mobility has a unit of m/N·s. Particle mobility is an indicator of how "mobile" the particle is. A higher mobility value indicates that the particle has a high velocity relative to the fluid, that a small drag force is exerted on the particle, or that there is a combination of both. The mobility of the particle decreases as the viscosity of the fluid increases, and decreases as the diameter of the particles increases. In analogy, an airplane is more mobile than a ship of similar size because the viscosity of air is lower than the viscosity of water.

A motor boat is more mobile than an oil tanker because the motor boat is smaller than the oil tanker. Combining Equation 4.38, Equation 4.39, and Equation 4.41, the terminal velocity of a particle is simply the product of its mobility and the external force exerted on the particle:

$$V_{TS} = F_e B \qquad (4.42)$$

where the external force $F_e = m_p g$, when the particle is settling in the air, or $F_e = m_p V_T^2/R$ as the centrifugal force when the particle is in circular motion.

Example 4.3: *A grain of concrete dust is falling down to the floor through room air. The particle diameter is 2 μm and the particle density is 2500 kg/m³. Assuming that the room air is still, determine the terminal settling velocity, drag force, and mobility of the particle. The room air is at standard conditions.*

Solution: In order to calculate the settling velocity, we first determine the slip correction factor for the 2 μm particle. At standard conditions, air viscosity $\eta = 1.81 \times 10^{-5}$ Pa·s (N·s/m²), air mean free path $\lambda = 0.066$ μm, and air density $\rho = 1.2$ kg/m³.

$$C_C = 1 + 2.52 \frac{\lambda}{d_p} = 1 + 2.52 \frac{0.066}{2} = 1.08$$

Substituting C_c values into Equation 4.30 gives

$$V_{TS} = \frac{\rho_p d_p^2 g C_c}{18\eta} = \frac{2,000 \times (2 \times 10^{-6})^2 \times 9.81 \times 1.08}{18 \times (1.81 \times 10^{-5})} = 0.00026 \ (m/s)$$

Before applying Stokes's law to calculate the drag force, the particle Reynolds number must be examined to ensure that the particle motion is in the Stokes region.

$$\text{Re}_p = \frac{\rho V_{TS} d_p}{\eta} = \frac{1.2 \times 0.00026 \times (2 \times 10^{-6})}{1.81 \times 10^{-5}} = 3.45 \times 10^{-5}$$

Because $Re_p \ll 1$, Stokes's law applies.

$$F_D = \frac{3\pi\eta V_{TS} d_p}{C_c} = \frac{3\pi(1.81 \times 10^{-5})(2.6 \times 10^{-4})(2 \times 10^{-6})}{1.08}$$

$$= 8.22 \times 10^{-14} \ N = 8.22 \times 10^{-9} \ dyn$$

Because the particle is free falling

$$F_e = m_p g = \frac{\pi d_p^3}{6} \rho_p g = \frac{\pi \times (2 \times 10^{-6})^3}{6} \times 2,000 \times 9.81 = 8.22 \times 10^{-14} \ N$$

The particle mobility is

$$B = \frac{V_{TS}}{F_e} = \frac{2.6 \times 10^{-4}}{8.22 \times 10^{-14}} = 3.16 \times 10^9 \ m/N \cdot s$$

4.6 NONSPHERICAL PARTICLES AND DYNAMIC SHAPE FACTOR

In practice, especially in an indoor environment, most particles are nonspherical. Particle *dynamic shape factor* is defined as the ratio of the actual resistance force of a nonspherical particle to the resistance force of a spherical particle that has the same equivalent volume diameter and the same settling velocity as the nonspherical particle.

$$\chi = \frac{F_D}{3\pi\eta V_{TS} d_e} \tag{4.43}$$

where F_D is the actual drag force exerted on the nonspherical particle, V_{TS} is the terminal velocity of the nonspherical particle, d_e is the equivalent volume diameter of the nonspherical particle, and C_c is the slip correction factor for d_e. Note that the denominator on the right side of Equation 4.43 uses V_{TS} of the nonspherical particle and thus is not the actual drag force of the spherical particle with d_e. The actual resistance force of the particle with d_e would be $3\pi\eta d_e V_r$, where $V_r > V_{TS}$. The dynamic shape factor is always greater than 1.0, except for certain streamlined shapes. Dynamic shape factors for some typical nonspherical particles can be found in references (Table 4.2).

The calculation or direct measurement of dynamic shape factors for irregular particles is very difficult. Such irregularly shaped particles include flakes (such as dander) and fibers (such as hair). Fibers have a relatively consistent dynamic shape factor of 1.06 (Table 4.2) with a variation of less than 5%. This constancy occurs because the fiber tends to keep its long axis in alignment with the moving direction, especially at low Reynolds numbers. For flake-shaped particles, dynamic shape factors are much larger than 1.0. This is because it is difficult to keep the long axes of the particles in alignment with the moving direction. An example of a flake-shaped particle with a large dynamic shape factor is a parachute, which falls in air much more slowly than other shapes, even those with the same equivalent volume diameter.

However, the dynamic shape factor for a specific particle can be measured indirectly. Substituting Equation 4.43 into Equation 4.36 gives

$$\chi = \frac{\rho_p d_e^2 g C_c}{18\eta V_{TS}} \tag{4.44}$$

Table 4.2 Dynamic Shape Factors for Typical Particles

Shape	Dynamic Shape Factor[a] χ Axial Ratio		
	2	5	10
Geometric shapes			
Sphere	1.00		
Cube[b]	1.08		
Cylinder[b]			
Vertical axis	1.01	1.06	1.0
Horizontal axis	1.14	1.34	1.58
Orientation averaged	1.09	1.23	1.43
Fiber ($L/d > 20$)[c,d]	1.06		
Clustered spheres[c,d]			
2 chain	1.12		
3 chain	1.27		
5 chain	1.35		
10 chain	1.68		
3 compact	1.15		
4 compact	1.17		
Dust			
Bituminous coal[e]	1.08		
Quartz[e]	1.36		
Sand[e]	1.57		
Talc[f]	1.88		

[a] Averaged over all orientations unless specified.
[b] *Source:* Johnson, D.L., Leith, D., and Reis, P.C., Drag on non-spherical, orthotropic aerosol particles, *J. Aerosol Sci.,* 18:87–97, 1987.
[c] *Source:* Dahneke, B.A., Slip correction factors for spherical bodies III: the form of general law, *J. Aerosol Sci.,* 4:163–170, 1973.
[d] *Source:* Dahneke, B.A. and Cheng, Y.S., Properties of continuum source particle beams. I. Calculation methods and results, *J. Aerosol. Sci.,* 10:257–274, 1979.
[e] *Source:* Davis, C.N., Particle fluid interaction, *J. Aerosol Sci.,* 10:477–513, 1979.
[f] *Source:* Cheng, Y.S., Yeh, H.C., and Allen, M.D., Dynamic shape factors of plate-like particles, *Aerosol Sci. Tech.,* 8:109–123, 1988.

The dynamic shape factor in Equation 4.44 is a function of its terminal settling velocity V_{TS}. The equivalent volume diameter (d_e) and particle density (ρ_p) and the terminal velocity of an irregularly shaped particle can be measured in a calm air chamber by recoding its falling distance and the traveling time. Thus χ can be determined. Such measurements require special instrumentation and stringent procedures.

4.7 AERODYNAMIC DIAMETER

Perhaps the most important physical property of a particle is its aerodynamic diameter. In reality, it is extremely difficult to measure or calculate equivalent volume diameters and shape factors because of the complexity of the geometric shapes of particles. Thus, we need an equivalent diameter that can be physically determined to characterize the particle property and behavior. This equivalent diameter is called the *aerodynamic diameter* d_a. Aerodynamic diameter is defined as the diameter of a sphere with a unit density and the same settling velocity as the particle of concern (Figure 4.6). From this definition and Equation 4.44

$$V_{TS} = \frac{\rho_p d_e^2 g C_{ce}}{18 \eta \chi} = \frac{\rho_0 d_a^2 g C_{ca}}{18 \eta} \tag{4.45}$$

where C_{ca} and C_{ce} are slip correction factors for d_a and d_e, respectively, and $\rho_0 = 1000$ kg/m^3. Solving Equation 4.45 for d_a gives

$$d_a = d_e \left(\frac{C_{ce} \rho_p}{C_{ca} \rho_0 \chi} \right)^{\frac{1}{2}} \quad \text{(for nonspherical particles)} \tag{4.46}$$

For spherical particles, $d_e = d_p$ and $\chi = 1$, then

$$d_a = d_p \left(\frac{C_c \rho_p}{C_{ca} \rho_0} \right)^{\frac{1}{2}} \quad \text{(for spherical particles)} \tag{4.47}$$

where d_p is the actual geometric diameter of the particle and C_c is the slip correction factor for d_p.

Aerodynamic diameter may be more easily understood and remembered through the following analogy. Drop the particle in question and a water droplet (spherical with a unity density) in still air from the same height. If the particle and the water droplet reach the ground at the same time (so that they have the same settling velocity), the aerodynamic diameter of the particle in question is equal to the diameter of the water droplet. In this analogy, we assume that the water droplet is a sphere. In reality, this is not true: The water droplet deforms when it is subjected to the air resistance; thus it is not a true sphere.

A more accurate description of aerodynamic diameter is shown in Figure 4.6, which demonstrates the equivalent volume and the aerodynamic diameter of the same particle.

For particles larger than 3 μm, the slip effect can be neglected. Thus, C_c for d_e, d_a, and d_p are equal unity, and Equations 4.46 and 4.47 can be written as follows:

$$d_a = d_e \left(\frac{\rho_p}{\rho_0 \chi} \right)^{\frac{1}{2}} \quad \text{(for nonspherical particles, } d_e \geq 3 \text{ μm)} \tag{4.48}$$

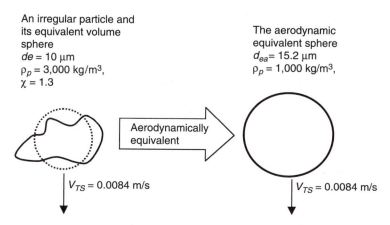

An irregular particle and
its equivalent volume
sphere
de = 10 μm
ρ_p = 3,000 kg/m³,
χ = 1.3

The aerodynamic
equivalent sphere
d_{ea} = 15.2 μm
ρ_p = 1,000 kg/m³,

Aerodynamically
equivalent

V_{TS} = 0.0084 m/s

V_{TS} = 0.0084 m/s

Figure 4.6 An irregularly shaped particle will behave similarly to a sphere with the same aerodynamic diameter.

$$d_a = d_p \left(\frac{\rho_p}{\rho_0} \right)^{\frac{1}{2}} \quad \text{(for spherical particles, } d_p \geq 3 \text{ μm)} \qquad (4.49)$$

For particles smaller than 3 μm, the slip effect must be considered. Equation 4.46 and Equation 4.47 each contains a slip correction factor C_{ca}, which is a function of the aerodynamic diameter (which is also in question). Aerodynamic diameter can be solved explicitly for particles smaller than 3 μm but larger than 0.1 μm. Substituting Equation 4.30 into Equations 4.46 and 4.47 and solving for d_a gives

$$d_a = \frac{1}{2} \left[\left(6.35\lambda^2 + 4d_e^2 C_{ce} \frac{\rho_p}{\rho_0 \chi} \right)^{\frac{1}{2}} - 2.52\lambda \right]$$

(for nonspherical particles and 0.1 μm < d_e < 3 μm) (4.50)

$$d_a = \frac{1}{2} \left[\left(6.35\lambda^2 + 4d_p^2 C_c \frac{\rho_p}{\rho_0} \right)^{\frac{1}{2}} - 2.52\lambda \right]$$

(for spherical particles and 0.1 μm < d_p < 3 μm) (4.51)

Aerodynamic diameter may be quite different from its equivalent volume or geometric diameters. For small particles, the slip effect can also significantly alter aerodynamic behavior, and thus the aerodynamic diameter. These differences can significantly change the output of aerodynamically based particle-measurement instruments. The following example gives a perspective on how the slip effect affects aerodynamic diameter.

Example 4.4. *Determine the ratio of the aerodynamic diameter to its actual geo-metric diameter. The particle is a 0.2 μm diameter sphere with a density of 2500 kg/m³. If the slip effect were ignored, what would the aerodynamic diameter be?*

Solution: Because the particle is smaller than 3μm, the slip effect must be considered:

$$C_c = 1 + \frac{2.52\lambda}{d_p} = 1 + \frac{2.52 \times 0.066}{0.2} = 1.83$$

Substituting the values above into Equation 4.51 gives

$$d_a = \frac{1}{2} \left[\left(6.35\lambda^2 + 4d_p^2 C_c \frac{\rho_p}{\rho_0} \right)^{1/2} - 2.52\lambda \right]$$

$$= \frac{1}{2} \left[\left(6.35 \times 0.066^2 + 4 \times 0.2^2 \times 1.83 \times \frac{2,500}{1,000} \right)^{1/2} - 2.52 \times 0.066 \right] = 0.353\,\mu m$$

The ratio of the aerodynamic diameter to the particle diameter is

$$\frac{d_a}{d_p} = \frac{0.353}{0.2} = 1.765$$

If the slip effect were ignored, the aerodynamic diameter would be

$$d_a = d_p \left(\frac{\rho_p}{\rho_0} \right)^{1/2} = 0.2 \times \left(\frac{2,500}{1,000} \right)^{1/2} = 0.316\,(\mu m)$$

Compared with the d_a, including the slip correction factor, ignoring the slip effect would introduce a 12% error into the aerodynamic diameter.

Aerodynamic diameter for particles smaller than 0.1 μm can be solved using iteration methods, by substituting Equation 4.31 into Equation 4.50 or Equation 4.51. In general, particles smaller than 0.1 μm tend to behave like gases, for which other particle transport mechanisms and behavior, such as diffusion, become more important than aerodynamic behavior. Gas properties are discussed in Chapter 5.

From Equation 4.45, the aerodynamic diameter standardizes both the density (to be the same as the water density) and the shape (to be a sphere) in accordance with the terminal settling velocity. Because the V_{TS} is relatively easy to obtain, the aerodynamic diameter becomes a very useful particle parameter. In this book, particle sizes are given by aerodynamic diameters unless specified.

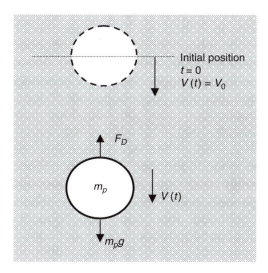

Figure 4.7 A particle with an initial velocity V_0 is falling down in air. Before it reaches its terminal velocity, the particle has an acceleration $dV(t)/dt$.

4.8 RELAXATION TIME

It has been assumed in the previous sections that a particle reaches its terminal velocity very quickly when it is released into a fluid or subjected to an external force. The quickness is said to be almost instantaneous. Thus, the terminal settling velocity can be determined experimentally by measuring the free-fall distance and the time of travel. Now is the time to examine how fast a particle accelerates in the air, and how long it takes a particle to achieve its terminal velocity.

Consider particle motion in still air along the vertical direction. The particle has an initial velocity V_0 when $t = 0$, and it is falling down in the air. A free body diagram of the particle is shown in Figure 4.7. There are two forces exerted on the particle: gravity $m_p g$, and drag force F_D. Applying Newton's second law and Stokes's law gives

$$m_p \frac{dV(t)}{dt} = m_p g - \frac{3\pi\eta d_p V(t)}{C_c} \tag{4.52}$$

where $V(t)$ is the particle velocity at any time. Substituting the mobility $B = C_c/3\pi\eta d_p$ (Equation 4.41) and $V_{TS} = m_p g B$ (Equation 4.42) into Equation 4.52 gives

$$m_p B \frac{dV(t)}{dt} = V_{TS} - V(t) \tag{4.53}$$

Separating variables, Equation 4.53 can be written as

$$\frac{d\left(V(t)-V_{TS}\right)}{V(t)-V_{TS}} = -\frac{dt}{m_p B} \tag{4.54}$$

Integrating Equation 4.54 with respective ranges of $V(t)$ and t,

$$\int_{V_0}^{V(t)} \frac{d\left(V(t)-V_{TS}\right)}{V(t)-V_{TS}} = -\int_0^t \frac{dt}{m_p B} \tag{4.55}$$

yields

$$V(t) = V_{TS} - (V_{TS} - V_0)\exp\left(-\frac{t}{m_p B}\right)$$

$$= V_{TS} - (V_{TS} - V_0)\exp\left(-\frac{t}{\tau}\right) \tag{4.56}$$

where

$$\tau = m_p B = \rho_p \frac{\pi d_p^3}{6}\left(\frac{C_c}{3\pi\eta d_p}\right) = \frac{\rho_p d_p^2 C_c}{18\eta} = \frac{\rho_0 d_a^2 C_{ca}}{18\eta} \tag{4.57}$$

The τ is referred to as the relaxation time of the particle. The term of the relaxation time implies the time period required for a particle to relax from a transient state (acceleration or deceleration) to a steady state (with a constant terminal velocity). Figure 4.8 shows the relationship between the terminal settling velocity and the time.

From Equation 4.57, τ is the time constant of the particle needed to reach its terminal velocity when it is subjected to an external force. When $t = 0$, $V(t) = V_0$. When $t \to \infty$, $V(t) = V_{TS}$. Theoretically, $V(t)$ never reaches its terminal velocity. Practically, $V(t)$ reaches 63.2% of $(V_{TS} - V_0)$ when $t = \tau$, 95% of $(V_{TS} - V_0)$ when t $= 3\tau$, and 99.3% of $(V_{TS} - V_0)$ when $t = 5\tau$.

Example 4.5: What is the relaxation time for a particle with an aerodynamic diameter of 3 μm? If the particle is injected from a nozzle into air and has an initial velocity of 5 m/s toward the ground, and then free-falls, what is the particle velocity after 0.001 second?

Solution: The slip correction effect can be neglected because $d_a = 3$ μm. The relaxation time of the particle is

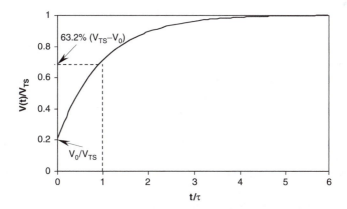

Figure 4.8 Relationship between the percentage of terminal settling velocity and the normalized time t/τ. The initial velocity is assumed to be zero.

$$\tau = m_p B = \frac{\rho_0 d_a^2 C_{ca}}{18\eta} = \frac{1,000 \times (3 \times 10^{-6})^2}{18 \times 1.81 \times 10^{-5}} = 2.76 \times 10^{-5}\ (s)$$

Because the relaxation time is so short, the particle reaches its terminal velocity almost instantaneously. To answer the second question, first determine the terminal settling velocity:

$$V_{TS} = \frac{\rho_0 d_a^2 g C_{ca}}{18\eta} = \frac{1,000 \times (3 \times 10^{-6})^2 \times 9.81}{18 \times 1.81 \times 10^{-5}} = 2.71 \times 10^{-4}\ (m/s)$$

Applying Equation 4.56 gives

$$V\big|_{t=0.001s} = V_{TS} - (V_{TS} - V_0)\exp\left(-\frac{t}{\tau}\right)$$

$$= 2.71 \times 10^{-4} - (2.71 \times 10^{-4} - 5)\exp\left(-\frac{0.001}{2.76 \times 10^{-5}}\right) = 2.71 \times 10^{-4}\ (m/s)$$

which is the same as the terminal settling velocity. This indicates that even with a high initial velocity of 5 m/s, the particle reaches its terminal velocity in less than 1 ms.

Although the relaxation time is derived using gravitational force, it can be derived when the particle is subjected to other types of external forces, such as centrifugal or electrical force, simply by substituting $B = V_T/F_e$ into Equation 4.57, where V_T is the terminal velocity of the particle and F_e is the external force exerted on the particle.

4.9 STOPPING DISTANCE

In addition to the terminal velocity of particles, the distance that a particle can travel is another important characteristic of particle mechanics. For example, how far will a particle generated from a grinding wheel travel in still air? The answer to this question involves the concept of *stopping distance*. Stopping distance is defined as the maximum distance a particle can travel with an initial velocity V_0 in still air in the absence of external forces. Such external forces include gravitational, centrifugal, and electrical forces. Drag force is not considered an external force in particle mechanics.

Considering the linear distance that a particle travels along one direction in still air as $S(t)$, then $V(t) = dS(t)/dt$. From Equation 4.56, the distance function $S(t)$ can be obtained by integrating the $dS(t)/dt$ over the time t:

$$S(t) = \int_0^t \left[V_{TS} - (V_{TS} - V_0) \exp(-\frac{t}{\tau}) \right] dt$$

$$= V_{TS}t - \tau(V_{TS} - V_0)\left(1 - \exp\left(-\frac{t}{\tau}\right)\right) \tag{4.58}$$

By the definition of the stopping distance, two factors must be considered. First, the maximum distance can only be achieved when the time $\to \infty$. Secondly, there is no external force exerted on the particle, that is, $V_{TS} \to 0$ when $t \to \infty$. Typically, we can consider $V_{TS} = 0$ when $t > 5\tau$ with 1% accuracy. Thus, when $t >> \tau$, $V_{TS} = 0$, and Equation 4.58 yields the stopping distance:

$$S = V_0 \tau = Bm_p V_0 = \frac{\rho_p d_p^2 C_c}{18\eta} V_0 \quad (Re_{p0} \le 1) \tag{4.59}$$

Similar to terminal velocity τ, the relaxation time is the time constant of the particle needed to reach its stopping distance when there are no external forces. Theoretically, $S(t)$ never reaches its stopping distance. Practically, $S(t)$ reaches 63.2% of its stopping distance when $t = \tau$, 95% of the stopping distance when $t = 3\tau$, and 99.3% of the stopping distance when $t = 5\tau$. Figure 4.9 shows the relationship between the stopping distance and the time.

Stopping distances calculated from Equation 4.59 are for particles moving in the Stokes region only, that is, when $Re_p < 1$. When $Re_p > 1$, the stopping distance of the particle is shorter than is predicted by Equation 4.59. This is because the drag force increases as the Re_p increases. Recalling Equation 4.16 and Equation 4.17, the drag coefficient C_D increases as the Re_p increases. The increase of the drag force attenuates the velocity of the particle, hence reducing the stopping distance. However, because the Re_p is proportional to the particle velocity and the drag force is proportional to the square of velocity, it is extremely difficult to obtain an analytical

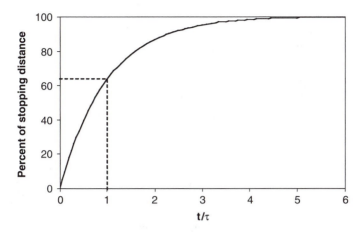

Figure 4.9 Relationship between the percentage of stopping distance and the normalized time, t/τ.

equation for stopping distance outside the Stokes region. Mercer [9] proposed an empirical equation to calculate the stopping distance for particles with $Re_p < 1500$:

$$S = \frac{\rho_p d_p}{\rho_g}\left[Re_{p0}^{1/3} - \sqrt{6} \times \tan^{-1}(\frac{Re_p^{1/3}}{\sqrt{6}})\right] \quad (1 < Re_{p0} \le 1{,}500) \qquad (4.60)$$

where Re_{p0} is the particle Reynolds number at the initial particle velocity V_0, and the term $\tan^{-1}\left[Re_p^{1/3}/\sqrt{6}\right]$ is in radians.[9] The accuracy of this estimation is within 3% of the calculated value.

Stopping distances for typical sizes of particles are listed in Table 4.3. The values are calculated using Equation 4.59 or Equation 4.60, according to the particle Reynolds number at the initial velocity. When Equation 4.59 was used, slip correction factors were calculated using Equation 4.30 or Equation 4.31, according to the particle size.

Table 4.3 Stopping Distances and Relaxation Times for Typical Sizes of Particles in the Stokes Region (Particle Sizes Are in Aerodynamic Diameter d_a)

d_a (µm)	τ (s)	$V_0 = 5$ m/s		$V_0 = 10$ m/s	
		Re_{p0}	S (m)	Re_{p0}	S (m)
0.01	7.05×10^{-9}	0.00331	3.53×10^{-8}	0.00663	7.05×10^{-8}
0.1	8.17×10^{-8}	0.0331	4.09×10^{-7}	0.0663	8.17×10^{-7}
1	3.58×10^{-6}	0.331	1.79×10^{-5}	0.663	3.58×10^{-5}
10	0.000312	3.31	0.00152	6.63	0.00276
100	0.0307	33.1	0.0960	66.3	0.153

Example 4.6: Calculate the stopping distance of a particle with an aerodynamic diameter of 20 μm in still air. Assume that the initial velocity of the particle is 5 m/s. What would the error be if the particle motion were assumed to be in the Stokes region?

Solution: The initial particle Reynolds number is

$$\text{Re}_{p0} = \frac{\rho d_a V_o}{\eta} = \frac{1.2 \times 20 \times 10^{-6} \times 5}{1.81 \times 10^{-5}} = 6.63$$

Because $Re_{p0} > 1$, Equation 4.60 should be used.

$$S = \frac{\rho_p d_p}{\rho_g} \left[\text{Re}_{p0}^{\frac{1}{3}} - 0.04276 \times \tan^{-1}(0.4082\,\text{Re}_{p0}^{\frac{1}{3}}) \right]$$

$$= \frac{1,000 \times 20 \times 10^{-6}}{1.2} \left[6.63^{\frac{1}{3}} - 0.04276 \times \tan^{-1}(0.4082 \times 6.63^{\frac{1}{3}}) \right]$$

$$= 0.0046\,(m)$$

If the particle were assumed to be in the Stokes region, the stopping distance would be calculated using Equation 4.59.

$$S = V_0 \tau = \frac{\rho_0 d_a^2}{18\eta} V_0 = \frac{1,000 \times (20 \times 10^{-6})^2}{18 \times 1.81 \times 10^{-5}} \times 5 = 0.00614\,(m)$$

The error then would be

$$\varepsilon = \frac{0.00614 - 0.0046}{0.0046} \times 100\% = 33\%$$

DISCUSSION TOPICS

1. How would the inertial force of a particle vary with the particle position in Equation 4.5? Can you name an example in real life for verification?
2. What does *slip* mean in particle mechanics? Is it the same slipping mechanism as something sliding on ice?
3. Why does the slip correction factor increase as ambient pressure decreases?
4. Why does the slip correction factor increase as the ambient temperature decreases?
5. A cotton mill needs to specify the properties of cotton fibers. How can you design an experiment to determine the shape factor χ for a piece of cotton fiber? Assume that you have already measured under a microscope the length L and diameter d of the cotton fiber.

6. How would you determine the dynamic shape factor of a particle with a known geometric shape?
7. What are the advantages of aerodynamic diameter over geometric diameter d_p and equivalent volume diameter d_e?
8. What are the other factors affecting the particle concentration in an indoor environment? Can the particles eventually settle down even if they are smaller than 100 μm in diameter?
9. Letting the particles settle down themselves seems a very economical means of air cleaning. How could you accelerate particle settling in practical air cleaning?
10. Under what circumstances are particle relaxation time and stopping distance important?

PROBLEMS

1. A pitcher throws a baseball at 120 km/h across a wind of 40 km/h. The wind direction is perpendicular to the ball direction. The baseball is 7.3 cm in diameter. The air density is 1.2 kg/m³. What is the Reynolds number for the baseball?
2. A roof ventilation fan exhausts 1.2 m³/s of air through a rectangular duct with a cross section of 0.7 × 0.5 m, as shown in the figure below. What is the largest particle that can be exhausted by the fan? Assume that the particles are spherical and density is 1000 kg/m³.

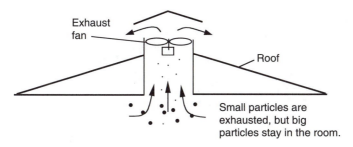

Schematic of an exhaust fan on a roof.

3. A circular supply air duct in an office is 15 cm in diameter. The flow rate is 0.015 m³/s. A dust particle of 20 μm diameter and 2000 kg/m³ density enters the air stream.
    a. Find the settling velocity of the particle.
    b. Find the Reynolds number of the air stream. Is the air flow laminar?
    c. Find the Reynolds number of the particle. Is the particle motion in the Stokes region?
4. A 0.3 μm diameter particle escapes through a vacuum cleaner filter and is tossed into the room air 2 m above the floor. Assume that the particle has a density of 1000 kg/m³ and free-falls from the air. How long will the particle take to settle on the floor?
5. Coal-burning particles from a power plant range from 1 to 100 μm in diameter. Assume that the particles are spherical and the density is 1200 kg/m³. The plant's chimney is 40 meters high. Assume that the ambient air is calm. The dust particles start to settle once they exit the chimney. Calculate the following:

    a. How long will it take for the smallest and the biggest particles to settle on the ground, neglecting the slip effect?

    b. If slip correction factor C_c is included, what is the settling velocity for the 1 μm diameter particles?

    c. If the wind is 10 km/h, how far can a 1 μm particle travel horizontally?

6. A glider jumps from a cliff 500 m high. The glider stays in the air for 5 minutes before touching down. The glider (including the glide) weighs 100 kg, with approximately a standard density. Find the dynamic shape factor of the glider.

7. Assume the airflow in a duct is laminar. The duct is 20 m long, 0.6 m wide, and 0.4 m high. The air flow rate is 0.5 m³/s. What is the percentage of 10 μm particles in aerodynamic diameter that settle in the duct? Assume that the particles never reenter the air once settled.

8. A 50 μm steel particle leaves a grinding wheel of 20 cm in diameter and 1700 rpm at an angle of 30° toward the floor, as shown in Figure 4.11. The particle is spherical with a density of 7800 kg/m³. The center of the wheel is 1.2 m above the floor. How far can the particle travel horizontally, and how long will the particle remain airborne before it touches down? (Hint: The particle trajectory can be approximated as a free fall after it reaches the terminal settling velocity. The trajectory of the particle is the dashed line shown below.)

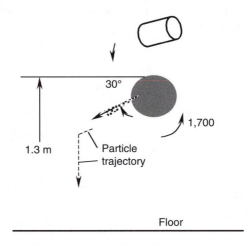

Particle trajectory of a sanding wheel.

9. A piece of asbestos fiber, with a density of 2400 kg/m³, a diameter of 1 μm, and a length of 30 μm, falls from the insulation of a pipe that is 2.8 m above the floor in a room. How long will the fiber remain airborne? Assume that the room air is still and the fiber free-falls.

10. A piece of dander is in the approximate shape of a circular plate, with a diameter of 300 μm and a thickness of 10 μm. The density is 800 kg/m³. The terminal settling velocity is measured at 0.05 m/s. Determine the aerodynamic diameter.

11. Calculate the aerodynamic diameter of a spherical particle of 0.5 μm in diameter. The density is 3000 kg/m³.

12. Plot the relationships between $V(t)/V_{TS}$ vs. t/τ on the same graph when $V_0 = 0$ and $V_0 = 2V_{TS}$, using Equation 4.56, and similar to Figure 4.8. Interpret the difference between the two curves as the time varies.

REFERENCES

1. Reynolds, O., An experimental investigation of the circumstances which determine whether the motion of water shall be direct or sinuous, and of the laws of resistance in parallel channels, *Trans. R. Soc. Lond.*, 174, 1883.
2. Streeter, V.L. and Wylie, E.B., *Fluid Mechanics*, 6th ed., McGraw-Hill, New York, 279, 1975.
3. Vincent, J.H., *Aerosol Sampling — Science and Practice*, John Wiley & Sons, New York, 1989.
4. Clift, R., Grace, J.R., and Weber, M.E., *Bubbles, Drops, and Particles*, Academic Press, New York, 122, 1978.
5. Bird, B.R., Stewart, W.E., and Lightfoot, E.N., *Transport Phenomena*, Wiley, New York, 1960.
6. Hinds, W.C., *Aerosol Technology*, John Wiley & Sons, New York, 1999.
7. Allen, M.D. and Raabe, O.G., Re-evaluation of Millikan's oil drop data for the motion of small particles in air, *J. Aerosol Sci.*, 6:537–547, 1982.
8. Allen, M.D. and Raabe, O.G., Slip correction measurements of spherical solid aerosol particles in an improved Millikan apparatus, *Aerosol Sci. Tech.*, 4:269–286, 1985.
9. Mercer, T.T., *Aerosol Technology in Hazard Evaluation*, Academic Press, New York, 1973.

CHAPTER 5

Gas Properties and Kinetics

Thermodynamics, such as psychrometric processes dealing with heating, ventilation, and refrigeration, use macroscopic variables, such as pressure, temperature, and humidity to describe energy and mass transfers. Nothing is mentioned about the fact that the air is made up of atoms, and the laws of mechanics are still applicable to these atoms. On the other hand, applying mechanics to every atom or molecule is an overwhelming task, even with the help of a supercomputer. For example, one beer can's worth of air would need 8.9×10^{21} differential equations to describe the motion of all the molecules in the beer can at any moment.

Fortunately, the detailed life histories of individual atoms or molecules in a gas are not important if we want to calculate only the macroscopic behavior of the gas. When the basic laws of mechanics are applied to individual gas molecules statistically, the macroscopic properties, such as temperature and pressure, can be expressed as averages of molecular properties. For example, the pressure exerted on a container wall is the average rate per unit area at which the gas molecules or atoms transfer momentum to the wall as they collide with it. Modern statistical mechanics includes quantum mechanics in addition to classical mechanics, and can be applied to many atomic systems, including liquids and solids. Gas kinetics is a branch of statistical mechanics applied to gases. Gas kinetics, particularly in its more advanced form, is the theory of applying statistical mechanics to gain a microscopic understanding of gas properties such as pressure, temperature, viscosity, and diffusivity. Because air and gaseous contaminants are all constituents of the indoor air environment, this chapter provides fundamentals of many aspects of indoor air quality, such as gas transportation and interaction between gases and particulates.

By completing this chapter, the reader will be able to

- Characterize gas kinetics by calculating the following gas properties:
  - Mean speed and root-mean-square speed of gas molecules
  - Kinetic energy
  - Mean free path of gas molecules
  - Collision frequency of gas molecules and molecules on a surface

- Molecule diameter
- Collision diameter of molecules
- Relate the gas macroproperties to the microproperties. The microproperties include the previously listed gas kinetics, all from a molecular point of view. The macroproperties are usually measurable using appropriate instrumentation. They include the following:
  - Temperature
  - Pressure
  - Viscosity
  - Diffusivity

5.1 IDEAL GAS LAW

Consider a gas sample with a mass m_g that is confined in a container of volume V. The density of the gas can be changed by removing some gas from the container or by putting the gas into a larger container (increasing the V). If the densities of the gases are low enough, the temperature T, pressure P, and volume tend to have a certain relationship. Densities of all gases in indoor environments can be considered low enough compared to liquid or solid substances. The ideal gas law is the result of two experiments and is stated as follows:

1. For a given mass of gas held at a constant temperature, the pressure is inversely proportional to the volume (Boyle's law).
2. For a given mass of gas held at a constant pressure, the volume is directly proportional to the temperature (law of Charles and Gay-Lussac).

The ideal gas law may be given in an equation:

$$PV = nRT \quad \text{(for a fixed mass of gas)} \tag{5.1}$$

where P is in pascals (N/m²), V is in m³, T is in K, n is the number of moles of gas in the gas sample, and R is called the *universal gas constant*, which is the same for all ideal gases:

$$R = 8.314 \; \frac{J}{mole \cdot K} = 1.986 \; \frac{cal}{mole \cdot K} \tag{5.2}$$

Gases that obey Equation 5.1 under all conditions are referred to as *ideal gases*. Equation 5.1 is called the equation of state of an ideal gas, or Boyle's law. Almost all gases in indoor environments can be considered ideal gases.

The equation of state of an ideal gas (Equation 5.1) gives a macroscopic description of gases. In order to apply statistical mechanics to ideal gases, it is appropriate to describe the ideal gases microscopically. The microscopic definition of an ideal gas includes the following:

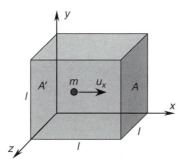

Figure 5.1 An ideal gas molecule is moving along the x-axis at the speed of u_x, in a cubical box with the side length of l.

1. A gas consists of particles, which are gas molecules.
2. The molecules are in random motion and obey Newton's law of motion.
3. The total number of molecules is large, so the statistical mechanics can be meaningful.
4. The volume of molecules is negligibly small compared to the volume occupied by the gas.
5. No appreciable forces act on the molecules except during the collision.
6. Collisions are elastic and are of negligible duration.

5.2 MOLECULAR SPEEDS

One microscopic description of a gas is its molecular speed, derived from kinetic calculations of pressure. Consider a gas molecule with a mass m in a cubical box whose walls are perfectly elastic (Figure 5.1). The length of each edge of the box is l, and the area of each side is $A = l^2$. The speed of the molecule is u. Let the molecule move along the x-axis with a speed u. We can resolve the speed u into three components, u_x, u_y, and u_z, in the direction of the box edges. Because the walls are perfectly elastic, this particle will rebound after it collides with the wall, without losing speed but exactly reversing direction. Based on the laws of mechanics, the change of momentum of the molecule is

$$\Delta mu = (\Delta mu_x)_f - (\Delta mu_x)_i = -(mu_x) - (mu_x) = -2mu_x \qquad (5.3)$$

where subscripts f and i represent final and initial states, respectively. This change of momentum is normal to the wall. Because the total momentum is conserved, the momentum imparted on the surface A will be $2mu_x$.

Suppose that this molecule reaches the wall A' after rebounding from wall A without striking any other molecules on the way. The time required to cross the cube will be l/u_x. The molecule will rebound back to A again. The round trip will take a time of $2l/u_x$. Therefore, the number of collisions per unit time with the surface A for this molecule is $u_x/2l$, so that the rate of momentum transferred to the surface A by the molecule is

$$\frac{\Delta mu}{\Delta t} = 2mu_x \frac{u_x}{2l} = \frac{mu_x^2}{l} \tag{5.4}$$

Note that this rate of momentum transferred to surface A is the force exerted on the surface by the molecule according to Newton's law of momentum. To obtain the total force exerted on the surface A, that is, the rate at which momentum is transferred to A by all gas molecules, we must sum up mu_x^2/l for all molecules.

$$F_t = \frac{m}{l}(u_{x1}^2 + u_{x2}^2 + \cdots) \tag{5.5}$$

Then, this total force is divided by the area $A = l^2$ to obtain the pressure:

$$P = \frac{F_t}{A} = \frac{m}{l^3}(u_{x1}^2 + u_{x2}^2 + \cdots) = \frac{mN}{V}\left(\frac{u_{x1}^2 + u_{x2}^2 + \cdots}{N}\right) \tag{5.6}$$

where N is the total number of gas molecules in the cubical box and V is the volume of the box. Since mN is the total mass in the box, mN/V is the mass per unit volume, that is, the density, ρ. The quantity of $(u_{x1}^2 + u_{x2}^2 + \cdots)/N$ is the average of u_x^2 for all the molecules in the box, called $\overline{u_x^2}$. Then, Equation 5.5, or the pressure of the gas along the x-axis, can be written as

$$P = \rho \overline{u_x^2} \tag{5.7}$$

For any molecules, $u^2 = u_x^2 + u_y^2 + u_z^2$. Because there are many molecules and because they are moving entirely at random, the average values of u_x^2, u_y^2, and u_z^2 are equal, and the value of each is exactly one third of u^2. There is no preference among the molecules for motion along any one of the three axes. Therefore, the pressure exerted on the wall is

$$P = \rho \overline{u_x^2} = \frac{1}{3}\rho \overline{u^2} \tag{5.8}$$

The square root of $\overline{u^2}$ is called the root-mean-square (rms) speed (u_{rms}) of the molecules and is a kind of average molecular speed. The result of Equation 5.8 is true when collision is considered, provided that the collision is elastic and occurs between identical molecules. The time spent during the elastic collision is also negligible compared to the time between the collisions. From Equation 5.8, the relationship between a macroscopic variable (the pressure P) and a microscopic variable (rms speed u_{rms}) of the gases can be established:

$$u_{rms} = \left(\overline{u^2}\right)^{1/2} = \left(\frac{3P}{\rho}\right)^{1/2} \tag{5.9}$$

Equation 5.9 shows that if the pressure and the density of a gas are measured, the root-mean-square speed of the gas molecules can be determined. In some extreme cases, such as when the time period is extremely short (as in nanoseconds), the volume of gas is extremely small (as in cubic nanometers), or the gas density is extremely low (near absolute vacuum), the u_{rms} value calculated could fluctuate. In most practical cases, u_{rms} values can be calculated with sufficient accuracy.

The rms speed of molecules is just one type of average of molecule speed. Different averages can be obtained using different averaging methods. The speeds of individual molecules vary over a wide range of magnitude. To obtain other types of averages of the molecules, molecular speed distribution $F(u)$ must be determined.

James Clark Maxwell first solved the problem of the most probable distribution of speed in a large number of gas molecules. This molecular speed distribution law, also called the Maxwell distribution, is used for a gas sample containing N molecules. The one-dimensional and three-dimensional molecular speed distributions are expressed as[1]

$$F(u_x) = N\left(\frac{m}{2\pi kT}\right)^{3/2} \exp\left(-\frac{mu_x^2}{2kT}\right) \text{ (for the x-direction only)} \tag{5.10}$$

$$F(u) = 4\pi N\left(\frac{m}{2\pi kT}\right)^{3/2} u^2 \exp\left(-\frac{mu^2}{2kT}\right) \text{ (for all directions)} \tag{5.11}$$

where k is the Boltzmann constant:

$$k = \frac{R}{N_a} = 1.381 \times 10^{-23} \frac{J}{molecule \cdot K}$$

and N_a is Avogadro's number:

$$N_a = 6.022 \times 10^{23} \frac{molecules}{mole}$$

Equation 5.10 shows that in any one of the three directions, the speed distribution is a normal distribution, symmetrical with respect to the zero velocity. It can be imagined that a molecule is bouncing back and forth along a string, so that one velocity forward will always have another velocity backward with the same magni-

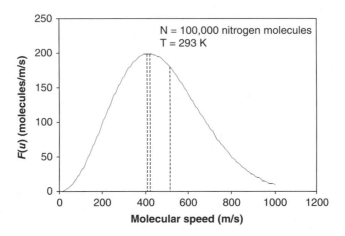

Figure 5.2   The Maxwell distribution of speeds for 100,000 nitrogen molecules at room temperature 20°C (293 K). The area under the curve is the total number of molecules.

tude. Equation 5.11 gives the three-dimensional speed distribution of molecules falling into the speed range of $u + du$. The total number of molecules, N, can be obtained by summing up $F(u)$ over the entire range of molecular speeds $(0,\infty)$.

$$N = \int_0^\infty F(u)du \qquad (5.12)$$

Figure 5.2 plots the Maxwell distribution of speeds of 100,000 nitrogen molecules at a room temperature of 20°C (293 K). As Equation 5.12 shows, the area under the speed distribution curve, which is the integral in that equation, is equal to the total number of molecules in the sample. At any given temperature, the number of molecules falling into the range $u + du$ increases, up to a maximum frequency (the most probable speed, or the mode, $\hat{u}$) and then decreases asymptotically toward zero. The speed distribution curve is not symmetrical about the most probable speed, because the lowest speed must be zero.

The *arithmetic mean speed*, or the *mean speed*, of molecules $\bar{u}$, can be obtained by integrating the product of speed u and its speed distribution $F(u)$ over the entire speed range, then dividing by the total number of molecules N.

$$\bar{u} = \frac{\displaystyle\int_0^\infty uF(u)du}{N} = \int_0^\infty 4\pi \left(\frac{m}{2\pi kT}\right)^{3/2} u^3 \exp\left(-\frac{mu^2}{2kT}\right)du$$

$$= \left(\frac{8kT}{\pi m}\right)^{1/2} = 1.59 \left(\frac{kT}{m}\right)^{1/2} \qquad (5.13)$$

Similarly, the root-mean-square speed of molecules, u_{rms}, can be obtained by integrating the product of u^2 and its speed distribution $F(u)$ over the entire speed range, dividing by the total number of molecules, N, and then taking the square root:

$$u_{rms} = (\overline{u^2})^{\frac{1}{2}} = \left\{ \frac{\int_0^\infty u^2 F(u)du}{N} \right\}^{\frac{1}{2}}$$

$$= \left\{ \int_0^\infty 4\pi \left(\frac{m}{2\pi kT} \right)^{\frac{3}{2}} u^4 \exp\left(-\frac{mu^2}{2kT} \right) du \right\}^{\frac{1}{2}} \tag{5.14}$$

$$= \left(\frac{3kT}{m} \right)^{\frac{1}{2}} = 1.73 \left(\frac{kT}{m} \right)^{\frac{1}{2}}$$

The most probable molecular speed $\hat{u}$ is the speed at which $F(u)$ has a maximum value. It is given by requiring that

$$\frac{dF(u)}{du} = 0 \tag{5.15}$$

Substituting Equation 5.11 into Equation 5.15 and solving for u gives the most probable molecular speed:

$$\hat{u} = \left(\frac{2kT}{m} \right)^{\frac{1}{2}} = 1.414 \left(\frac{kT}{m} \right)^{\frac{1}{2}} \tag{5.16}$$

From Equation 5.13, Equation 5.14, and Equation 5.16, the most probable speed does not occur at the mean speed or the rms speed, as shown in Figure 5.2. Of the three, the rms speed is perhaps the most important because it is related to the kinetic energy of the gas. Table 5.1 gives mean and rms speeds for several typical gases and vapors of indoor environments at 20°C.

We must distinguish the speeds of individual gas molecules, described as urms or $\overline{u}$, from the speed of the gas diffusion, which is much lower. For example, if we open a bottle of ammonia in one corner of the room, we smell the ammonia at the opposite corner only after a noticeably measurable time lag, say a couple of seconds. The diffusion speed is slow because large numbers of collisions with air molecules greatly reduce the tendency of ammonia molecules to spread themselves uniformly throughout the room.

Table 5.1 Molecular Speeds of Typical Gases and Vapors of Indoor Environments at 20°C

Gas or Vapor	Density (kg/m³)	Molecular Weight (kg/mole)	rms Speed (m/s)	Mean Speed (m/s)	Kinetic Energy (J/mole)
H_2	0.081	0.002	1909	1761	3644
CH_4	0.60	0.016	678	622	3677
NH_3	0.72	0.017	657	604	3669
H_2O	0.62	0.018	639	587	3675
CO	1.17	0.028	510	469	3641
N_2	1.19	0.028	510	469	3641
Air	1.20	0.029	501	460	3640
O_2	1.25	0.032	477	438	3640
H_2S	1.44	0.034	463	426	3644
CO_2	1.82	0.044	407	374	3644

5.3 KINETIC ENERGY

The kinetic energy E_g of a gas sample is defined as half of the product of the gas mass and the square of the root-mean-square speed of the gas molecules:

$$E_g = \frac{1}{2} M_g u_{rms}^2 = \frac{1}{2} \rho V u_{rms}^2 = \frac{1}{2} n M u_{rms}^2 \tag{5.17}$$

where M_g is the mass, ρ is the density, V is the volume, n is the number of moles, and M is the molar weight of the gas. Note that the kinetic energy is directly proportional to the average of the speed squared, that is $\overline{u^2} = u_{rms}^2$, rather than the mean speed square $\bar{u}^2$. If the kinetic energy were calculated based on the mean speed $\bar{u}$, the result would be 8% lower than its actual value.

Substituting Equation 5.8 into Boyle's law gives

$$\frac{1}{3} \rho u_{rms}^2 V = \frac{1}{3} n M u_{rms}^2 = nRT \tag{5.18}$$

Combining Equation 5.17 and Equation 5.18 gives

$$E_g = \frac{3nRT}{2} \tag{5.19}$$

Equation 5.19 shows that, for an ideal gas, the kinetic energy is independent of pressure, volume, and molecular weight. It depends only on temperature. Therefore, by simply measuring the temperature of a given gas sample, its kinetic energy can be calculated. Kinetic energies per mole for typical gases are calculated using Equation 5.19 and listed in Table 5.1. As shown in Table 5.1, there is very little difference

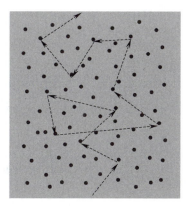

Figure 5.3 A molecule traveling through a gas, colliding with other molecules in its path. The mean path is the average of all straight lines (distances between two collisions). All other molecules are moving in a similar way.

in the molar kinetic energy at a given temperature for different gases. The mean value of E_g is 3652 ± 15.5 J/mole, which varies only $\pm 0.4\%$ from its mean value.

5.4 MEAN FREE PATH

We have assumed that the molecules in a gas sample move with a constant mean speed $\bar{u}$, regardless of direction. Because the molecular motion is completely random, a molecule has to wiggle to travel in a crowd of molecules (Figure 5.3). The degree of "crowdedness" and the molecular speed are important to the transportation of the gas molecules. Rather than using the average spacing between molecules, a more useful concept is the *mean free path*, which is defined as the average distance traveled by a molecule between successive collisions with the same species of molecules. The mean free path of a gas can be determined from the average number of collisions of a particular molecule, n_c, and the average distance it has traveled in the same period of time, which is the mean speed $\bar{u}$ times the t.

$$\lambda = \frac{\bar{u}t}{n_c} \qquad (5.20)$$

This definition of mean free path is analogous to driving a car at 100 km/h and having five (elastic) collisions every hour. The average distance between collisions is 20 kilometers. In order to calculate the mean free path, the collision frequency must be determined. Let us first consider one molecule, traveling in a gas, that collides with other molecules. For convenience in later discussion, we treat this molecule as a bullet. When the bullet molecule with a diameter d_1 travels at a speed u, another molecule with a diameter of d_2 (we treat it as a target) within the distance of d_c (center to center from bullet to target) will collide with the bullet molecule

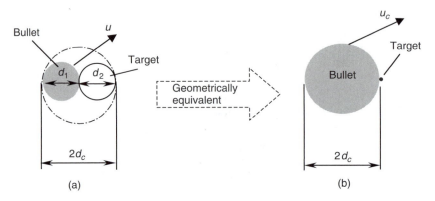

Figure 5.4 A molecule ("bullet") of d_1 in diameter with a speed u that collides with other molecules ("targets") of diameter d_2 (a) creates a sweeping area of πd_c^2 and (b) is geometrically equivalent to a molecule with diameter of $2d_c$ that collides with a molecule with a zero diameter.

(Figure 5.4). The distance between the centers of the two colliding molecules is called *collision diameter dc*:

$$d_c = \frac{d_1 + d_2}{2} \qquad (5.21)$$

where d_1 and d_2 are diameters of the two colliding molecules. The colliding cross-sectional area will be πd_c^2, shown in Figure 5.4a as the larger, dash-lined circle. Geometrically (hence the mass of each colliding molecule does not change), this is equivalent to a molecule with a diameter of $2d_c$ colliding with a point molecule with a diameter of zero, as shown in Figure 5.4b. That is, if the bullet and the target have a zero diameter, the chance of the bullet with a diameter of $2d_c$ hitting the target with a zero diameter is the same as the bullet with a diameter of d_1 hitting the target with a diameter of d_2.

When the equivalent molecule with a colliding area of πd_c^2 travels with a speed u_c, relative to its colliding partner, the volume that this colliding area sweeps over a time t will be

$$V_c = \pi d_c^2 u_c t \qquad (5.22)$$

As shown in Figure 5.5, this volume is a column with a cross-sectional area of πd_c^2 and a length of $u_c t$. All target molecules that fall into the swept volume will be considered to have collided with the bullet molecule. The cylinder in Figure 5.5 will be a zigzag shape, changing direction with every collision, and u_c will not be constant.

Furthermore, if other molecules are stationary, the collision speeds will be the same as the speed of the moving molecule u. However, other molecules are moving targets and move in all directions. Thus, the collision speed, which is the result of the speeds of the two colliding molecules, is not the same as the molecular speed

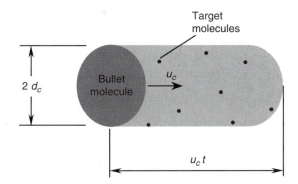

Figure 5.5   A colliding particle sweeps a volume as a column with a cross-sectional area of πd_c^2 and a length of $u_c t$.

u. Consider the same gas species, in which all molecules have the same mass and mean velocity. When two molecules of speed u move toward each other, the collision speed will be $2u$; two molecules moving in the same direction will have a collision speed of 0; two molecules moving at right angles will have a collision speed of $\sqrt{2}\, u$. Considering that the bullet molecule collides with target molecules from all directions, the partial sphere with $u_c > u$ is larger than the partial sphere with $u_c < u$. The mean collision speed over all colliding directions will be

$$\bar{u}_c = \sqrt{2}\, \bar{u} \tag{5.23}$$

A more rigorous derivation of the collision frequency, collision speed in general form (multiple species of gases with different molecular diameters, masses, and mean speeds) is given at the end of this chapter.

If the concentration of molecules is C_{gn} in number of molecules per volume, the number of collisions, or the total numbers of molecules contained in the column (as shown in Figure 5.5), will be

$$n_c = \sqrt{2}\, \pi d_c^2 C_{gn} \bar{u} t \tag{5.24}$$

Substituting Equation 5.24 in to Equation 5.20 gives

$$\lambda = \frac{1}{\sqrt{2}\, \pi d_c^2 C_{gn}} \tag{5.25}$$

Equation 5.25 gives the microscopic description of mean free path of gas molecules, as it involves the diameter of the molecule and the number concentration of gas molecules. For a given gas, that is, a fixed d_c, the mean free path depends only on gas density; it increases with increasing temperature and decreases with increasing pressure. In the atmosphere, the mean free path increases with altitude.

The collision frequency of a gas molecule with other molecules during one second, F_c, is simply the reciprocal of the mean free path. The mean free path and the collision frequency are analogous to wavelength and frequency, respectively, in radio broadcasting.

$$F_c = \sqrt{2} \pi d_c^2 C_{gn} \tag{5.26}$$

The molecular number concentration C_{gn} can be calculated using the Avogadro number (Na) and molar volume (V_M), or the molar weight (M) and density (ρ), of the ideal gas:

$$C_{gn} = \frac{N_a}{V_M} = \frac{N_a \rho}{M} \tag{5.27}$$

Note that the molar volume for an ideal gas is approximately a constant, 0.024 m^3/mole at standard conditions. Therefore, $C_{gn} = 2.5 \times 10^{25}$ (molecules/m^3) is a constant for all ideal gases at standard conditions. If the gas of concern is only part of a mixture, the molecular number concentration of the gas of concern must be multiplied by its volumetric concentration.

For example, the molecular number concentration of CO_2 in a mixture of 20% CO_2 and 80% air in volume is 0.5×10^{25} (molecules/m^3).

Example 5.1: *Calculate the mean free path and the collision frequency of an air molecule at $P_0 = 101.325$ kPa, $T = 21°C$. The density of air at these standard conditions is 1.2 kg/m^3, and the diameter of air molecules is 3.7 Å.*

Solution: The molar weight of air is 0.029 kg/mole, from Equation 5.27, the molecular concentration:

$$C_{gn} = \frac{N_a \rho}{M} = \frac{6.022 \times 10^{23} \times 1.2}{0.029} = 2.492 \times 10^{25} \left(\frac{molecules}{m^3} \right)$$

$$\lambda = \frac{1}{\sqrt{2} \, \pi d_c^2 C_{gn}} = \frac{1}{\sqrt{2} \, \pi (3.7 \times 10^{-10})^2 \times 2.492 \times 10^{25}}$$

$$= 6.6 \times 10^{-8} \, (m) = 0.066 \, \mu m$$

$$F_c = \frac{1}{\lambda} = \frac{1}{6.6 \times 10^{-8}} = 1.515 \times 10^7 \, (1/s)$$

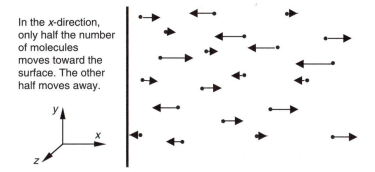

In the x-direction, only half the number of molecules moves toward the surface. The other half moves away.

Figure 5.6 Only half of the total number of molecules moves toward the surface; the other half moves away from the surface. The molecular speeds along the x-direction (u_x), denoted as arrows with different magnitudes, follow Maxwell's one-dimensional speed distribution.

5.5 COLLISION OF GAS MOLECULES WITH A SURFACE

There are two collision frequencies of concern in studying air quality:

- Collision frequency of gas molecules among themselves, F_c, in a gas, which has been defined by Equation 5.24.
- Collision frequency of gas molecules with a stationary surface, F_{cs}. The collision frequency of molecules with a surface is defined as the number of collisions that a unit area (m^2) received in a unit time (s).

The latter is a significant parameter in classical thermodynamics for determining heat conductivity and diffusivity. In air quality studies, this parameter is important in explaining processes such as diffusion, coagulation, and absorption. For example, in a gas detector with a flat sensing cell, how many times will the molecules of the gas of concern make contact with the sensing cell? Or, in an absorption device, the efficiency will depend on how often the gas molecules contact the adsorption surface.

Consider a surface with a unit area and the surrounding gas with a molecular concentration of C_{gn} (Figure 5.6), only half of the number of molecules, or $C_{gn}/2$, moving toward the surface. The other half of the molecules moves away from the surface. Only those molecules moving toward the surface could collide with the surface. The total number of collisions that the unit surface area experiences in one second, or the collision frequency of molecules with a surface, F_{cs}, is the total number of particles that potentially can collide with the surface times the mean molecular speed along the x-direction perpendicular to the surface:

$$F_{cs} = \frac{C_{gn}}{2} \bar{u}_x \qquad (5.28)$$

where $\bar{u}_x$ is the mean molecular speed along the x-direction perpendicular to the surface. From Maxwell's one-dimensional molecular speed distribution (Equation 5.10), $\bar{u}_x$ can be obtained using an equation similar to Equation 5.13:

$$\bar{u}_x = \frac{\int_0^\infty u_x F(u_x)du_x}{N} = \int_0^\infty \left(\frac{2m}{\pi kT}\right)^{1/2} u_x \exp\left(-\frac{mu_x^2}{2kT}\right)du_x$$

$$= \left(\frac{2kT}{\pi m}\right)^{1/2} = \frac{1}{2}\bar{u}$$

(5.29)

Substituting Equation 5.29 into Equation 5.28 gives the collision frequency of molecules with a unit area of surface:

$$F_{cs} = \frac{C_{gn}\bar{u}}{4}$$

(5.30)

where C_{gn} is the concentration of gas molecules in molecules/m³, and $\bar{u}$ is the molecular mean speed in m/s.

Although Equation 5.30 is derived for a flat surface, it can be used to make a reasonable estimate of collision frequency between air molecules and an airborne particle. Considering the relative sizes of air molecules and the 0.1 μm particle, these assumptions are reasonable for the purposes of this calculation (Figure 5.7). When an air molecule with a mean speed $\bar{u}$ collides with the particle, it rebounds and travels a distance on the order of the mean free path (0.066 Fm) since its last collision with another gas molecule.

For air at standard conditions, C_{gn} is 2.5×10^{25}/m³ and $\bar{u}$ is 460 m/s. Equation 5.30 gives the rate of molecular collisions with any surface as 2.87×10^{27} /m²·s. A

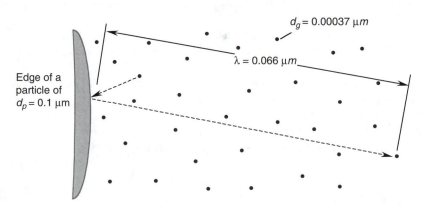

Figure 5.7   Relative size and spacing of air molecules and a 0.1 μm particle.

Table 5.2 Comparison of the Relative Size and Spacing of Air
Molecules with a 0.1 Fm Diameter Particle at Standard
Conditions (A Molecule Striking the Particle Will Have
Traveled a Distance)

Distance	Length (μm)	Ratio to Air Molecular Diameter
Molecular diameter	0.00037	1
Approximate molecular spacing	0.004	10
Mean free path	0.066	180
0.1-μm particle	0.1	260

1 m^3 cubicle has a surface area of 6 m², and will thus experience 1.72×10^{28} collisions per second!

Example 5.2: *Under standard conditions, how many times in one second will air molecules collide with a particle 1 μm in diameter? Assume that the particle surface is flat and stationary relative to the gas molecules.*

Solution: The concentration of air molecules under standard conditions has been calculated in Example 5.1, where $C_{gn} = 2.492 \times 10^{25}$ (molecule/m³). From Table 5.2, the mean air speed is 460 m/s. Substituting into Equation 5.30 gives the collision frequency:

$$F_{cs} = \frac{C_{gn}\bar{u}}{4} = \frac{2.492 \times 10^{25} \times 460}{4} = 2.87 \times 10^{27} \left(\frac{collisions}{s \cdot m^2} \right)$$

The surface area of a 1 μm diameter particle as a sphere is

$$A = \pi d_p^2 = \pi \times (10^{-6})^2 = 3.14 \times 10^{-12} \ (m^2)$$

The total number of collisions with the particle in one second is

$$n_t = F_{cs}A = 2.87 \times 10^{27} \times 3.14 \times 10^{-12} = 9.01 \times 10^{15} \left(\frac{collisions}{s} \right)$$

Compared to the collision frequency (151,515 collisions per second) of an air molecule in the previous example, a 1 μm particle experiences 60 billion times more collisions than an air molecule.

5.6 VISCOSITY

5.6.1 Viscosity for a Single Gas

Consider a gas contained between two large parallel plates, each with an area A. The two plates are separated by a small distance Y, so that the flow is laminar.

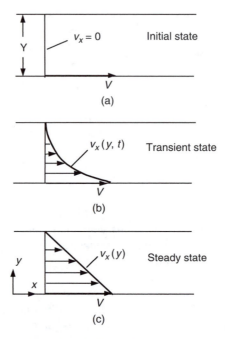

Figure 5.8 Establishment of steady-state flow velocity between two parallel plates.

The initial condition of each plate is stationary. The lower plate is set in motion along the x-direction at a constant velocity V (Figure 5.8a). The gas gains momentum and tends to follow the motion of the moving plate (Figure 5.8b) during a transient period of time. Eventually, the gas establishes a steady-state velocity profile, as shown in Figure 5.8c. At such a steady state, a constant force F (push or pull) on the lower plate is required to maintain the motion. This force is proportional to the area of the plate, A, and to the relative velocity between the two plates; it is inversely proportional to the distance Y. At the same time, a shear force F_τ, caused by the friction, is exerted on the lower plate at a direction opposite to force F.

$$F_\tau = -F = -\eta A \frac{\Delta v_x}{\Delta y} = -\eta A \frac{V}{Y} \qquad (5.31)$$

The coefficient of the proportionality, η, is called the *dynamic viscosity,* or simply the viscosity, in $N \cdot s/m2$, $kg/m \cdot s$, or $Pa \cdot s$. The viscosity of air at 20°C and one atmosphere pressure is 1.81×10^{-5} $Pa \cdot s$. It should be noted that v_x is the gas flow velocity, not the molecular speed.

Equation 5.31 can be expressed in a more explicit form of the shear stress τ_{yx} between the gas layers at a distance y from the moving plate:

$$\tau_{yx} = \frac{F_\tau}{A} = -\eta \frac{dv_x}{dy} \qquad (5.32)$$

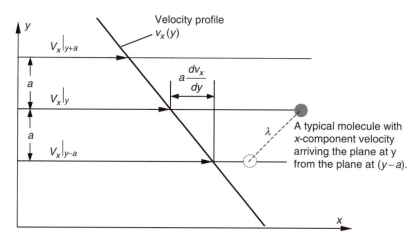

Figure 5.9   A typical molecular x-momentum transport from the plane at y-a to the plane at y.

Equation 5.32 states that the shear force per unit area is proportional to the negative of the local velocity gradient dv_x/dy. The viscous momentum flux (i.e., the shear stress) is in the direction of the negative velocity gradient, that is, the momentum tends to go downhill from a high-velocity region to a low-velocity region. This is analogous to heat transfer, where the heat flows from a high-temperature region to a low-temperature region. Equation 5.32 is known as Newton's law of viscosity. Gases that behave in this fashion are called *Newtonian fluids*. All gases and most simple liquids are Newtonian fluids. Fluids that do not obey Equation 5.32 include pastes, slurries, and high polymers.

Viscosity in Equation 5.32 is a macroscopic description in terms of the shear stress and the velocity gradient. The microscopic description of the viscosity must be started from the molecular motion and the momentum transfer among the molecules. Consider a gas composed of rigid spherical molecules with a diameter d, a mass m, and a concentration of C_{gn}. The steady flow established in Figure 5.8c can be amplified to a molecular scale, as shown in Figure 5.9. At any plane with a v_x, between and parallel to the two confinement plates, the molecules reaching the plane have, on the average, had their last collision at a distance a from the plane[3] :

$$a = \frac{2}{3}\lambda \qquad\qquad (5.33)$$

where λ is the mean free path of the gas molecules.

As shown in Figure 5.9, the flux of x-momentum across any plane y is the sum of the x-momenta of the molecules that cross the plane from the positive y-direction, subtracting the x-momenta of the molecules from the opposite direction. Because the total number of molecules in a unit volume involves such momenta transportation, the molecular collision frequency with the surface at y, F_{cs}, and all molecules that arrive at the y plane have experienced their last collision; the momentum flux, or the shear stress τ_{yx} can be written as:

$$\tau_{yx} = F_{cs} m v_x \big|_{y-a} - F_{cs} m v_x \big|_{y+a} \tag{5.34}$$

At the molecular level, viscosity of a gas represents a transfer of molecular momentum from a faster-moving layer to a slower-moving layer. This transfer is accomplished by the random thermal motion of molecules traveling between the layers. An appealing physical analogy for this process is that of two trains of flatcars traveling at slightly different speeds on parallel, adjacent tracks. The flatcars' passengers, each with a mass m, amuse themselves by jumping back and forth between the trains. Each jump from the faster train to the slower train imparts a momentum of $m\Delta v_x$ to speed up the slower train. Similarly, jumps the other way slow the faster train. The greater the difference in speed and the more frequent the jumping, the greater the effect. For a jumping rate of F_{cs} (collision frequency with a surface) per second, each train experiences a force of $F_{cs} m\Delta v_x$, accelerating the slower train and decelerating the faster train. At steady state, the velocity profile $v_x(y)$ is linear and the x-direction velocity can be determined from Figure 5.9, where

$$v_x \big|_{y-a} = v_x \big|_y - \frac{2}{3} \lambda \frac{dv_x}{dy} \tag{5.35}$$

$$v_x \big|_{y+a} = v_x \big|_y + \frac{2}{3} \lambda \frac{dv_x}{dy} \tag{5.36}$$

Substituting Equation 5.35, Equation 5.36, and Equation 5.30 into Equation 5.34 gives:

$$\tau_{yx} = -\frac{1}{3} C_{gn} m \bar{u} \lambda \frac{dv_x}{dy} \tag{5.37}$$

Equation 5.37, which was developed by Maxwell, gives a microscopic description of a fluid flow between two parallel plates. Combining Newton's law (Equation 5.32) and Maxwell's equation (Equation 5.37) gives

$$\eta = \frac{1}{3} C_{gn} m \, \bar{u} \lambda = \frac{1}{3} \rho \, \bar{u} \lambda \tag{5.38}$$

Substituting Equation 5.13 and Equation 5.25 into Equation 5.38 gives

$$\eta = \frac{2}{3\pi^{3/2} d_c^2} (mkT)^{1/2} \tag{5.39}$$

which represents the viscosity of a gas composed of *hard spheres*.

Error Analysis: Equation 5.39 gives an approximate prediction of viscosity vs. the collision diameter and temperature. One error is derived from the assumption that molecules are rigid spheres. Actually, the molecules are not perfectly rigid, thus the collisions are not perfectly elastic. Accurate prediction of viscosity must replace the rigid-sphere molecular model with a more realistic molecular force field. Such an analytical task is very difficult. Some semiempirical analyses have been described elsewhere.[2–4] A more rigorous description of viscosity is given by

$$\eta = \frac{5\pi}{32} C_{gn} m \bar{u} \lambda = \frac{5}{16\pi^{1/2} d_c^2} (mkT)^{1/2} \tag{5.40}$$

Equation 5.40 predicts viscosity for a gas and is especially recommended for applications where the collision diameter is in question. The viscosity of a gas is independent of pressure and has been verified experimentally over a pressure range of 0.001 to 100 atm. This equation provides a relationship between viscosity and collision diameter. To determine the collision diameter, the viscosity at one condition should be experimentally determined. Because the collision diameter remains the same, the viscosity at other conditions can then be predicted using Equation 5.40. The viscosity given by Equation 5.40 is independent of pressure and increases with temperature. This may be against our intuitive expectation, as the viscosities of many liquids, such as oils and honey, decrease when they are heated. This occurs because the liquid molecules are close to each other, thus the cohesive force, rather than the collisions among the molecules, primarily contributes to the shear stress. The gas molecules are usually too far apart to have a significant cohesive force among the molecules.

A second error source associated with Equation 5.39 and Equation 5.40 is that the true relationship of viscosity with temperature is not proportional to any power function of T. It varies between $T^{1/2}$ and T. In most indoor air quality problems, we are dealing with gas temperatures varying within a narrow range (e.g., between –40 and 40°C), and errors estimated using Equation 5.39 are less than 5%, which is sufficiently accurate for most applications.

Example 5.3: Viscosity of air at standard conditions (20°C) is measured as 1.81×10^{-5} Pa·s. Calculate the air viscosities at –23°C and 77°C.

Solution: From the experimental data and Equation 5.40, the collision diameter of air is

$$d_c = \left(\frac{5(mkT)^{1/2}}{16\eta\pi^{1/2}} \right)^{\frac{1}{2}} = \left(\frac{5 \times \left(\dfrac{0.029}{6.022 \times 10^{23}} \times 1.381 \times 10^{-23} \times 293 \right)^{1/2}}{16 \times 1.81 \times 10^{-5} \pi^{1/2}} \right)^{1/2}$$

$$= 3.69 \times 10^{-10} \ (m) = 3.69 \ \overset{o}{A}$$

Using the collision diameter of 3.69 $\approx$, and Equation 5.40, we have

$$\eta\big|_{T=250K} = \frac{5}{16\pi^{1/2}d_c^2}(mkT)^{1/2} = \frac{5\times\left(\dfrac{0.029}{6.022\times10^{23}}\times1.381\times10^{-23}\right)^{1/2}}{16\times\pi^{1/2}\times(3.69\times10^{-10})^2}\times T^{1/2}$$

$$= 1.056\times10^{-6}\times250^{1/2} = 1.67\times10^{-5}\ (Pa\cdot s)$$

$$\eta\big|_{T=350K} = 1.056\times10^{-6}\times350^{1/2} = 1.976\times10^{-5}\ (Pa\cdot s)$$

The measured viscosities of air at 250 K and 350 K are 1.596 $\times$ 10⁻⁵ and 2.082 $\times$ 10⁻⁵ Pa·s, respectively.[5] These represent a 4.6 % and a 5% error between the calculated and measured data.

Note that if Equation 5.39 were used, d_c would be 3.04 $\approx$, which is 17% smaller than the actual air molecule diameter (3.7 $\approx$). The viscosities would be the same as predicted by Equation 5.40. This demonstrates that when the collision diameter is in question, Equation 5.40, rather than Equation 5.39, should be used. Example 5.3 shows that the viscosity does not change exactly with the $T^{1/2}$. Using Equation 5.40 only gives a reasonable prediction of the viscosity of a gas.

5.6.2 Viscosity for a Mixture of Gases

For a mixture of gases, viscosity can be calculated using a semiempirical method given by Wilke:[6]

$$\eta_{mix} = \sum_{i=1}^{n}\frac{f_i\eta_i}{\displaystyle\sum_{j=1}^{n}f_j\Phi_{ij}} \tag{5.41}$$

in which

$$\Phi_{ij} = \frac{1}{\sqrt{8}}\left(1+\frac{M_i}{M_j}\right)^{-1/2}\left[1+\left(\frac{\eta_i}{\eta_j}\right)^{1/2}\left(\frac{M_i}{M_j}\right)^{1/4}\right]^2 \tag{5.42}$$

where n is the number of gas species in the mixture, f_i and f_j are the molar fractions of species i and j, η_i and η_j are the viscosities of species i and j at the system temperature and pressure, and M_i and M_j are the molar weights for species of i and j. Note that Φ_{ij} is dimensionless, and when $i = j$, $\Phi_{ij} = 1$. Provided that the viscosity for each species is accurate, the viscosity predicted by Equation 5.41 is within 2%

error. Generally, prediction errors using Equation 5.40 and Equation 5.41 are larger for polar molecules and for highly elongated molecules because of the highly angle-dependent force fields that exist between such molecules. Such polar molecules include H_2O, NH_3, CH_3OH, and NOCl.

Example 5.4: *Calculate the viscosity of a near-standard air sample at 300 K and 1 atm pressure. The air sample is composed of 78% nitrogen, 21% oxygen, and 1% argon, all in volume. Other given conditions are listed in the table below.*

Species n	Volume Fraction (Mole fraction) f	Mass Fraction	Molar Weight M (kg/mole)	Viscosity η (Pa·s)
1: N_2	0.78	0.7541	0.028	1.79×10^{-5}
2: O_2	0.21	0.2320	0.032	2.08×10^{-5}
3: Ar	0.01	0.0138	0.040	2.29×10^{-5}

Solution: Using Equation 5.41 and Equation 5.42, the main steps of calculation are shown in the table below.

i	j	M_i/M_j	η_i/η_j	Φ_{ij}	$\sum_{j=1}^{3} f_j \Phi_{ij}$
1	1	1.000	1.000	1.000	
	2	0.875	0.861	0.991	0.999
	3	0.700	0.782	1.048	
2	1	1.143	1.162	1.007	
	2	1.000	1.000	1.000	1.006
	3	0.800	0.908	1.028	
3	1	1.429	1.279	0.942	
	2	1.250	1.101	0.936	0.941
	3	1.000	1.000	1.000	

Substituting the values from the above table into Equation 5.41 gives

$$\eta_{mix} = \sum_{i=1}^{n} \frac{f_i \eta_i}{\sum_{j=1}^{n} f_j \Phi_{ij}}$$

$$= \frac{0.78 \times 1.79 \times 10^{-5}}{0.999} + \frac{0.21 \times 2.08 \times 10^{-5}}{1.006} + \frac{0.01 \times 2.29 \times 10^{-5}}{0.941}$$

$$= 1.86 \times 10^{-5} \quad (Pa \cdot s)$$

This value matches the measured viscosity of air at 300 K. This is not surprising, because Equation 5.41 was derived from empirical data.

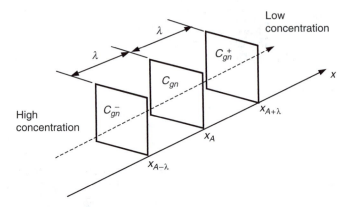

Figure 5.10 Mass diffusion field with a concentration gradient dC_{gn}/dx.

5.7 DIFFUSIVITY

Mass transfer of one gas through another in the absence of fluid flow is called *diffusion*. It is the result of the motion of gas molecules in a concentration gradient. For a gas b that is diffusing through air, the flux J, the quantity of gas b transferred per unit time through a unit area perpendicular to the direction of diffusion (molecules/m² @ s) under the influence of a concentration gradient dC_{gn}/dx, is given by Fick's first law of diffusion:

$$J = -D_{ba} \frac{dC_{gn}}{dx}$$ (5.43)

where the quantity D_{ba} is called the *diffusion coefficient* or *diffusivity* of gas b in gas a and has units of m²/s. The negative sign indicates that the mass transfer is always toward the region of lower concentration, similar to the heat transfer in which the heat always goes "downhill," that is, from a high-temperature region to a low-temperature region.

To determine the diffusivity D_{ba}, consider a plane at x_A with unit area perpendicular to the x-axis (Figure 5.10). The concentration gradient is along the x-axis. At plane x_A, the molecular concentration is C_{gn}. At the two adjacent planes, $x_{A-\lambda}$ and $x_{A+\lambda}$, the molecular concentrations are C_{gn}^- and C_{gn}^+ for the high and low concentration sides, respectively. The two adjacent planes are parallel to, and each has a distance of λ from, plane x_A. In the concentration gradient field, the concentrations C_{gn}^- and C_{gn}^+ are proportional to the concentration gradient dC_{gn}/dx.

$$C_{gn}^- = C_{gn} + \lambda \frac{dC_{gn}}{dx}$$ (5.44)

$$C_{gn}^{+} = C_{gn} - \lambda \frac{dC_{gn}}{dx} \tag{5.45}$$

The total number of molecules from plane $x_{A-\lambda}$ that arrive at plane x_A is one sixth of the molecules on the plane $x_{A-\lambda}$. The factor is one sixth because the molecular motion on plane $x_{A-\lambda}$ is moving in all three directional components, x, y, and z, and each component in two opposite directions along the axis. Therefore, only one sixth of the total number of molecules will arrive at plane x_A. Similarly, only one sixth of the total number of molecules on plane x_A will arrive at plane $x_{A-\lambda}$ from plane x_A. The molecular flux at plane x_A, which is defined as the total number of molecules passing through the unit area plane x_A in one second, is the sum of the molecules leaving plane x_A and the molecules arriving plane x_A:

$$J = \frac{1}{6}(C_{gn}^{+}\bar{u} - C_{gn}^{-}\bar{u}) = -\frac{1}{3}\lambda\bar{u}\frac{dC_{gn}}{dx} \tag{5.46}$$

Compare Equation 5.46 with Equation 5.43:

$$D_{ba} = \frac{1}{3}\lambda\bar{u} \tag{5.47}$$

Substituting Equation 5.25 and Equation 5.13 for λ and $\bar{u}$, respectively, into Equation 5.47, yields

$$D_{ba} = \frac{2}{3\pi^{3/2}C_{gn}d_c^2}\left(\frac{RT}{M}\right)^{1/2} \tag{5.48}$$

which represents the diffusivity of gas b in gas a, where all variables are associated with gas b only, that is, C_{gn}, d_c, and M are the molecular concentration, collision diameter, and molar weight of gas b. This is based on the fact that gas a does not have a concentration gradient. If a mixture of gases with different molar weights diffuses in air, the diffusivity becomes more complicated.[3] In most indoor air problems, gaseous contaminants diffusing in air can be treated as individual gases, because those gases are in low concentrations compared with air.

Error Analysis: Equation 5.48 gives an approximate prediction of diffusivity vs. the collision diameter and temperature, similar to viscosity. One error is derived from the assumption that molecules are rigid spheres.[2] A more rigorous prediction of diffusivity, taking into account a more realistic molecular force field, is given by

Table 5.3   Diffusivity of Typical Gases in Air at Standard Atmospheric Pressure (Assuming that the Content of Air Is Much Higher than the Gas)

System	Diffusivity × 10⁻⁵ (m²/s)	
	$T = 273.15$ K	$T = 293.15$ K
Ar–air	1.67	1.48
CH_4–air		1.06
CO–air		2.08
CO_2–air		1.60
H_2–air	6.68	6.27
H_2O–air		2.42
He–air	0.617	5.80
NH_3-air		2.47

Source: Lide, D.R., ed., *Handbook of Chemistry and Physics*, 79th ed., CRC Press, Boca Raton, FL, 1999, 6–168. With permission.

$$D_{ba} = \frac{3}{8\pi^{1/2} C_{gn} d_c^2} \left(\frac{RT}{M}\right)^{1/2} \tag{5.49}$$

Compare Equation 5.48 and Equation 5.49: The factor is 0.1197 ($2/3\pi^{3/2}$) for Equation 5.48 and 0.2116 ($3/8\pi^{1/2}$) for Equation 5.49. To illustrate the difference, the diffusivity of air at standard conditions predicted by Equation 5.48 is 1.02×10^{-5} m²/s; the diffusivity predicted by Equation 5.49 is 1.80×10^{-5} m²/s; and the measured value is 2.0×10^{-5} m²/s. Evidently, Equation 5.49 has a much smaller error in predicting the diffusivity than Equation 5.48. Therefore, Equation 5.49 should be used to calculate the diffusivity of gases when a measured value is not available. Diffusivities of typical gases in air under standard atmospheric pressure conditions are listed in Table 5.3. More typical gas-to-gas diffusivities at different temperatures are in Appendix 5.

5.7.1   Concentration Variation Caused by Diffusion

It is useful to determine the time and spatial variation of flux caused by diffusion. For example, how long does it take for a gas to diffuse from one side of a room to the other? Or how far must a gas be from an odorous gas source to be diluted enough to be below the smell threshold limit? Answers to these questions involve the concentration (flux) variation with time and distance from a source or an original state.

In the following discussion, we will discuss a simple situation: one-dimensional diffusion. Consider a thin slab of a mixture of gas b and air with thickness dx and a unit area (similar to Figure 5.10 by considering λ as dx). The x-axis is perpendicular to the slab face. Assume that the concentration of gas b is C_{gn} at the surface x, and the flux of gas b in the x-direction through the face at surface x is given by Fick's first law:

$$J\big|_x = -D_{ba} \frac{dC_{gn}}{dx} \tag{5.50}$$

The flux of gas b through the surface $x + dx$ is

$$J\big|_{x+dx} = -\left(J\big|_x + \frac{dJ}{dx}dx\right) = -\left(J\big|_x + \frac{d}{dx}\left(D_{ba}\frac{dC_{gn}}{dx}\right)dx\right) \tag{5.51}$$

The net accumulation of gas b in a time interval dt within the slab is

$$dN_b = \left(J\big|_x - J\big|_{x+dx}\right)dt = dx\,\frac{d}{dx}\left(D_{ba}\frac{dC_{gn}}{dx}\right)dt \tag{5.52}$$

Because the volume of the slab with a unit area is $V = dx$, and $C_{gn} = N_b/V$, Equation 5.52 becomes

$$\frac{dC_{gn}}{dt} = \frac{d}{dx}\left(D_{ba}\frac{dC_{gn}}{dx}\right) = D_{ba}\frac{d^2C_{gn}}{dx^2} \tag{5.53}$$

It is important to note that the derivation of the left side of Equation 5.53 is taken at a fixed location, and the right side is taken at a fixed time. This feature may be indicated by using the notation of partial differentiation:

$$\frac{\partial C_{gn}}{\partial t} = D_{ba}\frac{\partial^2 C_{gn}}{\partial x^2} \qquad \left\{ \begin{aligned} &C_{gn}(0,0) = C_{gn0} \\ &C_{gn}(x,\infty) = C_{gn}(\infty,t) = 0 \end{aligned} \right\} \tag{5.54}$$

Equation 5.54 is known as Fick's second law of diffusion. Consider a special situation in which the initial concentration of gas b is C_{gn0}, at $x = 0$ and $t = 0$ (Figure 5.11). The gas molecules will diffuse along the x-axis in both directions. Considered at the initial time $t = 0$, Equation 5.54 can be integrated and solved for C_{gn} as a function of time t and distance x.[2] The relationship of the concentration gradient of gas b at a distance x and time t becomes

$$\frac{dC_{gn}(x,t)}{dx} = \frac{C_{gn0}}{2\sqrt{\pi D_0 t}}\exp\left(\frac{-x^2}{4D_0 t}\right) \tag{5.55}$$

where C_{gn0} is the concentration of gas b introduced at the initial interface, x is the distance from the original position of the interface, and D_0 is the diffusion coefficient of gas b into gas a at the original position of interface.

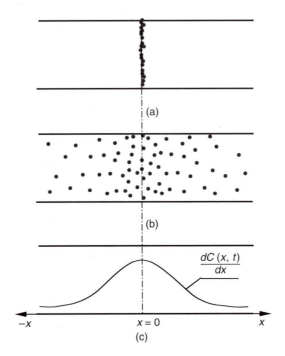

Figure 5.11 One-dimensional gas concentration distribution as a function of distance from the original interface ($x = 0$) and time t. (a) All molecules are assumed to be at the interface when $t = 0$. (b) Molecules diffuse along x-axis in both directions at a given time t. (c) Gas molecular concentration distribution with respect to x at a given time t.

Equation 5.55 also gives the particle concentration distribution at any distance from its origin ($x = 0$) and from the initial state ($t = 0$). As shown in Figure 5.11, assume that particles with an initial concentration C_{pn0} at $t = 0$ and the concentration along the x-axis in both directions, will decrease with t and x. If x is a constant, as in a confined chamber, the concentration gradient will eventually be zero. If x goes to infinity, as in an open space, the concentration will eventually be zero.

Substituting Equation 5.55 into Equation 5.43 gives the flux of gas b in gas a:

$$J(x,t) = -\frac{C_{gn0}}{2}\left(\frac{D_0}{\pi t}\right)^{1/2} \exp\left(\frac{-x^2}{4D_0 t}\right) \tag{5.56}$$

Note that Equation 5.56 gives the flux in one direction (either at $-x$ or at x, Figure 5.11c). Equation 5.56 is especially useful in finding the flow rate of a gas driven by diffusion only. The following example demonstrates such applications of concentration gradient and flux caused by diffusion. A more in-depth discussion of diffusion of gas, molecular, or particle flux, and deposition resulting from diffusion appears in Chapter 6.

Example 5.5: *A bottle contains pure carbon monoxide. A 0.5 m long tube is attached to the bottle, and the initial CO concentration at the intersection of the tube and the bottle is the same as in the bottle. Assume that there is no convection in the tube. The inside diameter of the tube is 5 mm. The tube's outlet (the end opposite to the bottle) is open to the room air. The collision diameter of CO is about 3.6×10^{-10} m. Both the CO in the bottle and the room air are at 1 atm pressure and 20°C. What will the flow rate of CO be at the outlet after 2 minutes?*

Solution: The molar volume at standard conditions is 0.0241 m³ for ideal gas. At the interface of the tube and the tank, the number concentration of CO is

$$C_{gn0} = \frac{N_a}{V_M} = \frac{6.022 \times 10^{23} \ molecules \ / \ mole}{0.0241 \ m^3 \ / \ mole} = 2.5 \times 10^{25} \ \frac{molecule}{m^3}$$

From Equation 5.49

$$D_0 = \frac{3}{8\sqrt{\pi} \ C_{gn0} d_c^2} \left(\frac{RT}{M} \right)^{\frac{1}{2}}$$

$$= \frac{3}{8\sqrt{\pi} \times 2.5 \times 10^{25} \times (3.6 \times 10^{-10})^2} \left(\frac{8.31 \times 293}{0.028} \right)^{\frac{1}{2}} = 1.93 \times 10^{-5} \ (m^2 \ / \ s)$$

At the outlet of the tube, $x = 0.5$ m and $t = 120$ s. Substituting values in Equation 5.56 gives

$$J_{gn}(x,t) = \frac{C_{gn0}}{2} \left(\frac{D_0}{\pi t} \right)^{\frac{1}{2}} \exp\left(\frac{-x^2}{4D_0 t} \right)$$

$$= \frac{2.5 \times 10^{25} \times \sqrt{1.92 \times 10^{-5}}}{2 \times \sqrt{\pi \times 120}} \exp\left(\frac{-0.5^2}{4 \times 1.92 \times 10^{-5} \times 120} \right)$$

$$= 4.69 \times 10^9 \ \frac{molecules}{m^2 \cdot s}$$

The flow rate of CO, Q_{co}, at the outlet is

$$Q_{co} = JA = 4.69 \times 10^9 \times \frac{\pi \times 0.005^2}{4} = 92,087 \ \frac{molecules}{s}$$

This is a very small flow rate for gas molecules, considering the initial concentration at the interface of tank and tube, indicating that the diffusion rate is very low. Typically, it takes several seconds for molecules of gas *b* to travel 1 cm in gas *a*. The diffusion rate of liquids is even slower than that of gases.

5.8 MOLECULAR COLLISION SPEED OF A MIXTURE OF GASES

Consider two species of gas molecules, A and B, each having N_A and N_B molecules in a unit volume. The molecular speed of species A has three components: u_{Ax}, u_{Ay}, and u_{Az}. The number of molecules for species A falling in a speed range $(u_{Ax}, u_{Ax+d} u_{Ax})$, $(u_{Ay}, u_{Ay+}du_{Ay})$, and $(u_{Az}, u_{Az+}du_{Az})$ can be calculated from Equation 5.10, the Maxwell–Boltzmann speed distribution:

$$dN_A = N_A \left(\frac{m_A}{2\pi kT} \right)^{3/2} \exp\left[\frac{-m_A\left(u_{Ax}^2 + u_{Ay}^2 + u_{Az}^2\right)}{2kT} \right] du_{Ax} du_{Ay} du_{Az} \qquad (5.57)$$

where m_A is the molecular mass of species A; u_{Ax}, u_{Ay}, and u_{Az} are three molecular velocity components; k is the Boltzmann constant; and T is absolute temperature.

Similarly, the number of molecules for species B falling in the domain of (x_B, x_B+dx_B), (y_B, y_B+dy_B), and (z_B, z_B+dz_B) can be calculated from the Maxwell–Boltzmann speed distribution:

$$dN_B = N_B \left(\frac{m_B}{2\pi kT} \right)^{3/2} \exp\left[\frac{-m_B\left(u_{Bx}^2 + u_{By}^2 + u_{Bz}^2\right)}{2kT} \right] du_{Bx} du_{By} du_{Bz} \qquad (5.58)$$

where m_B is the molecular mass of species B; u_{Bx}, u_{By}, and u_{Bz} are three molecular velocity components; k is the Boltzmann constant; and T is absolute temperature. The relative molecular speed, or the collision speed u_{AB} of species A and B is

$$u_{AB} = \left[\left(u_{Ax} - u_{Bx}\right)^2 + \left(u_{Ay} - u_{By}\right)^2 + \left(u_{Az} - u_{Bz}\right)^2 \right]^{1/2}$$

$$= (u_{cx}^2 + u_{cy}^2 + u_{cz}^2)^{1/2} \qquad (5.59)$$

where the subscript c refers to the collision (relative speeds of A and B).

Referring to Figure 5.5, the number of collisions, dn_{AB}, among the species A and B in one second, in the velocity domains $d\Omega_A$ [$(u_{Ax}, u_{Ax+d} u_{Ax})$, $(u_{Ay}, u_{Ay+}du_{Ay})$, $(u_{Az}, u_{Az+}du_{Az})$] and $d\Omega_B$ [(x_B, x_B+dx_B), (y_B, y_B+dy_B), (z_B, z_B+dz_B)], is given by

$$dn_{AB} = dN_A dN_B \pi d_{AB}^2 u_{AB} \qquad (5.60)$$

where

$$d_{AB} = \frac{d_A + d_B}{2} \qquad (5.61)$$

The total number of collisions n_{AB} within in the two velocity domains Ω_A (u_{Ax}, u_{Ay}, u_{Az}) and Ω_B (u_{Bx}, u_{By}, u_{Bz}) in one second can be obtained by integrating dn_{AB} over the entire two velocity domains.

$$n_{AB} = \iint_{\Omega_A, \Omega_B} dZ_{AB} = \pi d_{AB}^2 \iint_{\Omega_A, \Omega_B} u_{AB} dN_A dN_B \qquad (5.62)$$

Substituting Equation 5.57, through Equation 5.59 into Equation 5.60, and considering the speed range $(-\infty, \infty)$ in Maxwell–Boltzmann's distribution, gives

$$n_{AB} = \frac{\pi}{8} d_{AB}^2 N_A N_B \frac{(m_A m_B)^{3/2}}{(\pi kT)^3} \int_{-\infty}^{\infty} \int_{-\infty}^{\infty} \int_{-\infty}^{\infty} \int_{-\infty}^{\infty} \int_{-\infty}^{\infty} \int_{-\infty}^{\infty} (u_{cx}^2 + u_{cy}^2 + u_{cz}^2)^{1/2}$$

$$\exp\left(\frac{-m_A(u_{ax}^2 + u_{Ay}^2 + u_{Az}^2) - m_B(u_{Bx}^2 + u_{By}^2 + u_{Bz}^2)}{2kT}\right) \times \qquad (5.63)$$

$$du_{ax} du_{ay} du_{az} du_{Bx} du_{By} du_{Bz}$$

To solve Equation 5.63, let us change variables to

$$u_{Mx} = \frac{m_A u_{Ax} + m_B u_{Bx}}{m_A + m_B} \qquad (5.64)$$

$$u_{My} = \frac{m_A u_{Ay} + m_B u_{By}}{m_A + m_B} \qquad (5.65)$$

$$u_{Mz} = \frac{m_A u_{Az} + m_B u_{Bz}}{m_A + m_B} \qquad (5.66)$$

$$m_{AB} = \frac{m_A m_B}{m_A + m_B} \qquad (5.67)$$

where u_M can be considered the resultant speed of u_A and u_B, carried by a combined mass $(m_A + m_B)$. Substituting Equation 5.64 through Equation 5.67 into Equation 5.63 gives

$$n_{AB} = \frac{\pi}{8} d_{AB}^2 N_A N_B \frac{(m_A m_B)^{3/2}}{(\pi kT)^3} \int_{-\infty}^{\infty} \int \int \exp\left[\frac{-(m_A + m_B)(u_{Mx}^2 + u_{My}^2 + u_{Mz}^2)}{2kT}\right.$$

$$du_{Mx}\,du_{My}\,du_{Mz}\int\limits_{-\infty}^{\infty}\!\!\int\!\!\int (u_{cx}^2 + u_{cy}^2 + u_{cz}^2)^{1/2}$$

$$\exp\left[\frac{-m_{AB}\,(u_{cx}^2 + u_{cy}^2 + u_{cz}^2)}{2kT}\right] du_{cx}\,du_{cy}\,du_{cz} \qquad (5.68)$$

The first integral in Equation 5.68 is

$$\int\limits_{-\infty}^{\infty}\!\!\int\!\!\int \exp\left[\frac{-(m_A + m_B)(u_{mx}^2 + u_{my}^2 + u_{mz}^2)}{2kT}\right] du_{mx}\,du_{my}\,du_{mz}$$

$$=\left\{\int\limits_{-\infty}^{\infty} \exp\left[\frac{-(m_A + m_B)u_{mx}}{2kT}\right] du_{mx}\right\}^3$$

$$=\left(\frac{2\pi\,kT}{m_A + m_B}\right)^{3/2} \qquad (5.69)$$

The second integral can be solved by changing the cardinal coordinate into a spherical coordinate (Figure 5.12).

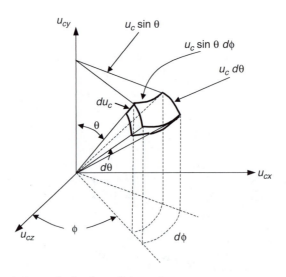

Figure 5.12 Spherical coordinates for collision volume.

$$du_{cx}\, du_{cy}\, du_{cz} = u_c^{\ 2} \sin\theta\, du_c\, d\theta\, d\phi \tag{5.70}$$

Substituting Equation 5.70 into the second integral in Equation 5.68 gives

$$\int\limits_{-\infty}^{\infty}\int\int (u_{cx}^2 + u_{cy}^2 + u_{cz}^2)^{1/2} \exp\left[\frac{-m_{AB}\,(u_{cx}^2 + u_{cy}^2 + u_{cz}^2)}{2kT}\right] du_{cx}\, du_{cy}\, du_{cz}$$

$$= \int\limits_{-\infty}^{\infty} du_c \int\limits_{0}^{2\pi} d\phi \int\limits_{0}^{\pi} u_c^{\ 3} \sin\theta \exp\left(\frac{-m_{AB}\,u_c^{\ 2}}{2kT}\right) d\theta$$

$$= 8\pi \left(\frac{kT}{m_{AB}}\right)^2 \tag{5.71}$$

Substituting Equation 5.63 and Equation5.65 into Equation 5.62 gives

$$n_{AB} = N_A\, N_B\, d_{AB}^2 \left(\frac{8\pi\, kT}{m_{AB}}\right)^{1/2} \tag{5.72}$$

Comparing Equation 5.72 with Equation 5.60 gives the average of the collision speed, or the relative velocity of molecule A and molecule B:

$$\bar{u}_c = \left(\frac{8kT}{\pi\, m_{AB}}\right)^{1/2} \tag{5.73}$$

When species A and B are identical

$$M_{AB} = \frac{m_A\, m_B}{m_A + m_B} = \frac{m}{2} \tag{5.74}$$

Note that the mean molecular speed $\bar{u} = \dfrac{8kT}{\pi\, m}$ gives

$$\bar{u}_c = \sqrt{2}\left(\frac{8kT}{\pi\, m}\right) = \sqrt{2}\,\bar{u} \tag{5.75}$$

DISCUSSION TOPICS

1. What variables are for *macroscopic* descriptions of gases? What variables are for *microscopic* descriptions of gases?
2. Explain why the kinetic energy, molar volume, and molecular number concentration for all ideal gases are approximately the same.
3. Discuss the error sources for equations that predict viscosity and diffusivity.
4. Why are predictions of viscosity and diffusivity for polar or elongated molecules more erroneous than for nonpolar molecules?
5. Why is viscosity independent of pressure, whereas diffusivity increases with pressure?
6. When collision diameter is calculated, there is a difference between results calculated using viscosity or diffusivity and those calculated using the actual molecule diameter. What are the possible error sources?
7. How do you explain that a calculated collision diameter is larger than the actual diameter?
8. If the diffusion rate of gases is so low, in the order of a few minutes or hours per meter, why can we smell spilled vinegar or ammonia in a room almost instantaneously?

PROBLEMS

1. Assume that a particle of 0.1 μm in aerodynamic diameter is a gas molecule. What would the molar weight of such a gas be?
2. A car tire contains 8 liters of air at a gauge pressure of 200 kPa (29 psi) when the temperature is 30°C. What will the gauge pressure be in the tire when air temperature drops to −10°C?
3. A tank of 100% hydrogen sulfide (H_2S) has a volume of 50 liters at a pressure of 10 atm. In an experiment, the H_2S is discharged at a constant rate of 0.01 l/min into room air at standard conditions (1 atm pressure and 20°C). How many hours can this tank of H_2S last?
4. The viscosity of ammonia (NH_3) at standard conditions (20°C, 1 atm pressure) is 1.02×10^{-5} Pa·s. The density is 0.7 kg/m3. Find the following:
   a. Kinetic energy per mole
   b. Diameter of an ammonia molecule
   c. Diffusion coefficient
   d. Collision frequency
   e. Mean free path
5. The viscosity of the air at standard conditions (20°C, 1 atm pressure) is 1.81×10^{-5} Pa·s, and the density is 1.2 kg/m³. Find the following:
   • Kinetic energy per mole
   • Collision diameter of a "molecule" of air
   • Diffusion coefficient
   • Collision frequency
   • Mean free path
6. In the development of a biosensor for measurement of offensive odors, a researcher selected skatole (C_9H_9N), a contributor to offensive odors, as an indicator. The biosensor surface area exposed to the measurement gas is a sphere of 100 μm in

diameter. The sensor starts to respond when the skatole molecules collide with the sensor surface 100 times per second. Find the lowest threshold limit of skatole concentration that the sensor can measure at standard room air conditions.

7. Calculate the mean center-to-center distance between two adjacent air molecules under standard indoor conditions (1 atm pressure and 20°C).

8. A calibration gas contains a mixture of 10% methane (CH_4) and 90% air in mass. At standard room conditions, (1 atm and 20°C), the viscosities for CH_4 and air are 10.8×10^{-5} Pa·s and 1.81×10^{-5} Pa·s, respectively. Find the viscosity of the mixture.

   The surface area of the olfactometry membrane (sensing cell area) in a human nose is approximately 0.5 cm². On the membrane, there are about 100 million receptor cells. All cells are the same size, that is, each cell is 5×10^{-9} cm². Assume that each receptor is reactive only to one specific type of gas. During one sniff, the inhalation takes 1 second. The hydrogen sulfide concentration in the air is 5 parts per billion (ppb). Under standard conditions (i.e., 20°C, 1 atm, and air density of 1.2 kg/m³), H_2S is 1.44 kg/m³.

   a. Find the mass concentration of H_2S in the air (g/ml).
   b. Find the molecular concentration of H_2S in the air (number of molecules/ml).
   c. How many H_2S molecules make contact (collide) with the smell membrane (the cell reactive to H_2S) during one breath?

10. A cylinder contains a mixture of 90% hydrogen sulfide (H_2S) and 10% air in volume (as shown in the following figure). Assume that the concentration of H_2S in the cylinder remains constant and that the mass transfer is by diffusion only. The diameter of an H_2S molecule is approximately 4 Å and an air molecule is 3.7 Å. An H_2S sensor with a lower detection limit of 0.1 ppbv is attached to the bottle via a 1 m long tube with an inside diameter of 6 mm. The sampling rate of the sensor is 100 ml/min. There is a valve at the interface of the bottle and tube. Assume that the tube is initially filled with air and that the valve is closed.

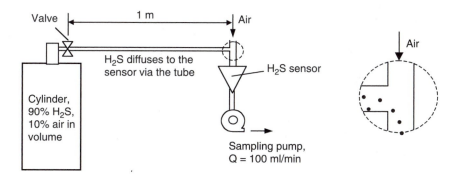

Gas is diffused from the cylinder to the air tube.

   a. How long will it take for the sensor to detect H_2S after the valve is opened?
   b. How long will it take for a sensor with a lower detection limit of 100 ppbv to detect H_2S after the valve is opened?
   (Hint: Derive from second Fick's law of diffusion.)

11. In question 10, assume that an initialization time is 20 minutes (the time period after the valve is opened). How long should the tube that connects the bottle and the sensor be in order to detect the H_2S?

REFERENCES

1. Wilkinson, F., *Chemical Kinetics and Reaction Mechanisms*, Van Mostrand Reinhold Company, New York, 1980, 97–102.
2. Kauzmann, W., *Kinetic Theory of Gases,* W.A. Benjamin Inc., New York, 1966.
3. Bird, R.B., Stewart, W.E., and Lightfoot, E.N., *Transport Phenomena,* John Wiley & Sons, Inc., New York, 1960.
4. Chapman, S. and Cowling, T.G., *Mathematical Theory of Non-Uniform Gases,* 2nd ed., Cambridge University Press, Cambridge, U.K., 1951.
5. Incropera, F.P. and De Witt, D.P., *Fundamentals of Heat and Mass Transfer,* John Wiley & Sons, New York, 1991.
6. Wilke, C.R., A viscosity equation for gas mixtures, *J. Chem. Physics,* 18:517–519, 1950.

CHAPTER **6**

Diffusion and Coagulation of Particles

Diffusion is the primary means of transportation and deposition of small airborne particles and gaseous molecules in a low air velocity field. For large particles and flow fields with high velocity, gravitational settling and convection become the primary means of transportation and deposition. A variety of applications of particle diffusion are found in air-quality instrumentation and control technologies. Typical applications include diffusive batteries for the study of size distribution of small particulate matters, and denuders to separate gaseous molecules from particles. This chapter will review the parameters that determine the particle diffusion coefficient, and then discuss the applications of particle diffusion for typical instrumentation and air-cleaning technologies.

Coagulation is a process of interaction among particles, through either diffusion or relative motion, to form larger particles. From the point of view of air-quality control, coagulation can be a particularly useful process, as it converts small particles to large ones that can be easily removed from an air stream. Examples are wet settling chambers and wet scrubbers, which allow small particles coagulated into larger particles to be separated from an air stream.

By completing this chapter, the reader will be able to

- Characterize particle properties related to diffusion:
  - Mean speed and root-mean-square (rms) speed of particles
  - Mean free path of particles
  - Diffusion coefficient of particles
  - Root-mean-square net displacement of particles in calm air
- Determine diffusive deposition:
  - Particle deposition for a single size and for total particle population
  - Deposition velocity, including the settling effect
- Understand diffusive denuders and batteries, including the four basic design geometries:
  - Cylinders
  - Parallel plates
  - Annular tubes

139

- Wire screen
- A combination of these designs and applications
- Understand coagulation. In the following three coagulation processes, the particle concentration follows the same equation (Equation 6.63), but the coagulation coefficients are different:
  - Monodisperse coagulation K
  - Polydisperse coagulation K_n
  - Kinematical coagulation K_c, (which is a function of the Stokes number)

6.1 THERMAL VELOCITY OF PARTICLES

Brownian motion originally described the wiggling motion of pollen particles at water surfaces. Similar wiggling motions of airborne particles were observed in the air in the late 19th century. This wiggling motion is caused by the impaction of air molecules onto the airborne particles. The particle velocity of each forward motion derived from the impaction is called *thermal velocity*. Whereas the wiggling motion is random and the average velocity in all directions is zero, the thermal velocity is the forward velocity for a step motion and thus not equal to zero. Note the difference in the terminologies of particle thermal velocity and molecular thermal velocity (often called *molecular thermal speed*). The former is a vector (step forward); the latter is a statistical value.

As shown in Figure 6.1, as the result of the bombardment of hundreds of air molecules, the airborne particle wiggles randomly in the air. Dotted spheres show the subsequent positions during the wiggling, and the solid-line curve with an arrow is the trajectory of the center of the particle. Each section of the curve between two

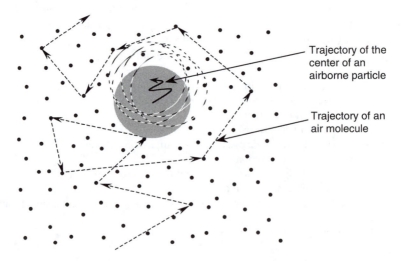

Trajectory of the center of an airborne particle

Trajectory of an air molecule

Figure 6.1   Wiggling motion of an airborne particle in air. The solid curve is the trajectory of the center of the particle, and the dotted straight lines are trajectories of air molecules.

turns of direction represents a step of forward motion of the particle. These forward-step motions can be used to calculate the particle's mean thermal velocity.

Because the wiggling motion of particles is the result of the impaction of air molecules, the particles and air molecules have the same kinetic energy, that is, $E_g = E_p = (3/2)RT$, for each mole of gas or particles, where E_g and E_p are kinetic energy for air and airborne particles at a given size, respectively; R is the gas constant; and T is the absolute temperature. With a diameter d_p and a mass m_p for each particle, the kinetic energy for 1 mole of particles is

$$E_p = \frac{1}{2} m_p N_a u_{rmsp}^2 \qquad (6.1)$$

where u_{rmsp} is the rms thermal velocity for the particle, and N_a is Avogadro's number. Equating Equation 6.1 and $E_p = (3/2)RT$ and solving for u_{rmsp} gives

$$u_{rmsp} = \left(\frac{3kT}{m_p} \right)^{1/2} = \left(\frac{18kT}{\pi \rho_p d_p^3} \right)^{1/2} \qquad (6.2)$$

where $k = R/N_a$ is Boltzmann's constant and ρ_p is the density of the particle.

Comparing Equation 6.2 with the root-mean-square thermal velocity of gases described in Equation 5.14, the rms thermal velocity for particles is in the same format as the rms speed for gases. For a given size of particles, the u_{rmsp} only depends on the temperature; that is why it is also called root-mean-square thermal velocity. Equation 6.2 indicates that particles' thermal velocity follows the Maxwell distribution (Equation 5.10 and Equation 5.11). Therefore, other averages of particle thermal velocities, such as arithmetic mean thermal velocity and the most probable thermal velocity (or mode), can be calculated in a similar way to a gas.

The *arithmetic mean thermal velocity*, or the *mean thermal velocity*, of particles $\bar{u}_p$, and the *most probable particle thermal velocity* $\hat{u}_p$ can be obtained from Equation 5.13 and Equation 5.16, respectively:

$$\bar{u}_p = = \left(\frac{8kT}{\pi m_p} \right)^{1/2} = \left(\frac{48kT}{\pi^2 \rho_p d_p^3} \right)^{1/2} \qquad (6.3)$$

$$\hat{u}_p = \left(\frac{2kT}{m_p} \right)^{1/2} = \left(\frac{12kT}{\pi \rho_p d_p^3} \right)^{1/2} \qquad (6.4)$$

Note that only the physical diameter and the actual density of the particles can be used in calculating the thermal velocities. Neither aerodynamic diameter nor Stokes diameter can be used. This is because the kinetic energy is mass dependent, whereas the drag force, from which the aerodynamic diameter and the Stokes

Table 6.1   **Thermal Velocities for Particles with a Density of 1000 kg/m³ under Standard Atmospheric Conditions (Values Calculated from Equations 6.2 through 6.4)**

Particle Size (μm)	Root-Mean-Square Thermal Velocity (m/s)	Mean Thermal Velocity (m/s)	Most Probable Thermal Velocity (m/s)
0.00037*	501	460	409
0.001	152	140	124
0.01	4.81	4.43	3.93
0.1	0.152	0.140	0.124
1	0.00481	0.00443	0.00393
10	0.000152	0.000140	0.000124
100	0.00000481	0.00000443	0.00000393

* Diameter of an equivalent "air molecule."

diameter are derived, is independent of mass. Thermal velocities for particles of typical sizes with a standard density (1000kg/m³) and under standard atmospheric conditions (20°C and 101.325 kPa) are listed in Table 6.1. All values are calculated from Equations 6.2 through 6.4.

6.2 MEAN FREE PATH OF PARTICLES

The small-scale motion of gas molecules can be characterized by the mean velocity and the mean free path. For airborne particles, it is useful to make a similar characterization. From the analysis of the previous section, the particle mean thermal velocity has the same formula as the gas mean molecular speed. The mean free path for particles, however, is quite different from that of gas molecules in many ways. The trajectory of a wiggling particle motion is a rather smooth curve (Figure 6.2), like a car traveling among roadblocks, whereas the trajectory of a gas molecule is a group of zigzagged straight lines, like a ball bouncing among surfaces (Figure 6.1). The mean free path of a gas molecule is defined as the average length traveled between two consecutive molecular collisions, that is, the total distance divided by the number of collisions (Equation 5.20).

Although each collision is clearly marked by the change of direction for gas molecules, the number of collisions and the change of direction of the trajectory are not so apparent for particles. As shown in Figure 6.2, there is no distinct turning point to identify the length between two consecutive collisions. The smoothly curved trajectory is the result of billions of gas molecule collisions with the particle. Therefore, unlike the mean velocity, gas molecular collision frequency is inappropriate for determining the mean free path of particles.

The *mean free path of particles* (λ_p) is defined as the average distance between an original position and the next position of the particle that has completely changed its direction (Figure 6.2). A complete direction change refers to 90° turn from its original direction, that is, the projection of the current velocity to the original velocity would be zero. This distance is equal to the straight-line distance that a particle travels in a given direction (i.e., λ_p) before it has lost all velocity in that direction.

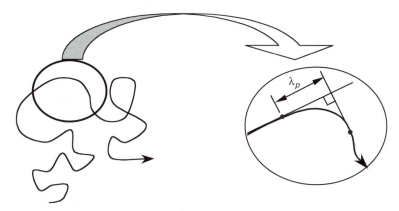

Figure 6.2   A two-dimensional trajectory of the center of a particle in a wiggling motion and a schematic description of particle mean free path.

Recalling that the stop distance of a particle is the initial particle velocity multiplied by the relaxation time, the mean free path of particles is simply the mean stopping distance of the particle at its mean thermal velocity:

$$\lambda_p = \tau \bar{u}_p \qquad (6.5)$$

where τ is the particle relaxation time defined by Equation 4.57 and $\bar{u}_p$ is the mean thermal velocity of particles defined by Equation 6.3.

Table 6.2 shows the mean free path values for typical sizes of particles with standard density under standard atmospheric conditions. Although the relaxation time and mean thermal velocity change rapidly with particle size, the mean free path changes only slightly, less than 12-fold for particles ranging from 0.001 μm to 100 μm. This is one of the very few particle properties that does not depend strongly on the particle size and is largely due to the combined effect of particle inertia, mean thermal velocity, and the slip correction coefficient. Small particles have smaller

Table 6.2   Mean Free Path for Typical Sizes of Particles with a Density of 1000 kg/m³ under Standard Atmospheric Conditions (Values Calculated from Equation 6.5)

Particle Size (μm)	Relaxation Time τ (s)	Mean Thermal Velocity, $\bar{u}_p$ (m/s)	Mean Free Path of Particles λ_p (μm)
0.00037*	—	460	0.066
0.001	6.89×10^{-10}	140	0.0966
0.01	7.05×10^{-9}	4.43	0.0313
0.1	8.17×10^{-8}	0.140	0.0115
1	3.58×10^{-6}	0.00443	0.0159
10	3.12×10^{-4}	0.000140	0.0438
100	0.0307	0.00000443	0.136

* Diameter of an equivalent "air molecule."

inertia than large particles and tend to travel a short distance (by having a shorter relaxation time). However, the mean thermal velocity and slip correction coefficient for small particles are greater than for large particles, allowing the small particles to travel at faster speeds and slip through among the air molecules. The particle mean free path increases with the particle density. Heavier particles have a greater mean free path. A closer look into Equation 6.5 reveals that the mean free path does not depend on the temperature. The kT term is canceled in mean thermal velocity (Equation 6.3) and air viscosity (Equation 5.38).

6.3 DIFFUSION COEFFICIENT OF PARTICLES

Diffusion of particles is the net mass transfer of these particles from a higher-concentration region to a lower-concentration region in still air (or in the absence of flow of other carrying gases). Diffusion is an important means of transportation for airborne particles, especially particles smaller than 1 micron. The diffusion of particles in air is the result of the random bombardment of air molecules against the particles. Similar to the diffusion of gas b into gas a, as discussed in Chapter 5, the diffusion of particles in air is given by Fick's first law of diffusion:

$$J_p = -D_p \frac{dC_{pn}}{dx} \qquad (6.6)$$

where J_p is the particle flux in particles/m^2·s, dC_{pn}/dx is the particle concentration gradient in particles/m^2, and D_p is the particle diffusion coefficient in m^2/s. The negative sign indicates that the mass transfer is always from a higher-concentration region to a lower-concentration region, similar to heat transfer, in which the heat always goes "downhill" — from a high-temperature region to a low-temperature region.

Comparing Equation 6.6 with Equation 5.43, the diffusion flux for particles takes the same formula as for gases. The main difference is the diffusion coefficient. The diffusion process of gases is driven by the partial pressure difference. Similarly, the diffusion process of particles is driven by the so-called osmotic pressure. The osmotic pressure P_{om} is defined by van't Hoff's law for n airborne particles per unit volume (i.e., particle number concentration C_{pn}) as

$$P_{om} = kTC_{pn} \qquad (6.7)$$

where k is Boltzmann's constant and T is the absolute temperature.

The osmotic pressure can be demonstrated experimentally by observing suspended solid particles in a liquid in two chambers separated by a semipermeable membrane (Figure 6.3). The semipermeable membrane allows the carrying liquid to pass through unimpeded but prevents the suspended particles from passing between chambers A and B. The membrane is free to slide along the chamber axis.

Assume that the membrane is initially at an equilibrium position, as shown in Figure 6.3a, and that the concentrations of suspended particles on both sides of the

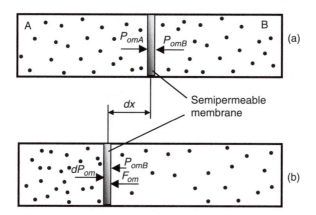

Figure 6.3   Illustration of osmotic pressure. (a) The membrane is in an equilibrium position when the particle concentrations (thus the osmotic pressure) in chambers A and B are equal. (b) A force must be exerted on the membrane to maintain the new position at which the particle concentrations are different in chambers A and B.

membrane are equal. If the membrane is pushed to the left by a distance dx, a force F_{om}, is needed (Figure 6.3b) to maintain the new position. This force is the same as the osmotic pressure difference. The force needed to balance the osmotic pressure difference, which is caused by the different particle concentrations in the two chambers, can be derived from van't Hoff's law:

$$F_{om} = dP_{om} = kTC_{pnB} - kTC_{pnA} = -kTdC_{pn} \qquad (6.8)$$

If it is assumed that the membrane has a unit area, then the volume change between the chambers A and B is dx. The number of particles in that volume is $C_{pn}dx$. Therefore, the diffusion force acting on each particle in this changed volume F_{omi} is

$$F_{omi} = -\frac{kT \, dC_{pn}}{C_{pn}dx} \qquad (6.9)$$

On the other hand, when a particle is driven by the diffusion force and moves in its carrying fluid, the diffusion force must be equal to the drag force. Recalling Stokes's drag force gives

$$F_{omi} = F_{D} = \frac{3\pi \eta V_{r}d_{p}}{C_{c}} \qquad (6.10)$$

Equating Equation 6.9 and Equation 6.10 and rearranging variables gives

$$C_{pn}V_r = -\frac{kTC_c}{3\pi\eta d_p}\frac{dC_{pn}}{dx}$$ (6.11)

Note that at the left side of Equation 6.11 is the particle speed relative to the carrying fluid multiplied by the particle concentration, which is the particle flux J_d. Comparing Equation 6.11 with Equation 6.6 gives

$$D_p = \frac{kTC_c}{3\pi\eta d_p}$$ (6.12)

Equation 6.12 is called the *Stokes–Einstein equation*. The derivation is based on the Stokes drag force and the osmotic pressure difference. The diffusion coefficient is one of the most important properties for particle transportation and air cleaning. Depending on the availability of other variables, the Stokes–Einstein equation can be written in other forms. Substituting the particle mobility, $B=C_c/3\pi\eta d_p$, into Equation 6.12 gives

$$D_p = kTB$$ (6.13)

The diffusion coefficient for particles can also be expressed in terms of particle mean free path and mean thermal velocity. Combining Equation 4.57, Equation 6.3, Equation 6.5, and Equation 6.12 gives

$$D_p = \frac{\pi}{8}\lambda_p\bar{u}_p$$ (6.14)

From Table 6.2, it is clear that the mean free path of particles for all sizes does not change rapidly, whereas the mean thermal velocity of particles increases rapidly as the particle size decreases. Therefore, the particle diffusion coefficient is primarily dependent on the mean thermal velocity.

It should be pointed out that Equations 6.12 through 6.14 are the same formula but are expressed in different variables. In summary of the diffusion coefficient for particles, it is appropriate to restate Einstein's findings:[1]

- The observable Brownian motion of an airborne particle is equivalent to that of a giant gas molecule. In other words, the scale of the wiggling motion of the particle, represented by its mean free path, is similar in magnitude to that of a gas molecule (Table 6.2).
- The kinetic energy of an airborne particle undergoing Brownian motion is the same as its carrying gas molecules. In other words, the energy conserves.
- The diffusion force of a particle is the osmotic pressure difference on that particle, as shown in the Stokes–Einstein equation.

Another useful expression of diffusion is the alternative statement of Fick's first law, in which the diffusion flux is written as

$$J_p = -\frac{(x_{rms})^2}{2t}\frac{dC_{pn}}{dx}$$ (6.15)

where x_{rms} is the root-mean-square net displacement of particles along any direction, t is time, and dC_{pn}/dx is the particle concentration gradient. Comparing Equation 6.15 with Equation 6.6 gives

$$D_p = -\frac{(x_{rms})^2}{2t}$$ (6.16)

or

$$x_{rms} = \sqrt{2D_p t}$$ (6.16b)

In the area of air quality, the net displacement, rather than wiggling motion or the thermal velocity of particles, is often of primary interest. For example, how fast a particle can be removed from an air stream depends on its net displacement x_{rms}. Because the Brownian motion of particles can be back and forth and three-dimensional, it is important to know that the x_{rms} is the net displacement and is much smaller than the total distance traveled at the rms thermal velocity over a time, $u_{rms}t$. In particle settling, the settling velocity is unidirectional and can be used to characterize the particle collection; in particle diffusion, u_{rms} is multidirectional and not suitable for characterizing particle collection. However, x_{rms} can be used for such characterization. Table 6.3 compares the net displacement driven by diffusive and gravitational forces for typical sizes of particles in one second.

Table 6.3 shows that diffusion is the predominant driving force in particle transportation for particles smaller than 0.1 µm with standard density. Gravitational

Table 6.3 Net Displacement Driven by Diffusive and Gravitational Forces for Typical Sizes of Particles with Standard Density in One Second under Standard Atmospheric Conditions

Particle Size (µm)	Diffusion Coefficient D_p (m²/s)	Net Displacement by Gravity in 1 s x_{TS} (m)	Net Displacement by Diffusion in 1 s x_{rms} (m)	Ratio of x_{rms}/x_{TS}
0.001	5.32×10^{-6}	6.76×10^{-9}	3.26×10^{-3}	4.83×10^5
0.01	5.45×10^{-8}	6.92×10^{-8}	3.30×10^{-4}	4770
0.1	6.31×10^{-10}	8.02×10^{-7}	3.55×10^{-5}	44.3
0.525	5.95×10^{-11}	1.09×10^{-5}	1.09×10^{-5}	1.00
1	2.76×10^{-11}	3.51×10^{-5}	7.44×10^{-6}	0.212
10	2.41×10^{-12}	0.0031	2.20×10^{-6}	7.17×10^{-4}
100	2.37×10^{-13}	0.302	6.89×10^{-7}	2.28×10^{-6}

settling is the predominant driving force in particle transportation for particles larger than 1 μm with standard density. Particles smaller than 0.1 μm tend to deposit onto surfaces via diffusive deposition or coagulation to form large particles. Particles larger than 1 μm tend to settle onto a surface because of the gravitational force. A particle size of interest is 0.525 μm, at which the net displacement caused by diffusion equals the net displacement caused by gravitational settling. For particles at this size range, neither diffusive nor gravitational forces prevail, leaving the particles in a relatively steady state (compared with other particle sizes). This may explain why there are a large number of particles smaller than 0.5 μm floating in the air. This is also the scientific basis on which diminutive particles are defined as particles smaller than 0.5 μm (see Chapter 2).

Example 6.1: *Determine the diffusion coefficient of particles of 0.2 μm diameter and with a density of 1000 kg/m³ at standard atmospheric conditions (20°C and 1 bar). What is the mean thermal velocity? How far can such a particle travel (i.e., what is the root-mean-square displacement) in one hour from its original position in still air by diffusion?*

Solution: The slip correction coefficient for a 0.2 μm diameter particle is

$$C_C = 1 + 2.52\frac{\lambda}{d_p} = 1 + 2.52\frac{0.066}{0.2} = 1.83$$

Using Equation 6.12

$$D_p = \frac{kTC_c}{3\pi\eta d_p} = \frac{1.38\times10^{-23}\times293\times1.83}{3\pi\times1.81\times10^{-5}\times0.2\times10^{-6}} = 2.17\times10^{-10}\left(\frac{m^2}{s}\right)$$

The mean thermal velocity and rms displacement can be calculated as

$$\bar{u}_p = \quad = \left(\frac{48kT}{\pi^2\rho_p d_p^3}\right)^{1/2} = \left(\frac{48\times1.38\times10^{-23}\times293}{\pi^2\times1000\times(0.2\times10^{-6})^3}\right)^{1/2} = 0.05\ (m/s)$$

$$x_{rms} = \sqrt{2D_p t} = \sqrt{2\times2.17\times10^{-10}\times3600} = 0.00125\ m$$

Note that the net displacement of the particle by diffusion in 1 hour is only 1.25 mm, whereas the mean thermal velocity of the same particle is 50 mm/s. The particle is virtually vibrating (wiggling) rather vigilantly but not making much net displacement.

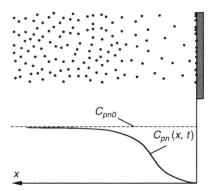

Figure 6.4   Particles deposited onto the surface cause a concentration gradient along the x-axis.

6.4 DIFFUSIVE DEPOSITION

Based on Einstein's study, the observable Brownian motion of an airborne particle is equivalent to that of a giant gas molecule. In other words, the scale of the wiggling motion of a particle, represented by its mean free path, is similar in magnitude to that of a gas molecule, as shown in Table 6.2. Therefore, the diffusion phenomenon of particles is in the same form as gas molecules, as stated by Fick's first law (Equation 5.50 and Equation 6.6 for gases and particles, respectively).

6.4.1   Interface with Concentration Gradient

Let us consider a simple but very common case of particle deposition by diffusion: deposition of particles onto a vertical wall from an infinite volume of airspace with a uniform initial particle concentration C_{pn0}. Unlike the gas molecules bouncing back and forth at the surface, particles deposit on the wall once they make contact with the surface during Brownian motion. This deposition process will gradually reduce the particle concentration along the direction perpendicular to the surface (Figure 6.4). The particle concentration at distance x from the surface at any time t, $C_{pn}(x,t)$, satisfies Fick's second law of diffusion:

$$\frac{\partial C_{pn}}{\partial t} = D_p \frac{\partial^2 C_{pn}}{\partial x^2} \qquad \left\{ \begin{array}{ll} C_{pn}(x,0) = C_{pn0}, & for\ x \to 0 \\[2mm] C_{pn}(0,t) = 0 & for\ t > 0 \end{array} \right\} \qquad (6.17)$$

where C_{pn} is the number concentration of particles and D_p is the diffusion coefficient for the particles at a given size.

Consider that at the initial time $t = 0$, Equation 6.17 can be integrated and solved for C_{pn} as a function of time t and distance x. The relationship of a particle's concentration gradient to the given size at a distance x and time t becomes

$$\frac{dC_{pn}(x,t)}{dx} = \frac{C_{pn0}}{\sqrt{\pi D_p t}} \exp\left(\frac{-x^2}{4D_p t}\right)$$ (6.18)

where C_{pn0} is the concentration of particles introduced at the interface, x is the distance from the original position, and D_p is the diffusion coefficient of the particles.

Note that Equation 6.18 is the same as Equation 5.55 but greater by a factor of two, yet both are results of Fick's second law of diffusion. This is because the boundary conditions are different. In the case of diffusion described by Equation 5.55, molecules or particles travel in both directions ($-x$ and x) along the x-axis, and the concentration at the interface ($x = 0$) is not zero. In the case of deposition, however, deposition velocity is only toward the deposition surface, and the concentration at the interface ($x = 0$) is zero.

Substituting Equation 6.18 into Equation 6.6 gives the one-dimensional particle flux at a distance x from the initial position at time t due to diffusion:

$$J_p(x,t) = -C_{pn0}\left(\frac{D_p}{\pi t}\right)^{1/2} \exp\left(\frac{-x^2}{4D_p t}\right)$$ (6.19)

Again, the deposition flux toward a surface is twice as great as that of diffusion in open space, expressed in Equation 5.56. The total number of particles deposited onto a unit area of surfaces, $N_{pd}(x,t)$, is the integral of the particle flux over a time period t:

$$N_{pd}(x,t) = \int_0^t J(x,t)dt$$

$$= -C_{pn0}\left(\frac{D_p}{\pi}\right)^{1/2} \int_0^t \left(\frac{1}{t}\right)^{1/2} \exp\left(\frac{-x^2}{4D_p t}\right)dt$$ (6.20)

To solve Equation 6.20, first let $t' = \sqrt{t}$, then let $t = 1/t'$, where t' is a dummy variable. If we change the integral domains accordingly, Equation 6.20 becomes

$$N_{pd}(x,t) = 2C_{pn0}\left(\frac{D_p}{\pi}\right)^{1/2} \int_{\infty}^{\frac{1}{\sqrt{t}}} \frac{1}{t^2} \exp\left(\frac{-x^2}{4D_p}t^2\right)dt$$ (6.21)

because

$$\int u\,dv = uv - \int v\,du$$ (6.22)

In Equation 6.21, let

$$u = \exp\left(\frac{-x^2}{4D_p}t^2\right) \tag{6.23}$$

$$v = -\frac{1}{t} \tag{6.24}$$

Substituting Equation 6.22 through Equation 6.24 into Equation 6.21 and changing the integral domain accordingly gives

$$N_{pd}(x,t) = 2C_{pn0}\left(\frac{D_p}{\pi}\right)^{1/2}\left[\begin{array}{l} -\sqrt{t}\exp\left(\frac{-x^2}{4D_p}t^2\right) + \\ \dfrac{x^2}{2D_p}\left(\displaystyle\int_0^\infty \exp\left(\frac{-x^2}{4D_p}t^2\right)dt - \int_0^{\frac{1}{\sqrt{t}}}\exp\left(\frac{-x^2}{4D_p}t^2\right)dt\right) \end{array}\right]$$

$$= 2C_{pn0}\left[-\left(\frac{D_p t}{\pi}\right)^{1/2}\exp\left(\frac{-x^2}{4D_p t}\right) + \frac{x}{2} - x\ erf\left(\frac{2}{x}\sqrt{\frac{D_p}{t}}\right)\right] \tag{6.25}$$

where *erf* is the error function, and Presnal integrals. Using dummy variables u and v, $erf(u)$ can be written as

$$erf(u) = \frac{2}{\sqrt{\pi}}\int_0^u \exp(-v^2)dv$$

$$= \frac{2}{\sqrt{\pi}}\sum_{n=0}^\infty \frac{\Gamma\left(\frac{3}{2}\right)\exp(-u^2)}{\Gamma\left(n+\frac{3}{2}\right)}u^{2n+1} \tag{6.26}$$

$$= \frac{2}{\sqrt{\pi}}\exp(-u^2)\left(u + \frac{2}{3}u^3 + \frac{4}{15}u^5 + \cdots\right)$$

6.4.2 Deposition Interface without Concentration Gradient

One simple but practical case for Equation 6.19 is to consider the particle deposition onto a surface in a constant concentration field (Figure 6.5). In many cases of indoor air quality, the assumption of a constant particle concentration field near a deposition surface is accurate. For example, particle concentration in a supply air duct or in a well-mixed room airspace can be considered relatively constant,

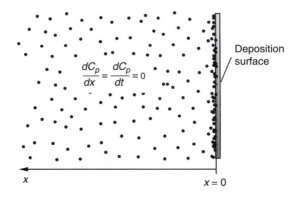

Figure 6.5   Particle deposition onto a surface in a uniform concentration field.

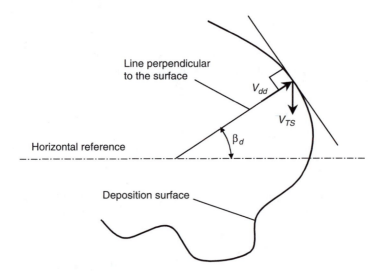

Figure 6.6   Illustration of diffusive deposition velocity and settling velocity at a deposition surface with an incline angle β_d.

because the air and the particles are continuously stirred and mixed. Even for small particles, the convective air velocity is usually several magnitudes higher than the particle diffusive velocity. Because the particle concentration near the surface is constant, there is no spatial variation. In this case, deposition is irrelevant to the distance x. By choosing the deposition surface as the original interface for x, the particle deposition occurs at $x = 0$ (Figure 6.6). If we substitute $x = 0$ into Equation 6.19, the particle deposition flux on a surface becomes

$$J_p(t) = -C_{pn0}\left(\frac{D_p}{\pi t_d}\right)^{1/2}$$

(6.27)

where the rate of particle deposition caused by diffusion at the interface can be also expressed in terms of deposition velocity V_{dd}, where the subscript $_{dd}$ indicates deposition by diffusion.

$$V_{dd} = \frac{J}{C_{pn0}} = \left(\frac{D_p}{\pi t_d}\right)^{1/2} \tag{6.28}$$

For a given deposition surface, there is no minus sign (−) in Equation 6.28. The diffusivity of the particle does not vary with the time for a constant concentration field. Unlike t in Equation 6.19, a time variable for accumulated particles N_{pd}, the presence of time t_d in Equation 6.27 and Equation 6.28 is for dimensional purposes, because the deposition velocity is defined as m/s. Therefore, t_d is the *unit time* for deposition velocity and should always equal one second. Note that the deposition velocity by diffusion is in all directions, whereas the minus sign (−) in the diffusion flux (Equation 6.19) indicates that the flux is "downhill"; there is no concentration gradient in Equation 6.28. Thus, the minus sign (−) has no physical meaning in either Equation 6.27 or Equation 6.28.

The diffusive deposition rate per unit area of surface for a given size of particles is the deposition flux multiplied by the deposition time t:

$$N_{pd}(t) = C_{pn0} \left(\frac{D_p}{\pi t_d}\right)^{1/2} t \tag{6.29}$$

Note that the deposition rate in Equation 6.29 is proportional to the deposition time. This is a much faster deposition process than the deposition rate when the deposition interface has a particle concentration gradient, shown in Equation 6.25. When a deposition interface has a particle concentration gradient, the surface can be considered to be "absorbing" the particles. This is particularly evident when two different fluids diffuse into each other. However, for airborne particles deposited onto a surface, such a concentration gradient may be negligible and result in a much faster deposition rate.

Up to this point, we have only discussed the deposition of monodisperse particles. For particles with n groups of sizes, the total number of particles deposited onto a surface, N_{pdt}, should be the sum of deposited particles of n size groups:

$$N_{pdt} = \sum_{1}^{n} N_{pdi} \quad (i = 1, 2, 3 \ldots, n) \tag{6.29b}$$

where N_{pdi} is the number of particles deposited onto the surface for i^{th} size group defined by Equation 6.29.

Example 6.2: *The concentration of particles in an exhaust fume hood duct at a restaurant is measured as follows: Oil fume particles have a concentration of 5 ×* 10^{17} *particles/m³ with a diameter of average mass of 0.1 μm and a density of 900 kg/m³. Other types of smoke particles are 7 × 10^{17} particles/m³ with a diameter of average mass of 0.05 μm and a density of 800 kg/m³. The duct is 0.2 m in diameter and 5 m long. Assume that the particle concentration in the duct is constant and temperature in the duct is 40°C. The restaurant is open 12 h per day. Gravitational settling is negligible. Find the following:*

    a. How many particles accumulated in the duct in 1 month (30 days) due to diffusion?

    b. What is the total mass of particles collected? Assuming that all fume particles are collected in a fume hood oil pan with a capacity of 1 kg oil, how often does the oil pan need to be cleaned?

Solution:

    a. For oil fume particles, $d_{p1} = 0.1$ μm

$$C_{C0} = 1 + 2.52\frac{\lambda}{d_p} = 1 + 2.52\frac{0.066}{0.1} = 1.83$$

For smoke particles, $d_{p2} = 0.05$ μm < 0.1 μm

$$C_c = 1 + \frac{\lambda}{d_p}\left[2.34 + 1.05\exp\left(-0.39\frac{d_p}{\lambda}\right)\right] = 5.12$$

Diffusion coefficients for fume and smoke particles are

$$D_{p1} = \frac{kTC_{c1}}{3\pi\eta d_{p1}} = \frac{1.38\times10^{-23}\times313\times1.83}{3\pi\times1.81\times10^{-5}\times0.1\times10^{-6}} = 2.32\times10^{-10}\left(\frac{m^2}{s}\right)$$

$$D_{p2} = \frac{kTC_{c2}}{3\pi\eta d_{p2}} = \frac{1.38\times10^{-23}\times313\times5.12}{3\pi\times1.81\times10^{-5}\times0.05\times10^{-6}} = 2.59\times10^{-9}\left(\frac{m^2}{s}\right)$$

Note that $C_{pn1} = 5 \times 10^{17}$ particles/m³, $C_{pn2} = 7 \times 10^{17}$ particles/m³, $t = 30 \times 12 \times 3,600 = 1,296,000$ s, and total surface area of the duct is $A = \pi \times 0.2 \times 5 = 3.14$ m². If we substitute these values into Equation 6.29 and Equation 6.29b, the number of particles deposited per m² in 1 month is

$$N_{pdt} = \sum_{1}^{n} N_{pdi} = C_{pn1} \left(\frac{D_{p1}}{\pi t_d} \right)^{1/2} t + C_{pn2} \left(\frac{D_{p2}}{\pi t_d} \right)^{1/2}$$

$$= 5 \times 10^{16} \times \left(\frac{2.32 \times 10^{-10}}{\pi} \right)^{1/2} \times 1,296,000 + 7 \times 10^{16} \times \left(\frac{2.59 \times 10^{-9}}{\pi} \right)^{1/2}$$

$$\times 1,296,000$$

$$= 5.57 \times 10^{17} + 2.60 \times 10^{18} = 3.16 \times 10^{18} \ (particles)$$

The total number of particles on the duct surface (noting that deposition area A = 3.14 m²) is

$$N_{pdt}A = 9.94 \times 10^{17} \ (particles)$$

b. The total mass of particles deposited on the duct in one month is

$$M_p = m_{p1} N_{pd1} A + m_{p2} N_{pd2} A$$

$$= 3.14 \times 5.57 \times 10^{17} \times \frac{\pi \rho_1 d_{p1}^3}{6} + 3.14 \times 2.60 \times 10^{18} \frac{\pi \rho_2 d_{p2}^2}{6}$$

$$= 1.25 \ (kg)$$

Because the oil pan capacity is 1 kg, the oil pan needs to be cleaned approximately every 0.8 month, or 24 days.

6.4.3 Total Deposition Velocity

Particles also deposit by gravitational settling, defined by the terminal settling velocity (Equation 4.38). The total deposition velocity V_d toward a surface is the combination of the settling terminal velocity V_{TS} and the diffusive deposition V_{dd}. Combining Equation 6.28 and Equation 4.38, and noting that the terminal settling velocity is always pointed down, gives

$$V_d = \left(\frac{D_p}{\pi t_d} \right)^{1/2} - \frac{\rho_p d_p^2 g C_c}{18\eta} \sin\beta_d \qquad \left(\frac{\pi}{2} \geq \beta_d \geq -\frac{\pi}{2} \right) \qquad (6.30)$$

where β_d is the incline angle between the normal line of the deposition surface of concern and the horizontal reference line (Figure 6.6).

The horizontal reference line can be located anywhere as long as it is horizontal. The first and second terms at the right side of Equation 6.30 represent the deposition velocity caused by diffusion and gravitational settling, respectively. When β_d is zero, the settling velocity component is zero, that is, particles do not deposit onto a vertical wall by gravitational settling. When β_d is $-90°$, the settling velocity becomes positive, that is, both the diffusive deposition and the gravitational settling contribute to the deposition onto a floor surface. When β_d is $90°$, the settling velocity is in the opposite direction to the diffusion deposition velocity. One situation of practical interest is when the settling velocity equals the diffusive velocity at a ceiling surface where $\beta_d = 90°$, that is, the total deposition velocity is zero:

$$\left(\frac{D_p}{\pi t_d}\right)^{\frac{1}{2}} = \frac{\rho_p d_p^2 g C_c}{18\eta} \tag{6.31}$$

Substituting Equation 6.12 for D_p in Equation 6.31 and solving for d_p gives the maximum particle diameter that can be deposited onto a ceiling surface, which is called the *critical deposition diameter of the particles, d_{pc}:*

$$d_{pc} = \left(\frac{108kT\eta}{\pi^2 t_d \rho_p^2 g^2 C_c}\right)^{\frac{1}{5}} \tag{6.32}$$

where k is Boltzmann's constant, T is the absolute temperature, η is the ambient air viscosity, $t_d = 1$ s (Equation 6.29), ρ_p is the density of particles, $g = 9.81$ m/s², and C_c is the slip correction coefficient.

Equation 6.32 can be solved using iterated procedures because C_c is a function of d_p. For standard particles at standard atmospheric conditions ($\rho = 1000$ kg/m³, T $= 293$ K), the critical deposition diameter is 0.355 µm, calculated using Equation 6.32. Note that the critical deposition particle size is only about two thirds of the particle size at which the net displacements caused by diffusion and gravitational settling are equal (Table 6.3). This is because the critical deposition diameter only considers the ceiling surfaces ($\beta_d = 90°$), or only the upward diffusive deposition velocity, whereas the net displacement of particles caused by diffusion considers the displacement in all directions. Particles larger than the critical deposition diameter will only settle down toward the floor but cannot be deposited onto a ceiling surface unless subjected to an external force.

Because diffusion is predominant for particles smaller than 0.1 µm and gravitational settling is predominant for particles larger than 1 µm with standard density, the deposition velocity for a given size of particles can be simplified to

$$V_d = \left(\frac{D_p}{\pi t_d}\right)^{\frac{1}{2}} \quad \text{(for } d_p \leq 0.1 \text{ µm)} \tag{6.33}$$

$$V_d = -\frac{\rho_p d_p^2 g C_c}{18\eta}\sin\beta_d \qquad \left(\frac{\pi}{2} \geq \beta_d \geq -\frac{\pi}{2}\right) \text{ (for } d_p \geq 1 \text{ }\mu m) \qquad (6.34)$$

Obviously, when $\beta_d \geq 0$, $V_d \leq 0$. When deposition velocity is zero or negative, there will be no deposition onto that surface. Therefore, particles larger than 1 μm with standard or greater densities will not deposit on a wall or a ceiling. For particles larger than 0.1 μm but smaller than 1 μm, Equation 6.30 should used to calculate the deposition velocity.

For a well mixed airspace (i.e., one in which particle concentrations for all n size groups are constant across the airspace), the total number of particles deposited onto a unit area surface by both diffusion and gravitational settling becomes

$$N_{pdt} = \sum_1^n C_{pni} V_{di} t \qquad (6.35)$$

Substituting Equation 6.30 into Equation 6.35 gives

$$N_{pdt} = \sum_1^n C_{pni} t \left(\left(\frac{D_{pi}}{\pi t_d}\right)^{1/2} - \frac{\rho_{pi} d_{pi}^2 g C_{ci}}{18\eta}\sin\beta_d \right) \qquad \left(\frac{\pi}{2} \geq \beta_d \geq -\frac{\pi}{2}\right) \quad (6.36)$$

6.5 DIFFUSIVE DENUDERS

One of the applications of particle diffusion theory is to separate small particles (<0.1 μm) or vapors from air. A diffusion denuder can be used for this purpose. Diffusion denuders are frequently used as scrubbers in an air sampling system to remove certain types of gases to avoid absorption of these gases on particles collected in the filter. This scrubbing characteristic is particularly useful in air-quality studies because, in many cases, gaseous contaminants must be separated from particulate contaminants for detailed chemical and biological analysis. The principle of a diffusion battery or a denuder is the same — allow the airborne particles or gases of concern to be collected by diffusion when the air flows through the device. Devices with different characteristics allow different sizes of particles to be collected.

Unlike the stagnant situations we considered in previous sections, a denuder requires airflow through it. As the airflow becomes dynamic, the diffusion deposition onto surfaces becomes more complicated than in stagnant situations. Solutions to Fick's second diffusion law for different boundary conditions strongly depend on the flow and boundary conditions. Explicit solutions to Fick's second law are available for only a few of many configurations. Further, a diffusion particle collector is always designed based on fully developed laminar flow conditions, because under turbulent conditions, the collection efficiency becomes overwhelmingly difficult to predict accurately. Therefore, flows in diffusion particle collectors must satisfy the condition of $Re < 2300$. When $Re > 2300$, the collection efficiency should be determined experimentally.

In the atmospheric sciences, *penetration* P_n is commonly used. The penetration rate refers to the number of particles that leave the device without being collected. The penetration is defined as the ratio of particle concentrations of exhaust air C_e to supply air C_s:

$$P_n = \frac{C_e}{C_s} \tag{6.37}$$

In the area of air quality, *collection* (or *separation*) *efficiency* ξ is often used. Collection efficiency is also simply called *efficiency* for many air-cleaning devices. The collection efficiency refers to the fraction of contaminants removed from an air stream by a device. For example, a filter of 90% efficiency refers to a filter that can remove 90% of particulate matter from the air passing through it. Obviously, the collection efficiency, or separation efficiency of particles, is the remainder of the penetration rate P_n:

$$\xi = 1 - P_n = \frac{C_s - C_e}{C_s} \tag{6.38}$$

The diffusion denuder technique was first conceived following the observation that the atmospheric nuclei lost in a tube are related to the diffusion coefficient.[2] Subsequently, particles lost through different geometries — including cylindrical and rectangular tubes, disks, and annular shapes — were either mathematically or experimentally derived to predict the particle collection efficiency accurately. Principles and operations of different diffusion denuders and batteries were reviewed by Cheng, where the sources of original research for different designs and equations were described.[3]

6.5.1   Coating Substrates for Denuders

In order to collect the gases or vapors of interest in an air stream onto the walls of a denuder, absorbent materials can be coated onto the tube wall of a denuder. Specific gases or vapors can be absorbed to specific materials. Table 6.4 lists substrates reported in the literature for removal of typical gases. Some materials absorb more than one gas. For example, sodium carbonate can absorb acidic gases in the ambient air, including HCl, HNO_2, HNO_3, and SO_2. The method of applying a material to the tube wall depends largely on the nature of the material. Most absorbent materials are first dissolved then applied to the tube wall. Solvents are allowed to evaporate, leaving the absorbent on the tube wall. In some cases, the glass denuder wall has been etched by sandblasting the surface to increase the wall's capacity to support the denuding chemical substrate. A nylon sheet has been used to line the

Table 6.4 Typical Absorbent Materials for Gases in Diffusion Denuders

Coating Material	Gas Absorbed
Oxalic acid	NH_3, aniline
Oleic acid	SO_3
H_3PO_3	NH_3
K_2CO_3	SO_2, H_2S
Na_2, CO_3	SO_2, HCl, HNO_3, HNO_2
$CuSO_4$	NH_3
PbO_2	SO_2, H_2S
WO_3	NH_3, HNO_3
MgO	HNO_3
NaF	HNO_3
NaOH and guaiacol	NO_2
Bisulfite-triethanolamine	Formaldehyde
Nylon sheet	SO_2HNO_3
Tenax powder	Chlorinated organics
Silica gel	Aniline
ICl	Tetra alkyl lead

Source: Cheng, Y.S., Diffusion batteries and denuders, in *Air Sampling Instruments for Evaluation of Atmospheric Contaminants*, 8th ed., American Conference of Governmental Industrial Hygienists, Cincinnati, OH, 1995. With permission.

inner wall of denuders.[4] Anodized aluminum surfaces have recently been found to be a good absorbing surface for nitric acid. Annular denuders made of anodized aluminum do not need coating.[5] Tenax or silica gel powder is difficult to apply as a coating on the tube wall, but these materials adhere to the glass wall with silicon grease and can be very effective absorbents.

6.5.2 Cylindrical Tubing Denuders

Penetration through a circular cylinder (Figure 6.7) at a flow rate Q for particles with a diffusion coefficient D_p has been developed by several researchers as a function of particle diffusion deposition parameter μ. The diffusion deposition parameter is a dimensionless number related to the device dimension, airflow rate, and particle diffusion coefficient. Penetrations at different μ values have been widely investigated.[6–9] Many equations are accurate only for a certain range of μ values. One set of the most accurate and versatile equations to satisfy the entire range of μ was developed by Soderholm based on previous studies:[10]

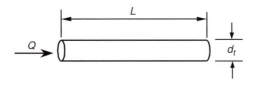

Figure 6.7 Schematic of a cylindrical denuder.

$$P_n = 0.81905 \exp(-3.6568\mu) + 0.09753 \exp(-22.305\mu)$$
$$\text{(for } \mu > 0.02) \quad (6.39)$$
$$+ 0.0325 \exp(-56.961\mu) + 0.01544 \exp(-107.62\mu)$$

$$P_n = 1 - 2.5638\mu^{\frac{2}{3}} + 1.2\mu + 0.1767\mu^{\frac{4}{3}} \quad \text{(for } \mu \leq 0.02) \qquad (6.40)$$

where

$$\mu = \frac{\pi L D_p}{Q} \qquad (6.41)$$

and where L is the length of the cylinder, D_p is the particle diffusion coefficient, and Q is the airflow rate through the cylinder. It may be surprising that there is no cylinder diameter (d_t) in Equation 6.41, which is intuitively thought to be an important geometric parameter for a cylinder denuder. This is because $Q = \pi d_t^2/4U$ already contains the effect of d_t, where U is the mean air velocity through the cylinder.

In cylindrical denuders for air sampling, a single cylindrical glass or Teflon tube is often used to collect gases or vapors.[2] The diameter and the length of the cylinder and the sampling flow rate are designed to have a collection efficiency greater than 99%. For example, a glass tube with a 3 mm inside diameter and a 35 mm length would have a collection efficiency of over 99% for ammonia at a flow rate of 3 l/min.[11] For higher sampling flow rate, parallel cylinder assemblies can be used. For example, 10 such cylinders would have a collection efficiency of 99% of ammonia at 30 l/min. In designing and selecting denuders, one must be sure to satisfy the laminar flow requirement — the Reynolds number must be smaller than 2300. Otherwise, collection efficiency will be sacrificed. In the case of air cleaning, a large airflow rate is usually required. Such large airflows are usually associated with high turbulent intensities. Therefore, air cleaning using diffusion techniques presents particular challenges and requires specific design considerations.

6.5.3 Rectangular Channels and Parallel Circular Plates

Penetration through a parallel narrow rectangular channel of a width W and a height H with the ratio of $H << W$ (e.g., $H < 0.1W$) (Figure 6.8a) and through inward flow parallel circular plates (Figure 6.8b) can be expressed as follows:[10]

$$P_n = 0.9104 \exp(-2.8278\mu) + 0.0531 \exp(-32.147\mu)$$
$$\text{(for } \mu > 0.05) \quad (6.42)$$
$$+ 0.01528 \exp(-93.475\mu) + 0.00681 \exp(-186.805\mu)$$

$$P_n = 1 - 1.526\mu^{\frac{2}{3}} + 0.15\mu + 0.0342\mu^{\frac{4}{3}}) \quad \text{(for } \mu \leq 0.05) \qquad (6.43)$$

where

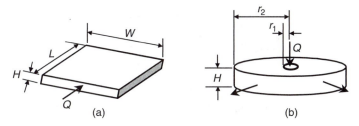

Figure 6.8 Schematic of diffusion denuders: (a) rectangular channels ($H << W$), (b) parallel circular plates.

$$\mu = \frac{8\,WLD_p}{3QH} \quad \text{(for rectangular channels)} \tag{6.44}$$

$$\mu = \frac{8\,\pi D_p (r_2^2 - r_1^2)}{3QH} \quad \text{(for parallel circular plates)} \tag{6.45}$$

and where L is the length of the rectangular channel and r_2 and r_1 are the outer and inner radii, respectively, of the circular plates.

In most applications, especially in the design process, the first term at the right side of Equation 6.39 and of Equation 6.42 is adequately accurate within an error of less than 5%. The greater the μ values, the smaller the error will be. For example, for a cylindrical tube with a μ value of 0.1, the error is about 2%, caused by using only the first term at the right side of Equation 6.39. In design, it is often required to estimate the μ value so that the air sampling rate and denuder length (and channel gap for parallel plates) can be determined. The first term at the right side of Equation 6.39 allows an explicit solution of μ, and an estimation of the denuder parameters can be made.

6.5.4 Annular Tubes

Penetration through an annular tube (Figure 6.9) has not been derived theoretically. One approach to estimating the penetration of particles through an annular tube is to approximate the annular tube as a rectangular channel, using Equation 6.42 and Equation 6.43, if the gap is much smaller than the inner radius of the annular tube. In this case, the airflow through the annular gap is similar to flow through the gap between two parallel plates. The gap between the inner and outer tubes is $H = r_2 - r_1$, and the width is $W = \pi(r_1 + r_2)$. Therefore, Equation 6.44 can be rewritten as

$$\mu = \frac{8\,\pi(r_1 + r_2)LD_p}{3Q(r_2 - r_1)} \quad \text{[for } (r_2 - r_1) < 0.1\pi(r_1 + r_2)] \tag{6.46}$$

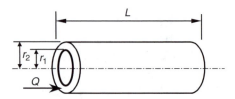

Figure 6.9   Schematic of an annular tube denuder.

where L is the length of the rectangular channel and r_2 and r_1 are the outer and inner radii, respectively, of the annular tube. Note that this approximation should satisfy the condition of $H < 0.1W$, that is, $(r_2 - r_1) < 0.1\pi(r_1 + r_2)$.

For very large μ values and low penetration, Possanzini et al. (1983) developed an empirical equation for the annular tube based on a sorption study of SO_2:[12]

$$P_n = (0.82 \pm 0.10)\exp[(-22.53 \pm 1.22)\mu] \quad \text{(for } \mu \geq 3.0 \text{ and } P_n < 0.1) \quad (6.47)$$

where

$$\mu = \frac{\pi LD_p(r_2 + r_1)}{4Q(r_2 - r_1)} \quad (6.48)$$

Note that Equation 6.47 can be used only for annular denuders with a large μ values ($\mu > 3.0$) and small penetration ($P_n < 0.1$).

It should be pointed out that an annular denuder is much more efficient in collection than a similar size of circular tubing denuder at the same sampling rate and length, as demonstrated in Example 6.3.

Example 6.3: Determine the ammonia collection efficiency under standard conditions for the following:

    a. A cylindrical tube denuder with a diameter of 30 mm
    b. An annular denuder with an outer diameter of 30 mm and an inner diameter of 24 mm

Both denuders have the same length (300 mm) and the same air sampling rate of 30 l/min.

Solution: The diffusion coefficient of ammonia in air, from Table 5.3, is $D = 2.47 \times 10^{-5}$ m²/s.

    a. The flow rate $Q = 30$ l/min $= 0.0005$ m³/s. For cylindrical tubing

$$\mu = \frac{\pi LD_p}{Q} = \frac{\pi \times 0.3 \times 2.47 \times 10^{-5}}{0.0005} = 0.04656$$

Because $\mu > 0.02$, substituting the μ value into Equation 6.39 gives

$$P_n = 0.81905\exp(-3.6568\mu) + 0.09753\exp(-22.305\mu) + 0.0325\exp(-56.961\mu)$$

$$+ 0.01544\exp(-107.62\mu)$$

$$= 0.81905\exp(-3.6568 \times 0.04656) + 0.09753\exp(-22.305 \times 0.04656)$$

$$+ 0.0325\exp(-56.961 \times 0.04656) + 0.01544\exp(-107.62 \times 0.04656) = 0.804$$

$$\xi = 1 - P_n = 0.196 = 19.6\%$$

b. For the annular tubing, $(r_2 - r_1) = 0.003 < 0.1\pi(r_1 + r_2) = 0.0085$, so

$$\mu = \frac{8\pi(r_1 + r_2)LD_p}{3Q(r_2 - r_1)} = \frac{8\pi \times (0.015 + 0.012) \times 0.3 \times 2.47 \times 10^{-5}}{3 \times 5 \times 10^{-4} \times (0.015 - 0.012)} = 1.117$$

Because $3.0 > \mu > 0.05$, Equation 6.42 should be used:

$$P_n = 0.9104\exp(-2.8278\mu) + 0.0531\exp(-32.147\mu) + 0.01528\exp(-93.475\mu)$$

$$+ 0.00681\exp(-186.805\mu)$$

$$= 0.0388$$

and

$$\xi = 1 - P_n = 0.961 = 96.1\%$$

Apparently, an annular denuder is much more effective at particle collection than a cylindrical denuder of the same size at the same sampling rate. In this example, the collection efficiency of the annular denuder is about five times more efficient than that of the cylindrical denuder. (Using Equation 6.47 and Equation 6.48 gives a collection efficiency of 92.2% for the annular denuder.)

6.5.5 Wire Screen

Penetration through a stack of fine mesh screens with circular fibers of uniform diameters (as shown in Figure 6.10) has been developed.[13, 14] The theoretical penetration is based on the particle filtration in the fan model because of their similarity in flow resistance and particle deposition characteristics.[15]

$$P_n = \exp\left[-B_f\, n_f\, (2.7Pe^{-2/3} + \frac{1}{B_h}B_i^2 + 1.24B_h^{1/2}Pe^{-1/2}B_i^{2/3}\right] \quad \text{(for } dp < 1\ \mu m) \quad (6.49)$$

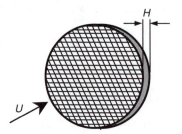

Figure 6.10 Schematic of a wire screen denuder.

where B_f is the solidity factor, B_i is the interception factor, B_h is the hydrodynamic factor, and Pe is the Peclet number, as defined in the following:

$$B_f = \frac{4\alpha_s H}{\pi(1-\alpha_s)d_f}$$ (6.50)

$$B_i = \frac{d_p}{d_f}$$ (6.51)

$$B_h = \left(\frac{2\alpha_s}{\pi}\right) - 0.5\ln\left(\frac{2\alpha_s}{\pi}\right) - 0.75 - 0.25\left(\frac{2\alpha_s}{\pi}\right)^2$$ (6.52)

$$Pe = \frac{Ud_f}{D_p}$$ (6.53)

where n_f is the number of screen filters, d_f is the screen fiber diameter, H is the thickness of a single screen, α_s is the solidity (which is the fraction of the solid volume of the screen or filter), and U is the face air velocity approaching the screen.

Equation 6.49 includes the effects of particle diffusion and interception. One application of Equation 6.49 is to estimate the filtration efficiency of small particles ($d_p < 1\ \mu m$) for fiber filters. For particles larger than 1 μm in diameter, Equation 6.49 may not be adequate, because the primary mechanism for particle collection is no longer diffusion but some other mechanism, such as impaction or interception.

6.5.6 Sampling Trains

It is sometimes necessary to collect gases and particulate contaminants separately. In this case, a sampling train consisting of an array of diffusion denuders and a filter pack with different characteristics has been used. A typical sampling train (as shown in Figure 6.11), which includes a cyclone precutter, two Na_2CO_3-coated annular denuders, and a filter pack with a Teflon and a nylon filter, has been used to collect acidic gases (HNO_3, HNO_2, SO_2, and HCl) separately from nitrate and

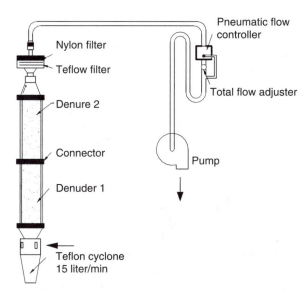

Figure 6.11 An ambient acidic particle sampling train consisting of a precutter, two annular denuders, and a filter pack. (From Stevens, R.K., Modern methods to measure air pollutants, in *Aerosols: Research, Risk Assessment, and Control Strategies*, Lee, S.D. ed., Lewis, Chelsea, MI, 1986. With permission.)

sulfate particles.[16] The first denuder removes gases quantitatively, whereas the second accounts for the interference from particulate material deposited on the wall under the assumption that particle deposition on each denuder is the same.[17] The denuders are placed vertically to avoid particle deposition on the walls by sedimentation. A diffusion scrubber can be connected to an ion chromatograph or other analytical instrument for real-time analysis of gases.[18, 19]

6.6 DIFFUSION BATTERIES

Diffusion batteries were originally developed to measure the diffusion coefficient of particles smaller than 0.1 μm in diameter. They have been used to measure size distribution by relating particle size and diffusion coefficient. Diffusion batteries are one of only a few instruments that can measure size distribution of particles between 0.001 and 0.2 μm in diameter. Several types of diffusion batteries have been designed and are commercially available.[3] The following section discusses the basic structures and types of diffusion batteries.

6.6.1 Basic Diffusion Battery Types

Basic types of diffusion batteries include a bundle of cylindrical tubes, parallel channels of equal length, and wire screens (Figure 6.12). The bundled tubes or channels are also referred to as *collimated hole* or *honeycomb structures* (CHSes). Tube-type diffusion batteries usually consist of a bundle of thin-walled tubes with

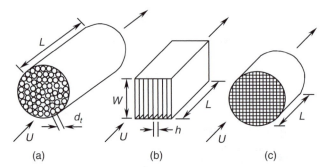

Figure 6.12 Basic types of diffusion screens: (a) cylindrical tube bundle, (b) stacked parallel
plates, (c) wire screen. (From Hinds, W.C., *Aerosol Technology*, John Wiley &
Sons, New York, 1999, Figure 7.10. With permission.)

an inside diameter smaller than 1 mm. Because diffusion coefficients for particles
are much smaller than those of gases, the length of the tube must be sufficiently
long, compared with the tube diameter or channel gap. There is no commercially
available CHS for construction of different diffusion batteries. However, the principle
can be used to construct such devices for air sampling or air cleaning.

One design criterion for diffusion batteries is that the flow through the diffusion
batteries must satisfy the laminar flow conditions, that is, $Re < 2300$. The penetration
(or collection efficiency) of a given particle size for each tube, channel, or wire mesh
can be calculated using the equations for calculating the corresponding denuders
described in previous section. For example, a diffusion battery consisting of a bundle
of 20 parallel cylindrical tubes will have the same penetration as a single-tube
denuder at an air sampling rate of 1/20 of the battery sampling rate. When diffusion
batteries with different characteristics are operated in parallel or in series, size
distribution can be measured.

6.6.2   Typical Diffusion Batteries

Depending on the arrangement of tube bundles, diffusion batteries can be generally categorized as parallel or serial batteries. A *parallel diffusion battery* consists
of bundles with different characteristics in a parallel fashion. Figure 6.13a gives a
schematic of a cluster-tube diffusion battery.[20] Three diffusion batteries with different
numbers of single tubes and different tube lengths were paralleled to collect three
different sizes of particles. The number of tubes and tube diameters in a bundle will
determine the flow rate through each tube, thus affecting the diffusion deposition
parameter and penetration. The length of the tubes is also proportional to the
diffusion deposition parameter. Penetration for each bundle can be determined using
Equation 6.39 and Equation 6.40. Tube-type diffusion batteries use commercially
available materials and are easy to construct. Light materials such as aluminum can
be used.

Figure 6.13b shows a schematic of a parallel wire-screen diffusion battery.[21] Different numbers of screens are stacked and housed in seven diffusion cells. The penetration through each cell can be determined using Equation 6.49. Because the bundles

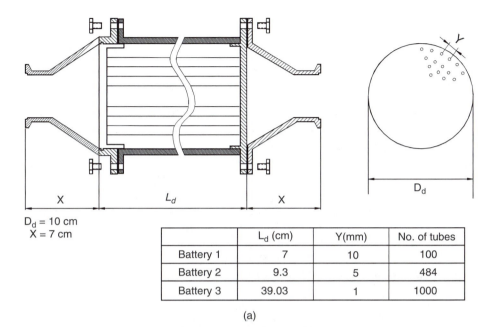

	L_d (cm)	Y(mm)	No. of tubes
Battery 1	7	10	100
Battery 2	9.3	5	484
Battery 3	39.03	1	1000

$D_d = 10$ cm
$X = 7$ cm

(a)

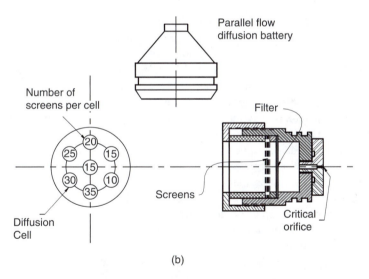

(b)

Figure 6.13 Parallel diffusion batteries. (a) Schematic of cluster-tube diffusion battery. (From Scheibel, H.G. and Porstendorfer, J., Penetration measurements for tube and screen-type diffusion batteries in the ultra fine particle size range, *J. Aerosol Sci.*, 15:133–145, 1984. With permission.). (b) Parallel-screen battery. (Adapted from Cheng, Y.S. et al., Characterization of diesel exhaust in a chronic inhalation study, *Am. Ind. Hyg. Assoc. J.*, 45:547–555, 1984.)

are parallel, a parallel diffusion battery is more flexible than denuders in its length variation to achieve a desirable collection efficiency. Parallel diffusion batteries often are heavy and bulky. Accurate flow control through each bundle can also be problematic.

Most diffusion batteries are arranged in a series of cells so that the concentration of particles in the flow is reduced gradually. One advantage of a *serial diffusion battery* is that the sampling flow rate through each cell can be controlled precisely. On the other hand, the concentrations of particles at the size of concern through each cell are not the true concentrations in the air sample, because the collection for each particle size is not clear cut but varies with a range of particle sizes. Figure 6.14a shows the schematic of a five-stage diffusion battery made of stainless-steel collimated holes. Hole lengths for different stages are different, to allow varying collection efficiency. The penetration of each stage is usually detected using a condensation nucleus counter (CNC). The difference in the particle count between any two stages can be used to calculate the penetration for a given size of particles. Figure 6.14b shows the schematic of a graded diffusion battery, including five stages of wire screens and a back-up filter. For certain particles with low concentrations and high diffusivity, care should be taken to attenuate concentration after each stage.

6.7 MONODISPERSE COAGULATION

One of the important applications of particle diffusion is coagulation. Coagulation, or agglomeration, of particles is a process in which airborne particles collide with one another, due to the relative motion between them, and adhere to form larger particles. When the relative motion is caused by Brownian motion, the coagulation is called *thermal coagulation*. When the relative motion is caused by external forces, such as gravitational settling or electrical force, the coagulation is called *kinematical coagulation*. Coagulation is the most important interparticle behavior that can be utilized in many air-sampling or air-cleaning processes.

Interparticle behavior is very complicated. Therefore, the exact theory of coagulation, especially for polydisperse particles, is also very complicated. However, in many indoor air quality problems, our concern is the change of concentration over a period of time for certain particle sizes, or how to change the airborne particle size distribution so that appropriate control strategies can be implemented. Therefore, some simplified coagulation theories can be useful in describing the changes of particle concentration and size.

Let us first consider the simplest coagulation — monodispersed coagulation for particles larger than 0.4 μm in diameter, also referred to as *Smoluchowski coagulation*. In this type of coagulation, particle size increases slowly, and particles adhere to each other after they have collided. When all particles have the same diameter d_p, the colliding diameter will be $2d_p$, and any two particles within the distance (center-to-center) of d_p from any direction will collide. The equivalent colliding area of a selected particle is then the surface area of an imaginary sphere with a diameter of $2d_p$, that is, $A_c = \pi(2d_p)^2 = 4\pi d_p^2$, shown in Figure 5.4 as the larger, dash-lined sphere. Because the diffusion flux is J_p, the rate of this particle colliding with other particles, dn_c/dt, is the product of the colliding area and the particle flux:

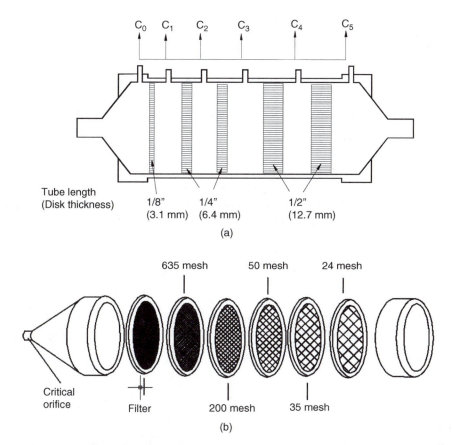

Figure 6.14 Multistage diffusion batteries: (a) schematic of a five-stage diffusion battery, (b) schematic of a graded diffusion battery with a series of wire screens and filters.

$$\frac{dn_{ci}}{dt} = A_c J_p = 4\pi d_p^2 J_p = -4\pi d_p^2 D_p \frac{dC_{pn}}{dx} \tag{6.54}$$

where x is the distance from and is perpendicular to the colliding surface, and n_{ci} indicates the collision rate for the ith particle. The rate of collision for the selected particle is proportional to the particle concentration gradient at the collision surface. Neglecting the transient change of particle concentration within a short period of time, the particle concentration gradient at a spherical collision surface for certain range of particle size, dCp/dx, was developed by Fuchs:[22]

$$\frac{dC_{pn}}{dx} = -\frac{2C_{pn}}{d_p} \quad \text{(for } d_p > \lambda_p) \tag{6.55}$$

where λ_p is the mean free path of the particle. Combining Equation 6.54 and Equation 6.55 gives the rate of collision for the ith particle:

$$\frac{dn_{ci}}{dt} = 8\pi d_p D_p C_{pn} \tag{6.56}$$

Because the total number of particles in a unit volume is C_{pn}, and each collision takes two particles to participate, the total rate of collision dn_c/dt becomes

$$\frac{dn_c}{dt} = \frac{C_{pn}}{2}(8\pi d_p D_p C_{pn}) = 4\pi d_p D_p C_{pn}^2 \tag{6.57}$$

Because after each collision the total number of particles in the unit volume will be reduced by one, the rate of particle number concentration reduction is numerically the same as the rate of collision:

$$\frac{dC_{pn}}{dt} = -4\pi d_p D_p C_{pn}^2 = -K_0 C_{pn}^2 \tag{6.58}$$

where the minus sign (−) indicates that the particle number concentration is always reducing during a coagulation process, and K_0 is the coagulation coefficient. Substituting D_p into Equation 6.58 gives

$$K_0 = 4\pi d_p D_p = \frac{4kTC_c}{3\eta} \quad \text{(for } d_p > 0.4 \text{ μm)} \tag{6.59}$$

where k is Boltzmann's constant, T is the absolute temperature, C_c is the slip correction coefficient, and η is the air viscosity.

Equation 6.59 can only apply to particles larger than about 0.4 μm, because Equation 6.55 does not correctly describe the concentration gradient within one particle mean free path of the particle. This error increases dramatically as the particle diameter decreases. For particles of 0.4 μm in diameter, the error is about 5%. For particles of 0.1 μm, the error is about 21%. For particles of 0.01 μm, the error is about 600%. Therefore, a coagulation coefficient correction factor β must be applied to obtain a corrected coagulation coefficient K for all particle sizes.

$$K = \beta K_0 \tag{6.60}$$

Table 6.5 lists β, K_0, and K values for typical particle sizes.[22] Based on the β values for the typical particle sizes from Hinds,[23] an equation can be developed for β vs. particle diameter. The relationship between the coagulation coefficient correction factor and the particle size follows a Gudermanian function pattern (Figure 6.15), which is a monotonic odd function asymptotic to $d_p = 0.35$ μm for the entire

Table 6.5 Coagulation Coefficients and Correction Factors for Typical Sizes of Particles

Particle Diameter d_p (µm)	Correction Factor β	Coagulation Coefficient K	
		$K_0 \times 10^{-16}$ (Equation 6.59), (m³/s)	$K = \beta K_0 \times 10^{-16}$ (m³/s)
0.004	0.037	168	6.2
0.01	0.14	68	9.5
0.04	0.58	19	10.7
0.1	0.82	8.7	7.2
0.4	0.95	4.2	4.0
1	0.97	3.4	3.4
4	0.99	3.1	3.1
10	0.99	3.0	3.0

Source: Hinds, W.C., *Aerosol Technology,* John Wiley & Sons, New York, 1999. With permission.

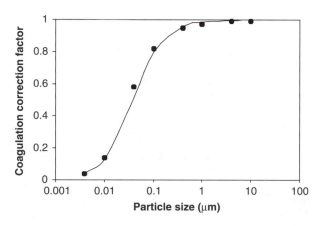

Figure 6.15 Coagulation coefficient correction factor vs. particle size. The solid curve is calculated from Equation 6.61, and data points are from Hinds.[23]

particle size range in logarithmic scale. Equation 6.61 can be used to calculate the β value within 7% for particles larger than 0.01 µm.

$$\beta = \frac{2}{\pi} \tan^{-1} \left\{ 45.3 \left[\exp(\log(d_p)) \right]^{2.694} \right\}$$ (6.61)

where d_p is in µm, and the term $tan^{-1}\{45.3[exp(log(d_p))]^{2.694}\}$ is in radians.

The particle concentration at any time, $C_{pn}(t)$, then can be solved from Equation 6.58, based on its initial concentration C_{pn0}, by integrating

$$\int_{C_{pn0}}^{C_{pn}(t)} \frac{dC_{pn}}{C_{pn}^2} = \int_0^t -K \, dt$$ (6.62)

$$C_{pn}(t) = \frac{C_{pn0}}{1 + C_{pn0} K t} \qquad (6.63)$$

where C_{p0} is in particles/m^3, K is in m^3/s, and t is in s.

Equation 6.63 describes monodisperse particle coagulation. It should be pointed out that the coagulation coefficient K changes slightly with particle size (Table 6.5) in a large range. Therefore, in many practical situations, such as during a short period of time or within small range change of particle sizes, K can be considered a constant. When the particle size changes substantially (typically in the order of 10 times or more), new K values, based on the coagulated particle diameter, should be used.

Particle size increases as a direct consequence of the decrease in particle number concentration for a given air space during a coagulation process. If there is no loss or gain of particle mass in the air space, the particle mass concentration C_{pm} conserves at the beginning and the end of the coagulation:

$$C_{pm} = C_{pn0} \frac{\pi}{6} \rho_p d_{p0}^3 = C_{pn}(t) \frac{\pi}{6} \rho_p (d_p(t))^3 \qquad (6.64)$$

Substituting Equation 6.63 into Equation 6.64 and solving for $d_{pn}(t)$ gives the particle diameter at any time t, with respect to the initial particle diameter and number concentration:

$$d_p(t) = d_{p0} (1 + C_{pn0} K t)^{\frac{1}{3}} \qquad (6.65)$$

Equation 6.63 and Equation 6.65 are approximately correct for solid particles. Coagulation of porous particles is faster than these equations predict. Although derived from monodisperse particles, Equation 6.63 and Equation 6.65 can be used for polydisperse particles, using K values calculated according to polydisperse particle coagulation (see next section). Equation 6.63 and Equation 6.65 can also be used directly for polydisperse particles with a count median diameter (CMD) larger than 0.1 μm and a geometric standard deviation (GSD) less than 2.5 within 30%. In this case, the CMD is used to calculate the K value.

A closer examination of Equation 6.58, Equation 6.63, and Equation 6.65 reveals that the rate of particle concentration is proportional to the square of the concentration. Therefore, the coagulation occurs vigorously in an airspace with a high particle concentration. At low particle concentration, the coagulation process is very slow. Figure 6.16 shows typical monodisperse particle concentration changes with time at an initial concentration of 5000 particles/ml, calculated using Equation 6.63. For example, for particles of 0.1 μm in diameter and with an initial concentration of 5000 particles/ml, it takes approximately 3 days to halve the concentration. Generally, the monodisperse coagulation effect will be negligible when the particle concentration is less than 10^6 particles/ml within a 5-minute observation period. Therefore, monodisperse particle coagulation can be neglected in most laboratory

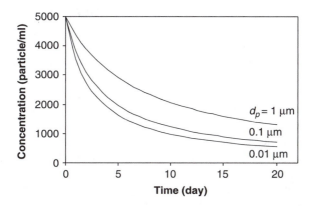

Figure 6.16 Particle number concentration change with time for typical particle sizes as the result of monodisperse coagulation, calculated using Equation 6.63.

experiments and indoor air quality problems. Smaller particles coagulate faster than larger particles because small particles have large coagulation coefficients in relation to the large diffusion coefficients.

Example 6.4*: What is the particle number concentration in an airspace after a 1-hour period for particles of 0.01 μm in diameter and with an initial concentration of 10^5 particles/ml? Assume that the process is monodisperse coagulation under standard atmospheric conditions.*

Solution: From Table 6.3 or Equation 6.12, the diffusion coefficient for 0.01 μm particles, $D_p = 5.45 \times 10^{-8}$ m²/s. From Table 6.4, the coagulation coefficient $K = 9.5 \times 10^{-16}$ m³/s. (Note that K can also be calculated using Equations 6.59 through 6.61).

Substituting these values into Equation 6.63, noting that the particle concentration should be in particles/m³, gives the particle concentration after one hour:

$$C_{pn}(t) = \frac{C_{pn0}}{1 + C_{pn0}Kt} = \frac{10^{11}\,(particle\,/\,m^3)}{1 + 10^{11}\,(particle\,/\,m^3) \times 9.5 \times 10^{-16}\,(m^3\,/\,s) \times 3600\,(s)}$$

$$= 7.45 \times 10^{10}\,(particle\,/\,m^3)$$

$$= 74500\,(particle\,/\,ml)$$

Note: The monodisperse coagulation process does not occur among gas molecules, although gas molecules are treated as very small particles within this context. For example, a cylinder of ammonia gas will remain ammonia gas, without coagulating into larger particles. However, particles in a cylinder of dust particle–laden air will coagulate and settle. Therefore, it is important that keep in mind that the monodisperse coagulation effect among gas molecules should be excluded when analyzing the interactions of particles and gases.

6.8 POLYDISPERSE COAGULATION

Particles in most air-quality problems are polydisperse. The rate of polydisperse coagulation depends on the particle size distribution, which itself is often a complex matter. Explicit mathematical solutions for polydisperse particles are not available. However, using the discussions of the previous section, the principles of coagulation of monodisperse particles can be extended to polydisperse particles, and Equation 6.63 can be used, provided that the coagulation coefficient K can be recalculated based on polydisperse particle interactions.

For particles with n size groups, the coagulation coefficient is based on the combination of products of d_{pi} and D_{pi}, ($i = 1, 2, 3, ..., n$). In this case, we must account for coagulations of a given size with any other size of particle. Further, the concentration fraction of each particle size also affects the coagulation process. The average coagulation coefficient for polydisperse particles, K_n, with n size groups can be calculated as

$$K_n = \sum_{i=1}^{n} \sum_{j=1}^{n} K_{i-j} f_i f_j \qquad (6.66)$$

where K_{i-j} is the coagulation coefficient for particles between the ith and j^{th} particle size groups, which will be discussed later, and f_i and f_j are the fractions of total number concentrations of particles for i^{th} and j^{th} size groups.

Equation 6.66 gives the coagulation coefficient of the polydisperse particles at a given time. In using Equation 6.66 for coagulation during a period of time, one should estimate that the K_n does not change appreciably during that period. Otherwise, the new coagulation coefficient must be calculated for a new time step because the particle size distribution varies with time.

In order to determine the average coagulation coefficient K_n, we must find each K_{i-j}. Let us consider the simplest case: An airspace contains particles of any two sizes, d_{pi} and d_{pj}, with diffusion coefficients D_{pi} and D_{pj}, respectively. Coagulations between the same sizes of particles are characterized by the product of $4\pi d_p D_p$ (Equation 6.59). Equation 6.59 can be viewed as the sum of coagulation between the two groups of particles. Coagulations between two different sizes of particles can be characterized by the combination of the effect of the products of d_{pi}, D_{pi}, d_{pj}, and D_{pj}. There are four combinations of products of d_p and D_p for these two sizes of particles: $d_{pi} D_{pi}$, $d_{pi} D_{pj}$, $d_{pj} D_{pi}$, and $d_{pj} D_{pj}$. Therefore, the coagulation coefficient for these two arbitrary sizes of particles, similar to Equation 6.59, can be written as

$$K_{i-j} = \pi \left[\beta(d_{pi}) d_{pi} D_{pi} + \beta(d_{pi}) d_{pi} D_{pj} + \beta(d_{pj}) d_{pj} D_{pi} + \beta(d_{pj}) d_{pj} D_{pj} \right] \quad (6.67)$$

where $\beta(d_{pi})$ and $\beta(d_{pj})$ are the coagulation coefficient correction factors defined by Equation 6.61.

Clearly, when $d_{pi} = d_{pj}$, Equation 6.67 becomes Equation 6.59, a monodisperse coagulation. If d_{pj} is large and d_{pi} is small, D_{pi} will be large and D_{pj} will be small

Table 6.6   Coagulation Coefficients for Typical Sizes of Particles under Polydisperse Coagulation (Values Calculated Using Equation 6.67)

d_{p1} (µm)	$K_{1-2} \times 10^{-16}$, (m³/s)			
	$d_{p2} = 0.01$ µm	$d_{p2} = 0.1$ µm	$d_{p2} = 1$ µm	$d_{p2} = 10$ µm
0.01	9.6	122	1700	17,000
0.1	122	7.2	24	220
1	1700	24	3.4	10.3
10	17,000	220	10.3	3.0

because the diffusion coefficient is inversely proportional to the particle diameter. Therefore, the product $d_{pj}D_{pi}$ will be much larger than the other three products. From a particle diffusion point of view, a large particle has a large surface to allow more particles to collide, especially for those small particles with a high diffusion coefficient. A small particle has a small surface but is very mobile (has a high diffusion coefficient) and has a greater chance of colliding with a large particle. The same sized particles have either a large mobility (high diffusion coefficient) and a small surface for small particles, or vice versa for large particles. In either case, the frequency of particle collisions is low. For example, if $d_{pj} = 1$ µm and $d_{pi} = 0.1$ µm, from Table 6.3, $D_{pj} = 2.76 \times 10^{-11}$ (m²/s) and $D_{pi} = 6.31 \times 10^{-10}$ (m²/s). The product $d_{pj}D_{pi}$ takes 87%, and the other three terms take 13%, of the total value of K_{i-j}. When $d_{pj} \geq 20d_{pi}$, the product of $d_{pj}D_{pi}$ takes approximately 95% of the total value of K_{i-j}. Therefore, Equation 6.67 can be simplified as follows:

$$K_{i-j} = \pi \beta(d_{pj})d_{pj}D_{pi} \ \ (\text{if } d_{pj} \geq 20d_{pi}) \tag{6.68}$$

Substituting Equation 6.67 or Equation 6.68 into Equation 6.66, one can obtain the average coagulation coefficient K_n and subsequently calculate the new particle concentration by substituting K_n into Equation 6.63.

Polydisperse particles coagulate much faster than monodisperse particles. Thus, air-quality control technologies using coagulation may become practical. Comparing Equation 6.68 with Equation 6.59, (although there is a factor of four missing in Equation 6.68), the coagulation coefficient of K_{i-j} can be much greater than K for the smaller particles, because the product of $d_{pj}D_{pi}$ is much greater than the product of $d_{pi}D_{pj}$. Coagulation coefficients for typical particle sizes are listed in Table 6.6. All data are calculated using Equation 6.67.

Although the coagulation coefficient for particles containing n size groups can be calculated using Equation 6.66, particles with continuous size distribution can be solved. For example, for particles with a lognormal size distribution, the coagulation coefficient has been derived explicitly by Lee and Chen for particles $d_p > \lambda$:[24]

$$K_n = \frac{2kT}{\eta}\left\{1 + \exp(\ln^2 \sigma_g) + \frac{2.49\lambda}{CMD}\left[\exp(0.5\ln^2 \sigma_g) + \exp(2.5\ln^2 \sigma_g)\right]\right\} \tag{6.69}$$

(for $d_p > 1$)

where k is Boltzmann's constant, T is the absolute temperature, η is the air viscosity, λ is the mean free path of air, CMD is the count median diameter, and σ_g is the geometric standard deviation of the particles.

For polydisperse particles, the diameter of average mass is directly related to the particle number concentration, regardless of the particle shape and size distribution. Assume that the total particle mass is conservative, that is, there is no gain or loss of particle mass before or after a coagulation process. Thus, the particle mass concentration can be calculated:

$$C_{pm} = C_{pn0} \frac{\pi}{6} \rho_p d_{\overline{m}0}^3 = C_{pn}(t) \frac{\pi}{6} \rho_p (d_{\overline{m}}(t))^3 \qquad (6.70)$$

where $d_{\overline{m}0}$ and $d_{\overline{m}}$ are the diameters of average mass of particles before and after the coagulation process. Solving for $d_{\overline{m}}(t)$ gives the particle diameter of average mass at any time t, with respect to the initial particle diameter of average mass and the change of particle number concentration:

$$d_{\overline{m}}(t) = d_{\overline{m}0} \left(\frac{C_{pn0}}{C_{pn}(t)} \right)^{1/3} \qquad (6.71)$$

Equation 6.71 predicts the change in diameters of average mass for polydisperse particles. If the particles have a wide size distribution, that is, a large GSD ($\sigma_g >$ 1.5), the coagulation will make the size distribution narrower. This is because the small particles will quickly coagulate with the larger ones, yet the sizes of larger ones do not change appreciably. For example, when a 0.1 µm particle coagulates with a 1 µm particle and forms a new, larger particle, the particle number is reduced by a factor of 2. The diameter of the new, larger particle is 1.001 µm, increased only by a factor of 0.001 over the original 1 µm particle. This characteristic may be useful in treating polydisperse particles for some air-cleaning processes, particularly for those processes that are efficient only for a narrow size range of particles.

Example 6.5: *Particles in an airspace have the following initial size distribution: 100,000 particles/ml for 0.01 µm, 5,000 particles/ml for 0.1 µm, and 200 particles/ml for 1 µm. Find the following after one hour:*

    a. Total particle number concentration
    b. The particle diameter of average mass

Solution:

    a. Using data from Table 6.6, the following list shows the coagulation coefficients for the three sizes of particles. (These values can also be calculated using Equation 6.67 if the table values are not available.)

$K_{1-1} = 9.6 \times 10^{-16}$ m³/s
$K_{1-2} = K_{2-1} = 122 \times 10^{-16}$ m³/s
$K_{1-3} = K_{3-1} = 1700 \times 10^{-16}$ m³/s
$K_{2-2} = 7.2 \times 10^{-16}$ m³/s
$K_{2-3} = K_{3-2} = 24 \times 10^{-16}$ m³/s
$K_{3-3} = 3.4 \times 10^{-16}$ m³/s

The fractions of total particle number concentration for each size group are calculated using the initial total particle number concentration.

$C_{pn0} = 100{,}000 + 5{,}000 + 200 = 105{,}200$ (particles/ml) $= 1.052 \times 10^{11}$ particles/m³,

$$f_1 = \frac{100{,}000}{105{,}200} = 0.95057$$

$$f_2 = \frac{5{,}000}{105{,}200} = 0.04753$$

$$f_3 = \frac{200}{105{,}200} = 0.0019$$

Substituting the preceding values into Equation 6.68 gives

$$K_n = \sum_{i=1}^{n} \sum_{j=1}^{n} K_{i-j} f_i f_j$$

$$= \begin{pmatrix} (K_{1-1}f_1 f_1 + K_{1-2}f_1 f_2 + K_{1-3}f_1 f_3) \\ +(K_{2-1}f_2 f_1 + K_{2-2}f_2 f_2 + K_{2-3}f_2 f_3) \\ +(K_{3-1}f_3 f_1 + K_{3-2}f_3 f_2 + K_{3-3}f_3 f_3) \end{pmatrix} \times 10^{-16}$$

$$= \begin{pmatrix} 9.6 \times 0.95057^2 + 122 \times 0.95057 \times 0.04753 + 1700 \times 0.95057 \times 0.0019) \\ +(122 \times 0.95057 \times 0.04753 + 7.2 \times 0.04753^2 + 24 \times 0.04753 \times 0.0019) \\ +(1700 \times 0.95057 \times 0.0019 + 24 \times 0.04753 \times 0.0019 + 3.4 \times 0.0019^2) \end{pmatrix} \times 10^{-16}$$

$$= 25.86 \times 10^{-16} \ (m^3 / s)$$

$$C_{pn}(t) = \frac{C_{pn0}}{1 + C_{pn0} K_n t}$$

$$= \frac{1.052 \times 10^{11} \ (particle \ / \ m^3)}{1 + 1.052 \times 10^{11} \ (particle \ / \ m^3) \times 25.86 \times 10^{-16} \ (m^3 \ / \ s) \times 3600 \ (s)}$$

$$= 5.3148 \times 10^{10} \ (particle \ / \ m^3)$$

$$= 53{,}148 \ (particle \ / \ ml)$$

The total particle number concentration decreases by about 50% in one hour.

b. The initial diameter of average mass can be found from the equation in Table 3.7. Note that the total particle number in this example is the particle number concentration, that is, $N = C_{pn0}$, and the frequency falling into the ith size group (F_i) is the number concentration of the same size group.

$$d_{\overline{m}0} = \left(\frac{\sum F_i d_{pi}^3}{N} \right)$$

$$= \frac{1}{1.052 \times 10^{11}} \left(10^{11} \times (0.01 \times 10^{-6})^3 + 5 \times 10^9 \times (0.1 \times 10^{-6})^3 + 2 \times 10^8 \times (1 \times 10^6)^3 \right)$$

$$= 1.25 \times 10^{-7} \ (m) = 0.125 \ (\mu m)$$

Equation 6.71 gives the new diameter of average mass after one hour:

$$d_{\overline{m}}(t) = d_{\overline{m}0} \left(\frac{C_{pn0}}{C_{pn}(t)} \right)^{1/3} = 0.125 \left(\frac{105{,}200}{53{,}148} \right)^{1/3} = 0.157 \ (\mu m)$$

6.9 KINEMATICAL COAGULATION

Kinematical, or *orthokinetical*, *coagulation* is the coagulation that results from the relative motion between particles caused by external forces other than Brownian motion. These external forces include electrostatic, centrifugal, and gravitational forces. In essence, particle kinematical coagulation does not belong to applications of particle diffusion. This topic is discussed briefly in this chapter, because we have already discussed other types of coagulation in previous sections.

Because kinematical coagulation is the direct result of relative motion between particles, relative velocity between particles is a direct measurement of the effect. Imagine that when a given particle (let us call it a droplet) with a diameter d_d travels in an airspace with a relative velocity to other particles, V_r, the droplet is "sweeping" the airspace, covering a volume within a unit time period, $\pi d_d^2 V_r/4$. If we assume that the number concentration of small particles is C_{pn}, the number of collisions during this sweeping of the droplet, or ith droplet, during the time period of one second, can be written as follows:

$$n_{ci} = \frac{\pi d_d^2}{4} V_r C_{pn} K_c \tag{6.72}$$

where K_c is the particle capture efficiency of the droplet.

The capture efficiency is low, except for particles larger than 1 μm in diameter. For particles smaller than 1 μm, the settling velocity is smaller than the mean thermal velocity; diffusive force is more significant. When the droplet is small, the sweep cross section is small and the collision frequency is low. Theoretical analysis of capture efficiency is very complicated because of the complexity of flow around the droplet and the motion of the small particles. An empirical expression of capture efficiency of particles with a diameter of d_p by a droplet with a diameter of d_d moving at a relative velocity V_r, neglecting particle interception,[23] is

$$K_c = \left(\frac{Stk_c}{Stk_c + 0.12} \right)^2 \quad \text{(for } Stk_c \geq 0.1) \tag{6.73}$$

where Stk_c is the Stokes number for particle capture efficiency, defined as

$$Stk_c = \frac{\rho_p d_p^2 C_c V_r}{18 \eta \, d_d} \tag{6.74}$$

where C_c is the slip correction factor for the particle, not for the droplet.

The capture efficiency decreases to zero when the sizes of particles and droplets are the same, because the relative velocity becomes zero. In the case of gravitational settling, $V_r = V_{TS}$ for large droplets with respect to small particles, such as in a raindrop and wet scrubbing process, and large droplets sweep small particles floating in the air. The settling velocity can be negligible compared with that of large droplets. In such a sweeping process, if $d_p \ll d_d$, the Stk_c is very small. When Stk_c is smaller than 0.1, as shown in Equation 6.73, the kinematical coagulation becomes small (less than 21%). In this case, Equation 6.73 cannot accurately predict the coagulation efficiency, and the polydisperse coagulation by diffusion becomes the major coagulation mechanism.

The total number of collisions for all droplets in a unit volume during a time period of t is the sum of the collisions for all droplets:

$$n_c = \sum n_{ci} t = \frac{\pi}{4} d_d^2 V_r C_{pn} C_{dn} K_c t \tag{6.75}$$

where C_{dn} is the concentration of droplets and t is the time period during which kinematical coagulation occurs.

Assume that a particle adheres to a droplet once it collides with that droplet, and the number of collisions among the droplets themselves is negligible (i.e., C_{dn} = constant); the total number of collisions is the number of particles collected by the droplets. This assumption is true when $C_{pn} \gg C_{dn}$. When $C_{pn} = C_{dn}$, the number of collisions among the droplets is about the same as the number of collisions between the particles and the droplets. Therefore, the total number of collisions during a unit time is equal to the rate of change in particle concentration;

$$\frac{d(C_{pn})}{dt} = -\frac{\pi}{4} d_d^2 V_r C_{pn} C_{dn} K_c \tag{6.76}$$

where the minus sign ($-$) indicates that the particle concentration is decreasing with time. Separating variables and integrating, noting that integration domains for C_{pn} and t are $[C_{pn0}, C_{pn}]$, and $[0, t]$, respectively, and C_{pn0} is the initial particle concentration, yields

$$C_{pn}(t) = C_{pn0} \exp\left(-\frac{\pi}{4} d_d^2 V_r C_{dn} K_c t\right) \tag{6.77}$$

Applications of kinematical and polydisperse coagulation are discussed in Chapter 11, where particle collection efficiencies of wet scrubbing devices are analyzed. In wet scrubbing devices, a liquid (usually water) is used to "scrub" the contaminated air. Particles or gases in the air are collected by the much larger liquid droplets. The large liquid droplets then can be collected easily.

DISCUSSION TOPICS

1. Thermal velocities for gases and airborne particles have the same characteristics. Can we treat airborne particles as ideal gases?
2. In diffusion, we can treat gas molecules as small particles, thus the principle of diffusion can be equally applied to gases and to particles. Why?
3. The kinetic energies for 1 mole of air and one mole of particles are the same. Assume that a car tire holds 1 mole of air. Can we replace the air in the tire with 1 mole of airborne particles, 0.1 μm in diameter, and expect normal performance?
4. Why do heavier particles have larger mean free paths than lighter particles of the same size?
5. How would you design an experiment to measure the diffusion coefficient for particles of 0.5 μm in diameter and with a density of 1000 kg/m³?
6. How would you plan to increase the air filtration efficiency for your home air-supply system? What are the primary factors that you can manipulate?

7. If the particle size distribution in a room is obtained and the particle density is measured, can you predict how much dust (in mass) accumulates on the floor, walls, and ceiling in a year? Schedule a cleaning plan based on your prediction.

8. Monodisperse particle coagulation has very little direct application for indoor air quality control. Why? Why might polydisperse coagulation be more practical in air cleaning?

9. Is a rainfall a process of monodisperse coagulation or polydisperse coagulation? With the same amount of rain, which is likely to be more efficient in cleaning the air: a drizzle (small droplets) or a storm (large droplets)? Why?

10. Kinematical coagulation is not effective either when $d_p << d_d$ or when $d_p \approx d_d$. However, water towers are effectively used to collect sulfur or ammonia gases, which are composed of very small particles (molecules). Why?

PROBLEMS

1. Derive a relationship between the mean free path and particle size, including the effect of the slip correction coefficient for particles.

2. Calculate the mean free paths for particles of 0.5 μm in diameter but with different densities: 1000 kg/m³ and 3000 kg/m³. Assume that all particles are spherical and under standard atmospheric conditions.

3. Calculate the diffusion coefficient for particles of 0.3 μm in diameter and the root-mean-square distance x_{rms}.

4. Determine the critical particle deposition diameter for particles with a density of 500 kg/m³, using the iteration method.

5. Particle size distribution in a room is measured in table below. Assuming that all particles are spherical and at standard densities, determine the deposition velocities at the surfaces of the ceiling, walls, and floor.

Particle Size (μm)	Number Concentration (particles/ml)
0.1	50
0.3	10
0.5	8
1	7
3	3
5	1.5
10	0.5

6. Consider the particle size distribution in the problem 5. Assume that all particles deposited on a surface are uniformly distributed without overlapping, that is, each particle is laid directly on the surface and occupies an area of $\pi d_p^2/4$. How long will it take to cover the entire floor surface with particles? How long will it take to cover the walls or the ceiling?

7. How long would it take for a return air shutter in a bedroom to be completely covered with dust? The shutter is on the wall and has an incline angle β_d of −70 degrees. Assume that no particles overlap on the shutter surface. The particle concentration in the room is 10 particles/ml. The average particle diameter is 0.7 μm, and the density is 600 kg/m³.

8. Using the equations provided in this chapter, plot the penetration of particles (in percentage) vs. the diffusion deposition parameter in the range of 0.001 to 1 for cylindrical tubes and for parallel plates. (Suggestion: P_n is the y-axis and μ is the x-axis and in log scale).

9. The smell threshold of ammonia for an average person is about 5 ppmv under standard conditions. Assume that the smell-sensing section of the human nose is similar to a cylindrical denuder with an inside diameter of 5 mm and a length of 6 mm. The air speed passing through the smell-sensing section is 25 m/s. How many ammonia molecules must be collected before one can detect the ammonia?

10. Using the equations provided in this chapter, plot the particle collection efficiency (in percentage) vs. the particle diameter in the range of 0.001 to 1 μm for a parallel-channel diffusion battery with 30 channels, $W = 50$ mm, $H = 1$ mm, and $L = 200$ mm. The air-sampling pump has three stages of flow rates: 0.1 l/min, 1 l/m, and 10 l/min. (Suggestion: ξ is the y-axis and d_p is the x-axis and in log scale).

11. In the design of a bundled-tube denuder to collect 0.1 μm particles at a 50% efficiency, the following components have been already preselected: The air-sampling pump is 1 l/min, tubes have an inside diameter of 2 mm, and 300 tubes will be used. Determine the length of the tubes.

12. In operating a parallel-channel diffusion battery with 30 rectangular channels, each with $W = 80$ mm, $H = 1$ mm, and $L = 200$ mm, the operator is interested in measuring particles of 0.05 and 0.2 μm with a collection efficiency of 50%. To achieve this, a variable-speed air pump is used to adjust the air sampling rate. What should the air sampling rates be for the 0.05 and 0.2 μm particles, respectively?

13. While smoking a cigarette, a smoker inhales particles of 0.01 μm in diameter with a number concentration of 3×10^9 particles/ml. After 5 seconds, the smoker exhales the smoke. What is the exhaled particle size? Assume that a simple monodisperse coagulation is occurring and that no particles are retained in the respiratory system.

14. In a room initially filled with smoke particles, with a number concentration of 10^6 particles/ml, assume that the particles are monodisperse with a diameter of 0.02 μm. Find the number concentration after one hour.

15. In a room air space, the concentration of smoke particles is 0.05 mg/m³. The smoke particles are monodisperse, with a diameter of 0.01 μm and a density of 900 kg/m³. Other dust concentration is 2 mg per cubic meter of air. Assume that the dust particles are monodisperse with a diameter of 12 μm. The density is 1000 kg/m³, and the monodisperse coagulation among the dust particles is negligible. What will the concentration of smoke particles be after 20 minutes?

16. A sample of polydisperse particles with a lognormal distribution has an initial number concentration of 10,000 particles/ml. The particles have a count median diameter of 1.5 μm and a geometric standard deviation of 1.8. Calculate the number concentration after 30 minutes.

17. In a wet scrubber system, particle- and ammonia-laden air passes through a chamber filled with water droplets. The water droplets have an average diameter of 50 μm, with a number concentration of 100 droplets/ml. The particles have an average diameter of 2 μm and a number concentration of 1000/ml. The ammonia concentration is 15 ppmv. The diameter of an ammonia molecule is approximately 0.00031 μm. Assuming that the number concentration reductions of water droplets and particles are negligible, how long would it take to reduce the ammonia concentration to 5 ppmv? (Hint: The effect of monodisperse coagulation of ammo-

nia should be excluded. The coagulation coefficient between ammonia molecules is zero.)

REFERENCES

1. Einstein, A., On the kinetic molecular theory of thermal movements of particles suspended in a quiescent fluid, *Ann. Physik*, 17:549–560, 1905. (English translation: *Investigation on the Theory of Brownian Movement*, Furth, R., ed., Dover, New York, 1956.)
2. Nolan, J.J. and Guerrini, V.H., The diffusion coefficient of condensation nuclei and velocity of fall in air of atmospheric nuclei, *Proc. R. Iri. Acad.*, 43:5–24, 1935.
3. Cheng, Y.S., Diffusion batteries and denuders, in *Air Sampling Instruments for Evaluation of Atmospheric Contaminants*, 8th ed., American Conference of Governmental Industrial Hygienists, Cincinnati, OH, 1995.
4. Durham, J.L., Ellestad, T.G., and Stockburger, L., A transition flow reactor tube for measuring trace gas concentrations, *J. Air Pollut. Control Assoc.*, 36:1228–1232, 1986.
5. John, W., Wall, S.M., and Ondo, J.L. A new method for nitric acid and nitrate aerosol measurement using the dichotomous sampler. *Atmos. Environ.*, 22:1627–1635, 1988.
6. Gormeley, P.G. and Kennedy, M., Diffusion from a stream flowing through a cylindrical tube, *Proc. R. Iri. Acad.*, A52:163–169, 1949.
7. Newman, J., Extension of the Leveque solution, *J. Heat Transfer*, 91:177–178, 1969.
8. Ingham, D.B., Diffusion of aerosols from a stream flowing through a cylindrical tube, *J. Aerosol Sci.*, 6:125–132, 1975.
9. Bowen, B.D., Levine, S., and Esptein, N., Fine particle deposition in laminar flow through parallel-plate and cylindrical channels, *J. Colloid Interface Sci.*, 54:375–390, 1976.
10. Soderholm, S.C., Analysis of diffusion battery data, *J. Aerosol Sci.*, 10:163–175, 1979.
11. Ferm, M., Method for determination of atmospheric ammonia, *Atmos. Eniron.*, 13:1385–1393, 1979.
12. Possanzini, M., Febo, A., and Aliberti, A., New design of a high-performance denuder for the sampling of atmospheric pollutants, *Atmos. Eniron.*, 17:2605–2610, 1983.
13. Cheng, Y.S. and Yeh, H.C., Theory of a screen-type diffusion battery, *J. Aerosol Sci.*, 11:313–320, 1980.
14. Cheng, Y.S., Keating, J.A., and Kanapilly, G.M., Theory and calibration of a screen-type diffusion battery, *J. Aerosol Sci.*, 11:549–556, 1980.
15. Cheng, Y.S., Yeh, H.C., and Brinsko, K.J., Use of wire screens as a fan model filter, *Aerosol Sci. Technol.*, 4:165–174, 1985.
16. Stevens, R.K., Modern methods to measure air pollutants, in *Aerosols: Research, Risk Assessment, and Control Strategies,* Lee, S.D., ed. Lewis, Chelsea, MI, 69–95, 1986.
17. Febo, A., DiPalo, V., and Possanzini, M., The determination of tetraalkyl lead air by a denuder diffusion technique, *Sci. Total Environ.*, 48:187–194, 1986.
18. Dasgupta, P.K. et al., Continuous liquid phase fluorometry coupled to a diffusion scrubber for the real-time determination of atmospheric formaldehyde, hydrogen peroxide and sulfur dioxide, *Atmos. Environ.*, 22:946–963, 1988.
19. Lindgren, P.F. and Dasgupta, P.K., Measurement of atmospheric sulfur dioxide by diffusion scrubber coupled ion chromatography, *Anal. Chem.*, 61:19–24, 1988.

20. Scheibel, H.G. and Porstendorfer, J., Penetration measurements for tube and screen-type diffusion batteries in the ultra fine particle size range, *J. Aerosol Sci.*, 15:133–145, 1984.
21. Cheng, Y.S. et al., Characterization of diesel exhaust in a chronic inhalation study, *Am. Ind. Hyg. Assoc. J.*, 45:547–555, 1984.
22. Fuchs, N.A., *The Mechanics of Aerosols,* Pergamon, Oxford, U.K., 1964.
23. Hinds, W.C., *Aerosol Technology,* John Wiley & Sons, New York, 1999.
24. Lee, K.W. and Chen, H., Coagulation rate of polydisperse particles, *Aerosol Sci. Technol.*, 3:327–334, 1984.

CHAPTER 7

Impaction

Impaction is particle motion that results in the collection or collision of particles at a surface due to particle inertia. All impactors are based on the same principle — the separation of particles from the carrying air by inertial force. In this chapter, jet and impaction theory is discussed first, based on an impaction model for a free circular jet. Analyses of impaction efficiencies and particle cutsizes establish the fundamentals for the design and evaluation of various particle collectors, impacting types of samplers and air cleaners. Typical impactors for particle collection, such as cascade impactors and virtual impactors, are discussed. Applications of impaction theory can be found in later chapters of this text.

By completing this chapter, the reader will be able to

- Characterize an air jet, including single jets and coalescing jets:
  - Jet zones
  - Divergence angles
  - Expansion
- Determine the particle impaction efficiency of a single free air jet impacted on a vertical wall (which is the fundamental for most impactors)
- Determine the particle cutsize and impaction efficiency for impactors (coalescing jets) using the following approaches:
  - Direct impaction
  - Stokes number
- Design a cascade particle impactor
- Design a virtual impactor

7.1 AIR JETS

One of the most important factors affecting particle impaction is the air jet property in which the particles are suspended. An *air jet* is an air stream with a boundary at which the velocity is substantially different from the ambient air. There

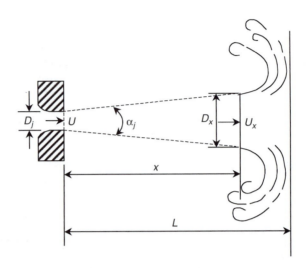

Figure 7.1 A free-isothermal circular air jet impact on a wall at an interception distance L.

are many types of air jets, depending on the discharging shapes and ambient boundary conditions. For example, if an air jet is not confined by a physical structure (such as a wall or a ceiling), it is considered a *free jet*. If air temperature in the jet is the same as the ambient air, it is considered an *isothermal jet*.

We will consider a free isothermal jet for the analysis of impaction theory. Figure 7.1 shows an isothermal free air jet, in which D_j is a characteristic diameter of the air jet and L is the distance between the discharging outlet and the interception wall. If the jet is circular, D_j is the diameter. If the jet is rectangular, D_j is the width of the shorter side. One reason for using the shorter side of a rectangular jet for its characteristic length is that the jet is much easier to expand at the shorter side and has behavior similar to a circular jet. An extreme example is a long, narrow slot air inlet, for which the expansion of the jet is primarily the jet width. In practice, most jets are circular, square, or nearly square in shape.

An air jet has four zones: core, transient, turbulent, and degradation,[1] defined as the distance from the discharging outlet (x) along the jet flow direction (Table 7.1). Particle impaction can occur effectively only in Zones I and II (primarily in Zone I) in most air sampling and air quality problems. Zones III and IV are important

Table 7.1 Jet Zone Classification and Characteristics

Jet Zone	Range	Major Characteristics
I: Core zone	$x \leq 4\,D_j$	Centerline velocity and temperature remain essentially unchanged from its discharging face
II: Transient zone	$4D_j < x \leq 25D_j$	Cross section of the jet expands and turbulence occurs
III: Turbulent zone	$25D_j < x \leq 100\,D_j$	Turbulent flow fully developed
IV: Degradation zone	$x > 100\,D_j$	Velocity and temperature degrade rapidly to the same as that of ambient air

Table 7.2   Divergence Angles for Four Types of Jets

Jet Type	Jet Diameter	Divergence Angle
Large single jet	$D_j \geq 12.5$ mm	$\alpha_j = 22 \pm 2°$
Coalescing large jets	$D_j \geq 12.5$ mm	$\alpha_j = 18 \pm 2°$
Small single jet	$D_j < 12.5$ mm	$\alpha_j = 14 \pm 2°$
Coalescing small jets	$D_j < 12.5$ mm	$\alpha_j = 10 \pm 2°$

for room air distribution but not for impaction or particle sampling. Therefore, characteristics of Zones III and IV will not be discussed in this chapter.

The angles of divergence, α_j, for free jets have been well defined for Zones I and II.[1, 2] According to the discharging diameter and its vicinity conditions, the diverging angle can be categorized for four types of jets:

- Large single jets
- Coalescing large jets
- Small single jets
- Coalescing small jets

Coalescing jets are a bank of parallel jets placed on the same plate. Generally, a single large free jet, such as an exhaust fan or a diffuser, expands much faster because of its large jet momentum and the pressure gradient around the jet. For coalescing jets, such as a bank of air nozzles in a wind tunnel, the pressure in the spaces among the jets "squeezes" the jets to expand at smaller angles. For very small air jets discharging into very small air spaces, such as those orifices in a cascade impactor plate, the divergence angle is even smaller.[3] Table 7.2 gives the divergence angles for these four types of jets.

From Figure 7.1, the average velocity of the jet at a distance, U_x, can be calculated by equating the flow rate discharged at the diameter D_j and at the diameter D_x, and solving for U_x, neglecting the effect of pressure change:

$$U_x = U_0 \left(\frac{D_j}{D_x} \right)^2 \tag{7.1}$$

and

$$D_x = D_j + 2x \tan\left(\frac{\alpha_j}{2} \right) \quad (x > 0) \tag{7.2}$$

where U_0 is the initial air velocity at the discharging outlet of the jet, α_j is the diverging angle of the jet, D_x is the diameter of the jet at the distance of x, and x is the distance from the discharging outlet to the bending point of the jet along the jet axis. The bending point is defined as $(D_x/2+H)$ from the impaction wall, which will be analyzed in the next section. The reason for defining x in this way is that the bending point is a critical parameter affecting the impaction and separation efficiency of particles, whereas the straight portion of the jet is of much less concern in impaction.

7.2 IMPACTION EFFICIENCY OF PARTICLES IN A FREE JET

An examination of the impaction area of a circular free jet where it approaches an interception wall results in a simplified impaction model, as shown in Figure 7.2. The jet starts to bend prior to reaching the interception wall and travels in a circular motion with a radius r, which is approximately equal to $D_x/2$. The jet at this space-shuttle launching pad was neither isothermal nor perfectly circular, but it offers an excellent visual to show how a jet expands and bends when impacted on an interception wall (Figure 7.2a). Figure 7.2b shows that the farthest exterior curved streamline reaches a separation contracta point, then bends upward. H is the height of the separation contracta, defined as the distance between the interception wall and the farthest exterior streamline at the contracta. The shaded area between two parallel lines is the particle separation zone; only in this zone do the flow streamlines change direction and particles separate from their original carrying flows. The separation zone consists of an array of quarter-circles with the same radius r and the same air velocity U_x. From Figure 7.2b, the distance between the jet discharging face and the bending point is

$$x = L - \left(\frac{D_x}{2} + H \right) \tag{7.3}$$

where L is the distance between the jet and the interception wall, D_x is the jet diameter at the distance x from the jet outlet, and H is the height of the contracta.

The terminal velocity of a particle relative to its carrying fluid is the product of its relaxation time and its acceleration (see Equation 4.32). When the particle moves

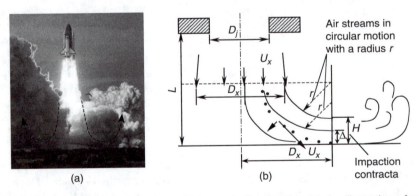

(a)   (b)

Figure 7.2   A simplified impaction model for a circular free jet. (a) An illustration of a jet impacting on the ground and bending with a diameter of approximately D_x. (b) Schematic of impaction of a free jet.

in the impaction region, as shown in Figure 7.2b, the acceleration of the particles is centrifugal acceleration U_x^2/r. The radial terminal velocity of a particle at any point relative to its carrying air, V_r, neglecting gravity, is

$$V_r = \tau \frac{U_x^2}{r} = \frac{\rho_p d_p^2 C_c U_x^2}{18 \eta \ r} \tag{7.4}$$

where U_x is defined in Equation 7.1. The time required for a particle to travel through the curved separation zone, t, is the ratio of distance of the quarter-circle, $\pi r/2$, to the jet velocity at the bending point, U_x:

$$t = \frac{\pi r}{2 U_x} \tag{7.5}$$

The radial distance that the particle traveled from the original streamline, Δ, can be obtained by combining Equation 7.4 and Equation 7.5:

$$\Delta = V_r t = \frac{\pi}{2} \tau U_x \tag{7.6}$$

At the end of each quarter-circle (i.e., at the contracta), each particle will have moved a radial distance of Δ from the original streamline toward the interception wall. If we assume that the particles that reach the interception wall will not be bounced back to the air stream, the impaction efficiency ξ_I is equal to the fraction of particles that have reached the interception wall. This fraction of interception is equal to the ratio of Δ to H:

$$\xi_I = \frac{\Delta}{H} \tag{7.7}$$

Assuming that the flow is incompressible in the impaction region, the volumetric flow rate at the diameter D_x must be equal to the volumetric flow rate at the separation contracta:

$$\frac{\pi D_x^2}{4} U_x = 2\pi D_x H U_x \tag{7.8}$$

Solving Equation 7.6 for H gives

$$H = \frac{D_x}{8} \tag{7.9}$$

Substituting Equation 7.9 into Equation 7.3 gives

$$x = L - \frac{5}{8} D_x \qquad (7.10)$$

Substituting Equation 7.10 into Equation 7.2 gives

$$D_x = \frac{D_j + 2L\tan(\alpha_j / 2)}{1 + 1.25\tan(\alpha_j / 2)} \qquad (7.11)$$

Substituting Equation 7.1, Equation 7.6 and Equation 7.9 through Equation 7.11 into Equation 7.7, and noting that the maximum impaction efficiency is 1, we have the impaction efficiency for a free jet:

$$\xi_I = \begin{cases} 4\pi\tau U_0 \left(\dfrac{D_j^2}{D_x^3} \right) = 4\pi\tau U_0 D_j^2 \left(\dfrac{1 + 1.25\tan(\alpha_j / 2)}{D_j + 2L\tan(\alpha_j / 2)} \right)^3 & (for\ d_p < d_{100}) \\[4mm] 1 & (for\ d_p \geq d_{100}) \end{cases} \qquad (7.12)$$

where U_0 is the initial air velocity from the jet, D_j is the diameter of the jet, α_j is the angle of divergence of the jet, L is the distance between the jet-discharging outlet and the interception wall, and d_{100} is the particle diameter at which the impaction efficiency reaches 100 percent. The d_{100} for given impaction conditions can be calculated using Equation 7.12 and Equation 4.57 by setting $\xi_I = 1$:

$$d_{100}\sqrt{C_c} = \frac{3}{D_j} \left(\frac{\eta}{2\pi\rho\ U_0} \right)^{\frac{1}{2}} \left(\frac{D_j + 2L\tan(\alpha_j / 2)}{1 + 1.25\tan(\alpha_j / 2)} \right)^{\frac{3}{2}} \qquad (7.13)$$

Because C_c is a function of d_{100}, it is not convenient to solve d_{100} explicitly. An empirical method can be used to estimate d_{100}, which is described in Equation 7.24. Theoretically, particles larger than d_{100} will have an impaction efficiency of 100%. Practically, some particles larger than d_{100} will remain in the air stream because of velocity distribution variation, turbulence, and particle bouncing. Figure 7.3 shows the particle impaction efficiency vs. particle size with a given set of conditions. An actual impaction efficiency curve diverges from the theoretical curve and reaches 100% only at a larger particle size than the d_{100}. Therefore, impaction efficiency calculated using Equation 7.12 is a theoretical estimation and can be considered the maximum impaction efficiency.

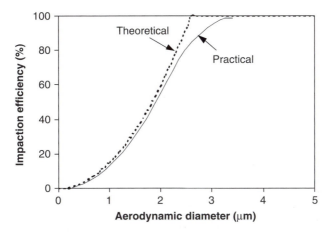

Figure 7.3   A typical impaction efficiency of an impactor as a function of particle aerodynamic diameter with a given set of jet parameters: D_j = 2 mm, L = 6 mm, U_0 = 20 m/s, and α_j = 10°.

Equation 7.12 shows that the impaction efficiency is independent of the particle concentration and is only a function of the particle relaxation time (i.e., particle size), jet configurations, and flow velocity. The impaction efficiency is proportional to the particle relaxation time and the initial air velocity of the jet. Because the particle relaxation time is proportional to the square of the particle size, the impaction efficiency increases rapidly with the increase of particle size (Figure 7.3). Impaction efficiency is inversely proportional to the jet diameter and the distance between the jet and the interception wall. Therefore, for a given size of particle, increasing U_0, decreasing D_j or L, or using a combination of these can improve impaction efficiency.

Impaction efficiency can be expressed in terms of flow rate of the air jet, Q_j, because the flow rate of the jet is often a given variable.

$$
\xi_I =
\begin{cases}
16\tau Q_j \left(\dfrac{1 + 1.25 \tan(\alpha_j / 2)}{D_j + 2L \tan(\alpha_j / 2)} \right)^3 & (for\ d_p < d_{100}) \\[4mm]
1 & (for\ d_p \geq d_{100})
\end{cases}
\tag{7.14}
$$

Example 7.1: *What is the impaction efficiency for 20 μm particles (in aerodynamic diameter) for an exhaust fan with a discharging diameter of 250 mm (see following figure)? Air leaves the fan outlet with a mean velocity of 7 m/s. Assume that the interception wall is 500 mm away from the fan and air is at standard conditions. What is the impaction efficiency for 10 μm particles?*

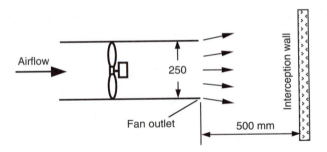

Interception wall for dust collection.

Solution: Because there are no other fans near the exhaust fan, and because the fan outlet diameter is greater than 12.5 mm, Table 7.1 shows that $\alpha_j = 22°$. The relaxation time for a 20 μm particle, noting that $C_{ca} = 1$, is

$$\tau = \frac{\rho_0 d_a^2 C_{ca}}{18\eta} = \frac{1,000 \times (20 \times 10^{-6})^2}{18 \times 1.81 \times 10^{-5}} = 0.00123\,(s)$$

$$\xi_I = 4\pi\tau U_0 D_j^2 \left(\frac{1 + 1.25\tan(\alpha_j / 2)}{D_j + 2L\tan(\alpha_j / 2)} \right)^3$$

$$= 4\pi \times 0.00123 \times 7 \times 0.25^2 \left(\frac{1 + 1.25\tan 11°}{0.25 + 2 \times 0.5\tan 11°} \right)^3$$

$$= 0.15 = 15\%$$

Because the impaction efficiency is proportional to the square of the particle size, the impaction efficiency for 10 μm particles is only 3.75%, or four times smaller than that for 20 μm particles.

Error Analysis: The simplified model, as described in Figure 7.2 and Equation 7.12, has three primary error sources:

- Air velocity distribution
- Turbulence intensity
- Particle bounce

The air velocity distribution of an air jet needs the solution of the Navier–Stokes equations for a particular jet geometry, which is often an impossible mathematical task. Take a circular jet, for example; the velocity profile along the centerline of the jet is illustrated in Figure 7.4. The mean jet velocity at a distance x varies along its radial position. The streamlines within the impaction region have some small eddies, which may prevent the particles from impacting or staying on the interception wall. A large D_x will have larger eddies, thus a lower impaction efficiency.

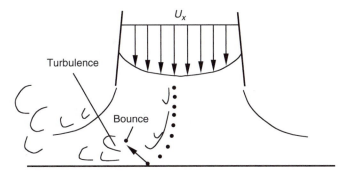

Figure 7.4 Airflow in the impaction region is very complicated, including complications in velocity variation, turbulent effect, and particle bounce.

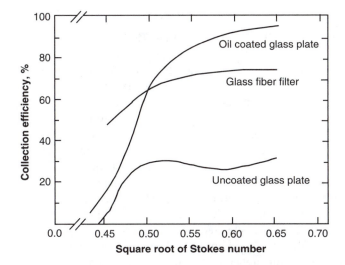

Figure 7.5 Effect of impaction surface on collection efficiency of a single-stage impactor. (From Rao, A.K. and Whitby, K.T., Non-ideal collection characteristics of inertial impactors — I: Single stage impactors and solid particles, *J. Aersol Sci.*, 9: 77–88, 1978. With permission.)

Particles that have impacted onto a wall may bounce back to the air stream, depending on the type of particle and the properties of the interception wall. Collection efficiencies for different interception wall surfaces in a single-stage impactor are demonstrated in Figure 7.5.[4] The collection efficiency varies largely with its surface adhesion property. Surfaces with rugged textures, such as a glass-fiber filter, have a much higher collection efficiency than smooth surfaces, such as a glass plate. An oil-coated surface can significantly improve particle collection efficiency. Many types of antibounce coatings are commercially available, including silicone stopcock, high-vacuum grease, petroleum jelly, and silicone oil. Impaction surfaces made up of porous materials or a coarse membrane filter saturated with a liquid can effectively reduce the particle bounce. The liquid serves as a reservoir and wicks through the deposited particles, maintaining an effective antibounce coating.

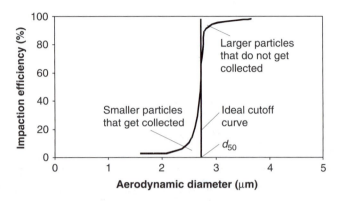

Figure 7.6   Schematic of ideal and actual particle cutsizes for an impactor.

7.3 CUTSIZE OF IMPACTED PARTICLES

From Figure 7.3, practical impaction efficiency approaches (but never reaches) zero for very small particles and approaches (but never reaches) 100% for very large particles. For most impactors, a complete curve of collection efficiency vs. particle size is not necessary. Impactors that have a sharp cutoff curve approach the ideal step-function efficiency curve (ideal from the standpoint of particle size classification), in which all particles greater than a certain aerodynamic size are collected and all particles less than that size pass through. The size in question is called the *cutsize, cutoff size, cutoff diameter, cut point, cutoff,* or d_{50}. The cutsize of particles impacted or collected is defined as the particle size at which the impaction efficiency is 50%. As a practical matter, most well designed impactors can be assumed to be ideal and their efficiency curves characterized by a particle cutsize d_{50}. As shown in Figure 7.6, this is equivalent to assuming that the mass of particles larger than the cutoff size that get through (the upper shaded area) equals the mass of particles below the cutoff size that are collected (the lower shaded area).

The recommended design criteria to produce a desirable sharp cutoff can be stated as follows.[5] The Reynolds number of the airflow in the jet nozzle throat should be between 500 and 3000. For a given cutoff size, the Reynolds number can be controlled by using multiple nozzles in parallel. The ratio of the separation distance (the distance between the nozzle and the impaction plate) to the jet diameter or width, L/D_j, should be 1 to 5 for circular nozzles and 1.5 to 5 for rectangular nozzles, with lower values preferred.

At 50% impaction efficiency, rewriting Equation 7.12 and noting that the slip correction factor C_c is a function of particle diameter, we can solve $d_{50}\sqrt{C_c}$:

$$d_{50}\sqrt{C_c} = \left(\frac{9\eta\, D_x^3}{4\pi\rho_p U_0 D_j^2} \right)^{\!\!1/2} \tag{7.15}$$

For particles larger than 0.1 μm, the slip correction factor has a linear relationship with its diameter. Assuming that the pressure effect on the slip correction factor is negligible under standard atmospheric conditions, Equation 4.22 can be rewritten in terms of d_{50}:

$$C_c = 1 + \frac{2.52\lambda}{d_{50}} \quad (d_{50} > 0.1 \ \mu m) \tag{7.16}$$

where λ is the mean free path of air. Substituting Equation 7.16 into Equation 7.15 and solving for d_{50} (noting that d_{50} cannot be a negative value) yields

$$d_{50} = 1.26\lambda + \frac{1}{2}\left(6.35\lambda^2 + \frac{9\eta D_x^3}{\pi \rho_p U_0 D_j^2}\right)^{\frac{1}{2}} \quad (d_{50} > 0.1 \ \mu m) \tag{7.17}$$

where D_x is defined in Equation 7.11. Under standard air conditions, $\lambda = 0.066$ μm, $\eta = 1.81 \times 10^{-5}$ Pa·s, and the standard density of the particle $\rho_p = 1000$ kg/m³.

$$d_{50} = 8.32 \times 10^{-8} + \frac{1}{2}\left(2.77 \times 10^{-14} + 5.185 \times 10^{-8} \frac{D_x^3}{U_0 D_j^2}\right)^{\frac{1}{2}} \quad (d_{50} > 0.1 \ \mu m) \tag{7.18}$$

where d_{50} calculated by Equation 7.18 is in meters.

An examination of Equation 7.18 reveals that a submicrometer cutsize requires a small D_x (which in turn requires a small air-jet diameter and a small impaction distance) operating at a high velocity. Practical limitations on these parameters for conventional impactors limit the smallest cutsizes to 0.2 to 0.3 μm. Micro-orifice impactors extend this limit to 0.06 μm by using large numbers of chemically etched nozzles as small as 50 μm in diameter. Another approach, which can have d_{50} values as small as 0.05 μm, is the low-pressure impactor. Low-pressure impactors operate at very low absolute pressures, 3 to 40 kPa. This has the effect of greatly increasing the slip correction factor according to Equation 4.27, which reduces the cutoff size. Volatilization of droplets at these low pressures can change the particle size. A large vacuum pump may be required because of the low exit pressure.

Example 7.2: *An impactor has 30 coalescing circular jets on a plate. Each jet has a discharging diameter of 1 mm. The sampling rate of the impactor is 30 L/min. The distance between the interception filter and the jet discharging outlet (L) is 3 mm. What is the particle cutsize, under normal atmospheric conditions?*

Solution: For coalescing small jets, from Table 7.2, $\alpha_j = 10°$. From Equation 7.11

$$D_x = \frac{D_j + 2L\tan(\alpha_j / 2)}{1 + 1.25\tan(\alpha_j / 2)} = 0.00137 \ (m)$$

The jet velocity at the outlet can be calculated from the flow rate for each jet, which is $Q/30$, and the jet opening area, which is $\pi D_j^2/4$:

$$U_0 = \frac{4Q}{30\pi D_j^2} = \frac{4 \times 0.0005}{30\pi \times 0.001^2} = 21.2 \ (m/s)$$

Substituting values of D_x, D_j, and U_0 into Equation 7.18 gives

$$d_{50} = 8.32 \times 10^{-8} + \frac{1}{2}\left(2.77 \times 10^{-14} + 5.185 \times 10^{-8} \frac{0.00137^3}{21.2 \times 0.001}\right)^{\frac{1}{2}}$$

$$= 1.34 \times 10^{-6} \ (m) = 1.34\mu m$$

7.4 STOKES NUMBER APPROACH

Similar to the Reynolds number for fluids, the Stokes number (Stk) is a dimensionless number used to characterize particle curvilinear motion and impaction. It is the ratio of the stopping distance of a particle to a characteristic length of the obstacle or the flow path:

$$Stk = \frac{S}{L} = \frac{\tau U_0}{L} \quad \text{(for } Re_{p0} < 1.0) \tag{7.19}$$

where S is the stopping distance, τ is relaxation time, U_0 is the undisturbed airflow velocity before the impaction occurs, and L is the characteristic length.

The initial particle Reynolds number Re_{p0} can be calculated from Equation 4.8. The characteristic length L can be different for different applications. Therefore, the Stokes number is application specific. For a particle impactor with a nozzle diameter of D_j and a jet that is impacting on a wall perpendicular to the jet, the characteristic length is defined as its radius, $Dj/2$. We use the subscript $_I$ to distinguish this specific Stokes number for impactors from other applications.

$$Stk_I = \frac{\tau U_0}{D_j / 2} = \frac{\rho_p d_p^2 U C_c}{9\eta D_j} \tag{7.20}$$

For a jet with a rectangular nozzle, the characteristic length should be the half-width of the nozzle instead of the radius in Equation 7.20.

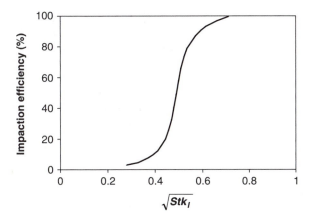

Figure 7.7 Typical impaction efficiency curve vs. the square root of the Stokes number.

Theoretically, particles with the same Stk_I have the same impaction efficiency, just as flows with the same Reynolds number have the same flow characteristics. In practice, however, the impaction efficiency cannot be explicitly expressed as a function of Stk_I because of the complexity of flow conditions around the nozzle and other variables in an impaction process. Impaction efficiency curves for impactors are usually plotted vs. the square root of the Stk_I, which is directly proportional to the particle size (Equation 7.20). To obtain such a curve, a series of monodisperse particle impaction efficiencies must be measured. By using a nondimensional variable, Stk_I, several design parameters and flow conditions can be varied to obtain the desired impaction efficiency (Figure 7.7).

The recommended design criteria for impactors that produce a sharp cutoff can be described as follows.[5] The Reynolds number of gas flow in the nozzle throat should be between 500 and 3000. For a given cutoff size, the Reynolds number can be controlled by using multiple nozzles in parallel. The ratio of impaction distance (from the nozzle face to the impaction interface) to the nozzle diameter should be 1 to 5 for circular nozzles and 1.5 to 5 for rectangular nozzles, with lower values preferred. For designs meeting the recommended criteria, the Stokes numbers for 50% impaction efficiency (Stk_{50}) for impactors are given in Table 7.3. All circular nozzles, regardless of diameter, will have the same Stokes number. Also, all rectangular nozzles meeting the design criteria, regardless of size, will have the same

Table 7.3   Design Criteria and Stokes Number for 50% Impaction Efficiency for Circular and Rectangular Impactor Nozzles

Impactor Type	Re	L/D_j[a]	Stk_{I50}
Circular nozzle	500–3000	1–5	0.24
Rectangular nozzle	500–3000	1.5–5	0.59

[a] L = distance between the nozzle outlet and the interception wall.

Stokes number. These design criteria substantially simplify the analysis of impaction efficiency and are sufficiently accurate for most practical applications.

When the criteria in Table 7.3 are satisfied, the particle cutsize can be derived from Equation 7.20. Let d_{50} be d_p and Stk_I be Stk_{I50}. Rearranging gives

$$d_{50}\sqrt{C_c} = \left(\frac{9\eta D_j Stk_{I50}}{\rho_p U}\right)^{1/2} \tag{7.21}$$

For many applications, the jet flow rate Q is often a known, and thus, Equation 7.21 can be also written as follows:

$$d_{50}\sqrt{C_c} = \left(\frac{9\pi\eta D_j^3 Stk_{I50}}{4\rho_p Q}\right)^{1/2} \tag{7.22}$$

For rectangular jet impactors with jet width W and length L

$$d_{50}\sqrt{C_c} = \left(\frac{9\eta W^2 L\, Stk_{I50}}{\rho_p Q}\right)^{1/2} \tag{7.23}$$

Because C_c is a function of d_{50}, Equations 7.21 through 7.23 cannot be solved conveniently for particle cutsize. Additionally, particle slip correction factor is affected by the air pressure downstream of the jet nozzle. For most conventional impactors, the effect of pressure change downstream of the jet can be negligible. For example, a jet nozzle with an air velocity of 30 m/s has a pressure decrease of 540 Pa, which has little effect on the slip correction factor. Therefore, for conventional impactors, d_{50} can be estimated using the following empirical equation:[5]

$$d_{50} = d_{50}\sqrt{C_c} - 0.078 \quad \text{(for } d_{50} \text{ in } \mu m) \tag{7.24}$$

This equation is accurate within 2% for $d_{50} > 0.2$ μm and pressures from 91 to 101 kPa. Impactor calibration (the cutoff diameter) is usually given in terms of aerodynamic diameters. If an impactor is operated at a flow rate that differs from its calibration flow rate, its cutoff size can be adjusted using Equation 7.22 or Equation 7.23.

Example 7.3: *Estimate the particle cutsize of the data given in Example 7.2, using the Stokes number approach.*

Solution: We first need to verify that the impactor jet in Example 7.2 satisfies the design criteria.

$$\text{Re} = \frac{\rho U D_j}{\eta} = \frac{1.2 \times 21.2 \times 0.001}{0.0000181} = 1406$$

Because of the Reynolds number for air flow, $500 < Re = 1406 < 3000$, and the ratio of $1 < L/D_j = 3 < 5$, the design criteria in Table 7.3 are satisfied. From Table 7.3, $Stk_{I50} = 0.24$. The particle cutsize can be estimated using Equation 7.21 and Equation 7.24:

$$d_{50}\sqrt{C_c} = \left(\frac{9 \times 0.0000181 \times 0.001 \times 0.24}{1000 \times 21.2} \right)^{1/2}$$

$$= 1.36 \times 10^{-6} \ (m) = 1.36 \ (\mu m)$$

$$d_{50} = d_{50}\sqrt{C_c} - 0.078 = 1.36 - 0.078 = 1.28 \ (\mu m)$$

The cutsize calculated in Example 7.2 is 1.34 μm, representing a 4.5% difference from the cutsize calculated using Stokes number approach.

7.5 CASCADE IMPACTORS

Because particle cutsizes of a given particle population depend primarily on the jet flow velocity, the jet diameter, and the distance of the impacting wall to the jet, these parameters of an impactor can be designed to have a specific particle cutsize. Several impactors with different particle cutsizes can measure the particle size distribution. Impactors can be arranged in parallel or serial. The use of parallel impactors is not common because of the difficulty in multiflow rate control. Serial impactors are more common in practice and are usually referred to as *cascade impactors*.

A cascade impactor is composed of several impactors arranged in series in order of decreasing particle cutsizes, as illustrated in Figure 7.8a. Particle-laden air passes through the nozzle and impinges on a collection plate oriented perpendicular to the nozzle axis. The airflow is laminar, and particles within the nozzle are accelerated to a nearly uniform velocity. At the nozzle exit, the streamlines of the air are deflected sharply by the collection plate. Larger particles are propelled across the air streamlines and deposited on the plate. Smaller particles follow the streamlines more closely and remain suspended in the air. In this serial arrangement, only one airflow rate control is needed. With the same airflow rate, the first stage, with a larger nozzle opening, allows a low air-jet velocity, hence a larger particle cutsize. The second stage has a smaller nozzle opening, allowing a higher air-jet velocity, hence a smaller particle cutsize than the first stage. Particle trajectories in a single-jet cascade impactor are illustrated in Figure 7.8a. The minimum size collected by an individual stage depends on the jet diameter and the air-stream velocity in the jet. Typically,

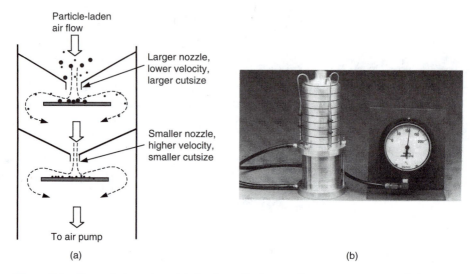

Particle-laden
air flow

Larger nozzle,
lower velocity,
larger cutsize

Smaller nozzle,
higher velocity,
smaller cutsize

To air pump

(a)   (b)

Figure 7.8   Cascade impactors. (a) A schematic diagram for a two-stage cascade impactor.
(b) An Andersen eight-stage cascade impaction air sampler.

the collection of smaller particles is achieved by using smaller-diameter jets with higher jet velocities. In operation, air enters at the top, passes through each of the impactor stages, and is exhausted through a vacuum air pump.

Figure 7.8b shows an example of a multistage cascade impactor. Each impactor stage consists of an array of coalescing jets followed by a collection plate. Successive stages are designed to collect smaller particles. Particles that penetrate the last impaction stage are collected by a backup filter. Airflow passing through the impactor is generated by means of an air pump and controlled by a valve or a critical orifice downstream of the backup filter.

Particle samples collected from cascade impactors are used to determine the distribution of particle mass and chemical composition with respect to particle size. When cascade impactor samples are analyzed gravimetrically, they provide an aerosol mass distribution. When cascade impactor samples are analyzed chemically, they yield species vs. particle size distributions. Simultaneous data on particle size and mass or chemical composition are important for assessing health effects and particle transport in indoor environments or in the atmosphere. Cascade impactors were introduced by the end of the second World War and have been used widely ever since.[6] More complete discussions and applications of the impacting instruments can be found in ACGIH and in Lodge and Chan.[7,8]

Cascade impactors have been designed for use with high-volume samplers. Many use multiple jets per stage to permit larger collection of particles at larger flow rates. Low flow rate impactors are used for personal and ambient sampling. Impactors are also used for stack sampling and viable particle sampling. The overall size range covered by an impactor depends on its design. A typical impaction efficiency graph for a cascade impactor is illustrated in Figure 7.9, in which the particle cutsizes for stages 1 through 6 are approximately 20, 10, 5, 1, 0.4, and 0.2 μm, respectively.

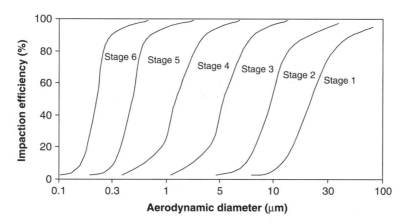

Figure 7.9   Illustration of impaction efficiencies for a typical six-stage cascade impactor. The cutsizes are 20, 10, 5, 1, 0.4, and 0.2 μm, respectively, for six stages.

Conventional cascade impactors can be designed to collect particles as small as 0.4 μm. Low-pressure and micro-orifice impactors can collect particles as small as 0.05 μm. Some impactors, such as the Andersen microbial sampler, are designed to collect very large particles, as much as 30 μm in diameter. Rotary impactors have been used with high efficiency to sample ambient air particles as large as 250 μm. In these rotary impactors, a rod moving through the ambient air impacts and collects particles larger than the characteristic cutsize for the sampler.

Impactors differ in the nature of the particle collection surface. Most particles are collected on a solid plate located immediately downstream of the accelerating jet, as shown in Figure 7.8a. However, unless the collection plate is greased, particles may bounce and be reentrained in the flow. To avoid this problem, the virtual impactor uses a nearly stagnant airflow to transport the size-fractionated sample to a filter. Although a virtual impactor does not have an impaction surface, the airflow streamlines are similar to those in conventional impactors.

7.6 VIRTUAL IMPACTORS

Virtual impactors, as the name implies, do not have a physical collection plate (Figure 7.10). Instead, an axial probe is placed below the impactor jet. Only a small fraction of the flow passes through the probe; the majority of the flow bends around the tip of the probe to pass to the next stage. The streamlines above the probe tip resemble those of a conventional impactor, and the particles are separated by size into the two air streams. One is the minor flow, which passes through the probe; the other is the major flow, which bypasses the probe. The minor flow through the probe carries with it all of the large particles from the total sample flow plus the small particles from the minor flow. The major airflow that bypasses the probe contains only smaller particles. Both the minor and major flows exit the impactor, and particles are collected by filtration.

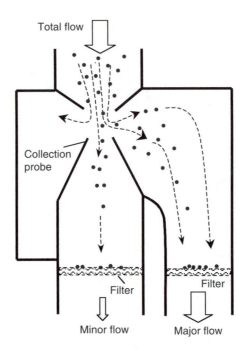

Figure 7.10 A schematic diagram of a virtual impactor.

A major advantage of virtual impactors is that they are not subject to errors resulting from particle bounce or reentrainment, and grease coatings are not required. Particles may be collected on whatever filter medium is best suited for the analysis to be performed. A limitation is that unless they are carefully designed and constructed, they are subject to significant wall losses for liquid particles near the cutpoint size. Experimentally determined criteria for minimizing these losses are given by Loo and Cork.[9] For particle sizes below the cutsize diameter, collection efficiencies reach a minimum value equal to the fraction of the total flow passing through the receiving probe. The cutoff diameter decreases as the fraction of the flow through the receiving probe is increased. Wall losses are most significant at the cutoff diameter. A critical factor in minimizing wall losses is the radius of curvature at the inlet of the receiving probe. John and Wall found that alignment of the jet and the receiving probe is critical and that deviations of more than 0.06 mm in concentricity can increase wall losses and affect the cutsize.[10]

The first type of virtual impactor was the aerosol centripeter, introduced by Hounam and Sherwood.[11] The most widely used virtual impactor is the dichotomous sampler introduced by Conner and developed by Dzubay and Stevens and Loo et al.[12-14] It operates at a sample rate of 16.7 L/min, or 1m³/hr, with aerosol collection onto two 37 mm diameter filters. The commercially available instrument provides a fine particle cut at 2.5 μm, although earlier versions had a 3.5 μm cut. The unit is generally operated with a PM_{10} sampler inlet, and it is most frequently used for ambient air monitoring. Calibration curves are given by McFarland et al. and by John and Wall.[10, 15]

Several other virtual impactors have been developed. Solomon et al. have developed a high-volume virtual dichotomous sampler that operates at 500 L/min and employs 100-mm diameter filters.[16] This sampler has the advantage of providing larger sample volumes, permitting analyses of trace species, or facilitating collection in cases of low airborne concentrations. Chen et al. have developed a virtual impactor that uses a particle-free air stream to eliminate fine particle collection in the minor (coarse) particle flow.[17, 18] Novick and Alvarez have designed a three-stage virtual impactor.[19] Noone et al. have developed a counterflow virtual impactor for separate sampling of cloud droplets and interstitial aerosol.[20] Theoretical analyses of virtual impactors are given by Forney and by Marple and Chien.[21–23]

DISCUSSION TOPICS

1. Why do coalescing jets have a smaller diverging angle than single jets of the same size?
2. Can you name a real-world example of a free jet impaction (as shown in Figure 7.2) from your own observation?
3. Why is the d_{100}, the particle size at which the impaction efficiency reaches 100%, usually larger than its theoretical values calculated using Equation 7.13? Is it possible to achieve the d_{100} value? If so, how?
4. What design parameters can be effective for the collection of very small particles (e.g., smaller than 0.1 μm) using the principle of impaction?
5. What are the major applications of particle impaction? Can impaction be an effective method of large-volume air cleaning?
6. During a particle sampling, if long lines of tubes or ducts are involved before the instrumentation, what are the critical design factors for these tubes and ducts to reduce impaction loss?
7. What are the major advantages and limitations for cascade impactors and virtual impactors? How can you overcome these limitations?

PROBLEMS

1. In an attempt to reduce dust emission to the atmosphere from a dusty building, a wall was placed 2 m from the air outlet, which is 1.2 m in diameter. The air jet has a mean velocity of 7 m/s and is perpendicular to the wall. The average diameter of mass of the particles from the air outlet is 12 μm in aerodynamic diameter. What is the particle removal efficiency of this impacting wall?
2. A rectangular air duct in a house has a cross section of 0.4 m by 0.4 m. There is a 90° elbow in the duct line with a centerline radius of 0.5 m. The airflow rate in the duct is 800 l/s. A 10 μm spherical particle with unit density travels along the centerline prior to entering the elbow. How far away from the centerline does the particle move at the exit of the elbow?
3. In order to have an 80% collection efficiency for 4 μm particles in aerodynamic diameter, a single small jet impactor (thus the diverging angle is 14°) is designed. The air velocity at the jet outlet is 10 m/s. The distance between the jet outlet and

the impaction wall is three times the nozzle diameter. What should the nozzle diameter be?

4. A single-stage impactor has 40 coalescing jets. Each jet nozzle has a diameter of 0.8 mm. The air pump has an airflow capacity of 20 l/min. The impaction wall is 3 mm from the nozzle outlet. What is the collection efficiency for particles 1 μm in aerodynamic diameter?

5. In problem 4, to what value should the airflow rate be adjusted to achieve a particle cutsize of 1 μm in aerodynamic diameter?

6. One stage of a cascade impactor has 400 coalescing holes, each with a diameter of 0.35 mm. The airflow rate of the sampling pump is 56.6 l/min. The impaction wall is 1.2 mm from the nozzle outlet. What is the cutsize of particles in aerodynamic diameter?

7. A cascade impactor has a cutsize of 3 μm in aerodynamic diameter when the sampling rate is 28.3 l/min. What will the particle cutsize be if the sampling rate is increased to 56.6 l/min?

8. What is the particle cutsize for a rectangular nozzle impactor under standard atmospheric conditions? The nozzle is square with a side length of 1 mm. The distance between the nozzle outlet and the impaction wall is 3 mm. The average velocity of air discharged from the nozzle is 15 m/s. Particles are in aerodynamic diameter.

9. One stage of a cascade impactor has 400 coalescing rectangular holes, each with a side length of 0.35 mm. The airflow rate of the sampling pump is 56.6 l/min. The impaction wall is 1.5 mm from the nozzle outlet. What is the cutsize of particles in aerodynamic diameter?

10. Assume that all design criteria in Table 7.3 are satisfied. Plot a graph of d_{50} (in aerodynamic diameter) vs. air jet velocity U, with nozzle diameters of 0.3, 1, 2, 3, and 4 mm, where d_{50} is in μm and on the y-axis, and where U is in m/s and on the x-axis. The ranges are 0.1 μm $< d_{50} <$ 20 μm, and $U <$ 50 m/s.

11. Compare Equation 7.18 and Equation 7.24 in terms of cutsize errors with respect to nozzle diameter and jet velocity. Use the data from problem 10.

12. Dust-laden air flows through a single-stage impactor with 40 coalescing jets. The particle size distribution has been measured as in Table 7.4. The impactor nozzle diameter is 1.5 mm, and the air velocity is 10 m/s at the nozzle outlet. The impacting wall is 5 mm from the nozzle outlet. Assuming that all particles are spherical and have unit density, find the following:
   a. What is the cutsize of the impactor?
   b. What is the percentage of particle mass collected that has a diameter smaller than the cutsize?

Size (μm)	Concentration (particle/ml)
0.3	30
1	10
4	7
10	1

REFERENCES

1. ASHRAE, *Handbook of Fundamentals*, American Society of Heating, Refrigeration and Air-conditioning Engineers, Atlanta, GA, 1997, ch. 31.

2. ISO, *International Standard ISO 5167: Measurement of Fluid Flow by Means of Orifice Plates, Nozzles and Venturi Tubes Inserted in Circular Cross-Section Conduits Running Full*, 1st ed., The International Organization for Standardization, Geneva, Switzerland, 1980.

3. McElroy, G.E., *Air Flow at Discharge of Fan-Pipe Lines in Mines*, U.S. Bureau of Mines Report of Investigation, Washington DC, 1943.

4. Rao, A.K. and Whitby, K.T., Non-ideal collection characteristics of inertial impactors — I: Single stage impactors and solid particles, *J. Aerosol Sci.*, 9: 77–88, 1978.

5. Hinds, C.W., *Aerosol Technology*, John Wiley & Sons, New York, 1999, 127.

6. May, K.R., The cascade impactor: an instrument for sampling coarse aerosols, *J. Sci. Instru.*, 22:187–195, 1945.

7. ACGIH, *Air Sampling Instruments — For Evaluation of Atmospheric Contaminants*, 8th ed., American Conference of Governmental Industrial Hygienists, Cincinnati, OH, 1995.

8. Lodge, J.P. and Chen, T.L., Cascade impactor sampling and data analysis, *Amer. Ind. Hyg. Assoc.*, Akron, OH, 1986.

9. Loo, B.W. and Cork, C.P., Development of high-efficiency virtual impactors, *Aerosol Sci. Technol.*, 9:167–176, 1988.

10. John, W. and Wall, S.M., Aerosol testing techniques for size-selective samplers, *J. Aerosol Sci.*, 14:713–727, 1983.

11. Hounam, R.F. and Sherwood, R.J., The cascade centripeter: a device for determining the concentration and size distribution of aerosols, *Am. Ind. Hyg. Assoc. J.*, 26:122–131, 1965.

12. Conner, W.D., An inertial-type particle separator for collecting large samples, *J. Air Pollut. Control Assoc.*, 16: 35–38, 1966.

13. Dzubay, T.G. and Stevens, R.D., Ambient air analysis with dichotomous sampler and X-ray fluorescence spectrometer, *Environ. Sci. Technol.*, 9:663–668, 1975.

14. Loo, B.W., Jaklevic, J.M., and Goulding, F.S., Dichotomous virtual impactors for large-scale monitoring of airborne particulate matter, in *Fine Particles, Aerosol Generation, Measurement, Sampling and Analysis*, Liu, B.Y.H., ed., Academic Press, New York, 1976, 311–350.

15. McFarland, A.R. and Bertch, R.W., Particle collection characteristics of a single-stage dichotomous sampler, *Environ. Sci. Technol.*, 12:379–382, 1978.

16. Solomon, P.A., Moyers, J.L., and Fletcher, R.A., High-volume dichotomous virtual impactor for the fractionation and collection of particles according to aerodynamic size, *Aerosol Sci. Technol.*, 2:455–465, 1983.

17. Chen, B.T., Yeh, H.C., and Cheng, Y.S., Performance of a modified virtual impactor, *Aerosol Sci. Technol.*, 5:369–376, 1986.

18. Chen, B.T. and Yeh, H.C., An improved virtual impactor: design and performance, *J. Aerosol Sci.*, 18:203–214, 1987.

19. Novick, V.J. and Alvarez, J.L., Design of a multistage virtual impactor, *Aerosol Sci. Technol.*, 6:63–70, 1987.

20. Noone, K.J. et al., Design and calibration of a counterflow virtual impactor for sampling of atmospheric fog and cloud droplets, *Aerosol Sci. Technol.*, 8:235–244, 1988.

21. Forney, L.J., Aerosol fractionator for large-scale sampling, *Rev. Sci. Instrum.*, 47:1264–1269, 1976.
22. Forney, L.J., Ravenhall, D.G., and Lee, S.S., Experimental and theoretical study of a two-dimensional virtual impactor, *Environ. Sci. Technol.*, 16:492–497, 1982.
23. Marple, V.A. and Chien, C.M., Virtual impactors: a theoretical study, *Environ. Sci. Technol.*, 14:976–985, 1980.

CHAPTER 8

Sampling Efficiency

Sampling efficiency, also referred to as *aspiration efficiency*, is an essential element in air-quality studies. In the measurement of airborne contaminants, it is critical to collect a sample of air containing representative contaminants in the air of concern, that is, contaminant concentration and size distribution in the sampled air must be the same as that of the air of concern. Factors affecting sampling efficiency vary, depending on the types of contaminants and size distribution. Sampler type and flow-field conditions are more important for particle sampling than for gas sampling. Material absorption characteristics are often more influential for gas sampling than for particle sampling. In this chapter, principles of air sampling efficiency and the factors affecting it will be discussed.

By completing this chapter, the reader will be able to

- Determine isokinetic sampling conditions, such as sampling rate or velocity and sampler diameter with respect to flow field
- Determine the particle sampling efficiencies under three anisokinetic sampling conditions:
  - Superisokinetic — When sampling velocity U_s is greater than flow field velocity U_0
  - Subisokinetic — $U_s < U_0$
  - Misalignment
- Determine calm-air sampling conditions, such as sampling rate or velocity and sampler diameter
- Perform an in-duct air sampling following corresponding standards
- Measure aerial pollutant emission from mechanically ventilated buildings

8.1 AIR SAMPLER AND SAMPLING EFFICIENCY

An *air sampler,* also referred to as a *sampling head,* is the physical inlet that allows air and airborne contaminants to enter an instrument or a system of concern, such as a human respiratory system or an analytical instrument. Figure 8.1 shows

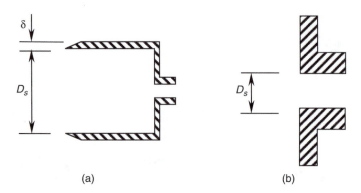

Figure 8.1   Two typical types of airborne particle samplers. (a) A thin-wall sampler commonly used for concentration measurement. (b) A blunt sampler to simulate the inhalation of a human respiration system.

two types of typical air samplers: a thin-wall sampler and a blunt-wall sampler. *Thin-wall samplers* have been widely used in particle instruments to quantify the particle concentration. Thin-wall samplers derive their name from the fact that the thickness of the sampler wall, δ, is much smaller than the diameter of the sampler D_s, typically $D_s/\delta > 10$. Many thin-wall samplers are sharp-edged to minimize disturbance to the airflow.

In many cases, occupational health and safety are the primary concern when airborne contaminants are inhaled by a human. Thus, a *blunt-wall sampler* is sometimes used to simulate a human head, mouth, or nose. Airflow around a blunt-wall sampler is substantially distorted compared to the surrounding airflow field, thus the sampling efficiency is more difficult to determine.

The first step in the study of airborne contaminants is to collect representative samples for analysis, whether concentration, biological, or chemical analysis. The sample must accurately represent the characteristics of the particles, such as concentration and size distribution in the concerned air. Because of inertia, an airborne particle tends to stay in its original direction of travel when airflow direction changes. This inertia effect results in a separation of the particle trajectory from the airflow. The departure of the particle from its original flow causes the difference in particle concentrations in the air, which in turn causes the error in measurement of particle concentrations. To describe this error in particle sampling, it is important to define the sampling efficiency of a sampler.

Particle sampling efficiency, also referred to as *aspiration efficiency*, ξ, is defined as the ratio of the particle concentration at the entrance of the sampler, C_s, and the particle concentration in the ambient air, C_0.

$$\xi = \frac{C_s}{C_0} \tag{8.1}$$

The value of ξ can vary widely in magnitude. The purpose of air sampling is to achieve a unit sampling efficiency. Figure 8.2 demonstrates how the particle con-

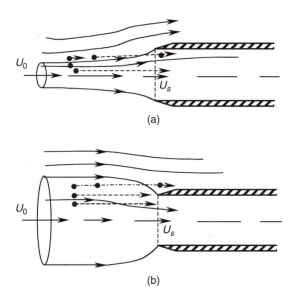

Figure 8.2 Particle concentration may vary between the air of concern and the sampled air due to the change of flow direction and the inertia effect. (a) Extra particles get into the sampler. (b) The sampled air has lost some particles.

centrations in the ambient air and in sampled air may be different for a tube sampler, thus changing the sampling efficiency. When the air velocity in the sampler, U_s, is different from the ambient air velocity U_0 and approaches the sampler head, the air streamlines change direction. When $U_s < U_0$, the air streamlines diverge from the sampling head and particles that have departed from the diverged air get into the sampler (Figure 8.2a), causing a higher particle concentration in the sampled air than in the ambient air. When $U_s > U_0$, the air streamlines converge into the sampling head, and particles that have departed from the converged air get away from the sampling head (Figure 8.2b), causing a lower particle concentration in the sampled air than in the ambient air. Particle sampling efficiencies vary with sampler type, airflow conditions, and sampling techniques.

8.2 ISOKINETIC SAMPLING

Thin-wall particle samplers are most commonly used in particle measurement instruments. The most important concept and practice in particle sampling is perhaps *isokinetic sampling*. Isokinetic sampling is a procedure used to ensure that the sampled particles are representative of those of concern in terms of concentration and size distribution. A thin-wall sampler has the least disturbance of airflow and can attain an isokinetic sampling more easily than other types of samplers. As shown in Figure 8.3, when the axis of the sampler is in alignment with and facing the free flow direction, and when the air velocity in the free-flow field is equal to the velocity at the face of the sampling head, the sampling is isokinetic. In an isokinetic sampling

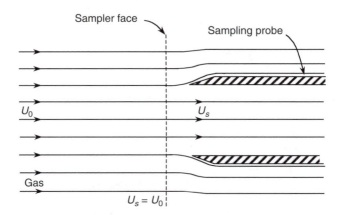

Figure 8.3   Isokinetic sampling with a thin-wall sampler. There is no change of flow direction
at the face of the sampler, thus there is no loss or gain of particles from beyond
the sampled air.

process, there is no change of flow direction at the face of the sampler, thus there
is no loss or gain of particles from beyond the sampled air.

To ensure isokinetic sampling, the following condition must be satisfied:

$$U_s = U_0 \qquad (8.2)$$

where U_s is the air velocity at the face of the sampling head and is called the *sampling
velocity,* and U_0 is the air velocity of the free-flow field.

In practice, the sampling rate is often a constant determined by the air pump,
but the air velocity of the free flow may vary widely. Therefore, it is important to
select a sampler head diameter that can attain the same sampling velocity as the
free-flow velocity, thus attaining an isokinetic sampling. The desirable sampler head
diameter for a specific sampling can be determined from the sampling rate and the
free-flow velocity. The sampling rate Q_s is

$$Q_s = \frac{\pi D_s^2}{4} U_s \qquad (8.3)$$

where D_s is the diameter of the sampler and U_s is the velocity at the sampler face
and equal to the velocity of the free-flow field. Rearranging Equation 8.3 and
substituting it into Equation 8.2 gives

$$D_s = \left(\frac{4Q_s}{\pi U_0} \right)^{1/2} \qquad (8.4)$$

Equation 8.4 shows that, with a given sampling rate and a free-flow air velocity, the diameter of the sampler can be designed accordingly to realize isokinetic sampling. It is practical to have an array of samplers with different diameters for a single sampling rate to achieve an isokinetic sampling. Alternatively, the sampling rate may be changed (by adjusting the flow rate of the sampling pump) to achieve an isokinetic sampling. Many air sampling systems are equipped with variable flow-rate controllers. Care should be taken in using flow-rate controllers, because the flow rate can vary with changes in power supply and pressure fluctuation across the sampling head.

Example 8.1. *Airflow velocity at the center of a duct is 2 m/s. A particle sampler has a variable flow-rate controller with a range of 0 to 20 l/min. The diameter of the sampling head is 5 mm. To what should the flow rate be adjusted to achieve an isokinetic sampling? Assume that the sampling head is facing and aligned with the free flow.*

Solution: Because the sampling flow rate can be adjusted, from Equation 8.4

$$Q_s = \frac{\pi D_s^2}{4} U_0 = \frac{\pi \times 0.005^2}{4} \times 2 = 0.00003925 \left(\frac{m^3}{s} \right) = 2.355 \left(\frac{l}{min} \right)$$

8.3 ANISOKINETIC SAMPLING

Sampling procedures that fail to sample isokinetically are called *anisokinetic sampling*. There are three general forms of anisokinetic sampling: superisokinetic, subisokinetic, and misalignment (Figure 8.4).

8.3.1 Superisokinetic Sampling

When the sampling velocity is higher than the free-flow velocity (i.e., $U_s > U_0$), it is called *superisokinetic sampling*. At superisokinetic sampling conditions, the free-flow streamlines converge, and some particles originally in the sampled air are lost outside the sampler because of their inertia effect (Figure 8.4a). The bigger the particles, the higher the loss will be. Therefore, the sampling efficiency for superisokinetic sampling is smaller than 1.

8.3.2 Subisokinetic Sampling

When the sampling velocity is lower than the free-flow velocity (i.e., $U_s < U_0$) it is called *subisokinetic sampling*. At subisokinetic sampling conditions, the free-flow streamlines diverge, and some particles originally not in the sampled air are impacted into the sampler because of their inertia effect (Figure 8.4b). Therefore, the sampling efficiency for subisokinetic sampling is greater than 1.

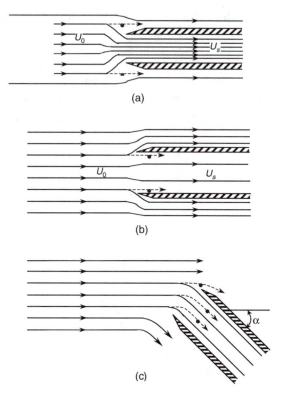

Figure 8.4 Three general forms of anisokinetic sampling. (a) Superisokinetic: $U_s > U_0$. (b) Subisokinetic: $U_s < U_0$. (c) Misalignment: $\alpha \neq 0$.

When the axis of the sampler is not aligned with the free flow directions, (i.e., $\alpha \neq 0$), it is called *misalignment sampling*. At misalignment sampling conditions, the free-flow direction changes at the face of the sampler with a misalignment angle α, and thus changes the sampling efficiency. Misalignment sampling will reduce the sampling efficiency, compared with alignment sampling. Therefore, under superisokinetic sampling conditions, misalignment sampling will further reduce the sampling efficiency, that is, the sampling efficiency will be much smaller than 1. Under subisokinetic sampling conditions, misalignment sampling can compensate for increased sampling efficiency. At an appropriate misalignment angle for a subisokinetic sampling, the sampling efficiency can be unity. Theoretically, the combination of subisokinetic and misalignment sampling conditions can achieve an equivalent isokinetic sampling. Practically, the misalignment of a sampler may change the flow patterns around the sampler and disturb the free-flow streamlines. These changes can introduce substantial errors in sampling efficiency. Therefore, one should use a true isokinetic sampling process rather than this "equivalent" isokinetic sampling approach, unless its use is necessary.

Sampling efficiency depends on particle size. When flow direction changes near the sampling head, larger particles tend to keep the original flow direction, while smaller ones tend to follow the changed airflow. The sampling efficiency for isoki-netic sampling is unity for all sizes of particles. However, sampling efficiencies for anisokinetic sampling are different for different sizes of particles. Before analyzing the sampling efficiencies for anisokinetic sampling, a useful number to characterize the sampler and particles — the *Stokes inlet number* — must be introduced. The Stokes inlet number is defined as follows:

$$Stk = \frac{\tau U_0}{D_s} = \frac{\rho_p d_p^2 C_c U_0}{18 \eta D_s} = \frac{\rho_0 d_a^2 C_{ca} U_0}{18 \eta D_s} \tag{8.5}$$

where τ is the relaxation time of the particle for a given diameter; U_0 is the free-flow air velocity; D_s is the diameter of the sampler; ρ_p and d_p are the density and diameter of the particle, respectively; $\rho_0 = 1000$ kg/m³; d_a is the aerodynamic diameter; and η is the viscosity of air.

The slip correction factor in Equation 8.5 should be determined if the particles of concern are smaller than 3 μm. Note the difference between the Stokes inlet number and the Stokes impaction number for a free jet, described in Chapter 7, which is denoted as Stk_J. The characteristic length for the inlet of a sampling head is the diameter of the sampler, D_s; the characteristic length for the impaction of a free air jet is the radius of the jet, $D_j/2$.

Similar to a Reynolds number, the Stokes inlet number characterizes particle behavior at the sampler face in a free-flow field with a velocity of U_0. The nominator represents the effect of the particles' inertia, and the denominator represents the effect of the sampler. A large value of *Stk* implies a large particle, a high free-flow velocity, a small sampling head diameter, or the combination of the three. A small *Stk* implies that the inertia effect is small with respect to the sampler diameter and may be negligible. In general, the inertia effect can be neglected when $Stk < 0.01$. Very small particles in the air always tend to follow the airflow, even when the flow changes its direction (Figure 8.5a). In that case, the sampling efficiency is always equal to 1, regardless of the sampling conditions and the sampler. Sampling effi-ciency is not a concern for gas sampling, because the gas molecules are so small that *Stk* is much smaller than 0.01.

When $Stk > 6$, the particle's inertia is so large that it maintains its original flow direction near the sampling head. Because the particle cannot change its direction at the face of the sampler, only the particles in the shaded volume shown in Figure 8.5 can get into the sampler. The total number of particles, N, in that shaded volume can be calculated as follows:

$$N = \frac{\pi D_s^2}{4} U_0 C_0 \tag{8.6}$$

The total number of the sampled particles is also equal to

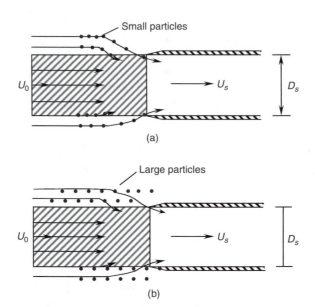

Figure 8.5 Physical implications of the Stokes inlet number. (a) When *Stk* < 0.01, the particles at the given size always follow the flow direction near the sampling head. (b) When *Stk* > 6, the particles at the given size maintain their original flow direction near the sampling head.

$$N = \frac{\pi D_s^2}{4} U_s C_s \qquad (8.7)$$

where C_s is the particle concentration in the sampler. Equating Equation 8.6 and Equation 8.7 gives the sampling efficiency for particles with *Stk* > 6:

$$\xi = \frac{C_s}{C_0} = \frac{U_0}{U_s} \quad \text{(for } Stk > 6 \text{ and } \alpha = 0) \qquad (8.8)$$

When 0.01 < *Stk* > 6, sampling efficiency for a given size of particle under anisokinetic sampling conditions is affected by both the inertia and the sampler diameter. Complete analysis of sampling efficiencies for sampling conditions of 0.01 < *Stk* < 6 is an overwhelming task, yet may still have large errors. Many efforts have been made to determine sampling efficiency experimentally under the given sampling conditions. When the sampling head is properly aligned with the free flow, the sampling efficiency can be calculated with the following empirical equation:[1]

$$\xi = \frac{C_s}{C_0} = 1 + \left(\frac{U_0}{U_s} - 1\right)\left(1 - \frac{1}{1 + Stk\,(2 + 0.62\,(U_s\,/\,U_0))}\right) \qquad (8.9)$$

(for *Stk* < 6 and $\alpha = 0$)

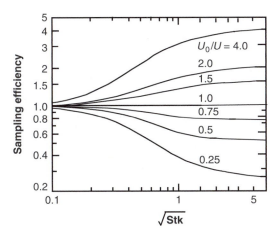

Figure 8.6   Sampling efficiency vs. *Stk*, with $\alpha = 0$.

The preceding analysis is appropriate for both superisokinetic and subisokinetic sampling. The sampling efficiency with respect to *Stk* is plotted in Figure 8.6. For Equation 8.8 and Equation 8.9 alike, $\xi < 1$ when $U_s > U_0$, and $\xi > 1$ when $U_s < U_0$. When $Stk \rightarrow 0$, $\xi \rightarrow 1$.

8.3.3   Misalignment Sampling

Misalignment is another type of anisokinetic sampling, as shown in Figure 8.4c. When the sampling velocity is isokinetic but the sampling head is misaligned with an angle of α, there will be a change of flow direction at the face of the sampling head and the concentration will be underestimated. We assume that particles will maintain the original flow direction with a large Stokes inlet number ($Stk > 6$) and that particles will completely follow the free flow when $Stk < 0.01$ (see Figure 8.5).

When $Stk > 6$, particles maintain their original flow direction near the sampling head. Because the particles cannot change their direction at the face of the sampler, only the particles in the shaded volume can get into the sampler. The shaded volume is $A_s \cos\alpha U_0$ (Figure 8.7), where A_s is the cross-sectional area of the sampler head. The total number of particles, N, in that shaded volume can be calculated as follows:

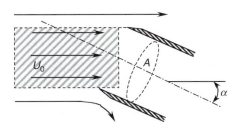

Figure 8.7   When the sampler head is misaligned with the flow direction but $U_0 = U_s$ and *Stk* > 6, only the particles contained in the shaded volume can get into the sampler.

$$N = A_s U_0 C_0 \cos\alpha = \frac{\pi D_s^2}{4} U_0 C_0 \cos\alpha \qquad (8.10)$$

because the total number of the sampled particles is also equal to

$$N = A_s U_s C_s = \frac{\pi D_s^2}{4} U_s C_s \qquad (8.11)$$

Noting that $U_s = U_0$ and equating Equation 8.6 and Equation 8.7 gives the sampling efficiency:

$$\xi = \frac{C_s}{C_0} = \cos\alpha \quad \text{(for } Stk > 6,\ 0 < \alpha < 90° \text{ and } U_s = U_0\text{)} \qquad (8.12)$$

When $0.01 < Stk < 6$ and the sampling velocity is isokinetic, the sampling efficiency is more complicated and very difficult to analyze. The relationship between the sampling efficiency and the misalignment angle has been studied extensively.[2,3] In general, the sampling efficiency can be expressed as

$$\xi = 1 + f(Stk)\left(\frac{U_0}{U_s}\cos\alpha - 1\right) \qquad (8.13)$$

where the function $f(Stk)$ must be determined experimentally. Durham and Lundgren (1980) evaluated sampling efficiency as a function of Stokes's inlet number and misalignment angles.[4] An empirical equation was developed:

$$\xi = \frac{C_s}{C_0} = 1 + (\cos\alpha - 1)\left(1 - \frac{1}{1 + 0.55\, Stk'\, \exp(0.25\, Stk')}\right) \qquad (8.14)$$

$$\text{(for } 0.01 < Stk < 6,\ U_0 = U_s\text{)}$$

where

$$Stk' = Stk\, \exp(0.022\alpha) \qquad (8.15)$$

where α is in degree. Note that Equation 8.14 only applies when $U_0 = U_s$. Apparently, the sampling efficiency is smaller than unity when a misalignment angle is present. Equation 8.14 also shows that when $Stk \to 0$, $\xi \to 1$ regardless of the misalignment angle. This implies that, for very small particles, the sampler orientation with respect to flow direction is irrelevant to the sampling efficiency. Sampling efficiency with respect to misalignment angle is plotted in Figure 8.8 using Equation 8.14.

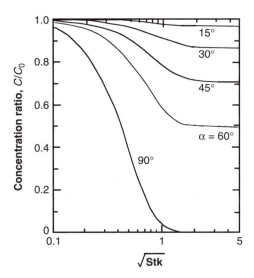

Figure 8.8 Sampling efficiency at different misalignment angles but at isokinetic sampling velocity, when 0.01 < *Stk* < 6, is plotted using Equation 8.14.

When $U_0 \neq U_s$ and $\alpha \neq 0$, sampling velocity and misalignment angles have a combined effect on the sampling efficiency. This combined efficiency is the product of two sampling efficiencies that are functions of (U_0/U_s) and α, respectively.

$$\xi = f_1\left(\frac{U_0}{U_s}\right) \cdot f_2(\alpha) \tag{8.16}$$

Equation 8.16 can be considered two steps in a sampling process. First, the sampling head is properly aligned and the sampling efficiency is $f_1((U_0/U_s))$. Second, a misalignment angle α is introduced and the overall sampling efficiency is $f_1((U_0/U_s)f_2(\alpha)$. For *Stk* < 0.01, the effect of anisokinetic sampling can be neglected; thus, $\xi = 1$. For *Stk* > 6, $f_1(Us/U_0)$ and $f_2(\alpha)$ are defined in Equation 8.8 and Equation 8.12, respectively. Substituting Equation 8.8 and Equation 8.12 into Equation 8.16 gives

$$\xi = \frac{U_0}{U_s}\cos\alpha \quad \text{(for } Stk > 6) \tag{8.17}$$

For 0.01 < *Stk* < 6, $f_1(Us/U_0)$ and $f_2(\alpha)$ are defined in Equation 8.9 and Equation 8.13, respectively. Substituting Equation 8.9 and Equation 8.13 into Equation 8.16 and simplifying gives

$$\xi = \left[1 + B_1\left(\frac{U_0}{U_s} - 1\right)\right]\left[1 + B_2(\cos\alpha - 1)\right] \quad \text{(for } 0.01 < Stk < 6) \tag{8.18}$$

where

$$B_1 = 1 - \frac{1}{1 + Stk\,[2 + 0.62\,(U_s\,/\,U_0)]} \qquad (8.19)$$

$$B_2 = 1 - \frac{1}{1 + 0.55\,Stk'\,\exp(0.25\,Stk')} \qquad (8.20)$$

where Stk' is defined in Equation 8.15.

Equation 8.17 and Equation 8.18 provide tools for achieving a 100% sampling efficiency when isokinetic sampling is not possible. When $U_0 > U_s$, there is a misalignment angle that can compensate for the anisokinetic effect of the sampling velocity. At a given U_0 and U_s, the misalignment angle for $\xi = 1$ can be solved from Equation 8.17:

$$\alpha = \cos^{-1}\left(\frac{U_s}{U_0}\right) \quad \text{(for } Stk > 6 \text{ and } U_0 > U_s) \qquad (8.21)$$

Because B_2 is a function of the misalignment angle, as shown in Equation 8.15 and Equation 8.20, α cannot be solved explicitly from Equation 8.18. However, Equation 8.18 can be solved using an iteration method by making $\xi = 1$. It is important to notice that this compensation only applies to the subisokinetic sampling. When the sampling is superisokinetic (i.e., $U_0 < U_s$) the sampling efficiency is less than 1, thus a misalignment angle will only reduce the sampling efficiency further.

Again, because the misalignment of a sampler with free-flow direction can change the flow patterns around the sampler and disturb the free-flow streamlines, a substantial amount of error in sampling efficiency can be introduced. Therefore, it is strongly recommended that isokinetic sampling should be used whenever possible. Apparently, one cannot make an infinite number of sampling heads to match all ratios of sampling velocity to free-flow velocity. In that case, the closest size of sampling head should be selected, so that the ratio of sampling velocity to free-flow velocity will be approximately unity.

An alternative approach to achieving isokinetic sampling conditions is to regulate the sampling rate so that the sampling velocity can be matched with the free-flow velocity. However, care should be exercised when adjusting the sampling flow rate, because it may be affected by many factors, such as power fluctuation. Precise flow-rate control is a critical factor in securing accurate air sampling.

Example 8.2: *Air velocity in a duct is 5 m/s. The sampling rate of a particle counter is 28.3 l/min. The diameter of the sampling head is 37 mm. Find the following:*

   a. *What is the sampling efficiency for particles with an aerodynamic diameter of 20 μm?*

b. What should the inlet diameter be to achieve isokinetic sampling?
c. *With the new sampling head, what is the sampling efficiency if the sampling head is misaligned by 20°?*

Solution:

a. For 20 μm particles, the slip correction factor $C_c = 1$.

$$Stk = \frac{\rho_0 d_a^2 C_c U_0}{18 \eta D_s} = \frac{1{,}000 \times (20 \times 10^{-6})^2 \times 1 \times 5}{18 \times 1.81 \times 10^{-5} \times 0.037} = 0.166$$

$$U_s = \frac{4 Q_s}{\pi D_s^2} = \frac{4 \times 0.000472}{\pi \times 0.037^2} = 0.439 \ (m/s)$$

Because $0.01 < Stk < 6$, and $\alpha = 0$, Equation 8.9 should be used.

$$\xi = 1 + \left(\frac{U_0}{U_s} - 1 \right) \left(1 - \frac{1}{1 + Stk\,(2 + 0.62\,(U_s / U_0))} \right)$$

$$= 1 + \left(\frac{5}{0.439} - 1 \right) \left(1 - \frac{1}{1 + 0.166\,(2 + 0.62\,(0.439 / 5))} \right) = 3.64$$

b. The correct inlet diameter for isokinetic sampling can be calculated using Equation 8.4.

$$D_s = \left(\frac{4 Q_s}{\pi U_0} \right)^{1/2} = \left(\frac{4 \times 0.000472}{\pi \times 5} \right)^{1/2} = 0.011 \ (m) = 11 \ mm$$

c. With the new sampling head, $D_s = 11$ mm, the sampling velocity is isokinetic. For $\alpha = 20°$, from Equation 8.15

$$Stk = \frac{\rho_0 d_a^2 C_c U_0}{18 \eta D_s} = \frac{1{,}000 \times (20 \times 10^{-6})^2 \times 1 \times 5}{18 \times 1.81 \times 10^{-5} \times 0.011} = 0.558$$

$$Stk' = Stk \exp(0.022\alpha) = 0.558 \times \exp(0.022 \times 20) = 0.868$$

$$\xi = 1 + (\cos \alpha - 1) \left(1 - \frac{1}{1 + 0.55\,Stk' \exp(0.25\,Stk')} \right)$$

$$= 1 + \left(\cos 20^{o} - 1\right)\left(1 - \frac{1}{1 + 0.55 \times 0.868 \times \exp(0.25 \times 0.868)}\right) = 0.977$$

8.4 SAMPLING IN CALM AIR

In the previous discussions, we assumed that the free stream always has a velocity of U_0 in a given direction. Based on this assumption, sampling efficiency can be measured or calculated for a typical sampler. However, free-flow velocities fluctuate and change direction in almost all practical situations. Airflow velocities and directions in a room, for example, vary greatly across the room and over a period of time. It is difficult to satisfy isokinetic sampling conditions in such varying situations. Further, isokinetic sampling makes no sense for sampling in calm air, because calm air requires a zero sampling rate. It is necessary to define a sampling procedure that is not sensitive to the free stream velocity and direction, depending only on the sampler size and sampling rate. Such a procedure involves the principles of sampling in calm air.

Truly calm air is rare in the real world. For very small particles, free flow, even with very high velocities, can be considered calm air. One the other hand, at very high sampling velocities, the relatively slow-moving, low free flow can be considered calm air. For practical purposes, Vincent (1989) gave the following criteria for calm air with respect to the sampler inlet Stokes number:[3]

$$U_0 < 0.1 U_s \frac{1}{Stk^{2/3}} \tag{8.22}$$

where U_0 is the free-flow velocity, U_s is the sampling velocity, and Stk is the Stokes inlet number defined by Equation 8.5. Substituting Equation 8.5 into Equation 8.22 and solving for U_0 gives

$$U_0 < 0.8 U_s^{3/5}\left(\frac{\eta D_s}{\rho_0 d_a^2 C_{ca}}\right)^{2/5} = 0.8 U_s^{3/5}\left(\frac{\eta D_s}{\rho_p d_p^2 C_c}\right)^{2/5} \tag{8.23}$$

For a given sampling system and a particle size of concern, Equation 8.23 defines the maximum free-flow velocity that can be considered calm air. Equation 8.23 also describes to what extent the sampling velocity and sampling head diameter affect the calm air threshold limit. Sampling velocity has a larger effect (with a 3/5 power) on calm-air criteria than does sampling head diameter (with a 2/5 power). The subsequent discussions are valid only when Equation 8.23 holds. Otherwise, sampling procedures described in previous sections should be exercised. Threshold limits for calm air are plotted in Figure 8.9 using Equation 8.23. If the free-flow air velocity falls into the lower-left area of a respected line, it can be treated as sampling in calm air.

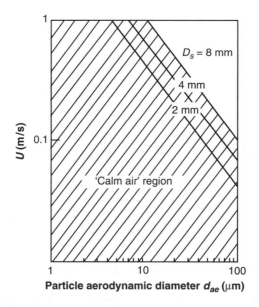

Figure 8.9 Criteria for calm air for different sampling velocities at standard air conditions. (Use Equation 8.23 and refer to Vincent, J.H., et al., On the aspiration characteristics of large diameter, thin-walled aerosol sampling probes at misalignment orientations with respect to the wind, *J. Aerosol Sci.*, 17:211–224, 1986. See also Figure 8.11).

If the free-flow velocity is below the calm-air criteria defined by Equation 8.23, procedures for sampling in calm air should be used. From Equation 8.23, the calm-air criteria are related to the sampling velocity U_s, which is in turn related to the sampling head diameter and the sampling rate. In the domain of "calm air," a sampler should meet certain criteria to minimize the error of sampling efficiency. In the context of obtaining representative sampling, a sampling efficiency $\xi = 1 \pm 0.1$ is considered an acceptable level.[5] The following criteria for sampler diameters are established based on this acceptable level.

Error sources in calm-air sampling are twofold:

- Effect of particle settling
- Effect of particle inertia

If a sampler with a large diameter is facing upward and has a low sampling rate, only a small column of air (as shown in the shaded area of Figure 8.10a) is allowed to enter the sampling head. In addition to the particles contained in that sampled air, particles — especially large particles — can settle into the sampling head by their gravity, thus introducing some gain of large particles in the sample. Loss of large particles from the sample can occur when the sampling head faces downward. On the other hand, if a sampler has a small diameter and a high sampling rate, a large column of air (the shaded area in Figure 8.10b) is allowed to enter the sampling head. Because the flow converges into the sampling head entrance suddenly, large

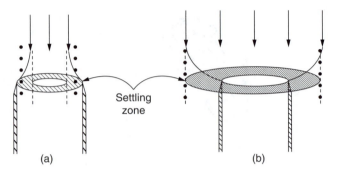

Figure 8.10 The two error sources in calm air sampling. (a) Settling effect: gain (if the sampler is facing upward) or loss (if the sampling head is facing downward) of large particles due to a large sampling head and a small sampling rate. (b) Inertia effect: loss of large particles due to a small sampling head and a large sampling rate.

particles tend to maintain their original position or motion due to the effect of inertia. This will cause a loss of large particles in the sample.

In order to minimize the sampling error caused by the settling effect, the sampling head diameter should be small enough so that the sampling velocity is high and prevents the gain or loss of large particles due to settling. Davis recommended a criterion for minimum sampling velocity:[6]

$$U_s > 25 V_{TS} \qquad (8.24)$$

where V_{TS} is the terminal settling velocity of the particles of concern.

Because the sampling velocity $U_s = 4Q_s/\pi D_s^2$ and $V_{TS} = \tau g$, the maximum diameter of the sampling head can be derived:

$$D_{s\,max} \le 0.4 \left(\frac{Q_s}{\pi \tau g} \right)^{\frac{1}{2}} \qquad (8.25)$$

where D_{smax} is the maximum diameter of the sampling head to minimize the particle settling effect, Q_s is the sampling rate, τ is the relaxation time of the particle of concern, and g is gravity.

On the other hand, if the sampling velocity is too high, the direction change of airflow at the sampling head may be sufficiently high and the flow converges to the sampler. As a result of this change of airflow direction, some large particles may be left out of the sampled air due to inertia. In order to minimize the sampling error caused by particle inertia effect, the sampling head diameter must be large enough so that the convergence of airflow at the sampler face will be small (Figure 8.10b) and the inertia effect of particles will be negligible. Davis derived the following criteria for minimum sampling head diameter:[6]

$$D_{s\min} \geq 10 \left(\frac{Q_s \tau}{4\pi} \right)^{\frac{1}{3}} \tag{8.26}$$

Combining Equation 8.25 and Equation 8.26 gives the range of sampling head diameters for particle sampling in calm air.

$$0.4 \left(\frac{Q_s}{\pi \tau g} \right)^{\frac{1}{2}} \geq D_s \geq 10 \left(\frac{Q_s \tau}{4\pi} \right)^{\frac{1}{3}} \tag{8.27}$$

When the sampling head diameter falls within the range defined by Equation 8.27, sampling efficiency will not be significantly affected by the direction of the sampling head, that is, the sampling head can face in any direction. This design criterion for sampling head diameter depends on the sampling rate and the particle size. For a given sampling rate, there will be a particle diameter at which $D_{s\min} = D_{s\max}$. For particles larger than that diameter, $D_{s\min}$ will be greater than $D_{s\max}$, which makes the design criterion invalid. This is because $D_{s\max}$ is inversely proportional to $(1/\tau)^{1/2}$ and $D_{s\min}$ is proportional to $\tau^{1/3}$. When the particle size increases (hence the τ increases), $D_{s\max}$ will decrease and $D_{s\min}$ will increase. For sampling large particles in calm air, Q_s must be sufficiently high to satisfy Equation 8.27 and the sampling efficiency requirement. If the sampling rate is too low, a large number of large particles will not follow the sampling airflow, causing loss in particle sampling.

For particles so large that a sampler diameter can no longer satisfy $D_{s\min}$ in Equation 8.27, one should use Equation 8.24 for the sampler design criteria. Hinds developed a graph to define the relationship between sampler head diameters and particle diameters.[16] Using this approach, sampling head diameter vs. particle aerodynamic diameter at different sampling rates for calm-air sampling is plotted in Figure 8.11.

When the diameter of the sampling head is larger than $D_{s\max}$, the settling effect may introduce a large error. In this case, the sampling head should be placed facing upward with a tilt angle β, to compensate for the error caused by the large sampling head diameter. As shown in Figure 8.10, the settling area for a sampler is A_s; if the desirable sampling head area is $A_{s\max}$, which is smaller than A_s, the sampler head can be tilted with an angle β, with respect to the vertical line to make

$$A_{s\max} = A_s \cos\beta \quad (0 < \beta < 90°) \tag{8.28}$$

$$\beta = \cos^{-1} \left(\frac{A_{s\max}}{A_s} \right) = \cos^{-1} \left(\frac{D_{s\max}^2}{D_s^2} \right) \quad (0 < \beta < 90°) \tag{8.29}$$

Example 8.3: *Determine whether a 37 mm diameter sampling head is appropriate for sampling in calm air. The room air velocity is 0.2 m/s, and the particle of concern*

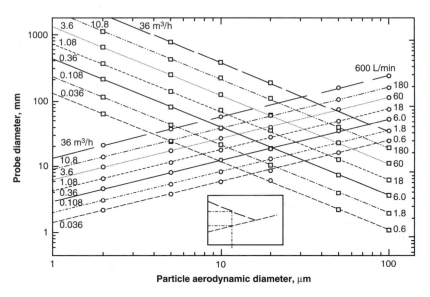

Figure 8.11 Sampling head diameter vs. particle aerodynamic diameter at different sampling rates, for standard air conditions.

is 10 μm or smaller in aerodynamic diameter. The sampling pump has a flow rate of 2 l/min.

Solution: First, determine whether the airflow satisfies the calm-air criteria. The sampling velocity at the face of the sampler is

$$U_s = \frac{4Q_s}{\pi D_s^2} = \frac{4 \times (3.33 \times 10^{-5})}{\pi \times 0.037^2} = 0.031\,(m/s)$$

$$U_0 < 0.8 U_s^{3/5} \left(\frac{\eta D_s}{\rho_0 d_a^2 C_{ca}} \right)^{2/5} = 0.8 \times 0.031^{3/5} \left(\frac{1.81 \times 10^{-5} \times 0.037}{1,000 \times (10 \times 10^{-6})^2} \right)^{2/5} = 0.213\,(m/s)$$

Because $U_0 = 0.2$ m/s < 0.212 m/s, the room air can be considered calm air.

Next, examine the sampling head diameter. For particles of 10 μm in aerodynamic diameter

$$\tau = \frac{\rho_0 d_a^2}{18\eta} = \frac{1,000 \times (10 \times 10^{-6})^2}{18 \times 1.81 \times 10^{-5}} = 0.00031\,(s)$$

Using Equation 8.25 and Equation 8.26

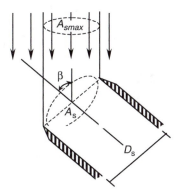

Figure 8.12 A tilt angle and a sampling head diameter equal to the required maximum diameter to minimize the particle sampling effect.

$$D_{s\,max} \leq 0.4\left(\frac{Q_s}{\pi\tau g}\right)^{1\!/2} = 0.4\times\left(\frac{3.33\times10^{-5}}{\pi\times0.00031\times9.81}\right)^{1\!/2} = 0.024\ (m) = 24\ mm$$

Because $D_s = 37$ mm $> D_{smax} = 24$ mm, the settling effect may introduce a larger error. The sampling head should be placed facing upward with a tilt angle (with respect to the vertical direction, as shown in Figure 8.12).

$$\beta = \cos^{-1}\left(\frac{D_{s\,max}^2}{D_s^2}\right) = \cos^{-1}\left(\frac{0.024^2}{0.037^2}\right) = 65^o$$

8.5 SAMPLING IN DUCTS

Airborne contaminants, especially particulate matter, may vary substantially at a duct cross section due to gravity settling, diffusion near the duct surface, flow characteristics, and other external forces, such as centrifugal force in a spiral flow after a blower. For example, the air velocity profile in a circular duct after an in-duct axial fan is shown in Figure 8.13. The concentration of very small particles near the duct surface may be much lower than that near the center of the duct, because many small particles are collected by the surface through diffusion. The concentration of very large particles near the duct surface can be much higher than near the center, because the spiral motion of the air can send large particles outward by centrifugal force.

To obtain a representative sample of the flow and the airborne contaminants, a series of samples must be taken at a transverse plane perpendicular to the duct axis. The number and locations of sampling points should be determined carefully. In determining the number and locations of sampling points, the following factors should be considered:

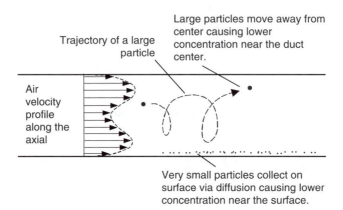

Figure 8.13 Concentration variation for different sizes of particles in a circular duct. Very small particles near the duct surface may be collected on the surface and cause lower small-particle concentration in the air near the surface. Large particles may be "thrown" outward by the spiral air motion caused by fan blades; thus, there is a lower large-particle concentration near the duct center.

- Shape of ducts (circular or rectangular)
- Types of contaminants of concern (particulate matter or gases)
- Distance from the flow disturbances (upstream or downstream)
- Duct cross-sectional area

8.5.1 Sampling Locations — Equal-Area Method

An *equal-area method* is to divide the total cross-sectional area of the duct into a number of equal subareas; the weighted center of each subarea represents the sampling point for that subarea. When the duct is square or rectangular, the equal subareas will be smaller squares or rectangles, and the sample will be taken from the center of each. Figure 8.14 shows an example of a 4 × 4 grid of a rectangular duct. The total duct cross-sectional area is divided into 16 equal rectangles, and 16 samples are taken — one from the center of each small rectangle.

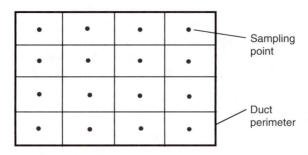

Figure 8.14 An example of 4 × 4 sampling points for a rectangular duct. The subareas on the traverse are equal.

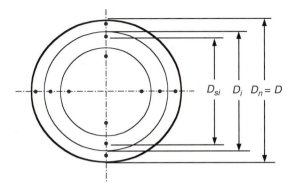

Figure 8.15 Equal-area method for a circular duct. Each annular and the inner circle must have the same area. The diameters at which the sampling points should be located are weighted diameters between two adjacent annular diameters.

When the duct is circular, the total cross-sectional area is divided into small, concentric annulars with the same area. As shown in Figure 8.15, a circular duct with a diameter D and an area A is divided into n annulars, each with an outer diameter D_i and a subarea A_i. The innermost subarea is circular with a diameter of D_1, and the outermost subarea has a diameter of $D_n = D$. The diameter at which the sampling point is located, D_{si}, should be at a weighted diameter that again divides the annular into two equal subareas:

$$\frac{\pi}{4}(D_i^2 - D_{si}^2) = \frac{\pi}{4}(D_{si}^2 - D_{i-1}^2) \quad (I = 1, 2, ..., n) \tag{8.30}$$

Solving for D_{si} from Equation 8.30 gives

$$D_{si} = \sqrt{\frac{D_i^2 + D_{i-1}^2}{2}} \tag{8.31}$$

This diameter D_{si} is greater than $(D_i + D_{i-1})/2$. Note that the outermost diameter in Equation 8.32 is the same as the duct diameter, that is, $D_n = D$ and $D_0 = 0$. Diameters of annulars with equal area can be determined from the duct diameter and the number of annulars divided.

$$A_i = \frac{\pi}{4}(D_i^2 - D_{i-1}^2) = \frac{A}{n} \tag{8.32}$$

$A = \pi D^2/4$ gives

$$D_{i-1} = \sqrt{D_i^2 - \frac{D^2}{n}} \tag{8.33}$$

Equation 8.33 can solved step-by-step to obtain each annular diameter.

Example 8.4: *What are the diameters at which the samplers should be located if the duct is divided into three equal area annulars? The duct is circular with a diameter of 0.9 m.*

Solution: From Equation 8.33, because $n = 3$, the outermost diameter $D_3 = D = 0.9$ (m).

$$D_2 = \sqrt{D_3^2 - \frac{D^2}{3}} = \sqrt{0.9^2 - \frac{0.9^2}{3}} = 0.735 \quad (m)$$

$$D_1 = \sqrt{D_2^2 - \frac{D^2}{3}} = \sqrt{0.735^2 - \frac{0.9^2}{3}} = 0.52 \quad (m)$$

The diameters at which the sampling locations should be located, D_{si}, are determined by applying Equation 8.31 and noting that $D_0 = 0$:

$$D_{s3} = \sqrt{\frac{D_3^2 + D_2^2}{2}} = \sqrt{\frac{0.9^2 + 0.735^2}{2}} = 0.822 \quad (m)$$

$$D_{s2} = \sqrt{\frac{D_2^2 + D_1^2}{2}} = \sqrt{\frac{0.735^2 + 0.52^2}{2}} = 0.637 \quad (m)$$

$$D_{s1} = \sqrt{\frac{D_1^2 + D_0^2}{2}} = \sqrt{\frac{0.52^2 + 0^2}{2}} = 0.37 \quad (m)$$

Once D_{si} are determined, the total number of sampling points can be selected. Note that the total number of sampling points is higher than the number of annulars. Usually, more than one sampling point is needed at the same diameter but at different locations (Figure 8.15). In each annular, four sampling points are selected at two perpendicular diameters. In this case, the total sampling points on a traverse will be $4 \times n$, where n is the number of annulars. The total number of sampling points is affected by the contaminant profile, which depends on the duct size and on distances for disturbances, such as an elbow or an in-duct blower. The U.S. EPA recommends a guideline to determine the total number of sampling points, as shown in Figure 8.16.[7] In using Figure 8.16, an equivalent diameter D_e for a rectangular duct must be determined. If a rectangular duct cross section has a width W and a height H, the equivalent diameter is defined as follows:

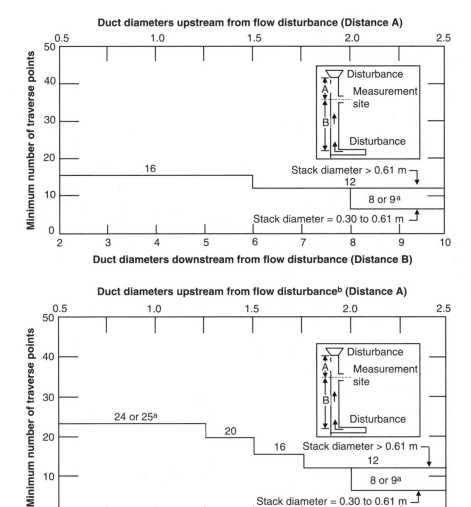

Figure 8.16 The minimum number of sampling points on a traverse for duct sampling (a) for gases and air velocity, (b) for particulate matter. (From U.S. EPA, *Standards of Performance for New Stationary Sources*, 40 CFR 60, U.S. Government Printing Services, Washington, DC, 1992. With permission.) [a] The higher number is for rectangular stacks of ducts. [b] From the point of any type of disturbance (bend, expansion, contraction, etc.).

$$D_e = \frac{2\,WH}{W + H} \tag{8.34}$$

where W is the width of duct, and H is the height of the duct. In many applications, use of EPA guidelines may lead to a conservatively high number of sampling points, especially for flows that are reasonably uniform and contaminants that are well mixed.[8]

8.5.2 Sampling Locations — Tchebycheff Method

An alternative method of air sampling in duct and stacks is the *log-Tchebycheff method*, or simply the *Tchebycheff method*. The method was originally developed for measurement of fluid velocities in ducts. Thus, it is particularly suitable for gaseous and air velocity sampling.[9]

To determine the profiles of air velocity or a contaminant concentration in the traverse plane, a straight average of individual point velocities will give satisfactory results when point velocities are determined by the log-Tchebycheff rule.[10] Figure 8.17 shows suggested sensor locations for traversing circular and rectangular ducts.[9] The log-Tchebycheff rule provides the greatest accuracy, because its location of traverse points accounts for the effect of wall friction and fall-off of velocity near the duct walls. For circular ducts, the log-Tchebycheff and log-linear traverse meth-

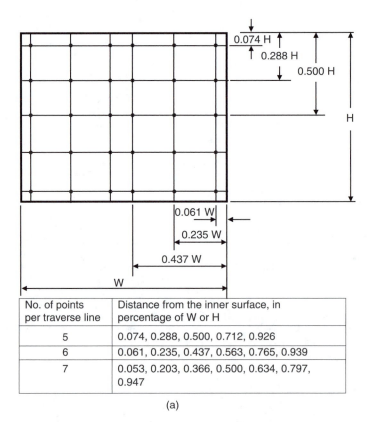

No. of points per traverse line	Distance from the inner surface, in percentage of W or H
5	0.074, 0.288, 0.500, 0.712, 0.926
6	0.061, 0.235, 0.437, 0.563, 0.765, 0.939
7	0.053, 0.203, 0.366, 0.500, 0.634, 0.797, 0.947

(a)

Figure 8.17 Sampling locations at a traverse, symmetric with centerlines for both *W* and *H* (a) for rectangular ducts, (b) for circular ducts. (From 2001 *ASHRAE Handbook — Fundamentals*, Chapter 14, Figure 6. © American Society of Heating, Refrigerating and Air-Conditioning Engineers, Inc. http://www.ashrae.org. With permission.)

ods are similar. Log-Tchebycheff is now recommended for rectangular ducts, as well. It minimizes the positive error (measured greater than actual) caused by the failure to account for losses at the duct wall. This error can occur when the method of equal subareas is used to traverse rectangular ducts.

For a rectangular duct traverse, a minimum of 25 points should be measured (Figure 8.16a). For a duct side of less than 450 mm, the points should be located at the center of equal areas not more than 150 mm apart, and a minimum of two points per side should be used. For a duct side greater than 1400 mm, the maximum distance between points is 200 mm. For a circular duct traverse, the log-linear rule and three symmetrically disposed diameters may be used (Figure 8.16b). Points on two perpendicular diameters may be used where access is limited.

If possible, measuring points should be located at least 7.5 diameters downstream and 3 diameters upstream from a disturbance (e.g., caused by an elbow). Compromised traverses as close as 2 diameters downstream and 1 diameter upstream can be performed with an increase in measurement error. Because field-measured airflows are rarely steady and uniform, particularly near disturbances, accuracy can be improved by increasing the number of measuring points. Straightening vanes located 1.5 duct diameters ahead of the traverse plane improve measurement precision.

When gas concentrations or air velocities at a traverse plane fluctuate, the readings should be averaged on a time-weighted basis. Two traverse readings in short succession also help to average out velocity variations that occur with time. If negative velocity pressure readings are encountered, they are considered a measurement value of zero and calculated in the average velocity pressure. ASHRAE Standard 111 has more detailed information on measuring flow in ducts.[11]

8.6 SAMPLING OF VAPOR AND GASES

Sampling of gases and vapors requires specific procedures and instruments. The words *gases* and *vapors* are frequently used interchangeably. However, they are not identical. The majority of gases of interest regarding indoor air quality are elements (e.g., chlorine) or inorganic compounds (e.g., hydrogen sulfide, ammonia, arsine, carbon dioxide, and carbon monoxide). Vapors of practical importance are primarily organic substances, such as methyl, ethyl, ketone, benzene, acetone, toluene, and toluene disocyanate, although some inorganic substances (e.g., mercury) are also encountered.

Although it is true that, at ordinary temperature and pressure, gases and vapors will both diffuse rapidly and form true solutions in air, they differ in other respects. Gases are generally understood to be noncondensable at room temperature, whereas vapors are derived from volatile liquids. Therefore, under ordinary conditions gases remain in the gaseous state, even when present at high concentrations. Vapors, on the other hand, may condense at high concentrations and coexist in both gas and aerosol forms. However, unless an aerosol is deliberately produced (as in a spray operation), concentrations of vapor pollutants in indoor environments rarely reach saturation conditions; gases and vapors can then be considered similar and the same devices used to collect them in indoor air studies. Therefore, in this textbook, the word *gas* (or *gaseous*) will refer to both gas and vapor, unless explicitly stated otherwise.

8.6.1 Sampling Procedures

There are two basic methods for collecting gaseous samples. In one, called *grab sampling,* an actual sample of air is taken in a flask, bottle, bag, or other suitable container. In the other, called *continuous* or *integrated sampling,* gases are removed from the air and concentrated by passage through an absorbing or adsorbing medium, such as a denuder or a filter.

The grabbing method usually involves the collection of instantaneous or short-term samples, usually within a few seconds or a minute, but similar methods can be used for sampling over longer periods. This type of sampling is acceptable when peak concentrations are sought or when concentrations are relatively constant. Grab samples were once used only for gross components of gases, such as methane, carbon monoxide, or oxygen, where the analysis was frequently performed volumetrically. The introduction of highly sensitive laboratory instruments, however, makes this technique limited only by the detection limit of the analytical methods available.

An important feature of grab samples is that their collection efficiency is normally 100%. However, it must be remembered that sample decay does occur for various reasons, such as reaction or adsorption on the inner surfaces of the collector or tubes through which the sample is collected, and grab sampling must be used with this clearly in mind.

Grab sampling is of questionable value when one of the following conditions pertains:

- The contaminant or contaminant concentration varies with time.
- There is a reaction with an absorbing solution (or reagent therein).
- There is collection onto a solid adsorbent.

Collection efficiency of active sampling devices used for these sampling procedures is frequently less than 100%. Therefore, individual efficiency percentages must be determined for each case. For more details of gas sampling procedures and instruments, see ACGIH.[12]

8.6.2 Selection of Sampling Devices

Selecting an appropriate sampling device for specific tasks is an important and complicated process. Criteria for selecting a sampling device may be different for different sampling purposes, such as research or trouble shooting. Although innovative devices may be a norm in research, established and commercially available devices should be used as much as possible in IAQ trouble-shooting processes.

The following general recommendations for sampling device selection are given by ACGIH.[12] The first step in selecting a sampling device and an analytical procedure is to search the available literature. Primary sources are the compendia methods recommended by regulatory authorities or governmental agencies, such as the

NIOSH *Manual of Analytical Methods* and the OSHA *Analytical Methods Manual.*[13,14] Secondary sources are published references in, for example, the *American Industrial Hygiene Association Journal* or *Applied Occupational and Environmental Hygiene,* or in books such as the Intersociety Committee's *Methods for Air Sampling and Analysis.*[15]

If a published procedure is not available, one can be devised from theoretical considerations. However, its suitability must be established experimentally before application. Important criteria for selecting sampling devices are solubility, volatility, and reactivity of the contaminant; the adsorptivity, reactivity, and permeability of the collection device; and the sensitivity of the analytical method.

Generally speaking, nonreactive and nonabsorbing gaseous substances may be collected as grab samples. Water-soluble gases and vapors and those that react rapidly with absorbing solutions can be collected in simple gas washing bottles. Volatile and less soluble gaseous substances and those that react slowly with absorbing solutions require more liquid contact. For the substances, more elaborate sampling devices may be required, such as gas washing bottles of the spiral type or fritted bubblers. Insoluble and nonreactive gases and vapors are collected by adsorption onto activated charcoal, silica gel, or another suitable adsorbent. Frequently, for a given contaminant, there may be several choices of sampling equipment.

8.7 SAMPLING RATE CONTROL

8.7.1 Conventional Flow-Rate Control Devices

Accurate measurement of the flow rate is very important in air sampling, because the contaminant concentration is determined by the ratio of the sampled contaminant quantity to the sampled air volume. One widely used conventional flow meter in air sampling is the rotameter. Rotameters are sensitive to pressure changes in upstream and downstream airflows.[17] Most flow meters are calibrated at atmospheric pressure, and many require pressure corrections when used at other pressures. When the flow meter is used in air sampling, it should be downstream of the filter to exclude the possibility of sample losses in the flow meter. Therefore, the sampled air is at a pressure below atmospheric due to the pressure drop across the filter. Furthermore, if the filter resistance increases due to the accumulation of dust, the pressure correction is not a constant factor. During the sampling period, the filter tends to get plugged, and the flow rate may decrease as filter resistance increases[18]. These factors make it difficult to measure the flow rate accurately.

Critical orifices[19] have been widely used in flow-rate control for air sampling because they are simple, reliable, and inexpensive. When the pressure drop across the critical orifice is more than 47% of the upstream pressure, the speed of sound is achieved in the throat, and the velocity will not change with a further reduction in downstream pressure. Under these conditions, the flow rate is kept constant if upstream conditions are constant. However, commercially available orifices were found to lack the required precision and accuracy: They differed from the nominal

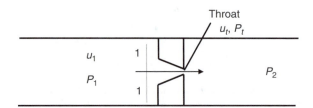

Figure 8.18 Gas flow through an orifice: v_1, upstream velocity; v_t, throat velocity; P_1, upstream pressure; P_t, throat pressure; P_2, downstream pressure.

flow rate by up to 15%.[20] Another disadvantage of most critical orifice designs is that a pressure drop in excess of 47 kPa is required to ensure a stable flow.[21] To achieve this pressure drop, a special high-power vacuum pump must be used. Some commercial flow-limiting orifices even require a vacuum as high as 72 kPa.

The electronic flow controller is another device used widely to measure and control the gas flow with high accuracy. However, compared with critical orifices, this device is much more expensive.

There is an increasing need for small critical airflow control devices with low pressure drop for air-quality studies. Many efforts have been made to develop critical flow devices.[20-23] For example, an experimental procedure was used to obtain a constant airflow by finishing disposable glass serological pipette tips.[20] The procedure and the products are both labor intensive and inaccurate. There is no existing design standard available for small orifices. The existing critical orifice design standard is available only for opening diameters larger than 12.5 mm.[24] In air-quality studies, sampling rates often require 20 l/min or lower, which represents a throat opening diameter of 1.6 mm or smaller at critical pressure drop. It is highly desirable to develop an accurate and reliable critical airflow control device to maintain a constant airflow rate at low pressure differentials for air sampling.

8.7.2 Critical Venturi

Figure 8.18 shows an example of gas flow through an orifice or a convergent nozzle.[25] Assume that the flow is isentropic, that is, the flow is adiabatic and frictionless. An *adiabatic* condition means that no heat is transferred between the system and its surroundings. The assumption of adiabatic flow can be considered true if the flow occurs very quickly and with a small amount of friction. Isentropic analysis can be applied to high velocity gas flows over short distances where friction and heat transfer are relatively small. This flow can be described with an energy conservation equation:[17]

$$\frac{u_t^2 - u_1^2}{2} = \left(\frac{\gamma}{\gamma - 1}\right)\frac{P_1}{\rho_1}\left[1 - \left(\frac{P_t}{P_1}\right)^{(\gamma-1)/\gamma}\right] \tag{8.35}$$

where u_1 is the upstream velocity of gas (cross section 1), ρ_1 is the upstream gas density, P_1 is the upstream pressure, u_t is the velocity of gas at the throat (minimum area), p_t is the pressure at the throat, and γ is the specific heat ratio.

Because the flow is assumed to be isentropic

$$\frac{P}{\rho^\gamma} = const \tag{8.36}$$

and

$$\frac{\rho_t}{\rho_1} = \left(\frac{P_t}{P_1}\right)^{1/\gamma} \tag{8.37}$$

where ρ_t is the gas density at the throat. The mass conservation equation shows

$$\rho_1 A_1 u_1 = \rho_t A_t u_t \tag{8.38}$$

where A_1 is the area at cross section 1 and A_t is the area at the throat (minimum area), assuming that the velocities are uniform across the flow area. Combining Equation 8.37 and Equation 8.38 results in

$$u_1 = \left(\frac{P_t}{P_1}\right)^{1/\gamma}\left(\frac{A_t}{A_1}\right) u_t \tag{8.39}$$

Substituting Equation 8.39 into Equation 8.35, u_t can be expressed as

$$u_t = C_a \sqrt{\left(\frac{2\gamma}{\gamma-1}\right)\frac{P_1}{\rho_1}\left[1-\left(\frac{P_t}{P_1}\right)^{(\gamma-1)/\gamma}\right]} \tag{8.40}$$

where C_a is the coefficient of velocity at the throat:

$$C_a = \frac{1}{\sqrt{1-\left(\frac{P_t}{P_1}\right)^{2/\gamma}\left(\frac{A_t}{A_1}\right)^2}} \tag{8.41}$$

The relationship between the coefficient of velocity at the throat C_a and the pressure ratio P_t/P_1 for the different values of diameter ratio d_t/d_1 is shown in Figure

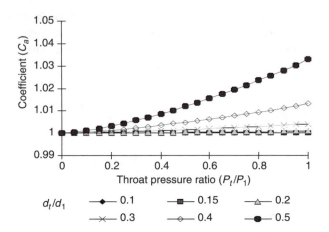

Figure 8.19 Relationship between coefficient C_a and throat-pressure to upstream-pressure ratio P_t/P_1 for different values of the throat diameter to upstream diameter ratio d_t/d_1. C_a can be considered as unity when $d_t/d_1 < 0.3$ with an error of less than 0.5%.

8.19, where d_t is the throat diameter and d_1 is the inlet diameter at cross section 1. These results show that C_a is near unity for all pressure ratios P_t/P_1 up to 1.0, provided $d_t/d_1 < 0.3$ (the error is less than 0.5%). Therefore, the velocity coefficient C_a can be considered to be unity, and Equation 8.40 can be simplified to

$$u_t = \sqrt{\left(\frac{2\gamma}{\gamma-1}\right)\frac{P_1}{\rho_1}\left[1-\left(\frac{P_t}{P_1}\right)^{(\gamma-1)/\gamma}\right]} \qquad (8.42)$$

The mass flow rate $\dot{m}$ can be obtained by combining Equation 8.37, Equation 8.38, and Equation 8.42:

$$\dot{m} = A_t\sqrt{\left(\frac{2\gamma}{\gamma-1}\right)P_1\rho_1\left(\frac{P_t}{P_1}\right)^{2/\gamma}\left[1-\left(\frac{P_t}{P_1}\right)^{(\gamma-1)/\gamma}\right]} \qquad (8.43)$$

Assume that the downstream pressure is P_2. When P_2 is greater than the critical pressure P_c, P_t is equal to P_2. Then

$$\dot{m} = A_t\sqrt{\left(\frac{2\gamma}{\gamma-1}\right)P_1\rho_1\left(\frac{P_2}{P_1}\right)^{2/\gamma}\left[1-\left(\frac{P_2}{P_1}\right)^{(\gamma-1)/\gamma}\right]} \qquad (8.44)$$

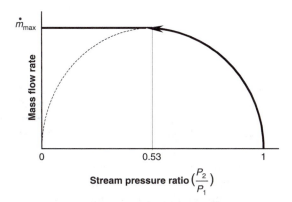

Figure 8.20 Relation between mass flow rate $\dot{m}$ and the downstream-pressure to upstream-pressure ratio P_2/P_1.

The variation of $\dot{m}$ with the pressure ratio P_2/P_1 in Equation 8.44 is illustrated by the curved line in Figure 8.20. When the pressure ratio P_2/P_1 decreases, the mass flow rate $\dot{m}$ increases. When the pressure ratio reaches a critical level (0.53), the flow rate $\dot{m}$ reaches a maximum value $\dot{m}_{max}$. The critical pressure P_c can be determined by differentiating $\dot{m}$ with respect to P_2 in Equation 8.44 and setting the result equal to zero. This operation gives

$$\frac{P_c}{P_1} = \left(\frac{2}{\gamma+1}\right)^{\frac{\gamma}{\gamma-1}} \qquad (8.45)$$

At this critical pressure ratio in Equation 8.45, $P_t = P_c$. If we substitute Equation 8.45 into Equation 8.42, the gas velocity u_t at the throat is

$$u_t = \sqrt{\frac{\gamma P_c}{\rho_c}} \qquad (8.46)$$

It equals the condition of the speed of sound at the throat, according to the definition of sound speed. When the downstream pressure is reduced below P_c, downstream pressure cannot transmitted back into the throat of the orifice, because the gas in the throat is moving with the same velocity of pressure propagation, that is, the speed of sound. When the downstream pressure P_2 is less than P_c, the pressure in the throat will not be affected by the downstream pressure and P_t is always equal to P_c. At this critical condition ($P_2 \le P_c$), P_t equals P_c and the mass flow rate reaches the maximum. Substituting Equation 8.45 into Equation 8.44, the maximum mass flow rate can be obtained with the following equation:

$$\dot{m}_{max} = \frac{A_t P_1}{\sqrt{T_1}} \sqrt{\frac{\gamma}{R} \left(\frac{2}{\gamma + 1} \right)^{\frac{\gamma+1}{\gamma-1}}} \qquad (8.47)$$

where T_1 is the absolute temperature of the air upstream and R is the gas constant.

Equation 8.47 shows that when the pressure ratio P_2/P_1 is less than the critical pressure ratio P_c/P_1, the mass flow rate depends only on the upstream conditions and will not be affected by the downstream pressure P_2. Therefore, the mass flow rate will always equal the maximum value $\dot{m}_{max}$ and remain constant, regardless of pressure changes downstream — as long as the pressure ratio between downstream and upstream is less than the critical pressure ratio.

For a convergent–divergent nozzle, the maximum flow rate can be reached when the pressure at the throat equals the critical pressure. However, the downstream pressure can be greater than the critical pressure, because the divergent section can recover some pressure (Figure 8.21). As the pressure ratio decreases from unity, the flow increases as the back pressure decreases. When the back pressure is reduced to a certain value, the flow rate reaches a limiting flow, because the throat is at the choked condition, that is, the velocity equals the speed of sound. This back pressure, at which the maximum flow rate is reached, is usually called the *first critical downstream pressure* P_{dfc}.[26] The flow rate will remain constant if the back pressure is less than the first critical pressure. That is, the first critical pressure is the highest downstream pressure at which the flow reaches a stable flow. It is obvious that the flow rate remains constant in a convergent–divergent venturi nozzle, provided that the back pressure is lower than the first critical pressure, because the throat is choked

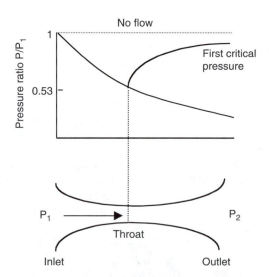

Figure 8.21 Pressure P vs. position in a convergent–divergent nozzle. P, pressure at different position in the nozzle; P_1, inlet pressure; P_2, exit pressure. (Revised from Robert and Benedict, *Fundamentals of Gas Dynamics,* John Wiley & Sons, New York, 1983.)

and the velocity in the throat remains sonic. In this case, the maximum flow rate will be the same as Equation 8.47.

According to the theoretical analysis and Equation 8.47, when an airflow venturi works at a state of critical condition, the volumetric flow rate Q can be determined with the following equation:

$$Q = \frac{\dot{m}}{\rho_1} = A_t \sqrt{\frac{\gamma P_1}{\rho_1}\left(\frac{2}{\gamma+1}\right)^{\frac{\gamma+1}{\gamma-1}}} \tag{8.48}$$

All equations are developed based on the assumptions of isentropic condition and perfect gas. However, there is friction in real situations; the discharge coefficient c_a is introduced in the volumetric flow rate.

$$Q = c_a A_t \sqrt{\frac{\gamma P_1}{\rho_1}\left(\frac{2}{\gamma+1}\right)^{\frac{\gamma+1}{\gamma-1}}} \tag{8.49}$$

From Equation 8.49, using $A_t = \pi d_t^2/4$ and $P_1/\rho_1 = RT_1$, the throat diameter d_t can be determined with the following equation:

$$d_t^2 = \frac{4Q}{c\pi}\sqrt{\frac{1}{R\gamma T_1}\left(\frac{\gamma+1}{2}\right)^{\frac{\gamma+1}{\gamma-1}}} \tag{8.50}$$

For example, when the gas to be measured is air and the assumed experimental conditions are $T_1 = 293$ K, $P_1 = 1.013 \times 10^5$ Pa, $\rho_1 = 1.19$ kg/m^3, and $\gamma = 1.41$, the required throat diameter d_t can be calculated from the volumetric flow rate Q and the discharge coefficient c_a:

$$d_t = \sqrt{\frac{Q}{157.6c_a}} \tag{8.51}$$

The discharge coefficient c depends on the configuration of the venturi. The discharge coefficient c is a function of venturi parameters and can only be determined experimentally. The process of the throat diameter design is iterative.

According to theoretical analysis, the pressure ratio P_2/P_1 for critical flow in the convergent–divergent nozzle can be greater than the theoretical critical pressure ratio in the convergent nozzle (0.53). It is desirable to keep a constant flow for a pressure drop as low as possible. Therefore, the design of the critical venturi is based on the convergent–divergent venturi nozzle.

Based on the preceding analysis, a low–critical pressure venturi for flow rate control was developed by Wang and Zhang.[28] Through an orthogonal experimental

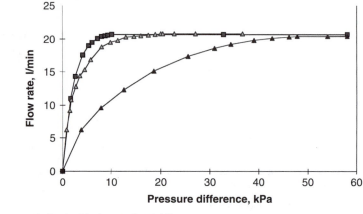

Figure 8.22 Comparison of optimized critical venturi with nonoptimized critical venturi and critical plate orifice.

design procedure, the critical pressure drop of the optimized critical venturi at a stable airflow is about 9 kPa and the pressure recovery efficiency is about 90%. Figure 8.22 shows the flow rate vs. pressure across a critical plate and two critical venturi.

Confirmation testing showed that the critical venturi with the optimized design gave the design flow rate at a low critical pressure drop. Twenty-seven critical venturi with the same specifications were fabricated and tested to check precision. The experimental results showed that the critical flow rates of all venturi nozzles are 21 ± 0.3 l/min. The critical pressure drops of all venturi nozzles are below 11 kPa. The pressure recovery efficiencies are between 0.89 and 0.92, and the discharge coefficients are from 0.93 to 0.96. The variation in flow rate is ±1.5% of the mean flow rate. That is, the maximum difference is 3%. This is consistent with the variation of discharge coefficient. The difference in performance is probably caused by the difference in the discharge coefficients and by inaccuracies in fabrication and experiment. For multipoint air sampling, the item of most concern is the consistency of the flow rate for each sampling head. These critical venturi are being used in the development of a multipoint air sampler for the study of spatial distribution of aerosol concentration.

8.8 SAMPLING OF PM EMISSION FROM BUILDINGS

For many ventilated buildings, such as animal production buildings, particulate matter (PM) emission often must be quantified. The PM from such sources usually has a wide and irregular size distribution, resulting in high variation in sampling efficiency. In addition, the airflow profile at the exits or near the fans can be irregular and complicated, adding more errors to sampling efficiency. All of these factors make fractional PM sampling, including PM2.5 and PM10, a very challenging task.

Other than fractional PM sampling, total suspended particulate (TSP) sampling is a reasonably accurate measurement (together with ventilation rate) for determining building emission rate. TSP is defined as all of the suspended particulate matter in a volume of air. This definition may still be ambiguous, considering the fact that a particle may settle out during sampling. Because there is no defined upper limit to the size, it is very difficult to design a TSP sampler and evaluate it properly, because large particles tend to settle much faster than small ones. The EPA reference method sampler for TSP had a wide range of cutsizes, depending on sampling procedure and on wind speed, etc., but the commonly accepted cutsize is about 35 μm, which includes most particles for a large variety of buildings.

Another measurement error arises from the sampling flow rate. Measurement and control of flow rate are critical to the accuracy of air sampling, because the PM concentration is determined by the ratio of PM collected to the air volume passed through the sampler. Unfortunately, sampling airflow rate measurements are often inaccurate because the flow rate varies.[29] Most flow meters are calibrated at atmospheric pressure, and many require pressure corrections when used at other pressures. Such corrections must be based on the static pressure measured at the inlet of the flow meter. The flow meter should be downstream of the filter to preclude sample losses in the flow meter. Therefore, the sampled air will be at a pressure below atmospheric pressure, due to the pressure drop across the filter. Furthermore, if the filter resistance increases due to dust load, as often occurs in practical sampling, the pressure correction will not be a constant factor. During the sampling period, the filter tends to become filled and the flow rate may decrease as filter resistance increases.[30] This makes it difficult to control the sampling flow rate accurately.

The following section describes a sampler that has been used successfully in PM concentration measurement for buildings with a free entry to the exhaust fans. Figure 8.23 shows a schematic of such a sampler, developed at the Bioenvironmental Engineering Systems and Structures (BESS) laboratory at the University of Illinois.

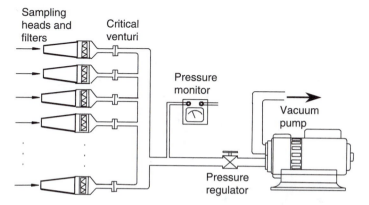

Figure 8.23 A schematic of a particle sampler for measurement of building particle emission or spatial concentration distribution when multiple sampling heads are used. The sampler can be used for TSP or fractional particle (PM2.5 and PM 10) sampling with corresponding sampling heads.

The sampler consists of an array of isokinetic sampling heads, each attached to a critical venturi to control the sampling rate accurately. Using the critical venturi described previously, the sampling rate will be consistent as long as the air pump maintains a pressure across the sampling head that is higher than the critical pressure. Each sampling head can have a flow rate ranging from 2 to 40 l /min. The central pump can provide simultaneous airflow for up to 50 sampling heads. Operating pressure for the air pump is recommended to be 50% higher than the critical pressure, to ensure that the critical venturi function properly even under conditions of dust loading and power fluctuations. The sampler can be used for TSP or fractional particle (PM2.5 and PM 10) sampling with corresponding sampling heads.

For particles with a large mass median diameter, there usually is a large spatial variation in concentrations along the building height. Therefore, it is desirable to measure several points. It was found that three samplers per fan are sufficient to measure the concentration variations accurately. In most situations, there is a greater variation in concentration vertically than horizontally; therefore, the three samplers should be arranged vertically on the centerline of the fan. One sampler can be located in the center of the fan and the others approximately 0.7 of the fan radius from the center. Exact locations may have to be adjusted due to physical constraints. If necessary, a horizontal arrangement can be used, with the same parameters used in the vertical arrangement just described, if the sampling locations are representative of room air.

As discussed in the section on sampling theory, it is important to achieve isokinetic sampling and to minimize the transport losses in the sampling line. To accomplish this, a simple, sharp-edged sampling head can be used. The diameter of the sampling head should match the velocity of the air at the sampling point. It is important to measure the sampling point flow-field air velocity so that an appropriately sized sampling head can be selected. Figure 8.24 shows the schematic side view of air velocity contours at the upstream of an exhaust fan on a large wall. The

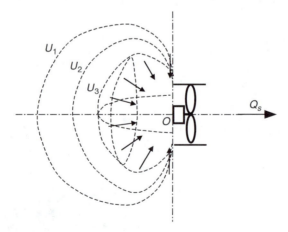

Figure 8.24 A schematic side view of air velocity contour at the upstream of an exhaust fan on a large wall. The contour may be asymmetric or distorted when the fan is adjacent to the floor, ceiling, or other obstruction.

air inhaled by the fan is free of obstruction, and the velocity contours are symmetric. The contour may be asymmetric or in rather distorted shapes when the fan is adjacent to the floor, ceiling, or some other obstruction. A field flow air velocity contour U_i equal to the sampling velocity U_s must be determined with an error of less than 20%. From the sampling efficiency analysis described in previous sections, if $U_i =$ 0.8 to 1.25 U_s, the sampling error will be smaller than 8%.

This sampling system is designed to be reasonably accurate and simple to use. Although the sampling head design is critical, much of the rest of the system is flexible. The filter used should have at least 99.9% efficiency and be appropriate for gravimetric analysis. Glass fiber filters are the most common and inexpensive, but Teflon or other appropriate membrane filters can be used if other types of analysis are needed in addition to gravimetric. Overall, the system provides stable flows even at fairly high particle loading and is much less expensive than an equivalent number of stand-alone samplers. The entrance and expansion slopes of the sampling head were designed based on existing EPA standards for method 5 stack sampling systems. The 15° slope of the entrance is satisfactory for the sharp-edged criterion while still being practical to manufacture. The 6° (one-sided) slope of the expansion was chosen because it fits within the range of EPA stack sampling head expansions and makes the total sampling head length reasonable. A more common design criterion is 3° to avoid eddy formation, but this was not considered necessary for sampling rates higher than 2 l/min.

After the sampling locations have been selected, to the method of supporting the sampling heads will depend on location and available materials. One option that has been used successfully is a ring stand resting on the ground or suspended from the ceiling to hold the sampling heads. This system allows for flexibility and the fine-tuning of sampling head positions. Another option involves installing rigid tubing immediately behind the filter holder, with the venturis near the pump. This option allows the rigid tubing to be attached to a pole, running vertically with adjustable zip ties. With this method, the distance from the fan can be easily adjusted. Alternative methods can be used, as long as they do not obstruct the airflow patterns near the sampling head.

Measuring the PM concentration (TSP, PM2, or PM10) is only half the task of PM emission measurement. The other half is to measure the ventilation rate of the building, because the pollutant emission rate equals the pollutant concentration multiplied by the ventilation rate. Ventilation rate measurement will be discussed in Chapter 14.

DISCUSSION TOPICS

1. According to the definition of sampling efficiency, sampling efficiency for particles may vary widely. Is this true for gaseous contaminants?
2. Sampling velocity is defined as the air velocity at the face of a sampling head. Why? Why is the definition not the sampling velocity in the sampling tube or another portion of the instrumentation?

3. Does sampling efficiency mean the accuracy of the instrumentation measuring the contaminant concentration?

4. Why is isokinetic sampling the most important concept in airborne particle sampling? Is it equally important for gas sampling?

5. Why should isokinetic sampling be exercised whenever possible, even though a misalignment arrangement can compensate for sampling efficiency?

6. What are the important variables affecting sampling efficiency? Can you prioritize these factors so that you can consider them in practice?

7. Does calm air apply to all room conditions or all samplers? What parameters affect the criteria for calm air?

8. If a sampler diameter does not satisfy the calm air criteria, is there any way to compensate for the sampling error?

9. Once calm-air criteria are established and the free flow is considered calm air, is the sampling efficiency always 100%, regardless of the sampler diameter or orientation?

10. In the design of a sampler head in calm-air sampling, there is a particle size at which the maximum equals the minimum of the sampler head diameter. For particles larger than that particle size, the calculated minimum sampler diameter will be larger than the maximum diameter. What criteria should be used under this circumstance?

11. There is no sampling efficiency analysis (equation) for calm-air sampling. Does that mean calm air sampling has no sampling efficiency concern? If there are sampling efficiency concerns, what are they?

12. Why is it recommended that the total number of sampling points be higher than the number of subannulars in a circular duct sampling?

13. If a long tube is used to connect the sampling head and the analytical instrument for the air sample, what factors should be considered for the gas of concern? What factors for the particle of concern?

PROBLEMS

1. Determine the diameter of a sampling head for isokinetic sampling. The air velocity in the free-flow field is 3 m/s, and the sampling pump is 20 l/min.

2. Intending to collect an average particle concentration in a large building, you carry a sampler and walk around the building. The sampler has a diameter of 30 mm and a sampling rate of 28.3 l/min. Assume that the sampler is in alignment with the walking direction and that the room air is calm. How fast should you walk?

3. Air velocities at two points in a duct are 2 and 5 m/s, respectively. A single-point sampler with a 10 mm diameter head is used to measure interchangeably the particle concentrations at the two points. The sampling head is facing and in alignment with the flow. At what sampling rates can isokinetic sampling be attained at each sampling point?

4. In order to measure the particle concentration in a fume hood, you use a sampling head of 10 mm in diameter. The sampling rate is 20 l/min. The mean air velocity in the hood is 1.2 m/s. Assuming that the sampling head is in alignment with the airflow in the hood, what is the sampling efficiency for particles with a diameter of 10 μm?

5. A laser particle counter has a constant sampling rate of 28.3 l/min and a sampler head with a diameter of 30 mm. Assume that you use this instrument to measure the particle emission rate at the outlet of an exhaust fan in a building. The air velocity at the outlet of the fan is 6 m/s. If the sampler is in alignment with the flow direction, what is the sampling efficiency for particles of 5 μm in diameter? To achieve a 100% sampling efficiency for particles of 5 μm in diameter, what should the tilt angle be?

6. A sampling head is misaligned with the airflow at an angle of 30°. The particles of concern are 10 μm. The free-flow velocity is 3 m/s, and the sampling rate is 20 l/s. Determine the sampling head diameter so that the sampling efficiency can be 1.

7. A commonly used particle mass sampler is 37 mm in diameter. In typical indoor environments, the largest particles of concern are about 20 μm in diameter and the air velocity is about 0.25 m/s. What should the range of sampling rate (minimum and maximum) be to satisfy the criteria of calm-air sampling?

8. A sampler has a diameter of 37 mm and a sampling rate of 2 l/min. What is the highest room air velocity that can be considered calm air? Assume that the particle size of concern is 20 μm.

9. Determine the orientation of a mass sampler in a room environment with calm air. The sampling rate is 10 l/min and the sampler diameter is 37 mm. The particles of concern are 20 μm in aerodynamic diameter.

10. In problem 9, the air sampler is facing sideways, that is, the axis of the sampler is horizontal. What is the sampling efficiency for 20 μm particles? When the sampler faces upward, more particles may be sampled due to the settling effect. What is the error when the sampler facing upward compared with that of sampler that faces sideways? Assume that a column of air is sampled as shown in the figure below. The particle concentration in the air is uniform. (Hint: During the sampling, the same amount of particles are falling into the column from above as are falling out from the bottom of the column.)

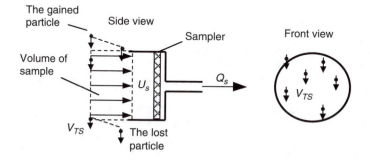

11. In calm-air sampling, assume that the sampling air velocity contour near the sampler is semispherical, as shown in the following figure. The center of the sphere is located at the center front of the sampling head. Derive the sampling efficiencies when the sampler is
   a. Horizontal
   b. Vertical and facing upward

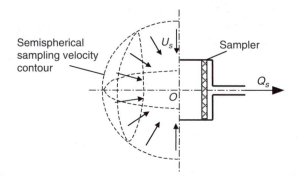

12. A sampler of 30 mm in diameter and with a sampling rate of 28.3 l/min measures a particle concentration of 20 particles/ml for particles of 10 μm. During the measurement, the sampler head is facing upward. Assume that the room air is calm. What is the actual particle concentration?

13. At a circular traverse, four equal-area annulars are divided. Determine the diameters at which the samplers should be located.

REFERENCES

1. Belyaev, S.P. and Levin, L.M., Techniques for collection of representative aerosol samples, *J. Aerosol Sci.*, 5:325–338, 1974.
2. Glauberman, H., The directional dependence of air samplers, *Am. Ind. Hyg. Assoc. J.*, 23:235–239, 1962.
3. Vincent, J.H., et al., On the aspiration characteristics of large diameter, thin-walled aerosol sampling probes at misalignment orientations with respect to the wind, *J. Aerosol Sci.*, 17:211–224, 1986.
4. Durham, M.D. and Lundgren, D.A., Evaluation of aspiration efficiency as a function of Stokes's number, velocity ratio and nozzle angle, *J. Aerosol Sci.*, 13:179–188, 1980.
5. Ogden, T.L., Inhalable, inspirable and total dust, in *Aerosols in the Mining and Industrial Work Environment*, Marple, V.A. and Liu, B.H.Y., eds., Ann Arbor Science, Ann Arbor, MI, 185–204, 1983.
6. Davis, C.N. The entry of aerosols into sampling tubes and heads, *Brit. J. Appl. Phys.*, 1:921–932, 1968.
7. U.S. EPA, *Standards of Performance for New Stationary Sources*, 40 CFR 60, U.S. Government Printing Service, Washington, DC, 1992, 145–1142.
8. ASME, *Flue and Exhaust Gas Analysis, Part 10: Instrumentation and Apparatus, Supplement to ASME Performance Test Code*, PTC 19.10-1981. ASME, United Engineering Ctr., New York, 1980, 9.
9. ASHRAE, *Handbook of Fundamentals*, Atlanta, GA, 1997.
10. ISO, *Standard 3966: Measurement of Fluid Flow in Closed Conduits — Velocity Area Using Pitot Static Tubes*, International Standards Organization, Geneva, 1977.
11. ASHRAE, *Standard 111: Practices for Measurement, Testing, Adjusting and Balancing of Building Heating, Ventilation, Air-Conditioning and Refrigeration Systems*, ASHRAE Atlanta, GA, 1988.

12. ACGIH, *Air Sampling Instruments – For Evaluation of Atmospheric Contaminants*, 8th ed., American Conference of Governmental Industrial Hygienists, Cincinnati, OH, 1995, ch. 21.

13. NIOSH, *NIOSH Manual of Analytical Methods*, 2nd ed., DHEW Publication No 94-113, National Institute for Occupational Safety and Health, Morgantown, WV, 1994.

14. OSHA, *OSHA Analytical Methods Manual*, OSHA Analytical Laboratories, Salt Lake City, UT, 1985. Available from ACGIH, 1330 Kemper Meadow Dr., Cincinnati, OH 45240-1634.

15. Intersociety Committee, *Methods of Air Sampling and Analysis*, 3rd ed., Lewis Publishers., Chelsea, MI, 1988.

16. Hinds, W.C., *Aerosol Technology*, John Wiley & Sons, New York, 1999, 206–216.

17. Doebelin, E.O., *Measurement Systems — Application and Design*, 4th ed., McGraw-Hill, New York, 1990.

18. ASTM, *ASTM Standards on Methods of Atmospheric Sampling and Analysis*, 2nd ed, American Society for Testing and Materials, Philadelphia, 1962.

19. Cohen, B.S. and Hering, S.V., *Air Sampling Instruments for Evaluation of Atmospheric Contaminants,* 8th ed., American Conference of Governmental Industrial Hygienists, Cincinnati, OH, 1995.

20. Vaughan, N.P., Construction of critical orifices for sampling applications, *3rd Annual Conference of the Aerosol Society*, 1989.

21. Zimmerman, N.J. and Reist, P.C., The critical orifice revisited: a novel low pressure drop critical orifice, *Am. Ind. Hyg. Assoc. J.,* 45(50):340–344, 1984.

22. Kotrappa, P., Pimpale, N.S., Subrahmanyam, P.S.S., and Joshi, P.P., Evaluation of critical orifices made from sections of hypodermic needles, *Ann. Occup. Hyg.,* 20:189–194, 1977.

23. Druett, H.A., The construction of critical orifices working with small pressure differences and their use in controlling airflow, *Br. J. Industr. Med.,* 12:65–70, 1955.

24. ISO, *International Standard ISO 5167: Measurement of Fluid Flow by Means of Orifice Plates, Nozzles and Venturi Tubes Inserted in Circular Cross-Section Conduits Running Full,* 1st ed., The International Organization for Standardization, Geneva, 1980.

25. Erett, J.B. and Liu, D.C., *Fundamentals of Fluid Mechanics*, McGraw-Hill, New York, 1987.

26. Binder, R.C., *Advanced Fluid Dynamics and Fluid Machinery*, Prentice Hall, New York, 1953.

27. Robert, P. and Benedict, P.E., *Fundamentals of Gas Dynamics*, John Wiley & Sons, New York, 1983.

28. Wang, X. and Zhang, Y., Development of a critical airflow venturi for air sampling, *J. Bio. Eng. Res.,* 73:257–264, 1999.

29. Cohen, B.S. and Hering, S.V., *Air Sampling Instruments for Evaluation of Atmospheric Contaminants*, 8th ed., American Conference of Governmental Industrial Hygienists, Cincinnati, OH, 1995.

30. ASTM, *ASTM Standards on Methods of Atmospheric Sampling and Analysis*, 2nd ed., American Society for Testing and Materials, Philadelphia, 1962.

CHAPTER 9

Deposition, Production, and Resuspension of Airborne Particles

In indoor air quality study, design, and trouble shooting, production and deposition rates of particulate contaminants often need to be estimated. *Particle production rate* refers to the rate of particles being generated within the airspace and becoming airborne. Particles not generated within the airspace or not airborne are not included in the production rate. *Particle deposition rate* refers to the rate of airborne particles settling onto the surfaces of the airspace. Quantification of production and deposition rates of particles within an airspace is important in at least two aspects. First, it is useful for implementing alternative particulate matter (PM) abatement strategies. For example, if the production rate is high yet the deposition rate is low, the airborne particle concentration will be high, and increasing the deposition of particles can be an effective air-cleaning strategy for reducing airborne particle concentration. Second, it provides key parameters for various models describing the mass balance and transportation of particles in airspaces. Such models can be used to predict or evaluate air quality control strategies and to design or improve ventilation systems.

By completing this chapter, the reader will be able to

- Determine deposition rates under different deposition modes for monodisperse particles:
  - Quiescent batch deposition
  - Perfect-mixing batch deposition
  - Continuous perfect-mixing deposition
  - Deposition in incomplete-mixing airspaces
- Determine the deposition rate for polydisperse particles
- Calculate and experimentally determine particle the production rate in ventilated airspaces
- Apply deposition principles to particle sampling and air cleaning:
  - Elutriators for particle size characterization

- • Settling chambers for air cleaning
- • Determine the range of air velocity that may cause particle resuspension in a flow field

9.1 QUIESCENT BATCH SETTLING FOR MONODISPERSE PARTICLES

In order to determine the deposition rate of particles for all sizes, we consider first the simplest case — deposition of monodisperse particles in a calm-air control volume in a batch mode (i.e., without flow-in or flow-out), also referred to as *quiescent batch settling* (Figure 9.1). The deposition on vertical surfaces, effect of diffusion, and resuspension are assumed to be negligible for quiescent batch settling. Assume that the control volume V has a height of H and a deposition area (projected area at the horizontal surface) A. The control volume is initially filled with monodisperse particles with a concentration of C_{pni}.

Assume that all particles settled on the deposition surfaces are removed and that the change of total particles in the controlled volume at any time is the same as the deposition rate of the particles, dN_{di}/dt. Therefore

$$\frac{dN_{pi}}{dt} = -\frac{\partial}{\partial t} \iiint_V C_{pni}\, dV = -\iint_A C_{pni} V_{di}\, dA \qquad (9.1)$$

where V_{di} and C_{pni} are the deposition velocity and number concentration of the particles in i^{th} size range, respectively, and A is the deposition surface area. The minus sign (−) indicates that the total number of airborne particles in the control volume is decreasing with time.

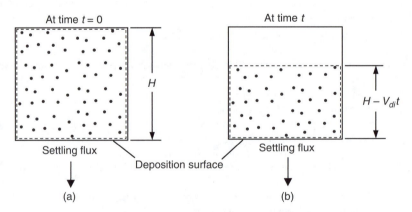

At time $t = 0$
H
Settling flux
Deposition surface
(a)

At time t
$H - V_{di}t$
Settling flux
(b)

Figure 9.1   A front view of a deposition of quiescent batch settling for monodisperse or i^{th} size range particles within a height of H and a deposition surface A. (a) At initial time $t = 0$. (b) At a time t.

Because in quiescent batch settling the concentration C_{pni} and the deposition area A remain constant, the only change is the height of deposition, which changes the deposition volume. At any time t, the deposition height becomes

$$H(t) = H - V_{di}t \tag{9.2}$$

and the total number of particles at that moment is

$$\iiint_V C_{pni}\, dV = C_{pni}A(H - V_{di}t) \tag{9.3}$$

Substituting Equation 9.3 into Equation 9.1 gives the particle deposition rate of ith size range:

$$\frac{dN_{pi}}{dt} = -\frac{\partial}{\partial t}C_{pni}A(H - V_{di}t) = C_{pni}AV_{di} \tag{9.4}$$

Similarly, Equation 9.4 can be used to calculate the mass deposition rate by substituting the particle number concentration for particle mass concentration for ith size range.

$$\dot{M}_{di} = -\frac{\partial}{\partial t}C_{pmi}A(H - V_{di}t) = C_{pmi}AV_{di} \tag{9.5}$$

The mass deposition rate for particles at the ith size range is proportional to the deposition velocity, particle concentration, and deposition area projected on a horizontal surface. The units of concentration [mass/length³] times the velocity term [length/time] equal the units of flux [mass/length²time]. *Flux* is the transport rate of mass to (or through) a surface. Then the deposition rate from Equation 9.5 is equal to the flux times the area, which is a mass transfer rate. The motion of particles in the box in this derivation was only due to settling, and the flux of particles will last for a time equal to H/V_{di}.

In calm air, the particle deposition velocity equals the terminal settling velocity of particles of the same size:

$$V_{di} = V_{TS} \tag{9.6}$$

In calm-air settling, particles settle down in a plug-flow fashion, and all particles will eventually settle down. Apparently, the minimum time required for particles at a given size to deposit completely onto a horizontal surface from a height H is

$$t_{min} = \frac{H}{V_{TS}}$$ \hfill (9.7)

where H is the vertical distance between the initial position of the particle and the settling surface. Equation 9.7 gives the relationship between time and the height of a settling chamber. To allow complete particle deposition, the settling time should be long, the settling height should be short, or both. Equation 9.7 provides a basic guideline for settling chamber design.

9.2 DEPOSITION OF PERFECT-MIXING BATCH SETTLING

In the previous section, we considered a deposition in batch mode without mixing, that is, the concentration remains constant during the entire settling period. Now, consider that the particles in the same control volume are continuously and perfectly mixed. The particles are still able to settle to the bottom of the box, and we still assume that the settled particles are removed from the system at a deposition rate, but because some particles are lost to sedimentation, we imagine that the particles remaining in suspension are still being mixed (Figure 9.2). At any time, the deposition rate is given by Equation 9.4 or Equation 9.5. However, the concentration is no longer constant in the perfect-mixing mode, as more and more particles settle out of the deposition surface of the control volume.

We first need to solve the particle concentration with respect to time. Note that the total volume involved in deposition remains constant in perfect-mixing batch settling; the change of total particles in the control volume is caused by the change in particle concentration. Assuming that the particle concentration is C_{pi0} initially

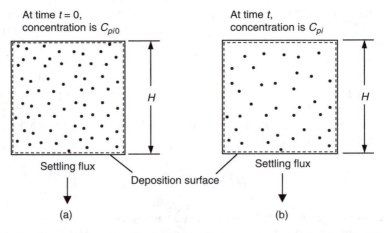

At time $t = 0$,
concentration is C_{pi0}

At time t,
concentration is C_{pi}

H

H

Settling flux

Settling flux

Deposition surface

(a)

(b)

Figure 9.2   A front view of a deposition of perfect-mixing batch settling for monodisperse or ith size range particles within a height of H and a deposition surface A. (a) At initial time $t = 0$. (b) At a time t.

and $C_{pi}(t)$ at any given time, from Equation 9.1 and the mass balance of particles within the control volume, we have

$$AH\frac{\partial C_{pi}}{\partial t} = -C_{pi}AV_{di} \qquad (9.8)$$

The minus sign $(-)$ indicates that the concentration is decreasing. Simplifying Equation 9.8 yields

$$\frac{dC_{pi}}{C_{pi}} = -\frac{1}{H}V_{di}\,dt \qquad (9.9)$$

Integrating with initial conditions $C_{pi} = C_{pi0}$ at $t = 0$ gives

$$C_{pi} = C_{pi0}\exp\left(-\frac{V_{di}\,t}{H}\right) \qquad (9.10)$$

Substituting Equation 9.10 into Equation 9.4 gives the particle number deposition rate for perfect-mixing batch settling:

$$\frac{dN_{pi}}{dt} = C_{pni}AV_{di} = AV_{di}C_{pi0}\exp\left(-\frac{V_{di}}{H}t\right) \qquad (9.11)$$

Notice that the particle deposition rate for quiescent batch settling does not depend on the height of the control volume, whereas the deposition rate for perfect-mixing batch settling does depend on the height H. The deposition rate for perfect-mixing batch settling is an exponentially decreasing function, whereas the flux for quiescent sedimentation was considered constant (only lasting a finite amount of time).

Example 9.1: A rectangular chamber is initially filled with particles 5 μm in aerodynamic diameter at an initial concentration of 10 particles/ml. The chamber measures 5 m long, 3 m wide, and 2.4 m high. Assume that the room air is completely mixing. What percentage of particles was removed by deposition after 8 minutes of settling?

Solution: This is a particle deposition under perfect-mixing batch settling. The average deposition velocity is the same as the terminal settling velocity.

$$V_d = V_{TS} = \frac{\rho_0 d_a^2 g}{18\eta} = \frac{1000\times(5\times10^{-6})^2\times9.81}{18\times1.81\times10^{-5}} = 0.000753\ (m\,/\,s)$$

From Equation 9.10, the particle concentration after 8 minutes is

$$C_{pi} = C_{pi0} \exp\left(-\frac{V_{di}\,t}{H}\right)$$

$$= 10 \times \exp\left(-\frac{0.000753 \times 480}{2.4}\right) = 8.64 \left(\frac{particles}{ml}\right)$$

Because the chamber is completely mixed, the total number of particles removed by deposition is the difference between the initial total number and the total number remaining airborne after 8 minutes. The percentage of particles removed, ξ_d, is

$$\xi_d = \frac{C_{pi0} - C_{pi}}{C_{pi0}} \times 100\% = 13.4\%$$

9.3 CONTINUOUS PERFECT-MIXING MODELS

In previous sections, we considered two simplified deposition models for monodisperse particles: quiescent batch settling and perfect-mixing batch settling. In the quiescent batch model, the particle concentration remains constant, and the volume involved in the settling process decreases by a rate of $V_{di}t$. In the perfect-mixing batch model, the volume involved in settling process remains constant, and the particle concentration decreases exponentially. Both models are in batch mode, that is, there is no flow-in or flow-out of the control volume.

Although these models are useful in some applications, such as particle measurement instrumentation and settling chambers, they have a major limitation in indoor air quality studies, because most indoor environments do not operate in a batch mode. Occupants, including humans, animals, plants, and even many stored goods, must be sustained by exchanging air, that is, by ventilation. A ventilated airspace is a *continuous-mixing model,* with flow-in and flow-out. In the continuous-mixing model, we first consider a simplified case — steady-state under perfect mixing conditions. That is, the deposition volume, particle concentration, and ventilation rate are all constant, and the air within the airspace is completely mixed.

9.3.1 Macro-Mixing Model without Source or Sink

We first consider, using the control volume approach, the mass balance of the PM in a simplified case — perfect mixing without a particle source or a sink within the airspace (Figure 9.3). Assume that the ventilation rate is Q. Particle concentrations at the supply air inlet and exhaust outlet are C_s and C_e, respectively, and the mass balance for airborne particles within the control volume can be written as

$$\frac{d(V\,C)}{dt} = QC_s - QC_e - V_d A C \tag{9.12}$$

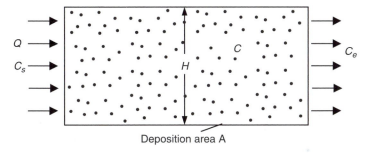

Figure 9.3 Schematic of continuous perfect-mixing deposition.

where V is the volume, A is the deposition area of the airspace, and C is the particle concentration within the airspace.

Because the airspace is completely mixed and the system is in steady state, $C_e = C$ and $d(VC)/dt = 0$. Therefore

$$QC_e = QC_s - V_d A C_e \qquad (9.13)$$

Rearranging Equation 9.13 gives us the particle deposition velocity in a perfectly mixed model:

$$V_d = \frac{Q(C_s - C_e)}{AC_e} \qquad (9.14)$$

When $C_s > C_e$, the deposition velocity is positive and particles will settle, thus it is a settling chamber. However, when $C_s < C_e$, the deposition velocity becomes negative. This indicates that there must be a particle source within the airspace. In practice, particle sources (such as cooking activities in a house) and sinks (such as a furnace filter) exist in an indoor environment. In this case, Equation 9.14 cannot predict the deposition velocity, because the left term is actually a "net" rate of particle change within the room, including the effect of a particle source and/or sink.

9.3.2   Average Vertical Velocity Model

To overcome the limitations of the macro-mixing model, the particle deposition velocity should be derived directly from the particle terminal settling velocity and its surrounding air velocity. Particle deposition velocity in calm air is the same as the terminal settling velocity. However, indoor air is rarely calm and has typical velocities of 0.1 m/s or higher, which are much higher than the settling velocities of most particles. Obviously, at locations where the upward air velocity is higher than the particle settling velocity, the particles will remain airborne. On the other hand, at locations where the room's air velocity is downward, particle deposition velocity will be higher than settling velocity. If the room air has an average air

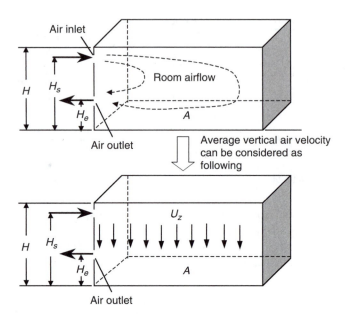

Figure 9.4   In the sense of statistical particle settling, the average vertical air velocity of a completely mixed airspace can be considered a plug flow with an average velocity proportional to the ventilation rate and the heights of the air inlet and outlet.

velocity downward, the overall deposition velocity will be higher than the settling velocity, in a statistical sense. When the effect of this average downward air velocity is not included, the deposition velocity tends to be underestimated. When the effect of average upward air velocity is not included, the deposition velocity tends to be overestimated.

In order to derive equations to calculate particle depositions in ventilated airspaces, we again assume that the room air is completely mixed. The particle deposition velocity V_{di}, as shown in Figure 9.4, is the sum of the particle settling velocity V_{TS} and the vertical component of the room air, U_{zi}. For the entire airspace, U_{zi} can be replaced by the mean vertical air velocity U_z. Thus

$$V_{di} = V_{TS} + U_z \tag{9.15}$$

Because the room air is completely mixed, we can treat the entire room as a control volume. In one extreme case, the air is supplied from the top and exhausted from the bottom. The airflow is a plug flow from the top to the bottom, and the average vertical air velocity over the entire room height is Q_e/A, where Q_e is the flow rate of the exhaust air and A is the floor area of the airspace. In another extreme case, the air inlet is at the same height as that of the exhaust, and the mean vertical air velocity over the entire room height H is zero, because the upward and downward air velocities will cancel each other. Therefore, the mean vertical room air velocity is proportional to the height difference between the air inlet and outlet

$$U_z = \frac{1}{N} \sum_{i=1}^{N} U_{zi} = \frac{Q_e(H_s - H_e)}{A\,H} \tag{9.16}$$

where H_s and H_e are the distances from the floor to the midpoints of the supply air inlet and the exhaust air outlet, respectively. H is the height of the airspace. Substituting Equation 9.16 into Equation 9.15 gives

$$V_{di} = V_{TS} + \frac{Q_e(H_s - H_e)}{A\,H} \tag{9.17}$$

From Equation 9.17, it is desirable to position supply the air inlet higher than the exhaust air outlet so that particle deposition velocities are increased and airborne particles can be reduced. Note that the deposition velocity may be negative when $H_e > H_s$ and the ventilation rate is high. In that case, the particle may not settle and thus remains airborne. In reality, deposition velocity in some zones will be still greater than zero, and particle deposition will occur in these zones at a lower rate.

Equation 9.17 agrees reasonably well with experimental data collected in rooms with $H_s > H_e$. In one study, settling dust and airborne dust were measured in 20 locations of $12 \times 12 \times 3.3$ m $(L \times W \times H)$.[1] The measured V_d was 60% higher than the calculated V_d when U_z was not considered. When U_z was considered, the measured V_d agreed with the predicted V_d, with approximately 10% of error.

Example 9.2: *A room measures $5 \times 4 \times 3$ m $(L \times W \times H)$. The supply air inlet and exhaust air outlet are 2.5 m and 0.5 m from the floor, respectively. The air exchange rate of the room is 0.01 m³/s. The initial concentration of 5 μm (in aerodynamic diameter) particles is 0.01mg/m³. Find the particle deposition rate.*

Solution:

$$U_z = \frac{Q_e(H_s - H_e)}{A\,H} = \frac{0.01 \times (2.5 - 0.5)}{5 \times 4 \times 3} = 3.33 \times 10^{-4} \ (m/s)$$

Because the particle is larger than 3 μm, the slip effect can be neglected.

$$V_{TS} = \frac{\rho_0 d_a^2\, g\, C_{ca}}{18\eta} = \frac{1{,}000 \times (5 \times 10^{-6}) \times 9.81}{18 \times 1.81 \times 10^{-5}} = 7.53 \times 10^{-4} \ (m/s)$$

$$V_{di} = U_z + V_{TS} = 1.086 \times 10^{-3} \ (m/s)$$

The deposition rate is

$$\dot{M}_{di} = V_{di}AC = 1.086 \times 10^{-3} \times (5 \times 4) \times 0.01 = 0.00217 \ (mg/s)$$

Very high ventilation may create high air velocities and turbulence intensity within the airspace, which may cause reentrainment of settled particles. Air velocity at the particle deposition surfaces rarely exceeds the resuspension threshold limit.[2] For indoor environments, the threshold limit to resuspend particles from a surface is higher than 1 m/s.[3] In general, the particles are assumed remain settled unless they are disturbed by other external forces, such as occupants' activities.

9.4 DEPOSITION IN INCOMPLETE MIXING AIRSPACES

Most ventilated airspaces are incompletely mixed to some extent. Therefore, solving for deposition velocities for specific locations relies on the knowledge of the local air velocity profile, which is often difficult to obtain. Mathematical models of particle deposition under incomplete mixing conditions can be overwhelmingly complex, involving descriptions of the entire air flow field, particle size distribution, and detailed boundary conditions of the airspace.

One practical approach for incomplete mixing airspaces is the *multizone model*. In many indoor environments, the airspace of concern can be divided into n zones, and each zone can be considered a perfectly mixed space (Figure 9.5). Each zone can have a different volume or deposition area. The total deposition is the sum of the depositions in n zones. One important step in applying the multizone model is to divide the zones appropriately to ensure that the characteristics, including air velocity, particle concentration, particle source, and sink, are similar within the same zone. A division that is too coarse will introduce large errors, and a fine division may increase unnecessarily the workload of calculation. The following example is intended to help the reader to generate a simple multizone model and provide rationale for analysis during the process.

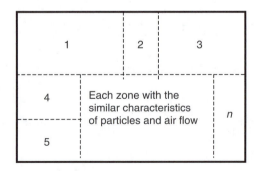

Figure 9.5 An incompletely mixed airspace can be divided into n zones, each zone having similar characteristics and being treated as a completely mixed space.

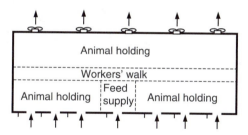

Figure 9.6 Subzones in an animal building.

Example 9.3: In determining the total particle deposition rate in an animal building as shown in Figure 9.6, the floor plan must be divided into a multizone environment. The animal room includes an animal holding area, a workers' walkway, and a feed supply area. All air inlets are on one side wall, and all exhaust fans are on the opposite side wall.

Solution: The room is not completely mixed and should be treated using the multizone model. Because animal activity is a major source of airborne particles, the three animal holding areas should be three different zones. The feed supply area may have a very different particle size distribution and concentration from other locations in the room; therefore, the feed supply area should be a separate zone. The workers' walkway, used occasionally, is very different in terms of disturbance of dust on the floor from the animal holding area. Therefore, the room is divided into five zones, as shown by dashed lines in the floor plan.

9.5 DEPOSITION OF POLYDISPERSE PARTICLES

The particle deposition rate for all sizes of particles is the sum of deposition rates for each particle size. From Equation 9.5, we have

$$\frac{dM_d}{dt} = \sum (V_{di} C_{pmi} A) = V_d C_{pm} A \tag{9.18}$$

where C_{pm} is the concentration of particles of all sizes and V_d is the mean deposition velocity for all sizes of particles, defined as

$$V_d = \frac{\sum V_{di} C_i}{\sum C_i} \tag{9.19}$$

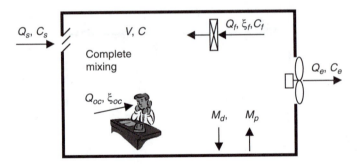

Figure 9.7   Variables and mass balance of airborne pollutants in a ventilated airspace.

9.6 PARTICLE PRODUCTION RATE IN A VENTILATED AIRSPACE

From Equation 9.5, the particle deposition rate is proportional to the particle concentration. Thus, when the particle concentration varies, the deposition rate changes.
Other factors, including ventilation, dust production, and particle filtration through
occupants, also affect the particle concentration. We now consider a ventilated
airspace as control volume with particle sources and sinks (Figure 9.7).

   In order to derive a meaningful and solvable equation, the following assumptions
are used. The ventilation rate, particle size distribution, and thermal conditions within
the airspace do not vary with time. The airspace is completely mixed, and there is
no infiltration or exfiltration. When the airspace is completely mixed, the dust
concentrations are the same for exhaust air and air within the airspace. The changes
of total particle mass with time within the airspace can be expressed as

$$V\frac{dC}{dt} = M_p + Q_s C_s - Q_e C_e - \xi_f Q_f C_f - \xi_{oc} Q_{oc} C_{oc} - M_d \qquad (9.20)$$

where V is the volume of the airspace in m³; C_e, C_f, and C_{oc} are particle mass
concentrations in mg/m³ for exhaust air, recirculation air, and inspiration air of
occupants, respectively; C_s is the particle concentration of the supply air in mg/m³;
M_p is the production rate of particles in mg/s; Q_s and Q_e are air exchange rates for
supply air and exhaust air in m³/s, respectively; Q_f is the airflow rate passing a
recirculation or an air filter in m³/s; ξ_f is the particle removal efficiency of the
recirculation or an air filter; Q_{oc} is the inspiration rate of the occupants within the
airspace in m³/s; ξ_{oc} is the particle removal efficiency of the respiration systems of
the occupants; and M_d is the particle deposition rate in mg/s. Note that the mass
balance expressed by Equation 9.20 is valid for any other airborne contaminants,
such as gases.

   In most biological structures, such as residential and animal buildings, the inspiration rate is much less than the minimum ventilation rate. The filtration efficiency ξ_{oc}
is equal to the aspiration efficiency of the respiratory system. Unless otherwise spec-

Table 9.1   **Inspiration Rate and Minimum Ventilation Rates for Typical Indoor Occupants**

Occupants	Q_{min}[a] (l/kg·h)	Q_{oc} (l/kg·h)
Human[a]	450	20
Production animals[b]	200	20
Laboratory animals[a]	900	20

[a] Based on ASHRAE standard (ASHRAE, Environmental Control for Animals and Plants, *ASHRAE 2001 Fundamentals Handbook*, American Society of Heating, Refrigerating and Air Conditioning Engineers, Atlanta, 2001, 10.1–10.21.)
[b] Based on ASAE Standard (ASAE, Structures, Livestock and Environments, *Standards Handbook*, ASAE, St. Joseph, MI, 2000, 595–698.)

ified, the effect of inspiration on particle removal can be neglected. In the case of a heavy density of occupants, the following linear equation can be used to estimate ξ_{oc}:

$$\xi_{oc} = 1 - \frac{d^2_{mm}}{200} \quad (\text{for } d_{mm} < 14 \; \mu m) \tag{9.21}$$

where d_{mm} is the mass mean diameter of particles in aerodynamic diameter and in μm.

Equation 9.21 approximates the aspiration efficiency of a human nose. When particles are small, the aspiration efficiency is high, and when particles are large, the aspiration efficiency is low. The cutsize (at 50% of aspiration efficiency) for human aspiration is about 10 μm. Although aspiration efficiency varies widely, Equation 9.21 is sufficiently accurate for mass balance analysis in most indoor air environments. Equation 9.21 closely represents the human aspiration efficiency standard.[4, 5] The minimum ventilation rate Q_{min} and the inspiration rate Q_{oc} for humans and animals, in terms of liters per kilogram of body weight per hour, are listed in Table 9.1.

Because the airspace is completely mixed

$$C_e = C_f = C_{oc} = C \tag{9.22}$$

Further, we assume that the room air is isothermal, so $Q_s = Q_e = Q$. Substituting Equation 9.21 and Equation 9.22 into Equation 9.20 gives

$$V\frac{dC}{dt} = \dot{M}_p + QC_s - QC - \xi_f Q_f C - \xi_{oc} Q_{oc} C - V_d AC \tag{9.23}$$

Equation 9.23 describes the mass balance of contaminants within a ventilated airspace. Assuming that the contaminants' production, ventilation, recirculation, and inspiration rates do not vary with time, Equation 9.23 can be solved. Separating variables in Equation 9.12 and integrating gives

$$\int_{C_0}^{C} \frac{dC}{C - \dfrac{M_p + QC_s}{Q + \xi_f Q_f + \xi_{oc} Q_{oc} + V_d A}} = \int_0^t -\frac{Q + \xi_f Q_f + \xi_{oc} Q_{oc} + V_d A}{V} \, dt \quad (9.24)$$

Solving Equation 9.24 for C yields the concentration at any time:

$$C(t) = (C_0 - C_\infty) \exp\left(-\frac{t}{\tau_c}\right) + C_\infty \quad (9.25)$$

and

$$C_\infty = \frac{M_p + QC_s}{Q + \xi_f Q_f + \xi_{oc} Q_{oc} + V_d A} \quad (9.26)$$

$$\tau_c = \frac{V}{Q + \xi_f Q_f + \xi_{oc} Q_{oc} + V_d A} \quad (9.27)$$

where C_0 is the initial particle concentration in the airspace, C_∞ is the steady-state particle concentration in the airspace, and τ_c is the time constant for transient particle concentration when the airspace is subjected to a step change. Apparently, Equation 9.25 becomes Equation 9.26 when $t \to \infty$, which is the steady-state particle concentration. The symbol τ_c is used to distinguish this time constant from the particle relaxation time τ.

At steady state, the particle deposition rate may be measured by sampling the settled particles on surfaces. When V_d and C are measured, M_p can be calculated using Equation 9.23:

$$\dot{M}_p = (Q + \xi_f Q_f + \xi_{oc} Q_{oc} + V_d A) C - QC_s \quad (9.28)$$

In many practical problems, particle deposition and production rates may not be measured or calculated directly. Thus, an indirect method to determine the particle production and deposition rates is desirable. Rewrite Equation 9.28:

$$\dot{M}_p = aV_d + b \quad (9.29)$$

where

$$a = AC \tag{9.30}$$

$$b = (Q + \xi_f Q_f + \xi_{oc} Q_{oc})C - QC_s \tag{9.31}$$

Equation 9.29 is linear and contains two unknowns: M_p and V_d. If we have two sets of values of a and b, M_p and V_d can be solved. Changing one or more of the following parameters — Q, Q_{oc}, or ξ_r — will result in a change of particle concentration within the airspace, and two sets of equations can be obtained.

$$\dot{M}_p = a_1 V_d + b_1 \tag{9.32}$$

$$\dot{M}_p = a_2 V_d + b_2 \tag{9.33}$$

Solving Equation 9.32 and Equation 9.33 simultaneously gives M_p and V_d. Note that changing the ventilation rate is not applicable unless the air inlet is at the same height as the air outlet, because V_d is a function of Q (Equation 9.17). Only when the air inlet and the outlet are at the same height (i.e., $U_z = 0$) is changing Q applicable, because in this case V_d is the same as the terminal settling velocity and is independent of the ventilation rate.

Example 9.4: A room measures $5 \times 4 \times 3$ m $(L \times W \times H)$. Steady-state dust concentration in the room is 0.5 mg/m³ at a ventilation rate of 0.02 m³/s. When an air cleaner with a cleaning efficiency of 0.8 is used to clean room air at a rate of 0.01 m³/s, the steady-state dust concentration becomes 0.35 mg/m³. Assuming that the dust concentration in the supply air is negligible and that there are no occupants within the room, estimate the dust production rate and deposition velocity.

Solution: Because Q_{oc} and C_s are zero and $Q_{f1} = 0$ m³/s, $C_1 = 0.5$ mg/m³, $Q_{f2} = 0.01$ m³/s, and $C_2 = 0.3$ mg/m³

$$a_1 = AC_1 = 5 \times 4 \times 0.5 = 10 \ (mg/m)$$

$$b_1 = (Q + Q_{f1})C_1 = 0.02 \times 0.5 = 0.01 \ (mg/s)$$

$$a_2 = AC_2 = 5 \times 4 \times 0.3 = 6 \ (mg/m)$$

$$b_2 = (Q + \xi_f Q_{f2})C_2 = 0.028 \times 0.38 = 0.01064 \ (mg/s)$$

Substituting these values into Equation. 9.21 and Equation 9.22 gives

$$\dot{M}_p = 10\,V_d + 0.01$$

$$\dot{M}_p = 6\,V_d + 0.01064$$

Solving for M_p and V_d yields

$$V_d = 0.00016 \text{ m/s}$$

$$\dot{M}_p = 0.0116 \text{ mg/s}$$

Alternative indirect methods to determine the particle production are available. Parameter estimation methods allow determination of $\dot{M}_p$ and V_d at both steady-state and transient conditions.[6]

9.6.1 Net Particle Production in a Ventilated Airspace

The net production rate $\dot{M}_{pn}$ is the sum of the production and deposition rates. In many practical indoor air quality cases, it is not necessary to know both deposition and production. Rather, we are interested in how much airborne contaminant is produced within the room, which is the net production.

$$\dot{M}_{pn} = \dot{M}_p - \dot{M}_d \tag{9.34}$$

9.7 APPLICATION OF PARTICLE DEPOSITION

Particle deposition by gravity is a useful property found in many applications. Several instruments rely on particle settling to measure the aerodynamic diameter of a particle or particle size distribution.[7] The simplest is the deposition cell, similar to the Millikan oil-droplet experiment, in which particles are introduced into a sealed volume of less than 1 cm^3 and illuminated through windows by an intense beam of light. The particles are viewed by a horizontal microscope in a direction perpendicular to the light beam. The particles, which appear as tiny specks of light, are timed individually as they settle between calibrated lines marked on the eyepiece of the microscope. The true settling distance is known from calibration and is usually less than 1.0 mm. The aerodynamic diameter is calculated directly from the measured settling velocity.

The Brownian motion of particles less than 0.3 μm in diameter causes significant variation in the measured settling velocity, and such particles may wander out of the field of view. Particles larger than 5 μm settle too fast for accurate measurement. Care must be taken to minimize thermal convection caused by the heat of the light

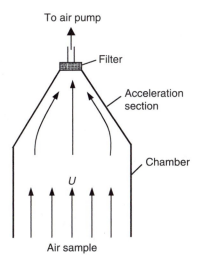

Figure 9.8   A schematic of a vertical elutriator for collection of particles smaller than a specific aerodynamic diameter.

beam. Care must also be taken to prevent operator bias toward the selection of larger and brighter particles.

For particles of 5 to 50 μm, the aerodynamic diameter can be measured directly using a settling tube.[8] Typically, this is a vertical glass tube about 7 mm in diameter and 0.3 to 0.8 m long, illuminated along its axis by a low-power laser. The tube is marked off in 10-cm or other convenient intervals. Particles are seen as points of light, and their settling time over a known distance is measured with a stopwatch. The choice of tube diameter is a compromise between reducing convection currents inside the tube and reducing tube wall effects. This method provides an absolute measurement of the aerodynamic diameter and requires no calibration. For measurement of irregularly shaped particles larger than 50 μm, the particle may not fall vertically and thus may require a larger tube diameter. For example, a flake of dander may flip-flop, glide in the air, and hit the tube wall.

A *vertical elutriator* is a device used to remove particles larger than a certain aerodynamic diameter from a particle-laden air stream (Figure 9.8). The air sample flows upward at a low velocity U in a vertical chamber or duct. Particles having a V_{TS} greater than the duct velocity cannot be carried out of the duct and are thereby removed from the air stream. Particles having a V_{TS} smaller than U will be collected on a filter. The acceleration section has an air velocity higher than U, thus ensuring that all particles that have passed the vertical chamber will be collected on the filter. The situation can be thought of as a contest between the particle and the air stream, with the faster one controlling the outcome (winning). These devices work satisfactorily for rough separation of large particles, but the distribution of gas velocities in the duct makes it difficult to obtain a precise cutoff. The large particles, having a net downward velocity, may "filter out" smaller particles moving upward. One critical factor for this device is to maintain a constant airflow rate so that U does not fluctuate.

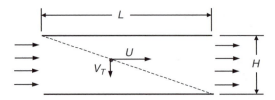

Figure 9.9   A side view of a schematic horizontal elutriator. The dashed line shows the trajectory of a particle for 100% collection.

The largest particle that can be collected by the filter of an elutriator is determined by the airstream velocity in the vertical chamber. In order for a particle to be collected by the filter, the particle terminal settling velocity must be smaller than the airstream velocity U in the vertical chamber. Because $U = Q/A$, where A is the cross-sectional area of the vertical chamber and Q is the airflow rate of the pump, the maximum particle size of a vertical elutriator must satisfy

$$U = V_{TS} = \frac{\rho_0 d_{a\max}^2 g}{18\eta} \qquad (9.35)$$

Noting that $U = Q/A$ and solving for d_{amax} gives the maximum aerodynamic diameter of the particle that the filter of a vertical elutriator can collect:

$$d_{a\max} = \left(\frac{18\eta Q}{\rho_0 g A}\right)^{\frac{1}{2}} \qquad (9.36)$$

The horizontal elutriator can be used either as a separator to fractionate particles in an air stream or as a particle spectrometer to measure the distribution of particle size. For separation, a particle-laden air stream is passed at low velocity through a horizontal duct with a rectangular cross section. Particle settling is perpendicular to the gas streamlines. Particles reaching the floor of the duct are removed from the air stream. This situation also may be viewed as a contest that results in an attenuation in particle concentration that is greater for the larger particles than for the smaller ones.

As shown in Figure 9.9, particles having a $V_{TS} > HU/L$ will be completely removed if the flow is laminar. Smaller particles will have a fraction $V_{TS}L/UH$ of their number removed. Horizontal plates evenly spaced in the duct will reduce H and increase the number collected of a given particle size.

A similar approach is utilized in an elutriation spectrometer to measure the particle size distribution. There are two flow streams used in an elutriation spectrometer: a clean air stream and a particle stream. A horizontal laminar flow of clean air at an average velocity U, called the *winnowing stream* or the *carrier gas stream,* is established in the chamber (Figure 9.10). At the beginning of the separation chamber, the particles are introduced as a thin stream or film along the upper surface of the duct. Particles of different sizes settle at different rates and deposit at different

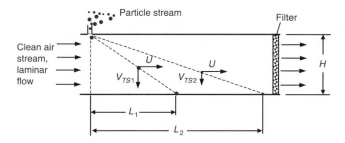

Figure 9.10 A side view of an elutriation spectrometer to measure particle size distribution. Particles of different sizes are deposited at different lengths along the settling chamber.

locations along the duct floor. There is a unique location along the floor where each particle size will deposit. The distance from the inlet to that location for the ith size of particle with a settling velocity V_{TSi} is

$$L_i = \frac{HU}{V_{TSi}} \qquad (9.37)$$

where H is the height of the chamber.

The bottom of the chamber is lined with glass slides or foil that can be removed in sections and analyzed for the number or mass of particles deposited. Each section corresponds to a range of V_{TSi} that defines a range of aerodynamic diameters. The distribution of number (or mass) as a function of aerodynamic diameter is determined from the fraction of the total number (or mass) in each size range. A filter can be placed at the end of the chamber to capture particles smaller than the limiting size. Because $V_{TS} \propto d_a^2$, the chamber must be excessively long to collect small particles. One approach is to use a duct whose width increases along its length to reduce U and permit smaller particles to be separated in an instrument of convenient length. Because gravitational settling of aerosol particles is a slow process, the carrier gas flow rate must be low. Consequently, these instruments are sensitive to convection currents.

The principle of the aerosol centrifuge is similar to that of the horizontal elutriator, except that the force of gravity is replaced by centrifugal force.[9] The elutriation duct is wrapped around a cylinder, with the collection surface at the periphery, and is rotated. The duct is often arranged in a spiral to permit the aerosol and the carrier gas air to be introduced along the rotational axis and to exit at the periphery. The outer side of the channel is lined with a deposition of foil, which is removed, segmented, and analyzed for particle mass to determine the mass distribution as a function of aerodynamic diameter. Because centrifugal force can easily be made much greater than the force of gravity, these devices can operate at a higher sampling flow rate and separate smaller particle sizes than can the horizontal elutriator. They are practical for particle sizes of 0.1 to 15 μm aerodynamic diameter. The radial component of flow velocity causes a secondary flow (eddies), which distorts the

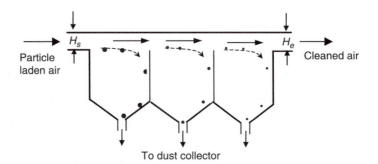

Figure 9.11 A cascade particle settling chamber. Large particles are deposited first in the left-hand bin, and fine particles are deposited in the right-hand bin.

flow stream and limits the resolution for small particle sizes. Some aerosol centrifuges are high-resolution instruments that can separate particles that differ in aerodynamic diameter by only a few percent.

In air cleaning, the settling chamber is an important means to separate particulate matter from an air stream. As shown in Figure 9.11, dust-laden air flows into a cascade chamber. Large particles settle faster than small particles. Therefore, the first bin will collect primarily large particles, the second bin will collect smaller particles than the first one, and the third will collect the smallest particles of the three. Together with diffusion and coagulation, separation of gases in a settling chamber is applicable.

The collection efficiency of such a particle settling chamber can be calculated using the same principle as that of a horizontal elutriator. It is important to maintain a laminar flow along the chamber.

9.8 PARTICLE ADHESION AND RESUSPENSION

In previous sections, we have assumed that once a particle touches a surface, it will deposit on the surface and will not reenter the air stream. In many practical situations, particles do touch a surface but reenter the air stream through bounce or a disturbance such as human activity, a gust of air, or a vibration. This section examines the detachment mechanism and the resuspension of particles.

9.8.1 Particle Balance under Adhesive Force

When a particle on a surface is detached and reenters the air stream, a detachment force is needed to overcome the adhesion force between the particle and the surface. In order to analyze the balance of a particle under adhesive force, let us consider a particle deposited on a vertical surface. Assume that the particle and the surface are electrostatic-free and that the particle is subjected to a flow field with a bulk air velocity U. Thus, the particle is subjected to three forces (Figure 9.12):

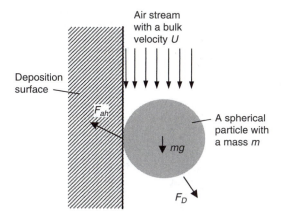

Figure 9.12 A particle deposited on a surface is subjected to three forces: adhesion, drag, and gravity.

- Adhesive force F_{ah}
- Gravity mg
- Drag force caused by air stream, F_D

Adhesive force F_{ah} between a surface and a particle depends on many factors, such as particle diameter, materials of the surface and the particle, shape of the particle, roughness of the surfaces, relative humidity of the air, and electrical charging of the surface and the particle. A complete analytical prediction of adhesive force is prohibitively difficult. However, an empirical expression of adhesive force for clean surfaces and particles based on the direct measurement of glass and quartz particles can be useful in many practical applications.[10]

$$F_{ah} = 0.063\,d_p(1+0.009RH) \tag{9.38}$$

where the force is in newtons, the particle diameter in meters, and the relative humidity RH in percentage (i.e., the RH value in Equation 9.38 should be 50 if the relative humidity is 50%). Adhesive forces for particles of typical sizes are listed in Table 9.2. The adhesive force for a given particle in saturated air is 90% greater than in dry air. Although Equation 9.38 was based on air temperature at 25°C, it can be applied to a wide range of temperatures in indoor environments.

From Equation 9.38, adhesive forces are proportional to the particle diameter. The drag force must be calculated using Newton's resistance law instead of Stokes's law, because the detaching process is out of the Stokes region ($Re_p \gg 1$). Because the particle is stationary on the surface, the relative velocity of the particle to the air stream, V_r, is the same as the bulk air velocity U, which is usually much higher than V_r in a free air stream. From Equation 4.13

$$F_D = C_D \frac{\pi}{8}\,d_p^2\rho_g\,U^2 \tag{9.39}$$

Table 9.2 Adhesive, Drag, and Gravitational Forces of Typical Sizes of Spherical Particles with Density of 1000 kg/m³ and Bulk Air Velocity of 10 m/s at 50% Relative Humidity (Calculated from Equation 9.38)

d_p (μm)	F_{ah}	F_D	mg
0.1	6.33E-09	1.71E-10	5.24E-19
0.2	1.27E-08	3.41E-10	4.19E-18
0.3	1.9E-08	5.12E-10	1.41E-17
0.5	3.16E-08	8.53E-10	6.55E-17
0.7	4.43E-08	1.19E-09	1.8E-16
1	6.33E-08	1.71E-09	5.24E-16
2	1.27E-07	4.26E-09	4.19E-15
3	1.9E-07	7.03E-09	1.41E-14
5	3.16E-07	1.38E-08	6.55E-14
10	6.33E-07	2.64E-08	5.24E-13
20	1.27E-06	6.43E-08	4.19E-12
30	1.9E-06	1.11E-07	1.41E-11
50	3.16E-06	2.27E-07	6.55E-11
100	6.33E-06	6.27E-07	5.24E-10

Note that here $V_r = U$, ρ_g is the density of air, and C_D is the drag coefficient determined by Equations 4.15 through 4.18.

The gravitational force of the particle can be written as

$$mg = \frac{\pi}{6} d_p^3 \rho_p \qquad (9.40)$$

Comparing Equation 9.38, Equation 9.39, and Equation 9.40, it is seen that adhesive force is proportional to the particle diameter d_p, whereas the drag force is proportional to d_p^2 and the gravitational force is proportional to d_p^3. These relationships indicate that the adhesive force is predominant for small particles and that gravitational and drag forces are more important for large particles. Therefore, smaller particles are much more difficult to remove from surfaces than large particles. For example, when you blow across a dusty surface, large particles can be blown away, but a thin layer of fine particles remains, which must be washed or wiped off. Figure 9.13 shows three forces of particles in different diameters with standard density and under a bulk airflow velocity of 10 m/s at 50% relative humidity. In general, gravitational forces for all airborne particles are very small compared to adhesive and drag forces. Adhesive force is more than 10 times greater than drag force for the same size of particles. For example, adhesive forces are 37 and 10 times greater than the drag forces for 0.1 and 100 μm particles, respectively.

On the other hand, small particles can agglomerate into larger particles by the strong adhesive force. These large particles can easily be removed from the surface. That is why a crust of dust (> 0.1 mm thickness) can be shaken off more easily than a thin layer of the same dust. Adhesive forces can be measured using a centrifugal method. When the particles deposited on a surface are subjected to a centrifuge, the

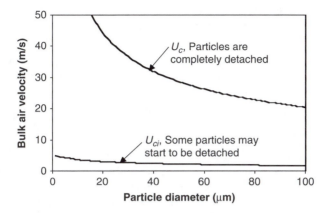

Figure 9.13 Critical air velocities for particle detachment from a surface for typical particles. U_{ci} shows that some particles that are rolling or sliding may be detached at very low air velocities.

particles will detach at a specific rotational speed. Centrifugal force is the measurement of the detachment force, or the adhesive force.

9.8.2 Particle Resuspension

Resuspension is a process in which particles detach from the surface and become airborne again. Particle resuspension may occur as a result of disturbances such as activity, vibration, or other mechanical or electrostatic forces. In the literature, a term closely related to resuspension is *reentrainment*. Reentrainment is resuspension caused by an air jet. Reentrainment sometimes refers only to the reentry of air during an expansion of an air jet. An example of particle reentrainment is the dust in a supply air duct. When the airflow in the duct (which can be considered an air jet) is high enough, some particles will reenter the air.

Let us first consider a particle in static conditions subject to the drag and gravity forces; both of these are opposite to adhesive force. Such a case can happen when a particle is attached to a ceiling and the bulk air flow direction is downward; thus, the gravitational and drag forces are both detaching the particle from the ceiling. To detach the particle from the surface, the sum of drag and gravitational forces must be equal to or greater than the adhesive force:

$$F_{ah} \leq F_D + mg \qquad (9.41)$$

Substituting Equations 9.38 through 9.40 into Equation 9.41, and simplifying, gives

$$\frac{\pi}{8} C_D \rho d_p U_c^2 + \frac{\pi}{6} d_p^2 \rho_p - 0.063(1 + 0.009 RH) = 0 \qquad (9.42)$$

where U_c is referred to as the *critical velocity* for particle detachment. For a given particle, if the bulk air velocity exceeds its critical velocity U_c, the particle will detach completely from the surface and become airborne. Note that Equation 9.42 includes an empirical expression of adhesive force, and thus the units do not cancel out. Therefore, Equation 9.42 only applies to one set of units: d_p in meters, ρ in kg/m^3, U_c in m/s, and RH in percentage.

Equation 9.42 contains the drag coefficient C_D, which is a function of the particle Reynolds number. The particle Reynolds number is a function of d_p and U. The particle drag coefficient is described in Equations 4.15 through 4.18 for different Reynolds number ranges. We can substitute Equations 4.15 through 4.18 into Equation 9.42 to solve the critical velocity for each Reynolds number range. As far as particle resuspension is concerned, not all ranges of Reynolds numbers are meaningful for the typical indoor environment. For $Re_p \leq 1$ (i.e., airflow around the particles is in the Stokes region), particle reentrainment will not occur because the adhesive force is much greater than the sum of drag and gravity. Small Re_p means that the particle is small, the air velocity is low, or both. Small particles mean a stronger adhesive force than its drag and gravity, and low air velocity means a weak drag force. This will prevent the particle from reentering the air stream. When $Re_p > 1$, airflow around the particle becomes turbulent and the particle is likely to be detached. However, the drag coefficient at a large Re_p becomes smaller than that at a small Re_p. Thus, particle reentrainment does not increase proportionally with Re_p. Therefore, we can only consider airflow with an $Re_p > 1$ when determining particle detachment for typical indoor environmental conditions.

Substituting the drag coefficient for $Re_p > 1$ (Equation 4.16) into Equation 9.42, noting that $U = U_c$, gives

$$\frac{9}{16}\pi\eta\rho_g U_c^2 + 3\pi\eta U_c + \frac{\pi}{6}\rho_p d_p^2 - 0.063(1 + 0.009RH) = 0 \qquad (9.43)$$

Solving for U_c gives

$$U_c = \frac{8}{3\rho_g d_p}\left[-\eta \pm \left(\eta^2 - \frac{1}{24}\rho_g\rho_p d_p^3 + \frac{0.063(1 + 0.009RH)\rho_g d_p}{4\pi}\right)^{\!\!1/2}\right] \qquad (9.44)$$

Because $U_c > 0$, the plus-or-minus sign ($\pm$) in Equation 9.44 must be a plus sign (+). The terms with a high order of d_p can be neglected, because $d_p \ll 1$ for typical indoor airborne particles. Even for particles of 100 μm, d_p^3 is only 10^{-12}, which is much smaller than the other terms. Therefore, Equation 9.44 can be simplified as

$$U_c = \frac{8\eta}{3\rho_g d_p}\left[\left(1 + \frac{0.063(1 + 0.009RH)\rho_g d_p}{4\pi\eta^2}\right)^{\!\!1/2} - 1\right] \qquad (9.45)$$

The critical velocity for particle detachment is a function of the air density and viscosity and of the particle diameter. It does not contain the density ρ_p, indicating that gravity has a negligible effect. For air at standard conditions ($\eta = 1.81 \times 10^{-5}$ Pa·s, $\rho_g = 1.2$ kg/m³), the critical velocity for particle detachment becomes

$$U_c = \frac{4.02 \times 10^{-5}}{d_p} \left[\left(1 + 18.36 \times 10^6 (1 + 0.009 RH) d_p \right)^{\frac{1}{2}} - 1 \right] \tag{9.46}$$

Equation 9.46 gives the critical air velocity at which the particle will be completely detached from the surface for spherical particles. Particle detachment is a very complicated process, involving variables such as particle shape, surface roughness, and the air velocity profile around the particle. Actually, particle detachment may occur at air velocities lower than the critical velocity. This may be especially true when a particle is in a dynamic condition, such as rolling or sliding on a surface. The rolling or sliding can substantially reduce the adhesive force of the particle. The dynamic adhesive force F_{ahi} is approximately 1% of the static adhesive force for a particle.[3] Although only a small portion of particles may be detached, we need to define this minimum adhesive force and determine the corresponding initial detaching air velocity U_{ci}. From Equation 9.38 and the minimum adhesive force

$$F_{ahi} = 0.00063 d_p (1 + 0.009 RH) \tag{9.47}$$

Substituting Equation 9.47 into Equation 9.43, noting that $U = U_{ci}$ and solving for U_{ci}, we have

$$U_{ci} = \frac{8\eta}{3\rho_g d_p} \left[\left(1 + \frac{0.00063(1 + 0.009 RH)\rho_g d_p}{4\pi\eta^2} \right)^{\frac{1}{2}} - 1 \right] \tag{9.48}$$

For air at standard conditions ($\eta = 1.81 \times 10^{-5}$ Pa·s, $\rho_g = 1.2$ kg/m³), the critical velocity for initial particle detachment becomes

$$U_{ci} = \frac{4.02 \times 10^{-5}}{d_p} \left[\left(1 + 0.1836 \times 10^6 (1 + 0.009 RH) d_p \right)^{\frac{1}{2}} - 1 \right] \tag{9.49}$$

Note that U_{ci} is not proportional to the adhesive force. For example, for a 100 μm particle, the initial detaching air velocity is 1.7 m/s, whereas the complete detaching air velocity is 21 m/s. This represents a changing factor of only 12, compared with the adhesive force's changing factor of 100. Figure 9.13 shows the initial critical air velocities vs. particle diameter under standard air conditions. The relative humidity of air is assumed to be 50%. When bulk air velocity is lower than U_{ci}, there will be no detachment for that given size of particle. When the bulk air

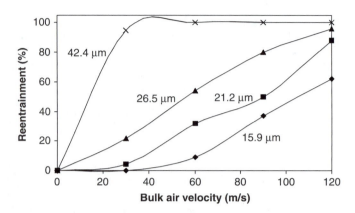

Figure 9.14 Fraction of glass bead particle reentrainment at different bulk air velocities. Data from Corn, M. and Stein, F., Re-entrainment of particles from a plane surface, *Am. Ind. Hyg. Assoc. J.*, 26:325–336, 1965.

velocity exceeds U_{ci}, some particles in a rolling or sliding condition may start to detach. The fraction of such stochastic detachment will increase as the bulk air velocity increases. When the bulk air velocity reaches U_c, all particles at that given size will be completely detached.

Figure 9.14 shows representative data of particle reentrainment for spherical glass beads on a flat surface under different bulk air velocities.[2] These experimental data are within the ranges of calculated U_{ci} and U_c shown in Figure 9.13.

Example 9.5: A bulk airflow with a velocity of 10 m/s passes through a cooling plate. What is the minimum particle diameter that will be completely detached from the cooling plate? Assume that the particles are spheres. Air is at standard conditions and 45% relative humidity.

Solution: The question is to find the particle with a diameter that has a critical air velocity of 10 m/s. From Equation 9.46

$$U_c = \frac{4.02 \times 10^{-5}}{d_p}\left[\left(1+18.36\times 10^6 (1+0.009RH)d_p\right)^{\frac{1}{2}} -1\right]$$

Solving for d_p gives

$$d_p = \frac{4.02 \times 10^{-5}(4.02 \times 10^{-5} \times 18.36 \times 10^6(1+0.009RH)-2U_c)}{U_c^2}$$

$$= \frac{4.02 \times 10^{-5}(4.02 \times 10^{-5} \times 18.36 \times 10^6(1+0.009 \times 50)-2 \times 20)}{10^2}$$

$$= 4.22 \times 10^{-4}\ (m) = 422\ \mu m$$

Reentrainment is a stochastic process in which one can estimate the fraction of particles of a given size that will be removed from the surface under a given air velocity. For most indoor environment air-handling systems, airflow is highly turbulent. There are always thin boundary layers that can be considered laminar flow in ducts or at room surfaces. These boundary layers protect the particles deposited on the surfaces and prevent reentrainment.

When a surface already has a layer of particles, the resuspension process is different from detachment from a clean surface. Two processes may occur. First, particles may be detached from the top surface of the particle layer as individual particles or as small clusters. This process is called *erosion*. For example, erosion can occur in an air supply duct. A layer of dust settles on the surface when the supply fan is off and then erodes when the fan is on again. Secondly, a whole section of the particle layer may be detached. Erosion can last for a long period of time, whereas detachment may only last a few seconds.

DISCUSSION TOPICS

1. What are the differences between deposition velocity and the terminal settling velocity of a particle? When can the two be considered the same?
2. In the quiescent batch settling model, what assumptions are made for key variables such as volume, particle concentration, and deposition air flow in which particles suspend?
3. Can you think of any examples in the real world to which the quiescent batch settling model may be applied?
4. Under which condition do particles of a given size settle down faster: quiescent batch settling or perfect-mixing batch settling? Why?
5. In the design of a settling chamber, what design parameters should be considered? Can you prioritize the factors of consideration?
6. Almost all ventilated indoor environments are not completely mixed, especially when they contain airborne particulate matter. However, complete (or perfect) mixing can be assumed in many applications. Name an example in which the airspace can be considered a perfect-mixing environment with reasonable accuracy.
7. The technique for calculating deposition and production in a ventilated airspace in Section 9.6 can be applied in many real-world situations. Can you design an alternative experiment to obtain particle deposition and production rates?

PROBLEMS

1. Particle size distribution in calm air is shown in Table 9.3. Calculate the following:
   a. Deposition velocities for each particle size group
   b. Total particle deposition rates of all sizes

Particle Size d_a (μm)	Concentration (mg/m³)
3	0.001
5	0.002
10	0.008
20	0.01
30	0.005

2. A settling chamber 3 m high is initially filled with particles. Assume that the chamber is in quiescent batch settling mode. How long it will take to remove 90% of particles 10 μm in aerodynamic diameter?

3. A settling chamber, as shown in Figure 9.9, is 1 m wide. The flow rate is 10 l/s. In order to obtain a 50% collection efficiency for particles 10 μm and larger, what is the minimum length of the chamber, assuming that the flow is laminar?

4. Prove that the deposition rate under quiescent batch settling is faster than for perfect-mixing batch settling. Assume that the initial particle concentration, settling area, and volume are the same. (Hint: Use the deposition rate equations and their derivatives).

5. In a dust explosion test, a chamber 3 m high is used. Two dust concentrations are measured at 20 mg/m³ and 5 mg/m³ after 1 h and 3 h of explosion, respectively. Assume that the settling process is in perfectly mixed batch mode and that all particles are of the same diameter. What is the initial particle concentration at the explosion, and what is the particle diameter?

6. Rocks and ashes were shot up 2,460 m from the Mount Mayon volcano in the Philippines as it erupted on June 29, 2001. Assuming that the air is calm, how long would it take for 4 μm particles to settle down completely?

7. In problem 6, assume that the dust is settling in a perfect-mixing batch mode. After two days, the dust concentration was measured at one location as 10 particles/ml for 4 μm particles. Assume that the background atmospheric particle concentration is negligible. What is the initial dust concentration for 4 μm particles at that location?

8. A rectangular supply air duct is 20 m long, 0.6 m wide, and 0.3 m high, with an airflow rate of 1 m³/s. Assume that the air in the duct is continuously mixing and that there is no particle source or sink within the duct. The concentrations of particles are measured at 0.08 mg/m³ and 0.07 mg/m³ at the inlet and the outlet of the duct, respectively. Find the average deposition velocity. What is the equivalent particle aerodynamic diameter for the settled particles, if all those settled particles have the same diameter?

9. A classroom contains 40 students. The particle concentration in the room is measured as listed in problem 1. What is the students' particle filtration efficiency?

10. An assembling shop measures 10 m long, 8 m wide, and 3 m high. The air supply inlet and outlet are at the same height but on opposite walls. The room ventilation rate is 600 l/s. The particle mass concentrations at the supply air inlet and the exhaust outlet are 0.08 mg/m³ and 0.3 mg/m³, respectively. Assume that the room is continuously mixing. When the ventilation rate decreases to 300 l/s, the particle mass concentration at the outlet increases to 0.5 mg/m³. Determine the particle deposition and production rates within the shop.

11. An experimental room has a ventilation rate of 200 l/s and is perfectly mixed. The room measures 5 m long, 3 m wide, and 2.4 m high. In one test, the room must be filled with airborne particles. The particle concentration in the supply air is negligible. A turnaround table is used to supply monodisperse particles of 5 μm in aerodynamic diameter at a rate of 1g/min. What is the particle concentration after 3 minutes? How long does it take to reach 99% of steady-state concentration in the room?

12. In the design of a vertical elutriator for collecting respirable particles (smaller than 4 μm in aerodynamic diameter), as shown in Figure 9.11, a 37 mm filter is used and the flow rate is 2 l/min. What should the diameter of the vertical chamber

be? If the airflow rate increased by 10%, what is the error in diameter of particles collected by the filter?

13. Air velocity approaching a smooth floor from the exhaust outlet of a vacuum cleaner is 8 m/s. What size particle can be initially detached, and what size will be completely resuspended? Particle size is in aerodynamic diameter. Room air is at standard conditions, and relative humidity is 40%.

14. Assume that the fraction of particle detachment is proportional to air velocity between U_{ci} and U_c, that is, 0% resuspension at U_{ci} and 100% at U_c. What is the resuspension fraction of particles of 50 μm at an air conditioner outlet with a velocity of 5 m/s? The temperature is 15°C and the relative humidity is 100%. The viscosity is 1.76×10^{-5} Pa·s.

REFERENCES

1. Barber E.M. et al., Spatial variability of airborne and settled dust in a piggery, *J. Agr. Eng. Res.*, 50:107–127, 1991.

2. Corn, M. and Stein, F., Re-entrainment of particles from a plane surface, *Am. Ind. Hyg. Assoc. J.*, 26:325–336, 1965.

3. Hintz, C.W., *Aerosol Technology — Properties, Behavior and Measurement of Airborne Particles*, John Wiley & Sons, New York, 1999.

4. ISO, *Air Quality — Particle Size Fraction Definitions for Health-Related Sampling*, Technical Report ISO/TR/7708-1983, International Standards Organization, Geneva, 1983.

5. ACGIH, Particle size-selective sampling in the workplace, *Report of the ACGIH Technical Committee on Air Sampling Procedures*, American Congress of Governmental and Industrial Hygienists, Cincinnati, OH, 1985.

6. Chen, Y., et al., Methods to measure dust production and deposition rates in buildings, *J. Agr. Eng. Res.*, 72:329–340, 1999.

7. ACGIH, *Air Sampling Instruments — For Evaluation of Atmospheric Contaminants*, 8th ed., American Conference of Governmental Industrial Hygienists, Cincinnati, OH, 1995.

8. Wall, S.J.W. and Rogers, D., Laser settling velocimeter: aerodynamic size measurement of large particles, *Aerosol Sci. Tech.*, 4:81–87, 1985.

9. Marple, V.A., Rubow, K.L., and Olsen, B.A., Inertial, gravitational, centrifugal and thermal collection techniques, in *Aerosol Measurements*, Willeke, K. and Barton, P.A., eds., Van Nostrand Reinhold, New York, 1993.

10. Corn, M., The adhesion of solid particles to surfaces, II, *J. Aerosol Sci.*, 11:566–684, 1961.

11. ASAE. Structures, Livestock and Environments, *Standards Handbook*, ASAE, St. Joseph, MI, 2000, 595–698.

12. ASHRAE, Environmental Control for Animals and Plants, *ASHRAE 2001 Fundamentals Handbook*, American Society of Heating, Refrigerating and Air Conditioning Engineers, Atlanta, 2001, 10.1–10.21.

CHAPTER 10

Filtration

Filtration is the most common and most economical means, in many applications, of air cleaning and air sampling. Fibrous or other particle-capturing materials can be easily made in different shapes with high efficiency of particle collection. Air filters are part of our daily life. They are in home furnaces, clothes driers, vacuum cleaners, cars, and facial masks or respirators. Almost all mechanically ventilated buildings are equipped with some kind of filter. In some special cases, such as a medical surgery room or an electronic fabrication shop, high-efficiency particle attenuation (HEPA) filters are necessary.

Although air filtration is a daily-life phenomenon and seemingly simple to understand, the process is very complex, involving almost all principles of particle mechanics, gas kinetics, and other properties, such as electrostatics. There is a wealth of literature regarding the filter research, and there are still many aspects that remain unclear. This chapter will review the known properties of filters and provide the reader with a basic understanding of the filtration process, as well as typical applications and limitations of filters.

By completing this chapter, the reader will be able to

- Understand five basic filtration mechanisms:
  - Interception
  - Impaction
  - Diffusion
  - Gravitational settling
  - Electrostatic deposition
- Design a fiber filter based on the analysis of total single-fiber efficiency and the total filter efficiency
- Apply filter performance criteria to determine the filter quality:
  - Filtration efficiency
  - Pressure drop across the filter
  - Dust-holding capacity

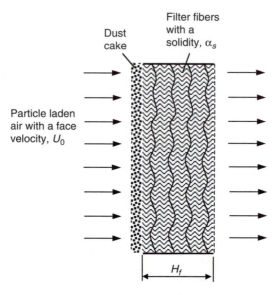

Figure 10.1 Schematic of a filter and its parameters used in performance evaluation.

- Design other types of fiber filters, such as bag houses, and evaluate filters using standardized filter testing methods:
  - Filter efficiency tests
  - Dust-holding capacity tests
  - Leakage tests
  - Environmental tests
- Apply cleanroom standards (both U.S. and ISO) to select filters and design an air-cleaning system for cleanrooms

10.1 FILTRATION MECHANISMS

When an air stream passes through a fibrous filter, particles or gases in the air stream may be collected by one or more of the following mechanisms: interception, impaction, diffusion, settling, or electrostatic deposition. The five basic filtration mechanisms form the basis for particle separation through all physical filter media, including filters, lungs, or a sampling tube. Physical characteristics of a filter include its fiber diameter d_f, filter thickness H_f, and filter solidity α_s, which is defined as the volumetric ratio of fibers and the total volume of the filter (Figure 10.1). A 0.05 solidity means that the fiber takes 5% of the total volume of the filter. Porosity of the filter is $1-\alpha_s$, representing the air volume in the filter.

Interactions occur among different filtration mechanisms, such as interception and impaction. In many cases, there is a competition among different filtration mechanisms and fibers for the same particle. Therefore, it is difficult to analyze the entire filter without knowledge of single-fiber collection efficiency. The following

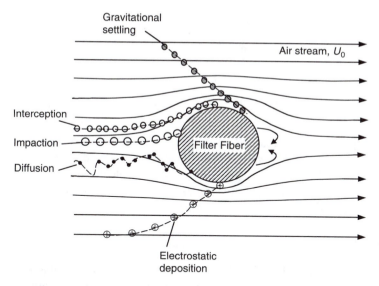

Air stream, U_0

Figure 10.2 Schematic of filtration mechanisms at a single-filter fiber. When a particle makes contact with the filter fiber, it is separated from the air stream.

discussions are focused on single-fiber collection efficiency (Figure 10.2). Subsequent sections then discuss the macroscopic filter efficiency and performance.

10.1.1 Interception

Interception is a particle separation process during which the particle in the air stream is intercepted and clings to the particle fiber (Figure 10.2). The interception is especially effective for the nonspherical particles, such as fibrous and flake-type particles. These particles usually have large surface areas and a large cross-sectional path. The particle collection efficiency for a single fiber due to interception, ξ_{fp}, is given by Lee and Ramamurthi as follows:[1]

$$\xi_{fp} = \frac{(1-\alpha_s)d_p^2}{Ku(1+\dfrac{d_p}{d_f})d_f^2}$$ (10.1)

where Ku is the Kuwabara hydrodynamic factor that compensates the effect of distortion of the flow field around a fiber because of its proximity to other fibers. The Ku number is dimensionless and depends only on the solidity α_s for fiber diameters $d_f \geq 2$ μm.

For filters with a fiber diameter smaller than 2 μm, the slip effect becomes significant. Kirsch and Stechkina recommended adding $2\lambda/d_f$ to compensate for the effect of slip, where λ is the free path of air and d_f is the diameter of the filter fiber.[2]

$$Ku = -\frac{\ln \alpha_s}{2} - \frac{3}{4} + \alpha_s - \frac{\alpha_s^2}{4} \quad \text{(for } d_f \geq 2 \text{ } \mu\text{m)} \tag{10.2}$$

$$Ku = \frac{2\lambda}{d_f} - \frac{\ln \alpha_s}{2} - \frac{3}{4} + \alpha_s - \frac{\alpha_s^2}{4} \quad \text{(for } d_f < 2 \text{ } \mu\text{m)} \tag{10.3}$$

Interception is an important mechanism of filtration. Interception efficiency increases as the ratio of d_p/d_f increases, but it cannot exceed a maximum value of 1, based on the definition of filter efficiency. If the calculated ξ_{fp} in Equation 10.1 is greater than 1 when the ratio of $d_p/d_f >> 1$, then $\xi_{fp} = 1$ because the maximum collection efficiency cannot exceed 100 percent.

10.1.2 Impaction

Impaction occurs when a particle, because of its inertia, is unable to follow the air stream to bypass the filter fiber. This is particularly easy for large and heavy particles (Figure 10.2). The most important parameter affecting the impaction efficiency of a filter fiber is the Stokes number, which is the ratio of the particle's stopping distance to the fiber diameter. The single-fiber impaction efficiency ξ_{fI} is given by Yeh and Liu:[3]

$$\xi_{fI} = \frac{(Stk)a}{2Ku^2} \tag{10.4}$$

where

$$Stk = \frac{\tau U_0}{d_f} = \frac{\rho_p d_p^2 C_c U_0}{18\eta d_f} \tag{10.5}$$

$$a = (29.6 - 28\alpha_s^{0.62})\left(\frac{d_p}{d_f}\right)^2 - 27.5\left(\frac{d_p}{d_f}\right)^{2.8} \quad \text{(for } d_p/d_f < 0.4) \tag{10.6}$$

$$a = 2.0 \quad \text{(for } d_p/d_f \geq 0.4) \tag{10.7}$$

Impaction efficiency of a single filter fiber increases as the Stokes number increases. Unlike interception, the impaction efficiency will decrease as the ratio of d_p/d_f increases for small particles and thick fibers. This effect disappears when the particle diameter approaches the diameter of the fiber. If the calculated ξ_{fp} in Equation 10.1 is greater than 1 when the ratio of $d_p/d_f >> 1$, then $\xi_{fp} = 1$ because the maximum collection efficiency cannot exceed 100 percent.

10.1.3 Diffusion

Filtration efficiency due to diffusion, or Brownian motion, of particles is a significant part of the overall filtration efficiency. The zigzag trajectories of small particles, as shown in Figure 10.2, greatly increase the chance of hitting a filter fiber or a surface while particles are passing through the filter in a nonintercepting streamline. The principles of a small particle hitting a fiber or a surface were described in Chapter 6. The single-fiber filtration efficiency due to diffusion, ξ_{fD}, depends on the ratio of particle diameter to fiber diameter, d_p/d_f, and two dimensionless numbers: the Peclet number Pe and the Kuwabara hydrodynamic factor Ku.[4] This is especially true when the filtration efficiency is low.

$$\xi_{fD} = 2Pe^{-2/3} + 1.24\left(\frac{d_p}{d_f}\right)^{2/3}(KuPe)^{-1/2} \tag{10.8}$$

where

$$Pe = \frac{d_f U_0}{D_p} \tag{10.9}$$

and where d_f is the diameter of the fiber, U_0 is the air velocity, and D_p is the particle diffusion coefficient.

The first term on the right side of Equation 10.8 is caused directly by diffusion, experimentally determined by Kirsch and Fuchs.[5] The second term represents the enhanced filtration efficiency induced by interception during the Brownian motion of particles. This term is relatively small for very small particles compared with the first term, but it becomes relatively significant for the larger particles. The explanation for this is that for very small particles, diffusion is the primary mechanism of particle collection. As particle size increases, however, diffusive collection decreases rapidly, because the particles are larger and so are more likely to be intercepted by a fiber.

Intuitively, particles smaller than a mesh opening will go through the filter, just as a grain of sand passes through a sieve. In reality, a large portion of particles smaller than the porosity of a filter will be collected due the diffusion effect. In air sampling, porosity sometime refers the mesh opening size of a filter. For example, a cellulous fiber filter has a porosity of 0.8 µm, meaning that the mesh opening size of the woven fiber is about 0.8 µm. A large portion of particles much smaller than 0.8 µm can be collected.

10.1.4 Gravitational Settling

Filtration efficiency due to gravitational settling, ξ_{fG}, is proportional to the terminal settling velocity and inversely proportional to the air velocity. This effect can

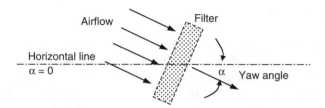

Figure 10.3 Yaw angle between airflow direction and the horizontal line affecting the gravitational settling of particles of a filter.

be described by a filtration efficiency factor due to gravitational settling G_f, defined as the ratio of particle terminal settling velocity and the face air velocity:

$$G_f = \frac{V_{TS}}{U_0} = \frac{\rho_p d_p^2 C_c g}{18\eta U_0} \tag{10.10}$$

ξ_{fG} also depends on the direction of the airflow with respect to particle settling velocity. If the airflow is downward, the filtration efficiency due to gravitational settling is greater than zero. If the airflow is upward, the filtration efficiency due to gravitational settling is negative and thus will reduce overall filtration efficiency. Assume that the yaw angle between the airflow direction and the horizontal line is α (Figure 10.3). When the airflow is horizontal, $\alpha = 0$. When the airflow is downward, $\alpha = 90°$, and when the airflow is upward, $\alpha = -90°$. The ξ_{fG} can be estimated as

$$\xi_{fG} = G_f (1 + \frac{d_p}{d_f}) \sin \alpha \tag{10.11}$$

Equation 10.11 shows that when the airflow is downward, $\sin\alpha = 1$, and the filtration efficiency due to gravitational settling is positive. When the airflow is upward, $\sin\alpha$ is negative. However, filtration efficiency due to gravitational settling cannot be negative. In this case, the ξ_{fG} should be zero. When the airflow is horizontal, $\sin\alpha = 0$; thus, the filtration efficiency due to gravitational settling is zero. This is true for most of practical filters unless the particle size is large and air velocity is very low. In fact, when air velocity is greater than 0.01 m/s, impaction is more important than gravitational settling.

10.1.5 Electrostatic Deposition

Electrostatic deposition can be an extremely important filtration mechanism if the particles or the fibers are charged. Quantifying the filtration efficiency due to electrostatic deposition requires knowledge of the charges of the particles and/or the fibers, which is often difficult to obtain. Brown gives a single-fiber collection efficiency based on an experimental measurement of a neutral glass fiber and a particle with a charge q:[4]

$$\xi_{fE} = 1.5 \left(\frac{(\varepsilon_f - 1)}{(\varepsilon_f + 1)} \frac{q^2}{12\pi^2 \eta U_0 \varepsilon_0 d_p d_f^2} \right)^{\frac{1}{2}} \tag{10.12}$$

where ε_f and ε_0 are the relative permittivities (dielectric constants) of the fiber and the vacuum, respectively, and q is the electric charge in C (coulombs) on the particle. The permittivity of a vacuum $\varepsilon_0 = 8.85 \times 10^{-12}$ C^2/N·m^2.

Usually, the electrostatic deposition effect is negligible in filtration. However, charged particles or charged filters can greatly enhance particle collection efficiency. When both the particle and the filter fiber are charged, the charge for the particles should be opposite to the charge on filters. Otherwise, the charges on the particle and on the filter will repel each other and reduce the collection efficiency. A common problem in charged filters is that the filters lose their charge quickly after the initial charge and must be recharged periodically.

10.1.6 Total Single-Fiber Collection Efficiency

The total collection efficiency of a single fiber, $\xi_{f\Sigma}$, can be calculated from the efficiencies of all filtration mechanisms discussed previously, assuming that each mechanism acts independently.

$$\xi_{f\Sigma} = 1 - (1 - \xi_{fp})(1 - \xi_{fl})(1 - \xi_{fD})(1 - \xi_{fS})(1 - \xi_{fE}) \tag{10.13}$$

Note that the second term at the right side of Equation 10.13 is always greater than or equal to zero. Thus the total filtration efficiency is always smaller than or equal to unity. The five filtration mechanisms are not additive, because different mechanisms are competing for the same particle and its collection could be counted more than once.

A typical filter efficiency for individual single-fiber filtration mechanisms and total single-fiber filtration efficacy calculated from Equations 10.1 through 10.13 is shown in Figure 10.4. For a given fibrous filter, there is a particle size, usually between 0.05 and 0.5 μm, which has a minimum collection efficacy. For a given size of particle, there is also a velocity at which the collection efficiency is minimal.

The greatest value of filtration mechanisms and total single-fiber filtration efficiency is that it provides principles and an analytical tool for filter design. For example, if the particle of concern is 0.3 μm or larger, diffusive filtration will not be effective. For HEPA filters, diffusive mechanism can be an important factor in filter design.

10.2 FILTER EFFICIENCY

A more practical concern of filters is the overall filtration efficiency, or filter efficiency ξ_f, which represents the primary performance of the filter. The filtration

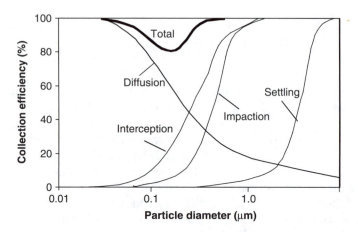

Figure 10.4 A typical filter efficiency for individual single-fiber filtration mechanisms and total single-fiber filtration efficiency, calculated from Equations 10.1 through 10.13: H_f = 1 mm, α = 0.05, U_0 = 0.1 m/s, and d_f = 2 μm.

efficiency of a filter is defined as the fraction of particles collected by the filter when particle-laden air passes through the filter. If the particle concentrations are C_{pin} and C_{pout} for the in-flow and out-flow of the filter, respectively, the filter efficiency can be expressed as

$$\xi_f = 1 - \frac{C_{pout}}{C_{pin}} \tag{10.14}$$

Now we can derive the filter efficiency from the single-fiber filtration efficiency $\xi_{f\Sigma}$. If we assume that all fibers in a filter have the same diameter d_f and that the solidity of the filter is α_s, the total length L of the fiber within a unit volume is

$$L = \frac{4\alpha_s}{\pi d_f^2} \tag{10.15}$$

When a particle-laden airflow with a particle concentration C_p passes through a filter layer with a finite thickness dH_f, the particle concentration changes by the amount of $d(C_p)$. In order to analyze $d(C_p)$, we must consider the physical meaning of total collection efficiency of a single fiber. The physical meaning of the total collection efficiency of a single fiber can be defined as the fraction of particles collected by a unit length of a single fiber when the airflow is perpendicular to the fiber axis. The incident area of the fiber is the unit length of the fiber times the fiber diameter. Consider a layer of filter with a unit area and a finite thickness dH_f; the total incident area of filter filters in this layer is $d_f \times L \times d(H_f)$, where L is the total length of fiber within a unit volume of a filter defined previously. The total number of particles captured by this incident area is proportional to the single-fiber total collection efficiency and the particle concentration. This total number of particles captured by this layer during a unit time period is also the change of the particle concentration in the particle-laden air, $d(C_p)$:

$$d(C_p) = -C_p \xi_{f\Sigma} d_f L dH_f \tag{10.16}$$

where the minus sign (−) indicates that the particle concentration is decreasing when passing through the filter.

Rearranging and integrating Equation 10.16, noting that the ranges for integration are 0 to H_f for filter thickness and C_{pin} to C_{pout} for particle concentration, gives

$$\int_{C_{pin}}^{C_{pout}} \frac{dC_p}{C_p} = \int_0^{H_f} -\xi_{f\Sigma} d_f L dH_f \tag{10.17}$$

Solving Equation 10.17 and substituting Equation 10.15 for L gives

$$\frac{C_{pout}}{C_{pin}} = \exp(-\xi_{f\Sigma} d_f L H_f) = \exp\left(-\frac{4\alpha_s \xi_{f\Sigma} H_f}{\pi d_f}\right) \tag{10.18}$$

Therefore, from Equation 10.14 and Equation 10.18, the relationship between the filter efficiency and the single-fiber filtration efficiency becomes

$$\xi_f = 1 - \exp\left(-\frac{4\alpha_s \xi_{f\Sigma} H_f}{\pi d_f}\right) \tag{10.19}$$

In many cases, the term of penetration P_n is used. Penetration is the fraction of particles passing through a filter of devices:

$$P_n = 1 - \xi_f = \exp\left(-\frac{4\alpha_s \xi_{f\Sigma} H_f}{\pi d_f}\right) \tag{10.20}$$

From Equation 10.19, filter efficiency increases exponentially with the single-fiber filtration efficiency, the solidity, and the thickness of the filter. It decreases exponentially as the diameter of the filter fiber increases. With a given solidity and filter thickness, the finer the fiber is, the higher the filter efficiency.

Example 10.1: *Determine the filter efficiency of a high-efficiency furnace filter for particles of 5 μm in aerodynamic diameter under standard room conditions and neglecting electrostatic effect. The filter is made of fiberglass with a solidity of 3%,*

2 mm thick. The average diameter of the fiber is 20 μm. The filter is perpendicular to the face airflow, and the airflow direction is upward, as shown in the diagram.

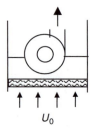

U_0

Solution: It is given that $\alpha_s = 0.03$, $H_f = 0.002$ m, $d_p = 5\times10^{-6}$ m, $d_f = 20\times10^{-6}$ m, $U_0 = 1.5$ m/s. First, we calculate the total collection efficiency of a single fiber.

$$Ku = -\frac{\ln \alpha_s}{2} - \frac{3}{4} + \alpha_s - \frac{\alpha_s^2}{4}$$

$$= -\frac{\ln 0.03}{2} - 0.75 + 0.03 - \frac{0.03^2}{4} = 1.033$$

$$\xi_{fp} = \frac{(1-\alpha_s)d_p^2}{Ku(1+\frac{d_p}{d_f})d_f^2} = \frac{(1-0.03)\times(5\times10^{-6})^2}{1.033\times(1+0.25)\times(20\times10^{-6})^2} = 0.047$$

From Equation 10.5 and Equation 10.6

$$Stk = \frac{\rho_p d_p^2 C_c U_0}{18\eta d_f} = \frac{1000\times(5\times10^{-6})^2\times1.5}{18\times1.81\times10^{-5}\times20\times10^{-6}} = 5.755$$

$$a = (29.6 - 28\alpha_s^{0.62})\left(\frac{d_p}{d_f}\right)^2 - 27.5\left(\frac{d_p}{d_f}\right)^{2.8}$$

$$= (29.6 - 28\times0.03^{0.62})(0.25)^2 - 27.5\times(0.25)^{2.8} = 1.083$$

$$\xi_{fl} = \frac{(Stk)a}{2Ku^2} = \frac{5.755\times1.083}{2\times1.033^2} = 1$$

The impaction efficiency is 1 if the calculated value is greater than 1. From Equation 10.9 and Equation 10.8

$$Pe = \frac{d_f U_0}{D_p} = \frac{3\pi d_f U_0 \eta d_p}{kTC_c}$$

$$= \frac{3\pi \times 20 \times 10^{-6} \times 1.5 \times 1.81 \times 10^{-5} \times 5 \times 10^{-6}}{1.38 \times 10^{-23} \times 293 \times 1} = 4.366 \times 10^8$$

$$\xi_{fD} = 2Pe^{-\frac{2}{3}} + 1.24 \left(\frac{d_p}{d_f}\right)^{\frac{2}{3}} (KuPe)^{-\frac{1}{2}}$$

$$= 2 \times (4.366 \times 10^8)^{-\frac{2}{3}} + 1.24 \times (0.25)^{\frac{2}{3}} \times (3.789 \times 4.366 \times 10^8)^{-\frac{1}{2}}$$

$$= 0.0000716$$

From Equation 10.11 and Equation 10.10 (note that the angle $\alpha = -90°$ because the flow is upward)

$$G_f = \frac{V_{TS}}{U_0} = \frac{\rho_p d_p^2 C_c g}{18\eta U_0} = 0.000502$$

$$\xi_{fG} = G_f (1 + \frac{d_p}{d_f}) \sin(-\alpha) = 0.000502 \times (1 + 0.25) \times \sin(-90°) = -0.000627$$

The total collection efficiency for a single filter fiber is from Equation 10.13:

$$\xi_{f\Sigma} = 1 - (1 - \xi_{fp})(1 - \xi_{fI})(1 - \xi_{fD})(1 - \xi_{fS})(1 - \xi_{fE})$$

$$= 1 - (1 - 0.0207)(1 - 1)(1 - 0.0000716((1 + 0.000627) = 1$$

The filter efficiency then can be obtained from Equation 10.19:

$$\xi_f = 1 - \exp\left(-\frac{4\alpha_s \xi_{f\Sigma} H_f}{\pi d_f}\right)$$

$$= 1 - \exp\left(-\frac{4 \times 0.03 \times 1 \times 0.002}{\pi \times 20 \times 10^{-6}}\right) = 0.978$$

The filter efficiency for 5 μm particles is 97.8%.

10.3 FILTER PERFORMANCE CRITERIA

In addition to filtration efficiency, a primary filter performance criterion, factors such as cost (initial capital and maintenance), space requirements, and airflow resistance (or pressure drop) are all important filter performance criteria. Establishment of filter performance criteria has encouraged the development of a wide variety of air filters and other types of air cleaners. Accurate comparisons of different air filters can be made only from data obtained by standardized test methods.

The three most commonly used operating characteristics that distinguish the various types of air filters are

- Filtration efficiency
- Pressure drop across the filter,
- Dust-holding capacity

Filtration efficiency measures the ability of the filter to remove particles from an air stream. Average filtration efficiency during the life of the filter is the most meaningful for most filters and applications. However, because the efficiency of many dry-type filters increases with dust load, the initial (clean filter) efficiency should be considered for design in applications with low dust concentrations. Pressure drop (or resistance to airflow) is the static pressure drop across the filter at a given airflow rate. The word *resistance* is used interchangeably with *pressure drop*. Dust-holding capacity defines the amount of a particular type of dust that a filter can hold when it is operated at a specified airflow rate to some maximum pressure drop.[6] Dust-holding capacity can vary substantially with the application and filter type. For example, a filter with a low solidity (such as a home furnace filter) has a higher dust-holding capacity than a filter with a high solidity (such as a HEPA filter).

Pressure drop across a filter represents the total drag force of all the fibers. One empirical equation gives pressure drop across a filter as follows:[7]

$$\Delta P = \frac{64 \eta H_f U_0 \alpha_s^{1.5} (1 + 56\alpha_s^3)}{d_f^2} \quad \text{(for } 0.006 < \alpha_s < 0.3\text{)} \qquad (10.21)$$

where ΔP is in Pa, η is the viscosity of air, H_f is the filter thickness, U_0 is the face velocity of air, α_s is the solidity, and d_f is the diameter of the filter fiber. From Equation 10.21, the pressure drop across a filter is proportional to air viscosity, filter thickness, and face velocity. It increases with the filter solidity and is inversely proportional to d_f^2.

In practice, the pressure drop can be measured accurately. Other parameters — H_f, α_s, and U_0 — can easily be measured. Therefore, Equation 10.21 can be used to determine the equivalent filter fiber diameter. Figure 10.5 shows the pressure drop vs. the fiber diameter at a given solidity, with thickness at standard conditions. Because the pressure drop is linearly proportional to the face velocity and the thickness of the filter, pressure drops at other face velocities and filter thicknesses

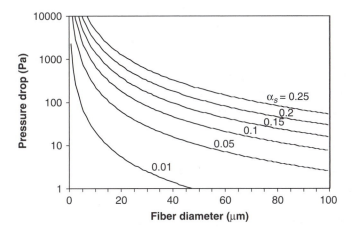

Figure 10.5 Pressure drop across filters at different face velocities. $U_0 = 1$ m/s, $H_f = 1$ mm.

can be easily calculated based on the value in Figure 10.5. For example, at $\alpha_s = 0.05$ and $d_f = 40$ μm, the pressure drop is about 15 Pa for $U_0 = 1$ m/s and $H_f = 1$ mm (Figure 10.5). For $U_0 = 2$ m/s and $H_f = 3$ mm, the pressure drop will be 90 Pa; the face velocity is twice and the filter thickness three times as high as that in Figure 10.5.

Complete evaluation of air filters, therefore, requires data on filtration efficiency, pressure drop, and dust-holding capacity. When applied to automatic renewable filtration devices (e.g., roll filters), the evaluation must include the rate at which the filtration medium is supplied to maintain constant pressure drop when standardized dust is fed at a specified rate. When applied to electronic air filters, the effect of dust buildup on efficiency should be evaluated.

A useful quantitative criterion to compare the performance of different filters is the filter quality q_f, which is logarithmically proportional to the filtration efficiency and inversely proportional to the pressure drop"

$$q_f = \frac{-\ln\left(1 - \xi_f\right)}{\Delta P} = \frac{4\alpha_s \xi_{f\Sigma} H_f}{\pi d_f \Delta P} \tag{10.22}$$

where ΔP is the pressure drop across the filter.

Apparently, q_f increases as ξ_f increases, as ΔP decreases, or both. The greater the value of q_f is, the better the performance of the filter. Comparison of q_f among filters must be made at the same face flow velocity and particle size. The filter quality value of a particular filter has little meaning, because it may be in a wide range, from as small as a fraction of percentage to as large as infinity. Filter quality is only appropriate for use in comparisons among filters, and it only takes two of the important parameters (filter efficiency and pressure drop) into consideration. Other factors, such as cost, maintenance requirements, and dust-loading capacity, may be more important factors to consider in particular applications.

10.4 STANDARDIZED FILTER TESTING METHODS

Air filter testing is complex, and no individual test adequately describes all characteristics of a filter. Ideally, performance testing of a filter should simulate the operation of the filter under actual conditions and evaluate the characteristics important to the user, such as the filtration efficiency, pressure drop, and dust-holding capacity. Wide variations in concentrations and particle types in indoor environments make evaluation difficult. Another complication is the difficulty of closely relating measurable performance to the specific requirements of users. Recirculated air tends to have a larger proportion of lint than outside air does. However, these difficulties should not obscure the principle that tests should simulate actual use as closely as possible.

Air filter test methods have been developed in several areas: the heating and air-conditioning industry, the automotive industry, the atomic energy industry, and government and military agencies. Several tests have become standard in general ventilation applications in the United States. In 1968, the test techniques developed by the U.S. National Bureau of Standards (now the National Institute of Standards and Technology, NIST) and the Air Filter Institute (AFI) were unified (with minor changes) into a single test procedure: ASHRAE Standard 52.2.[8]

10.4.1 Filter Efficiency Tests

In general four types of tests, together with certain variations, determine the filtration efficiency of a filter:[9]

- Arrestance
- Dust-spot efficiency
- Fraction efficiency
- Efficiency by particle size

These will be discussed in the sections that follow.

10.4.2 Arrestance

In this test, a standardized synthetic dust consisting of particles of various sizes and types is fed into the test air stream to the air cleaner, and the mass fraction of the dust removed is determined. In the ASHRAE Standard 52.1 test, the measurement is called *synthetic dust arrestance* to distinguish it from other efficiency values.[6] The synthetic dust is defined as a compounded test dust consisting of (by mass) 72% ISO 12 103-A2 fine test dust, 23% powdered carbon, and 5% No. 7 cotton linters. A known amount of the prepared test dust is fed into the test unit at a known and controlled rate. The concentration of dust in the air leaving the filter is determined by passing the entire airflow through a high-efficiency after-filter and measuring the gain in filter mass. The arrestance is calculated using the masses of the dust passing the tested filter and the total dust fed.

The indicated mass arrestance of air filters, as determined by the arrestance test, depends greatly on the particle size distribution of the test dust, which, in turn, is affected by its state of agglomeration. Therefore, this filter test requires a high degree of standardization of the test dust, the dust dispersion apparatus, and other elements of test equipment and procedures. This test is particularly suited to low- and medium-efficiency air filters, which are most commonly used on recirculating systems. It does not distinguish between filters of higher efficiency, because the dust mass-based test is not sensitive to small particles. Atmospheric dust particles range in size from a small fraction of a micrometer to 10s of micrometers in diameter. The artificially generated dust cloud used in the ASHRAE mass arrestance method is considerably coarser than typical atmospheric dust. It tests the ability of a filter to remove the largest atmospheric dust particles and gives little indication of the filter performance in removing the smallest particles. But where the mass of dust in the air is the primary concern, this is a valid test, because most of the mass is contained in the larger particles. When extremely small particles are involved, the mass arrestance method of rating does not differentiate between filters.

10.4.3 Dust-Spot Efficiency

One objectionable characteristic of finer airborne dust particles is their capacity to soil walls and other interior surfaces. The discoloring rate of a white, filter-paper target (microfine glass fiber HEPA filter media) filtering samples of air constitutes an accelerated simulation of this effect. By measuring the change in light transmitted by these targets, the efficiency of the filter in reducing the soiling of surfaces can be computed.

ASHRAE Standard 52.1 specifies two equivalent atmospheric dust-spot test procedures, each taking a different approach to correct for the nonlinearity of the relation between the discoloration of target papers and their dust load.[6] In the first procedure, called the *intermittent-flow method,* samples of conditioned atmospheric air are drawn upstream and downstream of the tested filter. These samples are drawn at equal flow rates through identical targets of glass fiber filter paper. The downstream sample is drawn continuously; the upstream sample is interrupted in a timed cycle so that the average rate of discoloration of the upstream and downstream targets is approximately equal. The percentage of off-time approximates the efficiency of the filter.

In the alternate procedure, called the *constant-flow method,* conditioned atmospheric air samples are also drawn at equal flow rates through equal-area glass fiber filter paper targets upstream and downstream, but without interrupting either sample. Discoloration of the upstream target is therefore greater than for the downstream target. Sampling is halted when the upstream target light transmission has dropped by at least 10% but no more than 40%. The opacities (the percent change in lighting transmission) are then calculated for the targets as defined for the intermittent-flow method. These opacities are next converted into opacity indices to correct for nonlinearity. The advantage of the constant-flow method is that it takes the same length of time to run, regardless of the efficiency of the filter, whereas the intermittent-flow method takes longer for higher-efficiency filters. For example, an efficiency test run of a 90% efficient filter using the intermittent-flow method takes 10 times as long as a test run using the constant-flow method.

The ASHRAE (1992) standard allows dust-spot efficiencies to be taken at intervals during an artificial dust-loading procedure.[6] This characterizes the change of dust-spot efficiency as dust builds up on the filter in service.

The dust-spot test measures the ability of a filter to reduce the soiling of fabrics and interior surfaces of buildings. Because these effects depend mostly on fine particles, this test is most useful for high-efficiency filters. The variety and variability of atmospheric dust may cause large difference in filtration efficiencies for the same filter at different test locations or different times.[10, 11] This discrepancy tends to an increase in low-efficiency filters.

10.4.4 Fractional Efficiency or Penetration

For high-efficiency filters of the type used in clean rooms and nuclear applications (HEPA filters), the normal test in the United States is the *thermal DOP method*, as outlined in U.S. Military Standard MIL-STD-282 and U.S. Army document 136-300-175A.[12, 13] DOP is dioctyl phthalate or bi- [2-ethylhexyl] phthalate, which is an oily liquid with a high boiling point. In this method, a smoke cloud of DOP droplets condenses to form a DOP vapor. The count median diameter for DOP aerosols is about 0.18 μm, and the mass median diameter is about 0.27 μm with a cloud concentration of approximately 80 mg/m^3 under properly controlled conditions. The procedure is sensitive to the mass median diameter, and DOP test results are commonly referred to as filter efficiency on 0.3 μm particles.

In this test, the DOP smoke cloud is fed into the filter, which is held in a special test chuck. Any smoke that penetrates the body of the filter or leaks through gasket cracks passes into the region downstream from the filter, where it is thoroughly mixed. The air leaving the chuck thus contains the average concentration of penetrating smoke. This concentration, as well as the upstream concentration, is measured by a light-scattering photometer. The filter penetration in percent is usually measured in the test procedure, because HEPA filters have filtration efficiencies so near 100%: for example, 99.97% or 99.99% on 0.3 μm particles. U.S. specifications frequently call for the testing of HEPA filters at both rated flow and 20% of rated flow. This procedure helps to detect gasket leaks and pinholes that would otherwise escape notice. Such defects, however, are not located by the DOP penetration test.

In fractional efficiency tests, the use of uniform-size particles has resulted in accurate measure of the particle size vs. efficiency characteristics of the filter over a wide atmospheric size spectrum. The method is time-consuming and has been used primarily in research. However, the dioctyl phthalate (DOP) or Emory 3000 test for HEPA filters is widely used for production testing at a narrow particle size range.

10.4.5 Efficiency by Particle Size

ASHRAE Standard 52.2 prescribes a method to test air-cleaning devices for removal efficiency by particle size while addressing two air filter performance characteristics of importance to users.[8] These characteristics are the ability of the device to remove particles from the air stream and its resistance to airflow.

In this standard method, air filter testing is conducted at a specific airflow, based on one of seven face velocities selected. Face air velocities may not be less than 0.60 m/s or greater than 3.80m/s in the test section. The test aerosol consists of laboratory-generated potassium chloride particles dispersed in the air stream. An optical particle counter measures the particle concentrations in 12 geometric, equally distributed particle size ranges both upstream and downstream for efficiency determinations. The size range encompassed by the test is 0.3 to 10 μm polystyrene latex–equivalent optical particle size. A method of loading the air filter with synthetic dust to simulate field conditions is also specified. The synthetic loading dust is the same as that used in ASHRAE Standard 52.1, consisting of (by mass) 72% ISO 12 103-A2 fine test dust, 23% powdered carbon, and 5% No. 7 cotton linters.[6]

In this test, a set of particle size removal efficiency performance curves is developed, together with an initial clean performance curve, which is the basis of a composite curve representing performance in the range of sizes. Points on the composite curve are averaged, and these averages are used to determine the minimum efficiency reporting value (MERV) of the air filter. A complete test report includes a summary section, removal efficiency curves of the clean filter at each of the loading steps, and a composite minimum removal efficiency curve.

10.4.6 Dust-Holding Capacity Tests

The exact measurement of true dust-holding capacity is complicated by the variability of atmospheric dust; therefore, standardized artificial loading dust is normally used. Such artificial dust also shortens the dust-loading cycle to hours instead of weeks or years. Under typical atmospheric conditions, a meaningful dust-holding capacity test would take months or years to collect enough dust mass.

Artificial dusts are not the same as atmospheric dusts, so dust-holding capacity measured by the accelerated test is different from that achieved by tests using atmospheric dust. The exact life of a filter in field use is impossible to determine by laboratory testing. However, tests of filters under standard conditions do provide a rough guide to the relative effect of dust on the performance of various units and can be used to compare different filters.

Reputable laboratories perform accurate and reproducible filter tests. Differences in reported values generally lie within the variability of test aerosols, measurement devices, and dusts. Because most filtration media are made of random air or water laid fibrous materials, the inherent media variations affect filter performance. Awareness of these variations prevents misunderstanding and specification of impossibly close performance tolerances. Caution must be exercised in interpreting published efficiency data, because the performance of two filters tested by different procedures generally cannot be compared. A value of air filter efficiency is only a guide to the rate of soiling of a space or of mechanical equipment.

In a dust-holding capacity test, the synthetic test dust is fed into the filter in accordance with the ASHRAE Standard 52.1 procedures.[6] The pressure drop across the filter (its resistance) rises as dust is fed. The test is normally terminated when the resistance reaches the maximum operating resistance set by the manufacturer. However, not all filters of the same type retain collected dust equally well. The test, therefore, requires that arrestance be measured at least four times during the dust-loading process and that the test be terminated when two consecutive arrestance values of less than 85%, or one value equal to or less than 75% of the maximum arrestance, has been measured. The ASHRAE dust-holding capacity is, then, the integrated amount of dust held by the filter up to the time the dust-loading test is terminated.

10.4.7 Leakage (Scan) Tests

ASHRAE (2000) describes the methods for filter leakage tests as follows.[9] In the case of HEPA filters, leakage tests are sometimes desirable to show that no "pinhole" leaks exist or to locate any that may exist, so that such leaks may be patched. A technique is employed that is essentially the same as that used in the DOP penetration test, except that the downstream concentration is measured by scanning the face of the filter and its gasketed perimeter with a moving probe. The exact point of smoke penetration can then be located and repaired. This same test can be performed after the filter is installed; in this case, a portable aspirator-type DOP generator is used instead of the bulky thermal generator. The smoke produced by a portable generator is not uniform in size, but its average diameter can be approximated as 0.6 μm. Particle diameter is less critical for leak location than for penetration measurement.

10.4.8 Environmental Tests

Some air filters may be subjected to fire, high humidity, wide temperature fluctuations, mechanical shock, vibration, and other environmental stresses. The performance of a filter under unusual environmental conditions must be tested. Several standardized tests exist for evaluating these environmental effects on air filters. U.S. Military Standard MIL-STD-282 includes shock tests (shipment rough handling) and filter media water-resistance tests.[12] The U.S. Department of Energy specifies humidity and temperature-resistance tests.[14, 15]

Underwriters Laboratories (UL) has two major standards for air filter flammability. The first, for commercial applications, determines flammability and smoke production. UL Standard 900 Class 1 filters are those that, when clean, do not contribute fuel when attacked by flame and emit only negligible amounts of smoke. UL Standard 900 Class 2 filters are those that, when clean, burn moderately when attacked by flame, emit moderate amounts of smoke, or both. In addition, UL Standard 586 for flammability of HEPA filters has been established. The UL tests do not evaluate the effect of collected dust on filter flammability: Depending on the dust, this effect may be severe. UL Standard 867 applies to electronic air filters.

The Air-Conditioning and Refrigeration Institute has published ARI Standard 680 and ARI Standard 850 for air filter equipment. These standards establish

- Definitions and classifications
- Requirements for testing and rating (performance test methods are per ASHRAE Standard 52.1)

Table 10.1 Selected Airborne Particulate Cleanliness Classes for Cleanrooms and Clean Zones[a]

Class Number	Concentration Limits, particles/m³					
	0.1 μm	0.2 μm	0.3 μm	0.5 μm	1 μm	5 μm
1	10	2				
2	100	24	10	4		
3	1,000	237	102	35	8	
4	10,000	2,370	1,020	352	83	
5	100,000	23,700	10,200	3,520	832	29
6	1,000,000	237,000	102,000	35,200	8,320	293
7				352,000	83,200	2,930
8				3,520,000	832,000	29,300
9				35,200,000	8,320,000	293,000

[a] Values shown in the table are the concentration limits for particles equal to or larger than the corresponding particle size.

- Specification of standard equipment
- Performance and safety requirements
- Proper marking
- Conformance conditions
- Literature and advertising requirements

10.5 CLEANROOM REQUIREMENTS

10.5.1 Cleanroom Criteria

The International Organization for Standardization (ISO) specifies particulate cleanliness for cleanrooms in Standard ISO 146441-1. The particle concentration limits is defined as a function of the class number and the particle diameter

$$C_{pn} = 10 \times CN \times \left(\frac{0.1}{d_p} \right)^{2.08} \tag{10.22}$$

where CN is the class number (1, 2, 3, ...) and d_p is in μm.

These specifications define the maximum particle concentrations for the given class of cleanrooms (Table 10.1 and Figure 10.6). The values in the table are the maximum concentrations of particles larger than the indicated particle size. For a given class, all limits for all particles sizes must be satisfied. For example, in a Class 1 cleanroom, particles larger than 0.1 μm should not exceed 10 particles per cubic meter of air, and particles larger than 0.2 μm should not exceed 2 particles per cubic meter. If the total concentration is 9 particles/m³ for particles larger than 0.1 μm, but particles larger than 0.2 μm have a concentration of 3 particles/m³, the Class 1 criteria have not been satisfied. The higher the class number, the higher particle concentration will be. The cleanest room is Class 1, which only allows 2 particles/m³ larger than 0.2 μm. A typical office building has a particle concentration of

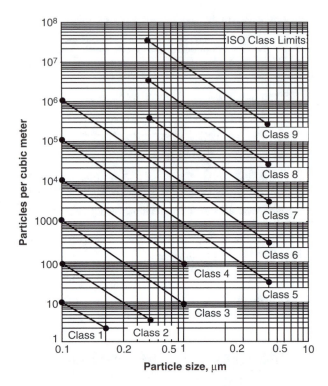

Figure 10.6 ISO cleanroom class limits calculated from Equation 10.22. (Modified from ASHRAE Handbook, *HVAC Applications*, American Society of Heating, Refrigerating and Air Conditioning Engineers, Atlanta, 1999, ch 15.)

10,000,000 particles/m³ or higher, or 10 particles/ml or higher for particles larger than 0.5 μm, which is approximately a Class 9 cleanroom.

The U.S. Federal Standard 209E (1992) specifies particulate cleanliness based on the maximum number of particles of 0.5 μm in diameter and larger per cubic foot of air.[16] For example, FS 209E Class 10 refers to rooms with concentrations less than 10 particles/ft³ for particles of 0.5 μm and larger; Class 1,000 refers to rooms with concentration be less than 1,000 particles/ft³ for particles of 0.5 μm and larger, and so forth. These FS 209E classes for cleanrooms can be expressed as a function of the class number and the particle diameter:

$$C_{pn} = CN \times \left(\frac{0.5}{d_p} \right)^{2.2} \tag{10.23}$$

where CN is the ISO class number (1, 2, 3, ...) and d_p is in μm.

These specifications define the maximum particle concentrations for the given class of cleanroom that are equivalent to the ISO cleanroom standard. FS 209E Class 1 is equivalent to ISO Class 3; FS 209E Class 10 is equivalent to ISO Class 4, and so on. The difference between the ISO and US FS 209E is negligible for most cleanroom applications.

10.5.2 Cleanroom Applications

Cleanrooms, or clean airspaces, have many applications, such as manufacturing and packaging. As technology advances, the need for cleaner work environments increases. The following major industries use clean airspaces for their products:[17]

Electronics — Advances in semiconductor microelectronics continue to drive cleanroom design. Semiconductor facilities account for a significant percentage of all cleanrooms in operation in the United States, with most the newer semiconductor cleanrooms being ISO Class 5 or cleaner.

Aerospace — Cleanrooms were first developed for aerospace applications to manufacture and assemble satellites, missiles, and aerospace electronics. Most applications involve clean airspaces of large volumes, with cleanliness levels of ISO Class 9 or cleaner.

Pharmaceuticals — Preparation of pharmaceutical, biological, and medical products requires clean airspaces to control viable particles (living organisms) that could produce undesirable bacteria growth and other contaminants. Many bacteria and viruses are airborne.

Hospitals — Some operating rooms may be classified as cleanrooms, yet their primary function is to limit particular types of contamination rather than the quantity of particles present. Cleanrooms are used in patient isolation and surgery, where risks of infection exist. Health care facilities usually have distinguished characteristics and design criteria.

Miscellaneous applications — Cleanrooms are also used in aseptic food processing and packaging, the manufacture of artificial limbs and joints, automatic paint booths, the crystal and laser/optic industries, and advanced materials research.

10.5.3 Air Cleaning for Cleanrooms

Airborne pollutants occur in nature as pollen, bacteria, miscellaneous living and dead organisms, windblown dust, and sea spray. Industry generates particles from combustion processes, chemical vapors, and friction in manufacturing equipment. Emissions from buildings are a major source of airborne pollutants. Occupants, including people, animals, and other living things, are prime sources of particles in the form of skin flakes, hair or feathers, fecal material, clothing lint, cosmetics, respiratory emissions, and bacteria from perspiration. These airborne particles vary in size form 0.001 µm to several hundred micrometers. Particles larger than 5 µm tend to settle quickly.

With respect to the clean space, particle sources are grouped into two general categories: external and internal. Particles from *external sources* are those particles that enter the clean airspace from the outside, normally via infiltration through doors, windows, and wall penetrations for pipes, ducts, etc. However, the largest external source is usually the outside air entering though the air supply system. In an operating cleanroom, external particle sources normally have little effect on overall cleanroom particle concentration, because HEPA filters clean the supply air. However, the particle concentration in clean spaces at rest relates directly to ambient particle concentrations. External sources are controlled primarily by air filtration, room pressurization, and airtightness of the room.

Internal sources of particles in cleanrooms include occupants, cleanroom surface shedding, process equipment, and the manufacturing process itself. Cleanroom occupants can be the largest source of internal particles. Workers may generate several thousand to several million particles per minute in a cleanroom. Personnel-generated particles are controlled with new cleanroom garments, proper gowning procedures, and airflow designed to shower the workers continuously with clean air. Polluted air is directed to air-cleaning devices or exhausted out. As personnel work in the cleanroom, their movements may reentrain airborne particles from other sources, including those settled on surfaces. Other activities, such as writing, may also cause higher particle concentrations.

Particle concentrations in the cleanroom have been used to define cleanroom class, but actual particle deposition on the product is of greater concern in cleanroom applications. Cleanroom designers may not be able to control or prevent internal particle generation completely, but they may anticipate internal sources and design control mechanisms and airflow patterns to limit the effect of these particles on the product. The following sections describe typical techniques used in obtaining a cleanroom environment.

10.5.4 Fibrous Air Filters

Proper air filtration prevents most externally generated particles from entering the cleanroom. Present technology for high-efficiency air filters centers around two types:

- High-efficiency particulate air (HEPA) filters
- Ultralow-penetration air (ULPA) filters

HEPA and ULPA filters primarily use glass fiber paper technology. Laminates and nonglass media for special applications have been developed. HEPA and ULPA filters are usually constructed in a deep pleated form with either aluminum, coated string, filter paper, or hot-melt adhesives as pleating separators. Filters may vary from 25 to 300 mm in depth; available media area increases with deeper filters and closer pleat spacing. Such filters operate at duct velocities near 1.3 m/s, with resistance rising from 120 to more than 500 Pa over their service life. These filters are the standard for cleanroom, nuclear, and toxic particulate applications.

Theories and models verified by empirical data indicate that interception and diffusion are the dominant particle collection mechanisms for HEPA filters. Fibrous filters have their lowest removal efficiency at the most penetrating particle size (MPPS), which is determined by filter fiber diameter, volume fraction or packing density, and air velocity. For most HEPA filters, the MPPS is between 0.1 and 0.3 μm. The rated efficiencies for HEPA and ULPA filters are based on 0.3 μm and 0.12 μm, respectively. For example, a HEPA filter must have a filter efficiency of 99.97% or higher for particles of 0.3 μm.

10.5.5 Airflow Pattern Control

Air turbulence in a cleanroom is strongly influenced by ventilation configurations, foot traffic, and process equipment layout. Selection of the air pattern configurations is the first step of good cleanroom design. Numerous air pattern configurations are in use, but in general they fall into two categories:

- Unidirectional airflow (often mistakenly referred to as laminar flow)
- Nonunidirectional airflow

Although not truly laminar airflow, unidirectional airflow, often called *displacement* airflow or ventilation, is characterized as air flowing in a single direction through a room or a clean zone with generally parallel streamlines. Ideally, the flow streamlines would be uninterrupted, and although personnel and equipment in the air stream do distort the streamlines, a state of constant velocity is approximated. Most particles that encounter an obstruction in a unidirectional airflow approach the obstruction and continue around it as the air stream reestablishes itself downstream of the obstruction.

Nonunidirectional airflow, often referred to as *turbulent* or *mixed* airflow or ventilation, does not meet the definition of unidirectional airflow because it has either multiple pass circulating characteristics or nonparallel flow. Variations of nonunidirectional airflow are based primarily on the location of supply air inlets and outlets and air filter locations. Airflow is typically supplied to the space though supply diffusers containing HEPA filters or through supply diffusers with HEPA filters located in the ductwork or air handler. In a mixed flow room, air is prefiltered in the supply section and is HEPA filtered at the workstations located in the clean space. Airflow pattern control will be discussed in more detail in Chapter 14.

10.6 TYPICAL FIBROUS FILTERS

Fibrous filters are made from woven, felted, and knitted materials. Fibers used in construction of filters include cotton, polypropylene, fiberglass, nylon, polyester, wool, Teflon, P84 polyamide, and ceramic, as well as variations of these fabrics, with Goretex laminates to improve dust release properties. The choice of fabric is based on the type of fabric filter collector, the cost of the media, the operating temperature, and the characteristics of the particulate matter and the carrying gas, such as corrosiveness, combustibility, and moisture content. These characteristics affect the useful life of the materials. Table 10.2 lists the highest operating temperatures for different fibrous materials.

Fibrous air filters can generally be divided into two common types: unit filters and renewable media filters. Fibrous unit filters have fixed filtration media that allow dust loads to accumulate and cause increased pressure drop to some maximum recommended value. During this dust accumulation period, filter efficiency normally increases. However, at high dust loads, dust may adhere poorly to filter fibers, and efficiency drops due to off-loading and reentrainment of dust to the air stream. Filters

Table 10.2 Highest Operating Temperatures for Different Fibrous Materials

Fibers	Highest Operating Temperature (°C)
Cotton	82
Polypropylene	88
Fiberglass	260
Nylon	93
Polyester	135
Wool	93
Teflon	232
P84 polyamide	190
Ceramic	980

in such a condition should be replaced or reconditioned, as should filters that have reached their final (maximum recommended) pressure drop. This type of filter includes viscous impingement and dry-type air filters, available in low-efficiency to ultrahigh-efficiency construction.

Viscous impingement filters are unit filters made up of coarse fibers with a high porosity. The filter media are coated with a viscous substance, such as oil (also known as an adhesive), which causes particles that impinge on the fibers to stick to them. Design air velocity through the media is usually in the range of 1 to 4 m/s. These filters are characterized by low pressure drop, low cost, and good filter efficiency on lint but low efficiency on normal atmospheric dust. They are commonly made 13 to 100 mm thick. Unit panels are available in standard and special sizes, up to about 610 mm × 610 mm. This type of filter is commonly used in residential furnaces and air conditioning; it is also often used as a prefilter for higher-efficiency filters.

A number of different materials are used as the filtering medium, including coarse (15 to 60 μm diameter) glass fibers, coated animal hair, vegetable fibers, synthetic fibers, metallic wools, expanded metals and foils, crimped screens, random-matted wire, and synthetic open-cell foams.

Although viscous impingement filters usually operate in the range of 1.5 to 3 m/s, they may be operated at higher velocities. The limiting factor, other than increased flow resistance, is the danger of blowing off agglomerates of collected dust and the viscous coating on the filter. The loading rate of a filter depends on the type and concentration of the dirt in the air being handled and the operating cycle of the system. Manometers, static pressure gauges, or pressure transducers are often installed to measure the pressure drop across the filter bank. From the pressure drop, it can be determined when the filter requires servicing. The final allowable pressure drop may vary from one installation to another; but, in general, unit filters are serviced when their operating resistance reaches 120 Pa. The decline in filter efficiency (which is caused by the absorption of the viscous coating by dust, rather than by increased resistance because of dust load) may be the limiting factor in operating life.

The manner of servicing unit filters depends on their construction and use. Disposable viscous impingement, panel-type filters are constructed of inexpensive materials and are discarded after one period of use. The cell sides of this design are usually a combination of cardboard and metal stiffeners. Permanent unit filters are generally constructed of metal to withstand repeated handling. Various cleaning methods have been recommended for permanent filters; the most widely used involves washing the filter with steam or water (frequently with detergent) and then recoating it with recommended adhesive by dipping or spraying. Unit viscous filters are also sometimes arranged for in-place washing and recoating.

The adhesive used on a viscous impingement filter requires careful engineering. Filter efficiency and dust-holding capacity depend on the specific type and quantity of adhesive used: This information is an essential part of test data and filter specifications. Desirable adhesive characteristics, in addition to efficiency and dust-holding capacity, include the following:

- A low percentage of volatiles to prevent excessive evaporation
- A viscosity that varies only slightly within the service temperature range
- The ability to inhibit growth of bacteria and mold spores
- A high capillarity or the ability to wet and retain the dust particles
- A high flash point and fire point
- Freedom from odorants and irritants

The media in dry-type air filters are random fiber mats or blankets of varying thickness, fiber sizes, and densities. Bonded glass fiber, cellulose fibers, wool felt, synthetics, and other materials have been used commercially. The media in filters of this class are frequently supported by a wired frame in the form of pockets, or V-shaped or radial pleat. In other designs, the media may be self-supporting because of inherent rigidity or because airflow inflates it into extended shape, as with bag filters. Pleating of the media provides a high ratio of media area to face area, thus allowing reasonable pressure drop and low media velocities.

In some designs, the filter media is replaceable and is held in position in permanent wire baskets. In most designs, the entire cell is discarded after it has accumulated its maximum dust load. The efficiency of dry-type air filters is usually higher than that of panel filters, and the variety of media available makes it possible to furnish almost any degree of cleaning efficiency desired. The dust-holding capacities of modern dry-type filter media and filter configurations are generally higher than those of panel filters.

The placement of coarse prefilters ahead of extended-surface filters is sometimes justified economically by the longer life of the main filters. Economic considerations should include the prefilter material cost, maintenance out labor, and increased fan power. Generally, prefilters should be considered only if they can reduce substantially the part of the dust that may plug the protected filter. A prefilter usually has a collection efficiency of 70% or more. Temporary prefilters are worthwhile during building construction to capture heavy loads of coarse dust. HEPA-type filters of 95% DOP efficiency and greater should always be protected by prefilters of 80% or greater. The initial resistance of an extended-surface filter varies with the choice of media and the filter geometry. Commercial designs typically have an initial resistance from 25 to 250 Pa. It is customary to replace the media when the final resistance of 125 Pa is reached for low-resistance units and 500 Pa for the highest-resistance units. Dry media providing higher orders of cleaning efficiency have a higher average resistance to airflow. The operating resistance of the fully dust-loaded filter must be considered in the design, because that is the maximum resistance against which the fan operates. Variable air volume and constant air volume system controls prevent abnormally high airflows or, possibly, fan motor overloading from occurring when filters are clean.

Flat-panel filters with media velocity equal to duct velocity are possible only in the lowest efficiency units of the dry type (open cell foams and textile denier nonwoven media). Initial resistance of this group, at rated airflow, is mainly between 10 and 60 Pa. They are usually operated to a final resistance of 120 to 175 Pa. In extended-surface filters of the intermediate efficiency ranges, the filter media area is much greater than the face area of the filter; hence, velocity through the filter media is substantially lower than the velocity approaching the filter face. Media velocities range from 0.03 to 0.5 m/s, although the approach velocities run to 4 m/s. Depth in direction of airflow varies from 50 to 900 mm.

Renewable media filters, in which fresh medium is introduced into the air stream as needed to maintain essentially constant resistance, are thus able to maintain a constant average efficiency. Automatic moving-curtain viscous filters are available in two main types. In one type, a random-fiber (nonwoven) medium is furnished in roll form. Fresh medium is fed manually or automatically across the face of the filter, while the dirty medium is rewound onto a roll at the bottom.

When the roll is exhausted, the tail of the medium is wound onto the take-up roll, and the entire roll is thrown away. A new roll is then installed, and the cycle is repeated. The moving-curtain filters may have the medium automatically advanced by motor drives on command from a pressure switch, timer, or media light-transmission control. A pressure switch control measures the pressure drop across the filtration medium and switches on and off at chosen upper and lower set points. Most pressure drop controls do not work well in practice. Timers and media light-transmission controls help to avoid these problems; their duty cycles can usually be adjusted to provide satisfactory operation with acceptable medium consumption. Filters of this replaceable roll design generally have a signal indicating when the roll is nearly exhausted. At the same time, the drive motor is deactivated so that the filter cannot run out of medium. The normal service requirements involve insertion of a clean roll at the top of the filter and disposal of the loaded dirty roll. Automatic filters for the design are not, however, limited in application to the vertical position. Horizontal arrangements are available for use with makeup air units and air-conditioning units. Adhesives must have qualities similar to those for panel-type viscous impingement filters, and they must withstand media compression and endure long storage.

The second type of automatic viscous impingement filter consists of linked metal mesh media panels installed on a traveling curtain that intermittently passes through an adhesive reservoir. In the reservoir, the panels give up their dust load and, at the same time, take on a new coating of adhesive. The panels thus form a continuous curtain that moves up one face and down the other face. The medium curtain, continually cleaned and renewed with fresh adhesive, lasts the life of the filter mechanism. The precipitated dirt must be removed periodically from the adhesive reservoir.

The resistance of both types of viscous impingement, automatically renewable filters remains approximately constant as long as proper operation is maintained. A resistance of 80 to 125 Pa at a face velocity of 2.5 m/s is typical of this class.

Moving-curtain dry-media filters with relatively high porosity are also used in general ventilation service. Operating duct velocities near 1 m/s are generally lower than those of viscous impingement filters. Special automatic dry filters are also available, which are designed for the removal of lint in textile mills and dry-cleaning establishments and for the collection of lint and ink mist in pressrooms. The medium used is extremely thin and serves only as a base for the buildup of lint, which then acts as a filter medium. The dirt-laden medium is discarded when the supply roll is used up. Another form of filter designed specifically for dry lint removal consists of a moving curtain of wire screen, which is vacuum cleaned automatically at a position out of the air stream. Recovery of the collected lint is sometimes possible with such a device.

Electret filters are composed of electrostatically charged fibers. The charges on the fibers augment the collection of smaller particles by Brownian diffusion with Coulomb forces caused by the charges on the fibers. There are three types of these filters: resin wool, electret, and electrostatically sprayed polymer. The charge on the resin wool fibers is produced by friction during the carding process. During production of the electrets, a corona discharge injects positive charges on one side of a thin polypropylene film and negative charges on the other side. These thin sheets are then shredded into fibers of rectangular cross-section. The third process spins a liquid polymer into fibers in the presence of a strong electric field, which produces the charge separation. The efficiency of charged-fiber filters is due both to the normal collection mechanisms of a media filter and to strong local electrostatic effects. These effects induce efficient preliminary loading of the filter to enhance the caking process. However, dust collected on the media can reduce the efficiency of electret filters.

Membrane filters are used predominately for air sampling and specialized small-scale applications where their particular characteristics compensate for their fragility, high resistance, and high cost. They are available in many pore diameters and resistances and in flat sheet and pleated forms.

Combination air cleaners exist that combine the previously mentioned types. For example, an electronic air filter may be used as an agglomerator with a fibrous medium downstream to catch agglomerated particles blown off of the plates. Electrode assemblies have been installed in air-handling systems, making the filtration system more efficient. A renewable media filter may be used upstream of a high-efficiency unit filter to extend its service life. Charged media filters can also increase particle deposition on media fibers by an induced electrostatic field. In this case, pressure loss increases as it does on a fibrous media filter. The benefits of combining different air cleaning processes vary. ASHRAE Standard 52.1 test methods may be used to compare the performance of combination air cleaners.[6]

10.6.1 Residential Air Filters

Filters used for residential applications are usually of the spun glass variety, which only filter out the largest of particles. These filters may prevent damage to downstream equipment, but they do little to improve air quality in the residence. In the absence of air-cleaning systems in line with the HVAC, console-type air cleaners can be used. Consoles equipped with HEPA filters or electrostatic precipitators can be effective in controlling particulates in a single room. For whole-house applications, in-line units are recommended.

10.7 DUST BAG HOUSES

Bag houses are widely used in dusty environments, such as power plants and cotton gins. Generally, the dirty gas enters the bag at the bottom and passes through the fabric, and the particulate matter is deposited on inside of the bag. Although there are many bag house designs, they may generally be classified by the method of cleaning and by whether the operation is periodic (or intermittent) or continuously automatic. Periodic operation requires shutdown of portions of the bag house at regular intervals for cleaning. Continuous automatic operation is required where periodic shutdown is not desirable. (Intermittent operation requires shutting down the entire process in order to clean the filter cloth. The dirty gas bypasses the equipment during the cleaning cue. This type of the design is of limited use, because very few effluents could be discharged to the atmosphere without cleaning, even on a temporary basis.) Cleaning is accomplished in a variety of ways, including mechanical vibration, shaking pulse jets, and reverse airflow.

A system with a mechanical vibration mechanism, such as shaking springs, is used for bag house cleaning (Figure 10.7a). When the dirty air passes through the filter bags, dust is collected on the inside of the bags. A mechanism can actuate the shaking springs to vibrate and shake off the dust at the bag surfaces. Dust will fall into the collection bin by gravity. All shaking springs can be actuated at the same time, or only some of them may be actuated. Typical bag height is 5 m or higher, and the operating air velocity through the filter is around 1 m/s.

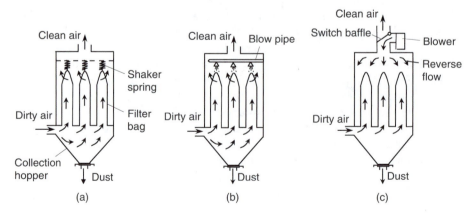

Figure 10.7 Schematic of cleaning mechanisms for bag houses: (a) mechanical vibration (b) shaking pulse jet (c) reverse airflow.

A pulse-jet system (Figure 10.7b) collects particulate matter on the outside of the bags. At an appropriate time, a short pulse of compressed air is directed downward through a venturi at the top of a row of bags. The pulse passes quickly down the bags, knocking the dust off. The dust settles by gravity to the bottom of the bag house. Because the time for cleaning a row of bags is very short and only a fraction of the bags can be cleaned at one time, continuous flow is maintained though the bag house. Because the rows of bags adjacent to the row being cleaned are in filtration mode, care must be taken to avoid reentrainment of the dust from the pulsed row. Typical filtering velocities in a pulse-jet bag house are in the range of 2 to 5 m/min, depending on the application. Bag heights in the latter case are usually less than 5 m, since the bottoms of extremely long bags are more difficult to clean using a pulse originating at the top of the bags.[18]

A reverse airflow collects particles on the inside of the bag, similar to the mechanical shaking method (Figure 10.7c). At the proper time, the flow of polluted air is cut off from the compartment, and cleaning air flows through the bag in the opposite direction. Typical dust bag houses are 6 m or taller. Bag diameters typically are in the 20 to 30 cm range, and recommended average filtering velocities (face velocity at the bag filter surfaces) are around 0.6 to 1.2 m/min.

10.8 LIMITATIONS OF FIBROUS FILTERS

There two major limitations of fibrous filters: the frequent cleaning requirement and pressure drop across the filter. Relatively frequent cleaning is necessary in order to avoid unreasonable pressure drops and desirable performance. As a result, the basic design of industrial filters is usually predicated on geometry that lends itself to relative ease of cleaning. One basic method for meeting these criteria is a collection system based on the bag house. Fibrous cylinders (supported internally in some manner), ranging from 10 to 35 cm in diameter and up to 12 meters long, are arranged in vertical rows. Many individual bags must be employed in one bag house when large gas volumes must be cleaned. One large bag house, capable of removing 6 tons of dust and fumes per hour from a large automotive foundry, consists of 16 compartments with a total of 4000 Dacron bags, each 20 cm in diameter and 57 cm long. This leads to one inherent disadvantage of bag house design: Namely, the overall size of the equipment is large in comparison to competitive types of removal devices.

Pressure drop across fabric filters is an important consideration. As the dust cake accumulates on the supporting fabric, the removal efficiency would be expected to increase. At the same time, however, the resistance to flow also increases. For a new filter that is to be loaded with particulate matter, the overall pressure drop can be approximated as the sum of that due to the new filter plus that contributed by the particles that form a dust layer or cake. For smooth, woven filters (in which the fibers are made into a yarn and the yarn is woven to form the filter), there is only a limited number of pore openings between the woven yarns in the warp and fill directions. In this case, there may be significant interaction between the filter and the particles, and the filtration results in pore blockage. For felted filters that have a random fiber orientation resulting in numerous pores in a variety of sizes, the initial interaction generally produces a somewhat linear initial rise in pressure drop with time. After the dust has filled or bridged the pores of the filter and begun to filter onto the previously deposited particles, the rate of rise in the pressure drop tends to become constant, resulting in a linear rise in pressure drop with time. The rise rate is a function of gas and particle characteristics.

DISCUSSION TOPICS

1. For a fibrous filter such as the one used in a home furnace, what filtration mechanism or mechanisms are most important, and why?
2. Generally, filtration due to gravitational settling can be negligibly small compared with other filtration mechanisms. However, gravitational settling may be significant in some specific applications. What factors, including design and operational factors, may increase the effect of gravitational settling?
3. What is the difference between single-fiber efficiency and filter efficiency? Is filter efficiency always higher than the total single-fiber efficiency?
4. A general criterion to select a filter is the filter quality q_f, which considers the filter efficiency and the pressure drop. What other factors should be considered when selecting a filter for a specific application?
5. In many dusty environments, such as a grain elevator, fibrous filters may not be economical for air cleaning because they require frequent filter replacement and high operating cost. Consider other air-cleaning options for such an environment.
6. List the applications of filters in your daily life. Estimate how much per year an average family with a 200 m² house may spend on products related to air cleaning (filters, air fresheners, vacuum cleaners, etc.).

PROBLEMS

1. A home furnace filter is 20 mm thick and has a solidity of 0.05. The average diameter of the filter fiber is 30 μm. The face air velocity is 2 m/s. The angle α = 90° for the filter. Calculate the following collection efficiencies of a single fiber in a filter for particles with an aerodynamic diameter of 5 μm:
   a. Collection efficiency by interception
   b. Collection efficiency by inertial impaction
   b. Collection efficiency by diffusion
   d. Collection efficiency by gravitation settling
   e. Total fiber collection efficiency
2. In problem 1, recalculate the particle collection efficiencies for particles in the size range of 0.1 to 100 μm in aerodynamic diameter. Plot the particle collection efficiencies for a single fiber (each filtration mechanism and the total single fiber) vs. particle size (in the x-axis).
3. Calculate the particle collection efficiency of a home furnace fiber filter for particles of 0.3 μm in diameter with standard density. The filter is 2.5 cm thick, made of fibers 100 μm in diameter, and with a solidity of 0.05. The average air velocity passing through the filter is 1 m/s.

4. What is the pressure drop across the filter described in problem 1? If a filter keeps all parameters the same as in problem 1, except that this filter is made of 10 μm fibers, what would the change in pressure drop be?

5. A prefilter for a cleanroom is required to capture most particles larger than 5 μm. The face flow direction is horizontal. Two brands of filters with similar costs are proposed for this application. Which filter should be selected based on the filter quality q_f? Specifications of the two brands are shown in the following table.

Parameter	Filter A	Filter B
Solidity	0.03	0.03
Thickness (mm)	10	5
Fiber diameter (μm)	70	50
Face velocity (m/s)	2	2

6. In one application, the filter needs to be replaced when its quality is reduced to 80% of its initial value, that is, when $q_f = 0.8q_{f0}$. Assume that the pressure drop across the filter proportionally increases with its operating time, t:

$$\Delta P = \Delta P_0 + C_f t$$

On the other hand, the filter efficiency increases with the time as follows:

$$\xi_f = 1 - (1 - \xi_{f0}) e^{-C_\xi t}$$

where ΔP_0 is the initial pressure drop of the filter at $t = 0$. C_f and C_ξ are proportional coefficients that can be determined experimentally. In one set of measurements at the same filter operating conditions, $\Delta P_0 = 100$ Pa and $\xi_{f0} = 90\%$. After one week, the pressure drop and the filter efficiency increase to 120 Pa and 93%, respectively. Determine the filter service life. (Hint: You may need to use iteration method to find the service life t.)

7. What will the service life be in problem 6, if we assume that the filter efficiency remains unchanged from its initial value? What is the error caused by this assumption?

REFERENCES

1. Lee, K.W. and Ramamurthi, M., Filter collection, in *Aerosol Measurement: Principles, Techniques, and Applications*, Willeke, K. and Baron, P.A., eds., Van Nostrand Reinhold, New York, 1993.

2. Kirsch, A.A. and Stechkina, I.B., The theory of aerosol filtration with fiber filters, in *Fundamentals of Aerosol Science*, Shaw, D.T., ed., John Wiley & Sons, New York, 1978.

3. Yeh, H.C. and Liu, B.Y.H., Aerosol filtration by fibrous filters, *J. Aerosol Sci.*, 5:191–217, 1974.

4. Brown, R.C., *Air Filtration: An Integrated Approach to the Theory and Applications of Fibrous Filters*, Pergamon, Oxford, 1993.

5. Kirsch, A.A. and Fuchs, N.A., Studies of fibrous filters — III: diffusional deposition of aerosol in fibrous filters, *Ann. Occup. Hyg.*, 11:299–304, 1968.

6. ASHRAE, Standard 52.1-1992, *Gravimetric and Dust-Spot Procedures for Testing Air-Cleaning Devices Used in General Ventilation for Removing Particulate Matter*, American Society of Heating, Refrigerating and Air Conditioning Engineers, Atlanta, 1992.

7. Davies, C.N., ed., *Air Filtration*, Academic Press, London, 1973.

8. ASHRAE, Standard 52.2-1999, *Method of Testing General Ventilation Air-Cleaning Devices for Removal Efficiency by Particle Size*, American Society of Heating, Refrigerating and Air Conditioning Engineers, Atlanta, 1999.

9. ASHRAE Handbook, *HVAC Systems and Equipment*, American Society of Heating, Refrigerating and Air Conditioning Engineers, Atlanta, 2000. ch. 24.

10. McCrone W.C., Draftz, R.G., and Delley, J.G., *The Particle Atlas*, Ann Arbor Science Publishers, Ann Arbor, MI, 1967.

11. Horvath, H., A comparison of natural and urban aerosol distribution measured with the aerosol spectrometer, *Environmental Science and Technology*, August: 651, 1967.

12. U.S. Military Standard MIL-STD-282, *DOP-Smoke Penetration and Air Resistance of Filters*, Department of Navy—Defense Printing Service, Philadelphia, 1956.

13. U.S. Army, *Instruction Manual for the Installation, Operation and Maintenance of Penetrometer, Filter Testing, DOP, Q107. Document 136-300-175A*, Edgewood Arsenal, MD, 1965.

14. Perters, A.H., *Application of Moisture Separators and Particulate Filtration in Reactor Contaminant*, USAEC-DP812, U.S. Department of Energy, Washington, DC, 1962.

15. Perters, A.H., Minimal specification for the fire-resistant high-efficiency filter unit, *USAEC Health and Safety Information (212)*, U.S. Department of Energy, Washington, DC, 1965.

16. Federal Standard 209E., *Airborne Particulate Cleanness Classes in Cleanrooms and Clean Zones*, Washington, DC, 1992.

17. ASHRAE Handbook, *HVAC Applications*, American Society of Heating, Refrigerating and Air Conditioning Engineers, Atlanta, 1999, ch. 15.

18. Buonicore, A. and Davies, W.T., eds., *Air Pollution Engineering Manual*, Van Nostrand Reinhold, New York, 1992.

CHAPTER 11

Aerodynamic Air Cleaners

Although fibrous filters are the most efficient and commonly used means of air cleaning, there are limitations often associated with this high filter efficiency, one of which is the relatively low dust-holding capacity. The primary problems associated with the fibrous filters stem from the direct contact of dust with a physical filtration media. This *contact* process accumulates dust onto the media quickly, and the media requires frequent maintenance or replacement. In many dusty environments, such as a grain elevator or an off-road vehicle, the dust concentration is typically tens or hundreds times higher than that of a residential building. As a result, filtration media of an air-cleaning device must be frequently maintained or replaced, which incurs high capital and operating costs.

Aerodynamic air cleaning, that is, the aerodynamic separation of dust particles from an air stream that does not pass through a physical filtration medium such as a fibrous filter, is desirable for many dusty environments. Such a *noncontact* air-cleaning process offers many advantages in daily life applications. This chapter provides a brief review of traditional cyclone technology and discusses the principles and applications of uniflow deduster (a type of uniflow cyclone to separate dust from an air stream). Other types of air-cleaning mechanisms and devices using a noncontact approach, such as wet scrubbers and venturi scrubbers, are also discussed. In the last section, the performance of an aerodynamic air cleaner is evaluated using the same criteria for filters: particle collection efficiency, pressure drop, and dust-holding capacity.

Although analytical methods are important for the design of different devices, many assumptions must be made for different design configurations and operating conditions. In this chapter, formulas for several devices are presented. It is important that the reader understand the principles, especially the forces and properties of the particles, and be able to apply them in different design or evaluation processes.

By completing this chapter, the reader will be able to

- Design cyclones:
  - Return-flow cyclones
  - Uniflow cyclones

• Analyze particle separation efficiency in laminar flow and complete-mixing flow in a cyclone
• Calculate the pressure drop across cyclones
• Design wet scrubbers — including spray chamber scrubbers, wet cyclones, and venturi scrubbers — and predict the particle separation efficiency
• Evaluate the overall performance (quality) of cyclones using particle collector performance criteria

11.1 RETURN-FLOW CYCLONES

Cyclones are particle separators that use centrifugal force generated from the spinning airflow to separate suspended particulate matter (solid or liquid) in the air stream. As the name implies, the device must create a cyclonic flow pattern with high air velocities. Typical initial velocities entering the cyclone range from 15 to 20 m/s. There are two major conventional return flow types of cyclones for dust separation: involute cyclones and vane-axial cyclones (Figure 11.1). For *involute cyclones,* the dust-laden air is introduced through a rectangular inlet that is tangential to the separation chamber. The air travels around the inner cylinder with a 180° involute and then exits. Dust separated is collected in a storage hopper (Figure 11.1a). The only difference between involute and vane-axial cyclones is the entrance of the cyclone. In a *vane-axial cyclone,* the dust-laden air is introduced through a set of vanes that force the air stream flows in a spiral motion (Figure 11.1b). The particle separation and collection processes for the vane-axial cyclones are the same as for involute cyclones.

11.1.1 Particle Separation Efficiency

Particle separation efficiency for a return flow type of cyclone depends the cyclone parameters, particle properties, and air flow characteristics. Therefore, different designs can substantially alter the particle separation efficiency. On the other hand, the complexity of airflow and cyclone configurations may prevent accurate analytical expressions of particle separation. The following analyses are based on a typical kind of return-flow cyclone found in many references.[1] As shown in Figure 11.2, assume that the particle-laden airflow has a velocity at the inlet, U_0, the height and width of the rectangular inlet are H and W, respectively, and the radii of the inner and outer cylinders are R_1 and R_2, respectively. As the particle-laden airflow enters the cyclone, it spins through N_e revolutions in the main outer chamber (Figure 11.1a) before it enters the inner chamber and travels upward to the exit of the cyclone. The number of revolutions N_e can be approximated as

$$N_e = \frac{1}{H}\left(L_1 + \frac{L_2}{2}\right) \tag{11.1}$$

where, L_1 is the height of the main upper cylinder and L_2 is the height of the lower cone.

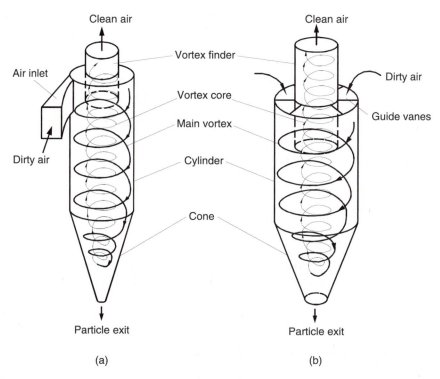

Figure 11.1 Sketch of return-flow cyclones: (a) involute type (b) vane-axial type. The only difference between the two common types of cyclones is the entry of air inlets.

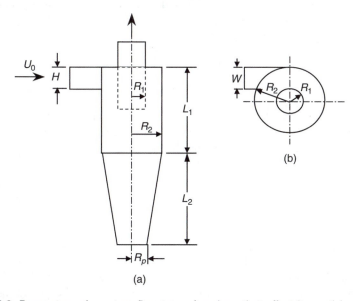

Figure 11.2 Parameters of a return flow type of cyclone that affect its particle separation efficiencies: (a) side view (b) top view.

Note that this approximation is based on the rationale that the airflow will turn one full revolution at one height of the inlet for the upper cylinder, and half a revolution at one height of the inlet for the lower cone. We assume that particles are uniformly distributed at the entrance of the cyclone. For particles with a diameter d_p to be collected at 100% efficiency, all particles at that size entering at the radius R_1 of the cyclone must reach the radius R_2 of the cyclone by the centrifugal force before the flow travels N_e revolutions. Subsequently, all particles of size d_p or larger entering the cyclone at some intermediate radius r_{sp} will also be completely collected. The minimum radius r_{sp} refers to the radius for which particles of size d_p will reach the outer wall during N_e revolutions. Therefore, the particle separation efficiency ξ_x can be expressed as

$$\xi_x = \frac{R_2 - r_{sp}}{R_2 - R_1} \quad (R_1 < r_{pl} < R_2) \tag{11.2}$$

In this design of a cyclone, the width of the cyclone rectangular inlet, W, is the same as the distance between the inner and outer cylinders:

$$W = R_2 - R_1 \tag{11.3}$$

Therefore, by developing a relationship of $(R_2\text{-}r_{sp})$, we can obtain the particle separation efficiency. The distance $(R_2\text{-}r_{sp})$ is directly related to the particle radial velocity V_r, which is normal to the airflow and moves toward the outer cylinder and the time the gas is in the separation chamber. This radial particle velocity can be found from its centrifugal force and drag force. For a spherical particles, if we assume that the particle motion is in the Stokes region and neglect the slip coefficient

$$\rho_p \frac{\pi d_p^3}{6} \frac{U_t^2}{r} = 3\pi\eta d_p V_r \tag{11.4}$$

where U_t is the tangential air velocity in the chamber, which can be considered the particle tangential velocity; V_r is the radial velocity of the particle; and r is the radius at which the particle is located. Solving the previous equation for V_r gives

$$V_r = \frac{\rho_p d_p^2 U_t^2}{18\eta r} \tag{11.5}$$

The tangential air velocity U_t is the entering air velocity U_0. The radial velocity V_r is a function of r. To simplify the solution, we assume that r is the average of R_1 and R_2. Thus V_r is essentially a constant for a particle moving outward. Note that the total distance needed for the particle to be separated is $(R_2\text{-}r_{sp})$ and the time is Δt. Between the two cylinders, the radial velocity of the particle can be also written as

$$V_r = \frac{R_2 - r_{sp}}{\Delta t} \tag{11.6}$$

Combining Equation 11.5 and Equation 11.6 gives

$$R_2 - r_{sp} = \frac{\rho_p d_p^2 U_0^2}{18\eta r} \Delta t \tag{11.7}$$

The time that the particle remains in the cyclone is also given by

$$\Delta t = \frac{2\pi r N_e}{U_0} \tag{11.8}$$

where N_e is the number of revolutions defined in Equation 11.1. Substituting Equation 11.7, Equation 11.8 and Equation 11.3 into Equation 11.2 gives the separation efficiency of the return-flow cyclone for particles with a size d_p:

$$\xi_x = \frac{\pi N_e \rho_p d_p^2 U_0}{9\eta W} \tag{11.9}$$

If the flow rate Q of the cyclone is known, the separation efficiency can be expressed as

$$\xi_x = \frac{\pi N_e \rho_p d_p^2 Q}{9\eta H W^2} \tag{11.9}$$

where H and W are the height and width, respectively, of the cyclone rectangular inlet.

The cutsize of the particles then can be obtained by making $\xi_x = 0.5$ in Equation 11.8 and Equation 11.9.

$$d_{50} = \left(\frac{9\eta W}{2\pi N_e U_0 \rho_p} \right)^{\frac{1}{2}} = \left(\frac{9\eta H W^2}{2\pi N_e U_0 \rho_p Q} \right)^{\frac{1}{2}} \tag{11.10}$$

Analysis: Note that the separation efficiency in Equation 11.9 reaches 1 at a certain particle diameter, and greater than 1 for particles larger than that certain size. This is usually not true in practice, because some particles larger than that size will pass through the cyclone due to turbulence, particle bounce, or other disturbance. Furthermore, Equation 11.9 shows that particle separation efficiency increases with d_p^2. Most cyclone separators have a particle separation efficiency with an exponential

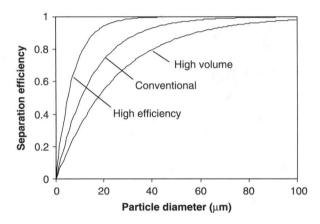

Figure 11.3 Particle separation efficiency for typical return-flow cyclones vs. particle sizes. Instead of increasing with d_p^2, the actual separation efficiency is somewhat exponential.

trend as shown in Figure 11.3. The separation efficiency does not really reach 100%. Regardless of the design, the separation efficiency of any cyclone drops rapidly beyond a certain particle size. A better cyclone is to achieve a 50% separation efficiency at a smaller particle size. Therefore, Equations 11.8 through 11.10 may cause large errors in predicting separations for a cyclone. The error sources are due to oversimplification of the mathematical model (Equation 11.9).

Return-flow cyclones are usually used to remove particles larger than 10 µm in diameter. However, conventional cyclones seldom remove particles with an efficiency greater than 90% unless the particles are larger than 25 µm. It is generally concluded that cyclones cannot be used practically in large-volume air-cleaning applications for indoor environments, where the particles of concern are smaller than 10 µm in diameter. In those cases, cyclones are primarily used as precleaners before other air-cleaning means, such as filtration.[2]

Equations 11.8 through 11.10 contain the basic cyclone design parameters. Although those equations may not able to predict particle separation efficiency or particle cutsize accurately, they indicate where improvement can be made. For example, increasing the number of turns (N_e) can increase the separation efficiency because they are proportionally related. At a given flow rate, reducing the width of the inlet can also improve the separation efficiency, because the ξ_x is inversely proportional to W^2. However, the height of the inlet has to be increased in order to maintain the same flow rate. A larger H means a smaller N_e. Therefore, specific designs must consider all parameters to achieve desirable particle separation efficiency. Conventional return-flow cyclones have standard parameters, as shown in Table 11.1.[1] All parameters are as shown in Figure 11.2.

The earliest application of centrifugal separation can be traced back many centuries, when Chinese people used this method to clarify tung oil. However, the exact date of the first application is not documented.[3] The first patent for a hydrocyclone was by Bretney.[4-6] In the U.S., early applications of centrifugal machines were primarily purging, hydro-extracting, clarifying, treating sewage, nitrating,

Table 11.1 Standard Return Flow Cyclone Parameters Based on the Outer Cylinder Radius R_2

Design Parameters	Symbol	Value
Length of cylinder	L_1	$4R_2$
Length of cone	L_2	$4R_2$
Height of inlet	H	R_2
Width of inlet	W	$R_2/2$
Radius of outlet for clean air	R_e	$R_2/2$
Radius of outlet for particles	R_p	$R_2/4$

dyeing, reclaiming oil and waste, and extracting oil.[7] Before 1950, centrifugal separations were applied mostly in liquid–liquid and liquid–solid separation.[8] Cyclones have more potential to stay where subsequent drying or pneumatic conveying are needed and where the equipment is exposed to high service temperatures, such as power generation, food and chemical processing, and building material production.[5]

Many models of commercial cyclones have been developed since the end of the 19th century.[9] Most commercial cyclones are limited to materials separation rather than air cleaning because of their low separation efficiency and high energy consumption for small particles. Different researchers have defined different classifications of cyclones for PM separation.[1, 2, 5, 10–12] Most cyclones fall into one of the following types, based on different inlets and plow passes: tangential, axial, and outflow; returned flow; and uniflow. No matter how a cyclone is classified, the inlet flow must be rotational in order to take advantage of the centrifugal forces. Usually, the rotational flow can be created in one of two ways: by a tangential inlet or by the guide vanes. Several studies have proved that the use of a deflector vane at the cyclone inlet reduces the cyclone pressure drop, but the separation efficiency is low.[5,13,14] High-efficiency cyclones, which are effective for particle sizes down to 3 μm, are available. Some cyclones are capable of collecting very small particles, but their flow rate is so small that they are only used in air-sampling instrumentation. These have not been widely used for large-volume air cleaning.[15–19] A high-volume airflow design sacrifices efficiency for high rates of collection.

11.1.2 Pressure Drop of Return-Flow Cyclones

The pressure drop of a cyclone, ΔP, is one of the most important criteria of cyclone performance. Most pressure drop equations for cyclones express the pressure loss as a function of the number of velocity heads H_v

$$\Delta P = H_v \frac{U^2}{2g} \frac{\rho_g}{\rho_L} \quad \text{(in meters of column of the liquid)} \qquad (11.11)$$

where ΔP is in meters of column of the liquid, U is the inlet velocity of the carrying gas in m/s, and the value of $U^2/2g$ is the value of one velocity head.

The ratio of gas density to liquid (i.e., water) density (ρ_g/ρ_L) accounts for the fact that the pressure loss in height of a fluid is usually reported in units of height of fluid (i.e., water) rather than height of the carrying gas. Substituting water density, 1000 kg/m³, and 10,000 Pa/m water column, the pressure drop can be expressed in Pa as

$$\Delta P = 10 \times H_v \frac{U^2 \rho_g}{2g} \text{ (in Pa)} \tag{11.12}$$

Although many equations have been developed to estimate the number of velocity heads of loss in cyclones, the empirical equations developed by Shepherd and Lapple in 1939 have been widely used and found to give a reasonable estimation:[14]

$$H_v = 3.75 \frac{HW}{R_e^2} \text{ (for involute cyclones)} \tag{11.13}$$

$$H_v = 1.875 \frac{HW}{R_e^2} \text{ (for vane axial cyclones)} \tag{11.14}$$

Example 11.1: *Calculate the aerodynamic particle cutsize and pressure drop of a standard involute cyclone with a cylinder diameter of 300 mm and an entry air velocity of 15 m/s. The ambient conditions are standard. The air density is 1.2 kg/m³.*

Solution: The radius of the cylinder is $R_2 = 0.15$ m. For a standard return-flow cyclone and using Table 11.2, other parameters are

$$L1 = 4R_2 = 0.6 \text{ m; } L2 = 4R_2 = 0.6 \text{ m; } H = R_2 = 0.15 \text{ m; } W = 0.5R_2 = 0.075 \text{ m;}$$
$$R_e = 0.5R_2 = 0.075 \text{ m}$$

From Equation 11.1, the number of revolutions (*Ne*) for the cyclone is

$$N_e = \frac{1}{H}\left(L_1 + \frac{L_2}{2}\right) = \frac{1}{0.15}\left(0.6 + \frac{0.6}{2}\right) = 6$$

At standard conditions, $\eta = 1.81 \times 10^{-5}$ Pa·s.

$$d_{50} = \left(\frac{9\eta W}{2\pi N_e U_0 \rho_p}\right)^{1/2} = \left(\frac{9 \times 0.0000181 \times 0.075}{2\pi \times 6 \times 15 \times 1000}\right)^{1/2}$$

$$= 1.14 \times 10^{-5} \text{ (}m\text{)} = 11.4 \text{ (}\mu m\text{)}$$

**Table 11.2 Performance and Limitation of
Conventional Cyclones**

Cyclone	Particle Cutsize (µm)	Pressure Drop (Pa)
High efficiency	> 5	> 1000
Conventional	> 10	> 125
High volume	> 20	> 500

For an involute cyclone, the number of velocity heads of loss, H_v, is

$$H_v = 3.75 \frac{HW}{R_e^2} = 3.75 \frac{0.15 \times 0.075}{0.075^2} = 7.5$$

Note that the carrying gas is air and $\rho_g = 1.2$ kg/m³.

$$\Delta P = 100 \times H_v \frac{U^2 \rho_g}{2g} = 10 \times 8 \frac{15^2 \times 1.2}{2 \times 9.81} = 1,032 \ (Pa)$$

Clearly, the pressure requirement for conventional cyclones presents a major limitation to applications of high-volume indoor-air cleaning. Most air delivery or recirculation fans have a cutoff pressure of approximately 250 Pa. The cutoff pressure is even lower than 100 Pa for many propeller fans used for large-volume air handling, as in animal production facilities. In conventional return-flow cyclones, the 180° change of the flow direction substantially increases the total pressure loss of the cyclone. Such designs are to increase the number of turns and reduce the diameter of the cyclone motion. The other major feature of conventional cyclones are to employ high-entry velocity, typically around 15 to 20 m/s, in an attempt to increase the centrifugal force. The high velocity, together with the high pressure in the particle separation chamber, is almost inevitably associated with high turbulence intensity. High turbulence intensity is not desirable for particle separation, because the higher the turbulence is, the higher the particle reentrainment.

The performance and limitations of conventional cyclones are summarized in Table 11.2. Although cyclone technology has been studied since early 1930s and has been widely used in many industrial applications, it has been concluded that it is impractical to separate particles smaller than 10 µm for high-volume air treatment, as in indoor air cleaning.[1, 20]

11.2 UNIFLOW CYCLONES

As described in the previous section, traditional cyclones have two major limitations:

- High energy consumption (due to high pressure drop)
- Low dust separation efficiency (due to high turbulence intensity)

These two limitations are due in turn primarily to one cause: high pressure across the cyclone. It was assumed that a high pressure drop (means high energy requirement) was required to create high-swirl air velocities to force the particle to separate from the air stream. The high air velocity was associated with high turbulence intensity and particle reentrainment means low particle separation efficiency. Some important factors, such as turbulence intensity, were also missing from existing separation theories, such as Equation 11.10. In addition, some assumptions, such as using the average diameter of the inner and outer cylinder to calculate centrifugal force, also contribute to errors in particle separation efficiency.

Figure 11.4 shows a sketch of a uniflow cyclone. The uniflow cyclone contains an air inlet, an outlet, and a dust bunker. The air inlet is made of two ducts, which form an annular tunnel. A set of guide vanes in front of the inlet is evenly distributed around the annular tunnel. Dust laden air enters the deduster chamber from guide vanes, and the airflow is forced to travel spirally. Dust particles travel gradually toward the exterior duct due to centrifugal force and are eventually separated from the air stream. The separated particles are deposited in a bunker, which can be cleaned periodically. The cleaned air will exit through the outlet. One major advantage of this uniflow cyclone is the low pressure drop. By properly designing the guide vanes, the airflow can be smoothly streamlined to increase particle separation efficiency at low air velocities. Such dust separators can be used for large-volume air cleaning with low energy consumption. The following analysis applies to both conventional vane-axial cyclones and low-pressure uniflow cyclones.

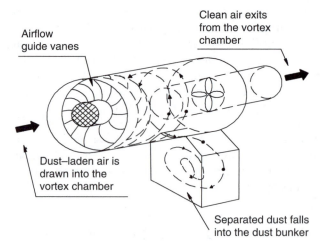

Figure 11.4 A sketch of a uniflow cyclone. Air enters the separation chamber through a set of guiding vanes to form a spiral motion. Particles are separated by centrifugal force and deposited in a dust bunker.

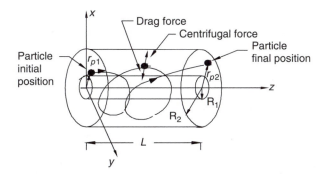

Figure 11.5 The locus of a particle in an annular tunnel. The particle travels toward the outer cylinder of the annular tunnel due to centrifugal force, thus separating from the air stream.

11.2.1 Particle Separation Efficiency in Laminar Flow

Consider that a dust particle with a diameter d_p is traveling with an air stream in an annular tunnel with a length L and a swirl angle α (Figure 11.5). The swirl angle is realized by inhaling air through a set of vanes around the annular tunnel having a discharging angle α at the outlet tip with respect to the axial centerline (z-direction) of the annular tunnel. The radii of the inner and outer cylinder of the tunnel are R_1 and R_2, respectively.

When the particle enters the tunnel, it has an initial radial position r_{p1}. When the particle leaves the tunnel, it has a final radial position r_{p2}. Air velocity in the tunnel consists of three components: tangential U_t, radial U_r, and axial U_z. The radial air velocity U_r is zero; assume that airflow in the annular tunnel is fully developed and symmetrical. The particle velocity consists of three components: tangential V_t, radial V_r, and axial V_z. At any point, the particle is balanced by three forces: Stokes drag force, centrifugal force, and gravity. Because the particle of concern is small, gravitational effect on the particle balance is negligible. The buoyancy of the air is also negligible, because the density of the particle is much greater than the density of the air. Thus, the centrifugal force equals the Stokes drag force acting on the particle:

$$\rho_p \frac{\pi d_p^3}{6} \frac{U_t^2}{r} = \frac{3\pi \eta d_p V_r}{C_c} \tag{11.15}$$

In Equation 11.15, r is the radius at which the particle is located and varies between R_1 and R_2; V_r is the radial velocity and a function of r that changes constantly. Because V_r varies all the time, it is not appropriate to use relaxation time and centrifugal acceleration to calculate the radial particle velocity. Note that in the conventional return-flow cyclone in the previous section, the V_r was treated as a constant. Because the velocity is the differential of the displacement with respect to the time, the radial velocity of the particle V_r can be expressed as follows:

$$V_r = \frac{dr}{dt} = \frac{dr}{dz}\frac{dz}{dt} = V_z\frac{dr}{dz} \tag{11.16}$$

where $V_z = dz/dt$ is the velocity component in the z-direction.

Introducing Equation 11.16 into Equation 11.15 and rearranging yields

$$rdr = \frac{\rho_p d_p^2 C_c}{18\eta}\frac{U_t^2}{V_z}dz \tag{11.17}$$

Assume that there is no axial slip for particles in the annular tunnel flow and that flow is steady, that is, $V_t = U_t$ and $V_z = U_z$. Integrating the two sides of Equation 11.17 with respect to radial transverse distance ($r_{p2} - r_{p1}$) and axial distance (L) gives:[21]

$$\int_{r_{p1}}^{r_{p2}} rdr = \int_0^L \frac{\rho_p d_p^2 C_c}{18\eta}\frac{U_t^2}{V_z}dz \tag{11.18}$$

which gives

$$(r_{p2}^2 - r_{p1}^2) = \frac{\rho_p C_c d_p^2 U_t^2 L}{9\eta U_z} \quad (R_1 \le r_p \le R_2) \tag{11.19}$$

The tangential and axial air velocities, U_t and U_z, can be either measured or calculated. If the size and density of the particle are known, and if either the initial or the final position of the particle is given, the radial displacement of the particle ($r_{p2} - r_{p1}$) can be calculated using Equation 11.19. If a particle reaches the inner surface of the outer cylinder, the particle is assumed to have been separated. Therefore, the smallest separated particle d_{sp}, at initial $r_{p2} - r_{p1}$, can be calculated using Equation 11.19:

$$d_{sp}^2 = \frac{9\eta(R_2^2 - r_{p1}^2)U_z}{\rho_p C_c U_t^2 \cdot L} \tag{11.20}$$

Separation efficiency ξ_x for a given particle diameter d_p at an entering radius of the cyclone, r_{p1}, can be calculated as the ratio of two annular areas A_r/A_T (Figure 11.6). A_r is the shaded area in Figure 11.6 that is divided by r_{p1}. Apparently, $A_r = \pi(R_2^2 - r_{p1}^2)$ and $A_T = \pi(R_2^2 - R_1^2)$. As shown in Figure 11.6, at the entering radius r_{p1}, there is a corresponding particle size, d_{sp}, which refers to a separated particle size, indicating that particles of size d_{sp} in the shaded area will be separated, and those not in the shaded area will not be separated. r_{p1} is the minimum radius at which particles of size d_{sp} will reach the outer cylinder surface during the axial distance L.

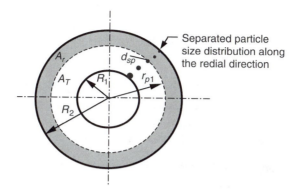

Figure 11.6 Particle separation efficiency for particles equal to or larger than the separated particle size d_{sp} can be calculated as the ratio of the shaded area between r_{p1} and R_2, A_r to the total annular area between R_1 and R_2, A_T.

$$\xi_x = \frac{A_r}{A_T} = \frac{(R_2^2 - r_{p1}^2)}{(R_2^2 - R_1^2)} \quad (R_1 \le r_{p1} \le R_2) \tag{11.21}$$

In reality, some particles that reach the outer cylinder surface may reenter the airflow due to particle bouncing and low adhesion between the particles and the surface. This reentrainment phenomenon is different from reentrainment caused by turbulent flow. The particle reentrainment factor k_{rp} can be defined as the ratio of particles at a given size at the collecting surface reentrained back to the airflow. Theoretically, the k_{rp} ranges from 0 to 1. The k_{rp} can be high when the air velocity is approaching to the resuspension velocity described in Chapter 9. When the collecting surface is coated with adhesive materials such as water or oil, the k_{rp} can be reduced to 0 and substantially improve the collection efficiency. To simplify the analysis, the value of k_{rp} is taken as zero when not specified. When the surface reentrainment is taken into account, the actual area of A_{rp} (Figure 11.5) should be multiplied by a factor of $(1 - k_{rp})$. Therefore, the separation efficiency for a given particle size becomes

$$\xi_x = \frac{(1 - k_{rp})(R_2^2 - r_{p1}^2)}{(R_2^2 - R_1^2)} \quad (0 \le k_{rp} \le 1) \tag{11.22}$$

Substituting Equation 11.19 into Equation 11.22, noting that $r_{p2} = R_2$ when a particle is separated, gives the particle separation efficiency with respect to particle size:

$$\xi_x = \frac{(1 - k_{rp})\rho_p C_c d_p^2 U_t^2 L}{9\eta U_z (R_2^2 - R_1^2)} \tag{11.23}$$

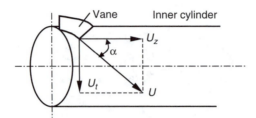

Figure 11.7 Vane angle with respect to the axial flow direction.

Substituting Equation 11.23 into Equation 11.20 with $\xi_x = 50\%$, the particle cutsize d_{50} can be calculated as follows:

$$d_{50}\sqrt{C_c} = \left(\frac{9\eta(R_2^2 - R_1^2)U_z}{2(1-k_{rp})\rho_p U_t^2 L} \right)^{\frac{1}{2}} \tag{11.24}$$

The axial air velocity U_z can be expressed in terms of the tangential air velocity U_t and the vane angle α, where the α is the angle between the U and U_z or the vane angle in degrees with respect to the axial direction (Figure 11.7):

$$U_t = U_z \tan \alpha \tag{11.25}$$

The axial velocity U_z can also be expressed in terms of the flow rate of the cyclone, Q:

$$U_z = \frac{Q}{\pi (R_2^2 - R_1^2)} \tag{11.26}$$

Substituting Equation 11.25 and Equation 11.26 into Equation 11.23 and Equation 11.24 gives

$$\xi_x = \frac{(1-k_{rp})\rho_p C_c d_p^2 L Q \tan^2 \alpha}{9\pi\eta\ (R_2^2 - R_1^2)^2} \tag{11.27}$$

$$d_{50}\sqrt{C_c} = 3(R_2^2 - R_1^2)\left(\frac{\pi\eta}{2(1-k_{rp})\rho_p L Q \tan^2 \alpha} \right)^{\frac{1}{2}} \tag{11.28}$$

If the particle cutsize is larger than 3 μm, the slip correction factor C_c can be considered as unity. If the particle cutsize calculated from Equation 11.28 is smaller than 3 μm, the slip correction factor C_c should be considered, and d_{50} can be solved by substituting Equation 4.23 into Equation 11.24:

$$d_{50} = \frac{1}{2}\left[\frac{18\eta(R_2^2 - R_1^2)U_z}{(1 - k_{rp})\rho_p U_t^2 L} + 6.35\lambda^2\right]^{\frac{1}{2}} - 1.26\lambda \quad (\text{for } d_{50} \leq 3 \text{ μm}) \qquad (11.29a)$$

where λ is the mean free path of gas. For air under standard conditions, $\lambda = 0.066$ μm, and Equation 11.29a becomes

$$d_{50} = \frac{1}{2}\left[\frac{3.26\times10^{-4}(R_2^2 - R_1^2)U_z}{(1 - k_{rp})\rho_p U_t^2 L} + 2.77\times10^{-14}\right]^{\frac{1}{2}} - 8.32\times10^{-8} \qquad (11.29b)$$

Example 11.2: *Calculate the particle cutsize (in aerodynamic diameter) of a uniflow cyclone within the main separation chamber. The chamber is 1.2 m long. The inner and outer diameters of the cyclone are 150 mm and 200 mm, respectively, with a flow rate of 100 l/s. The vane angle is 60°. Assume that there is no particle reentrainment and that the flow is laminar.*

Solution: The axial velocity is

$$U_z = \frac{Q}{\pi(R_2^2 - R_1^2)} = \frac{0.1(m/s)}{\pi(0.1^2 - 0.075^2)} = 7.28\,(m/s)$$

$$U_t = U_z \tan\alpha = 7.28\times\tan 60^o = 12.6\,(m/s)$$

Assume no reentrainment of particles once they are separated. We first try Equation 11.24 if the particle cutsize is larger than 3 μm. (If $d_{50} < 3$ μm, Equation 11.29 should be used.)

$$d_{50}\sqrt{C_c} = \left(\frac{9\eta(R_2^2 - R_1^2)U_z}{2\rho_p U_t^2 L}\right)^{\frac{1}{2}} = \left(\frac{9\times1.81\times10^{-5}\times(0.1^2 - 0.075^2)\times7.28}{2\times1,000\times12.6^2\times1..2}\right)^{\frac{1}{2}}$$

$$= 3.7\times10^{-6}\,(m) = 3.7\,(\mu m)$$

Because $d_{50} > 3$ μm, the slip correction factor can be approximately considered as unity, and the particle cutsize is 3.7 μm.

Note that the preceding analysis applies only to the main cylindrical separation chamber. In an actual design, other factors may have a significant effect on the particle separation efficiency. One critical section is the particle collection system where large turbulence may occur because of the structure's obstruction of the airflow. The dust collection and storage system should be carefully designed to minimize the reentrainment of separated particles back into the air stream. Although the particle separation efficiency is proportional to the air velocity entering the cyclone, as shown in Equation 11.23, the turbulence effect also increases substantially with air velocity. For a given design, one should seek an optimum air velocity that gives the highest particle separation efficiency. Such optimal conditions may be obtained through experimental design or through a sensitivity analysis.

11.2.2 Collection Efficiency in Complete-Mixing Flow

In reality, airflow in the annular tunnel is to some extent turbulent, and the trajectory of a particle is not unidirectional toward the outer cylinder surface. Turbulence in the radial direction can introduce reentrainment of particles and thus reduce the particle separation efficiency. Generally, the reentrainment of particles caused by turbulent mixing is much higher than the surface reentrainment caused by particle bounce and detachment. The higher the turbulence intensity is, the higher the reentrainment ratio of particles back into the airflow, hence the lower particle separation efficiency. The lowest separation efficiency happens when the airflow is completely mixing.

When particles and the carrying air are in a complete-mixing process, particle separation is analogous to the settling under complete-mixing conditions. As shown in Figure 11.8, only those particles very close to the receiving tube surface can be collected when the air is completely mixing. Assume that near the collection tube is a very thin layer of air, within which the airflow is laminar, and thus the particle is traveling unidirectionally toward the tube surface. Within an infinitesimal time interval dt, the particle will travel a small distance, $\delta = V_{TF}dt$, within this laminar flow layer, where V_{TF} is the terminal velocity caused by the centrifugal force. The area $2\pi R_2\delta$ can be defined as the particle collection zone, within which all particles with a velocity of V_{TF} are collected. The fraction of particles collected by the receiving tube is then the ratio of the area of $2\pi R_2\delta$ to the cross-sectional area of the separation chamber. On the other hand, the fraction of particles collected in a unit volume of air is the reduction of the particle concentration, dC_p/C_p. Therefore

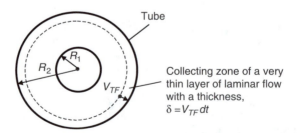

Figure 11.8 Particle collection of the uniflow cyclone under complete-mixing conditions.

$$\frac{dC_p}{C_p} = -\frac{2\pi R_2}{\pi(R_2^2 - R_1^2)}\delta = -\frac{2R_2 V_{TF}}{(R_2^2 - R_1^2)}dt \qquad (11.30)$$

where R_1 and R_2 are the radii of the inner and outer tubes, respectively, and δ is the distance traveled by the particle in the time interval dt. The minus sign $(-)$ indicates that the particle concentration is decreasing with the time during the separation process.

The particle terminal velocity V_{TF} at the outer tube surface, as shown in Figure 11.8, is constant and can be obtained from Equation 11.15:

$$V_{TF} = \frac{C_c \rho_p d_p^2}{18\eta}\frac{U_t^2}{R_2} \qquad (11.31)$$

where U_t is the tangential air velocity at the R_2 and equals $U_z \tan\alpha$, U_z is the axial air velocity, and α is the vane angle at the discharging tip. Substituting Equation 11.31 into Equation 11.30 gives the particle concentration at the time t in the separation chamber

$$C_p(t) = C_{p0}\exp\left(-\frac{C_c \rho_p d_p^2 U_t^2}{9\eta(R_2^2 - R_1^2)}t\right) \qquad (11.32)$$

where C_{p0} is the particle concentration in the supply air entering the separation chamber. The particle collection efficiency, according to the definition, can be written as

$$\xi_x = 1 - \frac{C_e}{C_{p0}} = 1 - \exp\left(-\frac{C_c \rho_p d_p^2 U_t^2}{9\eta(R_2^2 - R_1^2)}t\right) \qquad (11.33)$$

Let the particle concentration at the exhaust outlet of the separation chamber be C_e. Note that the retention time of the particle in precipitator, t, equals LA/Q or L/U_z, where L is the length, U_z is the average axial air velocity, A is the cross-sectional area and the annular area, and Q is the flow rate of cyclone.

$$\xi_x = 1 - \frac{C_e}{C_{p0}} = 1 - \exp\left(-\frac{C_c \rho_p d_p^2 Q \tan^2\alpha\, L}{9\pi\eta(R_2^2 - R_1^2)^2}\right) \qquad (11.34)$$

Or, in terms of axial air velocity

$$\xi_x = 1 - \exp\left(-\frac{C_c \rho_p d_p^2 U_z \tan^2\alpha L}{9\eta(R_2^2 - R_1^2)}\right) \qquad (11.35)$$

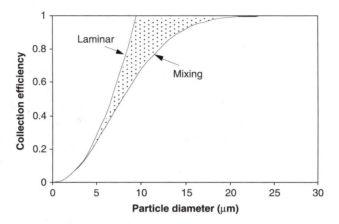

Figure 11.9 Collection efficiency for laminar and complete-mixing flow (calculated using Equation 11.27 and Equation 11.34) for a typical uniflow cyclone under standard room conditions: $R_2 = 0.15$ m, $R_1 = 0.1$ m, $\rho_p = 1000$ kg/m³, $\alpha = 60°$, $Q = 0.3$ m³/s, $L = 1$ m, and $k_{rp} = 0$.

For a given cyclone, particle collection efficiency depends on the airflow conditions. In most cases, airflow in the particle separation chamber is neither laminar nor completely mixing, but rather turbulent to some extent. Figure 11.9 illustrates the particle separation efficiency for a typical uniflow cyclone under laminar and complete-mixing conditions. The complexity of turbulent flow results in extreme difficulty in predicting the particle collection efficiency under actual turbulent conditions. However, it is reasonable to conclude that the particle separation efficiency falls into the shaded area in Figure 11.9. The shaded area has boundaries of the maximum separation efficiencies under laminar flow conditions, and minimum separation efficiencies under perfect-mixing flow conditions. Clearly, the particle collection efficiency for laminar flow is higher than the perfect-mixing flow. Therefore, one design consideration should be to minimize the turbulence intensity in the separation chamber. In reality, airflow in the separation chamber is usually neither laminar nor perfect-mixing, but rather to some extent turbulent. Therefore, the actual particle collection efficiency will be between the two curves shown in Figure 11.9.

It is worth noting that Figure 11.9 only considers the gravitational and drag forces during a particle separation process. Theoretically, the actual particle separation efficiency should be lower than the efficiency under laminar flow conditions, but higher than the efficiency under perfect-mixing flow conditions. In practice, other forces may be involved in the particle separation process. Such forces include diffusive force, electrostatic force, adhesive force, and bouncing. Generally, diffusive, electrostatic, and adhesive forces increase the separation efficiency, because they are likely to help particles get separated or stay separated. Bouncing force results in decrease in separation efficiency because some separated particles, especially large ones, may bounce back into the air stream.

Experimental data show that Equation 11.27 and Equation 11.34 can be used to predict the particle separation efficiency with reasonable accuracy. An aerodynamic uniflow cyclone was developed to validate the theory (Figure 11.10). An aerody-

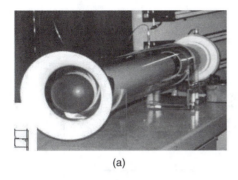 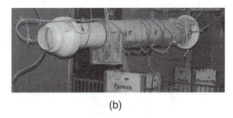

(a)   (b)

Figure 11.10   A uniflow cyclone was developed to validate Equation 11.27 and Equation 11.33, and the device has a flow rate of 135 l/s at pressure of 100 Pa. (a) The uniflow cyclone. (b) The experiment setup.

namic particle sizer (APS 3320, TSI) was used to measure particle separation efficiency. The particle separation efficiency was determined by measuring the particle concentrations upstream and downstream of the uniflow cyclone. The measurement was conducted in a building with a dust concentration of 1 ± 0.2 mg/m³, with a mass median diameter of 12 μm.

Figure 11.11 shows the measurement data and predicted data, using Equation 11.27 and Equation 11.34 for a uniflow cyclone. In this example, the uniflow cyclone has the following design and operating parameters: $R_1 = 0.065$ m, $R_2 = 0.103$ m, $Q = 0.15$ m3/s, $U_z = 7.5$ m/s, $L = 1.07$ m, and $\alpha = 60°$. The testing is under standard room air conditions, with a temperature of $20 \pm 2°C$, relative humidity of $40 \pm 5\%$ and the air viscosity $\eta = 1.81 \times 10^{-5}$ Pa·s. Each data point in Figure 11.7 is the mean value of 15 replications for the given particle size. For particles larger than the cutsize (4 μm in Figure 11.11), the standard error of the efficiency is smaller than 20% of its mean value for a given particle size. For particles smaller than 4 microns, the measured collection efficiency is higher than the predicted efficiency. Two possible forces may have contributed to this increase of collection efficiency. One is diffusive force: Small particles coagulate to form larger particles that can be separated easily. Second, the uniflow cyclone is made of plastic tubes. The electrostatic force at the surface of the plastic tubes attracts small particles, thus separating them from the airflow. For particles larger than 10 μm, the measured collection efficiency is lower than the predicted. The aerodynamic particle size has a low sampling efficiency for large particles. Particle bouncing and reentrainment may also contribute to the decrease in collection efficiency.

11.2.3   Pressure Drop in Uniflow Cyclones

Pressure drop in a uniflow cyclone, as shown in Figure 11.5, consists of three components: dynamic losses, friction losses, and losses of equipment fittings, such as elbows and nozzles:

$$\Delta P = P_v + P_f + P_e \tag{11.36}$$

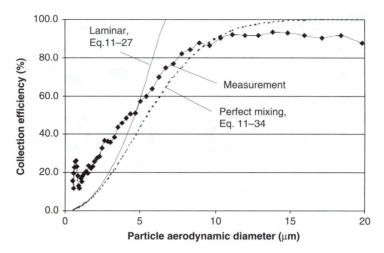

Figure 11.11 Comparison of particle collection efficiencies between those predicted by Equation 11.27 and Equation 11.34, and the measurement value for a uniflow cyclone
under standard room air conditions: $R_1 = 0.065$ m, $R_2 = 0.103$ m, $Q = 0.15$ m³/s,
$L = 1.07$ m and $\alpha = 60°$.

The dynamic pressure loss is caused by accelerating the gas velocity from a free
stream, U_0, to the velocity in the cyclone, U. The final steady flow velocity in the
cyclone is

$$P_v = \frac{U^2 - U_0^2}{2}\, \rho_g \tag{11.37}$$

where ρ_g is the density of the gas. When the vanes are designed properly, such as
curved in the shape of a flow nozzle, the pressure drop upstream and downstream
of the vanes will be approximately equal to the dynamic pressure loss.

Note that the air velocity in the cyclone, U, is the resultant velocity — not the
axial velocity U_z or the tangential velocity U_t. The free stream velocity U_0 is the
ambient air velocity near the cyclone inlet. If the cyclone were placed in an air
stream having a velocity of U_0 in the same direction as the U_z, U_0 would be positive
and the total dynamic pressure loss would decrease. If the free air stream were
opposite to the direction of U_z in the cyclone, U_0 would be negative and the total
dynamic pressure loss would increase.

Friction losses are due to fluid viscosity and are a result of momentum exchange
between molecules in laminar flow and between individual particles of adjacent fluid
layers moving at different velocities in turbulent flow. Friction losses for a uniflow
cyclone can be derived from the friction losses in a duct for straight direction flow.
Friction losses in a conduit can be calculated as follows:[22]

$$P_f = f\frac{L_f}{D_h}\frac{\rho_g U^2}{2} \tag{11.38}$$

where f is the friction coefficient, L_f is the length of the airflow trajectory, and D_h is the hydraulic diameter. In a uniflow cyclone chamber, L_f can be calculated from the chamber length L and the vane angle α:

$$L_f = \frac{L}{\cos \alpha} \tag{11.39}$$

The hydraulic diameter for the annular tunnel of the cyclone can be calculated as

$$D_h = \frac{4A}{L_w} = \frac{4\pi(R_2^2 - R_1^2)}{2\pi(R_2 + R_1)} = 2H_a \tag{11.40}$$

where A and L_w are the wetted area and perimeter of the annular tunnel, respectively, and H_a is the distance between the inner and outer cylinders, that is, $H_a = R_2 - R_1$.

The friction factor f is a function of the fluid Reynolds number for laminar flow. For a turbulent flow $(Re > 2000)$, f is a function of the duct surface roughness (also referred to as *absolute roughness*), hydraulic diameter, and the Reynolds number of the fluid. For air-cleaning or air-sampling cyclones, the Reynolds number in the cyclone is almost always turbulent. A simplified calculation for friction factor was developed by Altshul et al:[23]

$$f = 0.11\left(\frac{\varepsilon}{1000D_h} + \frac{68}{Re}\right)^{\frac{1}{4}} \tag{11.41}$$

where ε *is* the surface roughness in millimeters, D_h is the hydraulic diameter in meters, and Re is the Reynolds number for the gas flow. Degrees of surface roughness for typical duct materials are listed in Table 11.3 for use in Equation 11.41. Roughness for more materials can be found in Idelchik et al.[24]

Pressure losses due to equipment fittings depend on the connecting angles and shapes of specific components. ASHRAE (2001) offers a comprehensive summary for calculating the pressure losses for different equipment fittings, including elbows, bellmouths, screens, and nozzles.[22]

Example 11.3: *Calculate the total pressure loss a uniflow cyclone dust separator within the main separation chamber. The chamber is 1.5 m long. The inner and outer diameters of the duct are 400 mm and 300 mm, with a flow rate of 300 l/s. The separator is made of smooth PVC duct with an absolute roughness of 0.01 mm. The vane angle is 60°. Assume that the ambient air velocity is zero.*

Solution: The axial air velocity is

$$U_z = \frac{Q}{\pi(R_2^2 - R_1^2)} = \frac{0.3}{\pi(0.2^2 - 0.15^2)} = 5.46\ (m/s)$$

Table 11.3 Duct Roughness for Different Duct Materials

Duct Material	Roughness Category	Absolute Roughness (mm)
Uncoated carbon steel, clean (0.05 mm)	Smooth	0.03
PVC plastic pipe (0.01 to 0.05 mm)		0.01–0.05
Aluminum (0.04 to 0.06 mm)		
Galvanized steel, longitudinal seams, 1200 mm joints (0.05 to 0.10 mm)	Medium smooth	0.09
Galvanized steel, continuously rolled, spiral seams, 3000 mm joints (0.06 to 0.12mm)		
Galvanized steel, spiral seam with 1, 2, and 3 ribs, 3600 mm joints (0.09 to 0.12 mm)		
Galvanized steel, longitudinal seams, 760 mm joints (0.15 mm)	Average	0.15
Fibrous glass duct, rigid	Medium rough	0.9
Fibrous glass duct liner, air side with facing material (1.5 mm)		
Fibrous glass duct liner, air side spray coated (4.5 mm)	Rough	3.0

Source: ASHRAE, *Fundamentals Handbook*, American Society of Heating, Refrigeration and Air-Conditioning Engineers, Atlanta, 2001, Ch. 32. With permission.

$$U = \frac{U_z}{\cos \alpha} = \frac{5.46}{\cos 60^o} = 11.92 \ (m/s)$$

The Reynolds number of the separator, noting that $D_h = 2H_a = 2(R_2 - R_1) = 0.1$ m, is

$$Re = \frac{\rho U D_h}{\eta} = \frac{1.2 \times 11.92 \times 0.1}{1.81 \times 10^{-5}} = 79,027$$

$$L_f = \frac{L}{\cos \alpha} = \frac{1.5}{\cos 60} = 3 \ m$$

The friction loss coefficient of the duct, noting that the absolute roughness $\varepsilon = 0.01$ mm, is

$$f = 0.11 \left(\frac{\varepsilon}{1000 D_h} + \frac{68}{Re} \right)^{\frac{1}{4}} = 0.11 \left(\frac{0.01}{1000 \times 0.1} + \frac{68}{79027} \right)^{\frac{1}{4}} = 0.0194$$

The friction loss is

$$P_f = f \frac{L_f}{D_h} \frac{\rho_g U^2}{2} = 0.0194 \frac{3 \times 1.2 \times 10.92}{2 \times 0.1} = 41.6 \ (Pa)$$

The dynamic loss is

$$P_v = \frac{U^2 - U_0^2}{2}\rho_g = \frac{10.92^2}{2} \times 1.2 = 71.5\,(Pa)$$

The total pressure loss of the cyclone separator is

$$\Delta P = P_v + P_f = 113.1\ Pa$$

11.3 WET SCRUBBERS

In a *wet scrubbing device* a liquid, usually water, is used to capture particles or gases. There are two mechanisms in such a collection process: kinematical coagulation and diffusive coagulation. As discussed in Chapter 6, kinematical coagulation is caused by the relative motion of the water droplets and the particles. Diffusive coagulation is caused by the Brownian motion of particles or by the diffusivity of particles. Diffusive coagulation includes polydisperse and monodisperse coagulation.

As shown in Figure 11.12, when a droplet travels through a particle-laden airspace with a velocity V_r relative to the particle, particles may collide with, and thus be collected by, the droplet through two mechanisms: kinematical coagulation or polydisperse coagulation. In kinematical coagulation, we assume that all particles that fall into the column swept by the droplet (within the dotted lines) are collected. In polydisperse coagulation, particles can be collected by the droplet through their Brownian motion. In both cases, the number concentrations are decreased by the reduction of particle concentration rather than the reduction in droplets, because the droplets are assumed to be much larger than the particles.

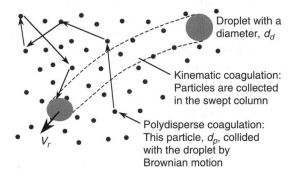

Figure 11.12 When a droplet travels through a particle-laden airspace, particles may collide with, and thus be collected by, the droplets through two mechanisms: kinematical coagulation and polydisperse coagulation.

Assume that particles have an initial number concentration of C_{pn0}, droplets have a concentration C_{dn}, and the relative velocity between the droplets and particles is V_r. The coagulation time period is t. Particle collection efficiency by kinematical coagulation, ξ_k, can be calculated from Equation 6.76:

$$\xi_k = \frac{C_{pn0} - C_{pn}(t)}{C_{pn0}} = 1 - \exp\left(-\frac{\pi}{4} d_d^2 V_r C_{dn} K_c t\right) \tag{11.42}$$

where d_d is the droplet diameter and K_c is the particle capture efficiency of the droplet defined by Equation 6.73.

The kinematical coagulation efficiency increases exponentially with d_d^2, V_r, C_{dn}, K_c, and t. Unlike diffusive coagulation, kinematical coagulation does not depend on the particle concentration. It depends on the particle size, though, by affecting Stk_c, which in turn affects K_c. One way to increase the ξ_k is to increase the relative velocity of the droplets to the particles, as in a venturi air scrubber, where a liquid stream is injected into the air stream with a very high velocity. A high V_r also increases the Stk_c and the K_c. Increasing C_{dn} means a high flow rate of the water. There are two methods to increase the droplet concentration: One is to increase the water flow rate; the other is to obtain a longer droplet retention time by selecting proper droplet size with respect to the air velocity in the tower. Increasing the droplet diameter may not necessarily increase the kinematical coagulation efficiency, because a large droplet diameter relative to the particle diameter will result in a small Stk_c, and in turn a lower particle capture efficiency of the droplet K_c. In general, if $d_p < 100 d_d$, the Stk_c will be much smaller than 0.1. From Equation 6.73 and Equation 6.74, the particle capture efficiency for the droplets is very low, even at high velocities. An explanation of this is that the airflow has been parted far ahead at the upstream of a large droplet, and small particles have less chance to impact on the droplet.

Note that the particle collection efficiency by the kinematical coagulation, ξ_k, is different from the particle capture efficiency of a droplet K_c. ξ_k is the particle collection efficiency of the entire wet scrubbing process, whereas K_c is the particle capture efficiency of a droplet due to the kinematical coagulation. If we view the kinematical coagulation process as a fiber filter filtration process, then ξ_k is equivalent to the filter efficiency and K_c is equivalent to the single-fiber collection efficiency.

Particle collection efficiency by polydisperse coagulation, ξ_c, can be determined by the following procedure. In a wet scrubbing process, consider the total concentration of particles and droplets as C_t, which is the sum of the concentrations of particles of concern (C_p) and the droplets (C_d):

$$C_t = C_p + C_d \tag{11.43}$$

From Equation 6.63, the total concentration of particles and droplets at a given elapse time can be expressed as

$$C_t(t) = \frac{C_{t0}}{1 + C_{t0}K_n t} \qquad (11.44)$$

where C_{t0} is the initial total concentration of particles and droplets, and K_n is the coagulation coefficient for polydisperse particles defined by Equation 6.66.

Because diffusive force becomes weak, compared with gravitational force, for particles larger than 0.5 μm, it is reasonable to neglect the coagulation effect for particles larger than 3 μm and particle concentrations lower than 500 particles/ml. For most scrubbing processes, the coagulation efficiency ξ_c is even smaller when the coagulation time is short (particularly in the order of less than a minute). Polydisperse coagulation of particles can be a more important particle-forming process in atmospheric studies, where coagulation time can last several days. For most man-made scrubbing systems, coagulation can be effective for submicron particles at high concentrations and for high concentrations of scrubbing droplets. Recalling that coagulation coefficient K_n is proportional to the droplet concentration fraction of the total concentration, the most effective concentration ratio for particles to droplets is 1:1, which is usually a very difficult and very expensive task.

In most practical problems, we are interested in the change of particle concentration, not in the change of droplet concentration. In fact, because the droplets are usually much larger than the particles of concern, it is reasonable to assume that the droplet concentration C_d remains unchanged during the concentration. Thus the change in C_t is solely caused by the change in C_p. Therefore, the particle collection efficiency due to the polydisperse coagulation can be calculated by combining Equation 11.43 and Equation 11.44; from its definition

$$\xi_c = \frac{C_{p0} - C_p(t)}{C_{p0}} = 1 - \frac{1}{C_{p0}}\left[\frac{C_{p0} + C_d}{1 + (C_{p0} + C_d)K_n t} - C_d\right] \qquad (11.45)$$

where C_{p0} is the initial concentration of particles.

Particle collection efficiency due to the kinematical coagulation and polydisperse coagulation is not additive, because many collected particles are counted twice. Particles within the swept volume are assumed to have been collected, and those particles are still considered in the polydisperse coagulation. Similarly, particles coagulated to the droplets are double-counted in the kinematical coagulation. The total particle collection efficiency can only have a maximum value of 100%. Similar to the definition of total particle capture efficiency of a single filter fiber, the total particle collection efficiency of the wet scrubbing is defined as

$$\xi_x = 1 - (1 - \xi_k)(1 - \xi_c) \qquad (11.46)$$

where ξ_k and ξ_c are defined in Equation 11.42 and Equation 11.45, respectively.

One potential error source for the preceding particle collection efficiency analysis may be V_r, which has been assumed to be a constant. For gravitational scrubbing in which the liquid droplets are falling down at their terminal settling velocity, this is true. For many spray systems, the smaller droplets have a short relaxation time and thus approach the gas flow velocity quickly. The loss of the relative velocity with the flow for the scrubbing droplets quickly reduces the kinematical coagulation efficiency. This effect leads to the important result that for a given particle size, there is an optimal droplet size that can maximize the collection efficiency. For many spraying scrubbing systems, such as venturi scrubbers, the liquid is injected at a high initial velocity. The scrubbing efficiency can be very high during the droplet relaxation time, even though this time period is short. Transient scrubbing efficiency is discussed in the next section.

Equation 11.46 can be applied to particles alone or to a mixture of particles and gases. In gas scrubbing, kinematical coagulation is negligible, because the molecules are very small compared with the droplets. Thus, the Stk_c is very small and the particle capture efficiency of a droplet is very small. The total gas collection efficiency is the efficiency of diffusive coagulation and its solubility β_s, with respect to the scrubbing liquid (usually water). As discussed in Chapter 6, gas molecules only coagulate to larger particles; they do not coagulate among themselves. A gas molecule may not be captured by the water droplet even when it collides with a droplet. Therefore, the solubility of the gas to the scrubbing liquid must be considered for gas scrubbing. Principles and applications of the scrubbing of gaseous pollutants, also commonly referred as *absorption*, is discussed in Chapter 12.

11.3.1 Spray Chamber Scrubber

Principles of wet scrubbing have been widely used in the air-cleaning industry and in numerous types of devices. One of the simplest wet scrubbing devices is the rectangular or circular *spray tower*. As shown in Figure 11.13, dirty air is pulled or pushed in from the bottom of the tower and moves upward against a shower of water spray. The water spray nozzles can be arranged in series or in parallel to achieve uniform coverage of the airflow passage. Desirable liquid droplet size and liquid flow rate can be achieved by selecting appropriate nozzles. Particles in the dirty air collide with the liquid droplets by kinematical and diffusive coagulation.

There are three scenarios for droplets within the tower:

First, if the upward velocity of dirty air in the tower is lower than the droplet terminal settling velocity, the liquid droplets will settle by gravity at the bottom of the tower. The dirty liquid is then drained from the tower for disposal or recycling.

Second, for some small droplets, the terminal settling velocity may be equal to the upward air velocity. In this case, the droplets will be suspended in the chamber until they change sizes by colliding with other droplets or by polydisperse coagulation. The extended retention time of those droplets in the tower allows a higher droplet concentration within the chamber and increased particle collection efficiency by polydisperse coagulation.

Third, the droplet terminal settling velocity may be lower than the upward air velocity. In this case, the droplets will follow the airflow and reach the air outlet. A mist eliminator is used to collect the liquid droplets, and only cleaned air is allowed to exit. The mist eliminator is usually a set of traps or fins that force the airflow to

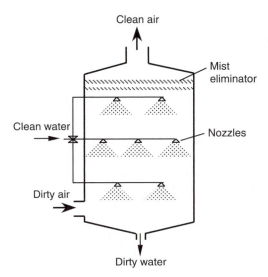

Figure 11.13 Sketch of a vertical spray tower.

change direction. The water mists are impacted on the traps or fins and removed from the airflow.

In order to analyze the particle collection efficiency of the water tower, as shown in Figure 11.13, the following assumptions are made: Water mists are perfectly mixed within the tower, and the average velocity of the droplets relative to the airflow is the terminal settling velocity of the droplets. This assumption may underestimate the V_t because the water droplets are sprayed at a high pressure with a high initial velocity at the nozzle outlet. However, droplets quickly attain their terminal velocities, especially in a large-volume tower with a longer retention time for dirty air.

Let us consider a water tower with an effective height (the height from the bottom to the mist eliminator) H and a cross-sectional area A. The volumetric flow rates for air and water are Q and Q_l, respectively, and the densities for air and water are D and D_l, respectively. The average retention time of dirty air within the tower can be determined as follows:

$$t = \frac{V}{Q} = \frac{HA}{Q} \tag{11.47}$$

where V is the volume of the tower.

Assume that the water droplets are spherical and have an average diameter d_d. The rate of water droplets sprayed into the tower, N_d, can be calculated from the water flow rate and the droplet diameter:

$$N_d = \frac{6Q_l}{\pi d_d^3} \tag{11.48}$$

where N_d is in droplets/s. When the droplets are perfectly mixed in the tower, the average retention time for droplets, t_l, is greater than the average retention time for air, because the air flows upward and the droplets flow downward.

$$t_l = \left| \frac{V}{Q - A\tau g} \right| \tag{11.49}$$

where Q is the airflow rate and τ is the relaxation time for the droplets.

When $Q \approx A\tau g$ $t_l \to \infty$, indicating that the droplets will suspend in the tower indefinitely and resulting a very high droplet concentration. When $Q < A\tau g$, t_l becomes an impossible negative. This indicates that the droplets will quickly settle out by gravity. In this case, the absolute value of retention time for droplets should be used. The concentration of water droplets in the tower is directly proportional to rate of droplet spray, the retention time of the droplets t_l, and the volume of the tower. Combining Equation 11.48 and Equation 11.49, we have

$$C_d = \frac{N_d t_l}{V} = \frac{6 Q_l}{\pi d_d^3 |Q - A\tau g|} \tag{11.50}$$

If we substitute Equations 11.48 and 11.50 into Equations 11.42 and 11.45 and then into Equation 11.46, the particle collection efficiency of the water tower, as shown in Figure 11.13, can be expressed as

$$\xi_x = 1 - \left(\frac{1}{C_{p0}} \left[\frac{C_{p0} + C_d}{1 + (C_{p0} + C_d) K_n t} - C_d \right] \right) \left(\exp(-\frac{3 Q_l V_r K_c V}{2 d_d Q |Q - A\tau g|}) \right) \tag{11.51}$$

where K_n is the coagulation coefficient between the droplets and particles defined by Equation 6.66, Q is the airflow rate in m^3/s, Q_l is the water flow rate in m^3/s, τ is the relaxation time of the droplets, K_c is the kinematical particle capture efficiency of droplet defined by Equation 6.73, V is the volume of the water tower, and A is the cross-sectional area of the tower.

Equation 11.51 may result in overestimation of the particle collection efficiency due to the following error sources:

First, part of the water droplets may impact on the tower walls, and the actual droplet concentration may be lower than predicted, resulting a lower kinematical coagulation efficiency.

Second, some particles (including gases) may not be water soluble, resulting in a low diffusive coagulation efficiency.

Third, air and water droplets may not be well mixed, resulting a lower overall particle collection efficiency.

Nevertheless, Equation 11.51 is a very useful design tool for determining water tower parameters.

Example 11.4: *In a cylindrical water tower, as shown in Figure 11.13, the effective scrubbing height (the height between the bottom and the mist eliminator) is 3 meters and the diameter is 1 meter. The airflow rate is 1 m³/s. The water supply rate for all spray nozzles is 0.8 l/min. Assume that the water droplets have a diameter of 100 μm. The concentration of 5μm particles in aerodynamic diameter is 50 particles/ml at the water tower inlet. What is the particle concentration at the tower outlet? All ambient conditions are standard (20°C temperature and 101.325 kPa atmospheric pressure).*

Solution: From Equation 6.12 and Table 6.3, the diffusion coefficients for the 3 μm particles and 100 μm droplets are 2.76×10^{-11} m²/s and 2.37×10^{-13} m²/s, respectively.

$$D_p = \frac{kTC_c}{3\pi\eta d_p} = \frac{1.38 \times 10^{-23} \times 293 \times 1}{3 \times 3.14 \times 1.81 \times 10^{-5} \times 5 \times 10^{-6}} = 4.74 \times 10^{-12} \left(m^2 \ / \ s \right)$$

From Equation 6.61 or Table 6.5

$$\beta(d_d) = \frac{2}{\pi} \tan^{-1} \left\{ 45.3 \left[\exp\left(\log(d_p) \right) \right]^{2.694} \right\} = 1$$

Because $d_d > 20d_p$, from Equation 6.68, denote the particles as *1* and the droplets as *2* in the subscript.

$$K_{1-2} = \pi\beta(d_d)d_d D_p = 3.14 \times 1 \times 100 \times 10^{-6} \times 4.74 \times 10^{-12} = 1.49 \times 10^{-15} \left(m^3 \ / \ s \right)$$

Note that the water flow rate $Q_l = 0.8$ l/min $= 1.33 \times 10^{-5}$ m³/s, the relaxation time for the droplets $\tau = 0.031$ s, the tower cross-sectional area $A = \pi D^2/4 = 0.785$ m², and the volume $V = 2.355$ m³. From Equation 11.50, the water droplet concentration is

$$C_d = \frac{6Q_l}{\pi d_d^3 |Q - A\tau g|} = \frac{6 \times 1.333 \times 10^{-5}}{3.14 \times \left(100 \times 10^{-6} \right)^3 \times |1 - 0.785 \times 0.031 \times 9.81|}$$

$$= 3.34 \times 10^7 \left(\frac{droplets}{m^3} \right)$$

The number concentration fractions for particles and droplets are

$$f_p = \frac{C_p}{C_p + C_d} = \frac{5 \times 10^7}{5 \times 10^7 + 3.34 \times 10^7} = 0.6$$

$$f_d = \frac{C_d}{C_p + C_d} = \frac{3.34 \times 10^7}{5 \times 10^7 + 3.34 \times 10^7} = 0.4$$

Noting that $K_{1-2} = K_{2-1}$, the polydisperse coagulation coefficient

$$K_n = \sum_{i=1}^{n} \sum_{j=1}^{n} K_{i-j} f_i f_j = (K_{1-2} + K_{2-1}) f_p f_d$$

$$= 2 \times 1.49 \times 10^{-15} \times 0.658 \times 0.342 = 2.28 \times 10^{-12} (m^3 / s)$$

The relative velocity is terminal settling velocity of the droplets:

$$V_r = V_{ts} = \tau g = 0.031 \times 9.81 = 0.25 \ (m/s)$$

The coagulation time is the height of the tower divided by the airflow velocity Q/A:

$$t = \frac{H}{Q/A} = \frac{3}{1/0.785} = 2.355 \ m/s$$

$$Stk_c = \frac{\rho_p d_p^2 C_c V_r}{18 \eta d_d} = \frac{1,000 \times (5 \times 10^{-6})^2 \times 1 \times 0.25}{18 \times 1.81 \times 10^{-5} \times 100 \times 10^{-6}} = 0.192$$

Because $S_{tkc} > 0.1$, from Equation 6.73

$$K_c = \left(\frac{Stk_c}{Stk_c + 0.12} \right)^2 = 0.379$$

Substituting the preceding values into Equation 11.51, noting that $V_r = V_{TS} = \tau g$ in this case

$$\xi = 1 - \left(\frac{1}{C_{p0}} \left[\frac{C_{p0} + C_d}{1 + (C_{p0} + C_d) K_n t} - C_d \right] \right) \left(\exp(- \frac{3 Q_l \tau g K_c V}{2 d_d Q|Q - A\tau g|}) \right)$$

$$= 1 - \left(\frac{1}{5 \times 10^7} \left[\frac{5 \times 10^7 + 3.34 \times 10^7}{1 + (5 \times 10^7 + 3.34 \times 10^7) \times 2.28 \times 10^{-12} \times 2.355} - 3.34 \times 10^7 \right] \right)$$

$$\times \left[\exp\left(- \frac{3 \times 1.33 \times 10^{-5} \times 0.25 \times 9.81 \times 0.379 \times 2.355}{2 \times 100 \times 10^{-6} \times |1 - 0.785 \times 0.25 \times 9.81|} \right) \right]$$

$$= 1 - (0.99942)(0.472) = 0.528$$

The particle concentration at the outlet is

$$C_{pout} = C_{pin}(1-\xi) = 50 \times 0.472 = 23.6\left(\frac{particles}{ml}\right)$$

The preceding analysis is based on the assumption of a steady state and perfect mixing. In a steady state we assume that $V_r = V_{TS}$ by ignoring the initial velocity of the droplets and its transient time to the terminal settling velocity. In reality, the droplet velocity is changing from V_0 to V_{TS} in a water tower. Because of high initial velocity, the particle collection efficiency for a droplet at its initial time period can be significantly higher than that of the rest of scrubbing time.

Let us divide the droplet–particle interaction time t into two stages. The first stage is transient. The droplet travels through the airflow with an initial velocity V_0 and has a duration time τ. The second stage is a steady state. The droplet travels in the airflow with its terminal settling velocity, t-τ. During the first stage, the droplet velocity relative to airflow can be calculated from Equation 4.56, and the stopping distance can be obtained from Equation 4.59 or Equation 4.60. The time-dependent V_r of the droplets can be substituted into Equation 6.76 to solve for kinematic coagulation efficiency. Because V_r also affects the Stokes number and the kinematic coagulation coefficient, explicit mathematical expression is very difficult. However, for most engineering applications, we can obtain the average droplet relative velocity during its relaxation time τ as follows:

$$V_r = \frac{V_0 + V_{TS}}{2} \tag{11.52}$$

where V_0 and V_{TS} are the initial velocity and terminal settling velocity, respectively, for the droplet.

During the second stage, $V_r = V_{TS}$. From Equation 6.77 and Equation 11.42, the particle concentration after a total particle–droplet interaction time t then becomes

$$\xi_k = 1 - \exp\left[-\frac{\pi}{4}d_d^2 C_{dn} K_c\left(\frac{V_0 + V_{TS}}{2}\tau + V_{TS}(t-\tau)\right)\right] \tag{11.53}$$

Equation 11.53 can be viewed as two kinematical coagulation processes in series: transient and steady state. In the transient coagulation, the droplet velocity is $(V_0+V_{TS})/2$, the coagulation time is τ, and the initial particle concentration is C_{pn0}. After time τ, the coagulation enters a steady state, with the new initial particle concentration at time τ, with a droplet velocity V_{TS} and coagulation time t-τ. For short retention time and large droplets, Equation 11.53 should be used to calculate the kinematic coagulation efficiency.

Water towers, as shown in Figure 11.13, are also called *counterflow towers*, for which the airflow direction and the droplet flow direction are opposite. An alternative design is the *crossflow water tower*. As shown in Figure 11.14, water is sprayed from the top of the chamber, and the dirty air is pulled horizontally across the

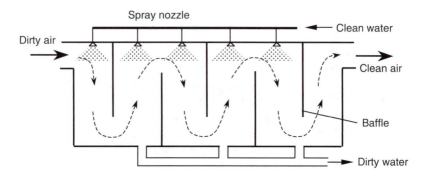

Figure 11.14 Sketch of a crossflow-type water scrubber.

chamber. The particles are captured by the impact on the vertical barriers and diffusion. Large droplets will fall to the bottom by gravity. A typical technique is to place vertical baffles after the spray section. Additional vertical baffles will improve the collection efficiency. Locating the baffles on both the top and the bottom of the tower and shaping a zigzag path of airflow increases the impaction and mixing (thus the chance of interaction between the particles and droplets).

General design guidelines for water flow rate are in the range of 1 to 5 m³/h for every 3600 m³/h of gas flow in industrial applications.[1] Supplemental water must be added to replenish that which evaporated into the gas stream. In most cases, the water used must be recycled to reduce the net water consumption. Thus some kind of water-cleaning system, such as a settling tank, is required in the water tower area. The collection efficiency for these types of water tower ranges from 94% for 5 μm particles to 99% for 25 μm particles. High collection efficiencies for particles of 1 μm can be attained by using high-pressure fog sprays.

11.4 VENTURI SCRUBBER

A *venturi scrubber* can have rectangular or circular cross-sectional flow channels. As shown in Figure 11.15, the dirty air enters the inlet with an airflow rate Q_g and converges in a throat (contracta) section, then diverges to the outlet. In the throat, a bank of nozzles on either side of the throat injects water into the gas stream though slots, or in some cases weirs, located on either side of the venturi throat. The water jets form a curtain that is perpendicular to the airflow direction. The high-velocity gas atomizes the liquid injected into the gas stream. Good atomization is essential if sufficient targets for inertial impaction are to be available. It is normally assumed that the fine particles enter the venturi throat with a velocity equal to that of the gas stream. The droplets of the scrubbing liquid, on the other hand, are assumed to have no initial axial velocity and to be accelerated though the venturi by the aerodynamic drag of the gas stream. An array of water nozzles or a water slot allows the water to spray into the throat to form a water curtain perpendicular to the airflow.

In the converging section, flow work associated with the fluid is converted into kinetic energy, with a concomitant decrease in static pressure and increase in velocity.

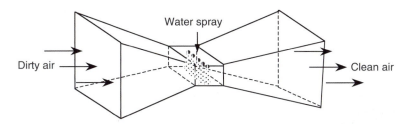

Figure 11.15 Sketch of venturi scrubber. Air is accelerated and reaches a high velocity at the throat. A water curtain perpendicular to the air flow "scrubs" off the particles in the airflow by impaction.

The velocity in the throat section can reach values of 50 to 180 m/s. The area ratio between the inlet and the throat cross sections typically is 4:1 in a venturi scrubber. The angle of divergence is around 4 to 7° in order to achieve good static either in the throat region (recommended) or at the beginning of the convergence section. The collection of fine particles by the liquid droplets is accomplished by inertial impact during the time the droplets are being accelerated. As the droplet velocity approaches that of the air, and the relative velocity between particle and droplet approaches zero, and the probability of inertial impaction downstream from the throat decreases rapidly. Because the impaction collection efficiency increases with an increase in relative velocity, high inlet air velocities are essential.

Because the energy of the high-velocity air stream in a venturi scrubber is used to increase liquid atomization and to accelerate the liquid droplets, it is not surprising that the pressure drop in the venturi is large in comparison to those of other dry and wet collectors. It has also been shown that the particle collection efficiency is positively related to the pressure drop of the venturi.

Although venturi scrubbers have enjoyed wide application in removing particles from gas streams, reliable design equations for the pressure drop and particle collection efficiency have been lacking. Several attempts have been made to solve the particle collection efficiency of venturi scrubbers analytically. Calvert[25] developed a particle penetration equation based on an analysis that takes into account droplet size, the inertial impaction parameter, and droplet concentration across the venturi throat, as well as the continuously changing relative velocity between particle and droplet. However, the predicted results often vary largely from experimental results.

One of the most widely accepted expressions for the pressure drop in venturi scrubbers was developed by Hesketh[26] based on a correlation of experimental data obtained from many different venturi scrubbers.

$$\Delta P = 0.532\, V_g^2 \rho_g A^{0.133}\left[0.56 + 16.6\left(\frac{Q_l}{Q}\right) + 40.7\left(\frac{Q_l}{Q}\right)^2\right] \qquad (11.54)$$

where ΔP is the pressure drop across the venturi in Pa, V_g is the gas velocity at the throat in m/s, ρ_g is the air density downstream from the venturi throat in kg/m³, A

is the cross-sectional area of the venturi throat in m², Q_l and Q are the flow rates of the liquid and gas, respectively, in m³/s. (Note that the original equation was in the imperial system.)

By examining experimental data, Hesketh (1974) concluded that the venturi scrubber is essentially 100% efficient in removing particles larger than 5 μm. Therefore, he studied the penetration for particles less than 5 μm in diameter.[26] On the basis of this study, he concluded that the overall mass collection efficiency of particles less than 5 μm in diameter is approximately related to the pressure drop across the venturi by this equation:

$$\xi = 1 - 9,319(\Delta P)^{-1.43} \tag{11.55}$$

where ΔP is the pressure drop in pascals. Apparently, when the pressure drops below 600 Pa, the wet venturi has a zero collection efficiency. This device should operate at pressure drops higher than 1 kPa.

The water circulation rate of a venturi scrubber varies from 0.25 to 1.6 l/cm³ of air. But more significant for this device, as compared with other dry and wet collectors, is the pressure loss. The pressure drop varies between 0.75 kPa and 25 kPa, depending on the removal efficiency desired. The collection efficiency is directly related to the energy expenditure, and hence to the pressure loss. The main advantage of the venturi scrubber, however, is that by taking a large draft loss, a very high efficiency can be attained even for very small particles. The efficiency may reach 99% in the submicron range, and 99.5% for 5 μm particles. In order for a collection device to attain these efficiencies, the dust-laden droplets must be removed from the air stream after passing through the venturi scrubber. A common method of achieving this is to move the gas stream through a cyclone separator in series with the venturi passage. Another technique is to move the gas from the venturi through a fluidized packed bed.

A major inference can be drawn from the previous discussions: If a high relative velocity between particles and liquid droplets is a basic requirement for high impaction efficiencies, then most of the impaction removal must occur within the first few inches of the divergent section. The added length is necessary for adequate pressure recovery, but it has little influence on the overall target efficiency.

Example 11.5: *Water is introduced into the throat of a venturi scrubber at a rate of 0.5 L/s. The airflow rate through the scrubber is 2 m³/s. The venturi throat has a rectangular cross section 0.2 m wide and 0.1 m high. The air density is 1.2 kg/m³. Determine the pressure drop and mass collection efficiency for particles smaller than 5 μm.*

Solution: The throat area $A = 0.2 \times 0.1 = 0.02$ m²; air velocity at the throat $V_g = Q/A = 2/0.02 = 100$ m/s. Note that $Q_l/Q = 0.5$ (liter)/2 m³ = 0.00025. From Equation 11.54

$$\Delta P = 0.532\, V_g^2 \rho_g A^{0.133} \left[0.56 + 16.6 \left(\frac{Q_l}{Q} \right) + 40.7 \left(\frac{Q_l}{Q} \right)^2 \right]$$

$$= 0.532 \times 100^2 \times 1.2 \times 0.02^{0.133} \left[0.56 + 16.6 \times 0.00025 + 40.7 \times 0.00025^2 \right]$$

$$= 2,142 \quad (Pa)$$

The mass collection efficiency for particles smaller than 5 μm is

$$\xi = 1 - 9,319 (\Delta P)^{-1.43} = 1 - 9,319 \times 2,142^{-1.43} = 0.84$$

11.5 OTHER NONCONTACT PARTICLE CLEANERS

The previously described techniques can be combined to form different air-cleaning devices. One example is the wet cyclone, which combines a dry cyclone and a wet scrubber to remove particles and water-soluble gases. As shown in Figure 11.16, dirty air from a confinement animal building passes through a wet scrubbing section, where gases and small particles are diffused and coagulated with water droplets, and the much larger water droplets are separated at the uniflow cyclone section. The device gives a 90% collection efficiency for particle mass and a 50% collection efficiency for ammonia.

The water circulation rate in wet cyclones runs from 0.1 to 1.2 l/m³ of treated air. The pressure drop typically ranges between 100 and 1000 P, depending on the internal arrangement of the equipment and the cyclone design. In general, wet cyclones have a collection efficiency of 100% for droplets of 100 μm and over, around 99% for droplets from 50 to 100 μm, and from 90 to 98% for droplets between 5 and 50 μm. The use of cyclones in series with a venturi scrubber as a collection device can also improve the particle collection efficiency and eliminate the water mists from the treated air.

11.6 PARTICLE COLLECTOR PERFORMANCE CRITERIA

In order to compare the overall performance of different particle collectors in different applications, some factors affecting the collector performance should be considered first. As with fibrous filters, particle collection efficiency and pressure drop are important criteria of particle collector performance. Unlike fibrous filters, dust-loading capacity is usually not a concern and thus is not considered as a criterion. On the other hand, the airflow capacity largely affects the particle collection efficiency and pressure drop, and thus should be considered an important criterion in particle collector performance. Other factors, such as space requirements and capital and operational costs, should also be considered. However, these factors are difficult to weight in the overall performance evaluation. From a technical point of

Figure 11.16   A wet uniflow cyclone was tested on an animal confinement building. It reduces
90% of dust and 50% of ammonia emissions from the building.

view, we consider the following three factors the primary criteria of particle collector performance:

- Particle collection efficiency
- Pressure drop across the collector
- Airflow capacity

We can define a quantitative criterion, the particle collector quality q_c, to compare the performance of different particle collectors (except for fibrous filters). The collector quality is logarithmically proportional to the particle collection efficiency and the airflow rate, and inversely proportional to the pressure drop.

$$q_c = \ln\left(\frac{1000Q}{1-\xi}\right)\frac{1}{\Delta P} \qquad (11.56)$$

where Q is the flow rate through the collector in m³/s, ξ is the particle collection efficiency for a given particle size (or the total efficiency of concern, such as the total particle mass collection efficiency), and ΔP is the pressure drop across the collector in pascals.

Apparently, collector quality increases as ξ increases, as ΔP decreases, or both. At the same particle collection efficiency and pressure drop, a larger airflow capacity gives a better collector quality. As discussed before, it is more difficult to achieve a high particle collection efficiency for high-volume collectors. The greater the value of q_c, the better the collector's performance is. In practice, the q_c value must be interpreted according to the application, because Q, ξ, and ΔP may have different weighing factors. For example, in a hospital facility, particle separation efficiency (cleanness) is more important than pressure drop (operating cost). In this case, selection of a particle collector selection should aimed for high particle separation efficiency.

DISCUSSION TOPICS

1. Why are return-flow cyclones not used in large-volume air cleaning?
2. In reality, the airflow in a uniflow cyclone is neither laminar nor complete mixing, but somewhat turbulent. How does this turbulent flow affect the particle separation efficiency?
3. In many conventional return-flow cyclones, the actual particle collection efficiency is much lower than the efficiency predicted using the complete-mixing model. Why?
4. What are the key parameters in improving the particle collection efficiency of a uniflow cyclone? How do you rank the importance of different design parameters?
5. List and justify the three most important parameters in a water tower design.
6. Discuss the advantages and disadvantages of dry and wet air cleaners. Give a real-life example to which each technology can be applied.
7. For a given wet scrubber, discuss the pros and cons of using fine water droplets. Is it true that the smaller the water droplets are, the higher particle collection efficiency?
8. How might you use quality criteria to compare different air cleaners? Does the value of air cleaner quality q_c have any physical meaning?

PROBLEMS

1. The average dust concentration at an exhaust fan is 2 mg/m^3 with an average diameter of 12 μm with standard density. A recommended dust emission threshold limit from such buildings is 0.23 mg/m^3. A dust collector is installed before the exhaust fan to meet this requirement. What should the dust collection efficiency be?
2. In an air-cleaning process, a cyclone serves as a precleaner in series with a fibrous filter. For particles of 5 μm in aerodynamic diameter, the collection efficiencies are 50% and 70% for the cyclone and the fiber filter, respectively. What is the overall collection efficiency of the cleaning system?
3. Calculate the particle separation efficiency for 10 μm particles (in aerodynamic diameter) and pressure drop of a standard involute cyclone with an outer cylinder radius of 500 mm and an entry air velocity of 15 m/s. Refer to Table 11.1 for standard cyclone parameters. The ambient air conditions are standard.
4. Calculate the particle separation efficiency for 10 μm particles (in aerodynamic diameter) and the pressure drop of a uniflow cyclone. The separation chamber is 2 m long. The cyclone is made of smooth PVC duct with an absolute roughness of 0.01 mm. The inner and outer diameters of the separation chamber are 400 mm and 500 mm, respectively, with an axial flow velocity of 10 m/s. The vane angle is 60°. The ambient air conditions are standard with a zero velocity. Assume that the particle reentrainment is negligible. What is the particle separation efficiency under each of the following conditions:
   a. Laminar flow in the chamber
   b. Perfect mixing in the chamber

5. Determine the particle collection efficiencies for particles of 10 μm in aerodynamic diameter for a given uniflow cyclone: $R_2 = 400$ mm, $R_1 = 300$ mm, $\alpha = 60°$, and $L = 2$ m. Assume that the air conditions are standard and that the air flow in the separation chamber is perfect mixing. What is the particle collection efficiency under the following conditions:
    a. The axial air velocity $U_{z1} = 3$ m/s
    b. The axial air velocity $U_{z2} = 8$ m/s

6. Plot the particle separation efficiency vs. $Q(R_2^2 - R_1^2)$ for $\alpha = 60°$. $L = 1$ m and particle aerodynamic diameters of 3, 10, 30 and 50 μm for a uniflow cyclone. Airflow in the separation chamber is perfect mixing and all air conditions are standard.

7. In a uniflow cyclone design, the following parameters have been chosen: airflow rate $Q = 0.5$ m³/s, chamber length $L = 0.6$ m, vane angle $\alpha = 45°$. Determine the outer and inner diameters of the cyclone to achieve a particle cutsize of 5 μm under perfect mixing conditions.

8. In an air supply duct with a diameter of 400 mm, a uniflow cyclone serves as an in-line precleaner. The airflow rate in the duct is 600 l/s. The cyclone outer diameter is chosen to be the same as the duct diameter. The length of the cyclone is 1.5 m. In order to achieve a particle cutsize of 10 μm, determine the relationship between the vane angle and the inner diameter. If the inner diameter is 300 mm, what should the vane angle be? Assume that the airflow is perfect mixing.

9. In a vertical wet scrubber, the air velocity is upward. The airflow rate is 2 m³/s. Water droplets are 100 μm in diameter. In order to increase the particle collection efficiency of the scrubber, it is desirable to have a high water droplet concentration. What should the scrubber cross-sectional area be to maximize the water droplet concentration?

10. A cylindrical wet scrubber has a diameter of 1.2 m and a height of 3 m. The airflow rate is 8 m³/s. The water flow rate is 0.15 l/s and the water droplet diameter is 70 μm. The concentration of particles of 3 μm is 10 particles/ml. The initial relative velocity of water droplets to the airflow is 20 m/s. Assume that the water droplets are perfectly mixed with the air in the scrubber. Determine the kinematic coagulation efficiency for particles of 3 μm in aerodynamic diameter, including both the transient and steady-state stages. What will the error be if transient kinematic coagulation is ignored?

11. Calculate the concentration of particles of 10 μm in aerodynamic diameter at the exit of a vertical cylindrical wet scrubber. The scrubber is 3 m high and 0.8 m in diameter. The airflow is upward with a rate of 1.2 m³/s, and the water flow rate is 0.25 l/s. The average water droplet diameter is 100 μm. Assume that the water droplets are perfectly mixed with airflow and the transient kinematic coagulation is negligible. The 1 μm particle concentration in the untreated air is 20 particles/ml. All air conditions are standard.

12. In problem 11, in order to achieve a particle cutsize of 5 μm, one method is to increase the water droplet concentration by increasing the droplet retention time in the scrubber. Determine the required water droplet concentration for 1 μm cutsize requirement. Determine the scrubber diameter to achieve the needed droplet concentration.

13. A uniflow cyclone is in series with a wet scrubber to serve as a mist eliminator. The flow rate is 0.8 m³/s. The outer and inner diameters are 0.4 m and 0.2 m, respectively. If the separation chamber is 0.8 m long, determine the vane angle

needed to collect 95% of the water droplets. If the vane angle is 60°, how long should the separation chamber be? Assume that the air is completely mixing.

14. Water is introduced into the throat of at a rate of 0.2 l/s. The airflow through a venturi scrubber is 0.8 m³/s. The venturi throat has a rectangular cross section that is 0.12 m wide and 0.08 m high. If a 90% collection efficiency is required for particles smaller than 5 μm, what should the minimum water flow rate be under standard air conditions?

REFERENCES

1. Wark, K., Warner, C.F., and Davis, W.T., *Air Pollution*, Addison-Wesley Longman, Berkeley, CA, 1988, 220–277.
2. ASHRAE, *Handbook of Systems and Equipment,* American Society of Heating, Refrigerating, and Air-Conditioning Engineers, Atlanta, 2000, ch. 25.
3. Stephens, H.H., The design, construction, and uses of centrifugals, *Metallurgical and Chemical Engineering*, vol. XI,(6):358–359, 1913.
4. Bretney, E., *U.S. Patent No. 453* 105, 1891.
5. Storch, O., *Industrial Separators for Gas Cleaning*, Elsevier Scientific Publishing Company, Oxford, U.K., 1979.
6. Svarovsky, L., *Hydrocyclones*, Holt, Rinehart & Winston, New York, 1984.
7. Stephens, H.H., The design, construction, and uses of centrifugals, *Metallurgical and Chemical Engineering*, vol. XI: 358–359, 1913.
8. Maloney, J.O. and Wilcox, A.C., *Centrifugation Bibliography*, University of Kansas Publications, Engineering Bulletin, no. 25, 1950.
9. Perry, R.H. and Green, D.W., *Perry's Chemical Engineer's Handbook*, McGraw-Hill, New York, 1984, ch. 20.
10. Ogawa, A., *Separation of Particles from Air and Gases*, CRC Press Inc., Boca Raton, FL, 1984, 42–45.
11. Leith, D. and Mehta, D., Cyclone performance and design, *Atmospheric Environment*,7:527–549, 1973.
12. Leith, D. and Licht, W. Collection efficiency of cyclone type particle collectors, a new theoretical approach, *A. I. Ch. E. Symposium Series: Air*, 1971.
13. Alden, J.L., *Heating and Ventilating, "Design of industrial exhaust systems,"* Industrial Press, New York, 48–53, 1938.
14. Shepherd, C.B. and Lapple, C.E., Flow pattern and pressure drop in cyclone dust collectors, *Ind. Eng. Chem.*,32:1246–1248, 1940.
15. Bernstein, D.M. et al., A high-volume sampler for the determination of particle distributions in ambient air, *JAPCA*, 26:1069, 1976.
16. John, W. and Reischl, G., A cyclone for size-selective sampling of air, *J. Air Pollut. Control Assoc.*, 30: 872–876, 1980.
17. Sitzmann, B.M., Kendall, J.W., and Williams, I., Characterization of airborne particles in London by computer-controlled scanning electron microscopy, *The Science of the Total Environment*, 241: 63–73, 1999.
18. Christoforou, C.S. et al., Trends in fine particle concentration and chemical composition in southern California, *J. Air Wastewater Manage. Assoc.*, 50: 43–52, 2000.
19. Kleeman, M.J., Schauer, J.J., and Cass, G.R., Size and composition distribution of fine particulate matter emitted from motor vehicles, *Env. Sci. Tech.*, 34, 2000.

20. Carpenter, G.A., Dust in livestock buildings — review of some aspects, *J. Agr. Eng. Res.*, 33: 227–241, 1986.

21. Licht, W., *Air Pollution Control Engineering — Basic Calculations for Particulate Collection,* 2nd ed. Marcel Dekker, New York, 1988.

22. ASHRAE, *Fundamentals Handbook*, American Society of Heating, Refrigeration and Air-Conditioning Engineers, Atlanta, 2001, ch. 32.

23. Altshul A.D., Zhivotovckiy, L.C., and Ivanov, L.P., *Hydraulics and Aerodynamics*, Stroisdat Publishing House, Moscow, 1987.

24. Idelchik, I.E. et al., *Handbook of Hydraulic Resistance*, CRC Press, Boca Raton, FL, 1994.

25. Calvert, S., *Handbook of Air Pollution,* John Wiley & Sons, New York, NY, 1984.

26. Hesketh, H.E., Fine particle collection efficiency related to pressure drop, scrubbant and particle properties and contact mechanics, *JAPCA,* 24(10):939–942, 1974.

CHAPTER 12

Electrostatic Precipitation

In particle mechanics, the most important factor affecting particle transport and fate is the force exerted on the particle, whether this force is gravitational, thermal (diffusion), or electrostatic. This chapter discusses the electrostatic effect on particles and various air-cleaning means utilizing the electrostatic effect. Most particles, as well as many other substances in the physical world, carry some electric charge. Some particles may be highly charged. For highly charged particles, the electrostatic force can be thousands of times greater than the force of gravity. The motion induced by electrostatic forces forms the basis for important types of air-cleaning equipment (electrostatic precipitators [ESPs]) and particle sampling and measuring instruments.

After completing this chapter, the reader will be able to

- Understand the principles of electrostatic precipitation, including particle electrical mobility and terminal velocities in an electrical field, in the Stokes region, and in the non-Stokes region
- Determine particle charging in different charging processes: field charging, diffusive charging, and combined charging of particles; charging limit; and corona discharge
- Design parallel-plate electrostatic precipitators — laminar flow model, complete mixing flow model
- Design tube-wire electrostatic precipitators — laminar flow model, complete mixing flow model

12.1 ELECTROSTATIC FORCE AND FIELD INTENSITY

A charged particle is acted upon by an electrostatic force near charged surfaces or other charged particles. The force acts remotely through air or a vacuum and does not require the flow of current. The charge on a particle can be negative or positive, depending on whether the particle has an excess or a deficiency, respectively, of electrons. In the following section, we will review the basic principles of electrostatics.

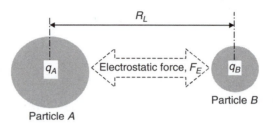

Figure 12.1 There is an electrical force between two charged particles, similar to the gravitational attractive force between two masses.

As shown in Figure 12.1, if two particles A and B are apart by a center-to-center distance R_L and have charges of q_A and q_B, respectively, a force is exerted on both particles. The force is attraction if the charges on the two particles are opposite; the force is repulsion if the charges on the two particles are the same in terms of polarity. The fundamental equation of electrostatics is Coulomb's law, which gives the electrostatic repulsive force F_E between two particles with charges of like signs separated by a distance R_L:

$$F_E = K_E \frac{q_A q_B}{R_L^2} \tag{12.1}$$

where q_A and q_B represent the amount of charge at the two particles and K_E is a constant of proportionality.

In the SI system of units, the ampere (A) is defined as the current required to produce a specified force between two parallel wires 1 m apart. The units of charge and potential difference are derived from the ampere. The unit of charge, coulomb (C), is defined as the amount of charge transported in 1 second by a current of 1 A. The unit of potential difference, volt (V), is defined as the potential difference between two points along a wire carrying 1 A and dissipating 1 watt (W) of power between the points. The unit and the value of K_E in Equation 12.1 are

$$K_E = \frac{1}{4\pi\varepsilon_0} = 9.0 \times 10^9 \left(\frac{Nm^2}{C^2} \right) \tag{12.2}$$

where ε_0 is the permittivity of a vacuum, 8.85×10^{-12} C²/N·A·m². Combining Equation 12.1 and Equation 12.2 gives

$$F_E = 9.0 \times 10^9 \frac{q_A q_B}{R_L^2} \tag{12.3}$$

where F_E is expressed in N, q_A and q_B are in C, and R_L is in m.

An electric field exists in the space around a charged object and causes a charged particle in this space to be acted on by the electrostatic force. We express the strength of such a field in terms of the magnitude of the force F_E produced per unit charge of the particle. The field intensity or strength E is

$$E = \frac{F_E}{q} \tag{12-4}$$

where q is the charge on the particle in C. The unit of field intensity E is in N/C. Field intensity is a vector that has the same direction as the force F_E. It is common to express the amount of charge q as n multiples of the smallest unit of charge, the charge on a single electron e, $e = 1.6 \times 10^{-19}$ C. The n is the number of electrons. Thus, the force on a particle with n elementary units of charge in an electric field E is

$$F_E = qE = neE = 1.6 \times 10^{-19} nE \tag{12.5}$$

Equation 12.5 is the basic equation for the electrostatic force acting on a particle. The application of the equation is straightforward if the values of n and E are known. The central problem in the application of electrostatic theory to aerosols is the determination of these two quantities. Both n and E can change with time and position, and the magnitude of one can affect the other.

Because Coulomb's law provides a relationship between charge, force, and geometry, it can be used to determine the field intensity at any point near a charged surface. An imaginary unit charge is positioned at the desired location, and the total electrostatic force on this unit charge is calculated by the vector sum of the Coulombic force due to each charge on the charged surface. The field intensity is then computed from the total force by Equation 12.4. The utility of this method is limited, because we usually do not know the location and magnitude of all the charges on a surface. In theory, the field intensity at any point could be determined by placing a real test charge at the point and measuring the electrostatic force on it, but there are many practical difficulties with such measurements.

An alternative definition of field intensity is based on voltage of potential, a relatively easy quantity to measure. The potential difference ΔW (in volts) between two points can be defined as the work required to move a unit charge between the two points. This work is equal to the force per unit charge (the field intensity) times the distance between the points, Δx:

$$\Delta W = \frac{F_E \Delta x}{q} \tag{12.6}$$

If we substitute Equation 12.6 into Equation 12.4 and rearrange, an alternative definition of field intensity is obtained by

$$E = \frac{\Delta W}{\Delta x} \tag{12.7}$$

where E is in V/m. The two units, N/C and V/m of the field intensity, are the same because V = N·m/C, work (N·m) equals the voltage (V) multiplied by the current (C is the current in 1 s at 1 V of voltage). From Equation 12.7, the field intensity in any direction at a point is equal to the potential gradient in that direction at the point. The determination of field intensity is primarily a problem of determining the potential gradient near charged surfaces.

The solution for an electric field (that is, an equation that gives the field intensity at any point), can be determined for simple geometries. Three simple geometries are considered here (Figure 12.2): single charged point, parallel plates, and a cylinder with a charged axial wire.

The field around a single point charge q_A can easily be determined by placing a charge q_B in the field and determining the force on it by Coulomb's law (Equation 12.1) and the field intensity by Equation 12.4:

$$E = \frac{F_E}{q_B} = \frac{K_E q_A}{R_L^2} \tag{12.8}$$

The field intensity between two oppositely charged, closely spaced parallel plates is uniform in the region between the plates (neglecting edge effects) and is given by

$$E = \frac{\Delta W}{\Delta x} \tag{12.9}$$

where E is in V/m, ΔW is the algebraic difference in voltage between the two plates, and Δx is the separation distance between the plates. In a uniform field, the electrostatic force on a charged particle is constant everywhere between the plates. For a positively charged particle, the direction of the force is toward the more negatively charged plate; for a negatively charge particle, the force is directed toward the more positively charged plate.

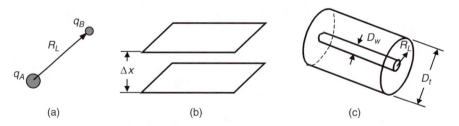

(a)   (b)   (c)

Figure 12.2 Three simple geometries for the calculation of electrical intensity: (a) single charged point, (b) parallel plates, and (c) cylinder with a charged axial wire.

The field intensity inside a cylindrical tube with a wire along its axis is given by

$$E = \frac{\Delta W}{R_L \ln(D_t / D_w)} \qquad (12.10)$$

where E is in V/m, ΔW is the algebraic difference in voltage between the wire and the tube, R_L is the radial position for which the field is being calculated, and D_t and D_w are the diameters of the tube and the wire, respectively. Note that in Equation 12.10, the radial position R_L varies with time as the particle is pulled toward the central wire as it passes through the cylinder. Equation 12.10 indicates that the field intensity goes to infinity as R_L goes to zero. The finite diameter of the conducting wire precludes a field at $R_L = 0$. The maximum field will be at the surface of the wire and will increase as the wire diameter decreases. Equation 12.10 can also be used for the calculation of field intensity in the space between two concentric cylinders. It becomes much more difficult to calculate the field intensity for complex geometries or when there is significant space charge (ions or charged particles in the field region).

In the troposphere near the earth's surface, there is an electric field caused by the difference in potential between the earth's surface (negative) and the upper layers of atmosphere (positive). In normal, clear weather, the average field intensity at sea level is 120 V/m (1.2 V/cm), but it can be 10,000 V/m (100 V/cm) beneath thunder clouds and much higher at the site of a lightning discharge. Electrostatic field intensity in an indoor enclosure can be treated as the average field intensity in sea-level atmosphere. However, field intensity near a dry surface (such as a wall) can be much higher than the average atmospheric field intensity at sea level.

Example 12.1: *A particle of 1 μm in aerodynamic diameter is positioned 0.05 m from a point charge of 10-12 C. What is the field intensity at the particle? What is the electrostatic force on the particle if it has 100 excess elementary charges? What is the ratio of the electrostatic force to the gravity force?*

Solution: The field intensity around the point charge qA = 10-12 C, the charge on the particle is qB = 100e, then

$$E = \frac{K_E q_A}{R_L^2} = \frac{9 \times 10^9 \times 10^{-12}}{(0.05)^2} = 3.6 \ (N/C)$$

$$F_E = q_B E = 100 \times 1.6 \times 10^{-19} \times 3.6 = 5.76 \times 10^{-17} \ (N)$$

The gravity force, *FG*

$$F_G = mg = \frac{\pi d_a^3}{6} \rho_p g = \frac{\pi \times (10^{-6})^3}{6} \times 1000 \times 9.81 = 5.14 \times 10^{-15} \ (N)$$

$$\frac{F_E}{F_G} = 0.011$$

In this example, the electrical force is only about 1% of the gravitational force, thus electrical force cannot be used for effective particle separation. Apparently, the electrical force needs a much stronger field intensity to be predominant. In practice, the field intensity is typically higher than 10,000 V/m for air-cleaning precipitators.

12.2 ELECTRICAL MOBILITY AND TERMINAL VELOCITY IN AN ELECTRICAL FIELD

A charged particle will reach a terminal velocity in an electrical field, which is similar to the settling velocity in a gravitational field. Such terminal velocity is usually much higher than the settling velocity. Thus, the electrical force governs the transport separation of the particle. Additionally, high terminal velocity for the particle often results in a particle's Reynolds number being greater than one. Thus, Stokes's law is not valid and Newton's law must be applied.

12.2.1 Terminal Velocity in the Stokes Region ($Re_p \leq 1$)

For particle motion within the Stokes region, the electrostatic force equals the Stokes drag force, that is

$$neE = \frac{3\pi\eta V_r d_p}{C_c} \tag{12.11}$$

Noting that the terminal velocity caused by the electrical field (V_{TE}) is the particle velocity relative to the fluid velocity, from Equation 12.11 we have

$$V_{TE} = \frac{neEC_c}{3\pi\eta d_p} = neEB \tag{12.12}$$

where E is the electrical field intensity in V/m or N/C, and B is the mechanical mobility of the particle.

12.2.2 Electrical Mobility

The electrical mobility B_E is defined as the velocity of the particle with a charge of ne in an electrical field of a unit intensity

$$B_E = \frac{V_{TE}}{E} = \frac{neC_c}{3\pi\eta d_p} = neB \tag{12.13}$$

**Table 12.1 Electrical Mobility for Typical Particles
with a Single Charge[a]**

Particle Diameter d_p (μm)	Electrical Mobility with a Single Charge, B_E/n, (m²/V·s)
Electron	6.7×10^{-2}
Negative air ion	1.6×10^{-4}
Positive air ion	1.4×10^{-4}
0.01	2.1×10^{-6}
0.1	2.7×10^{-8}
1	1.1×10^{-9}
10	9.7×10^{-11}
100	9.3×10^{-12}

[a] Values are calculated from Equation 12.13.

From Equation 12.13, the unit of electrical mobility B_E is in m²/V·s if E is in V/m (or in m·C/N·s if E is in N/C). The terminal velocity of a particle in an electrical field can be calculated directly from its electrical mobility and the field intensity:

$$V_{TE} = B_E E \qquad (12\text{-}14)$$

Electrical mobilities for typical sizes of particles with a single charge are listed in Table 12.1. When the particle motion is within the Stokes region, the electrical mobility is simply the product of the corresponding value in Table 12.1 and ne. However, for particles larger than 0.5 μm and with maximum charge, the terminal velocity can be high and the particle motion can be beyond the Stokes region. In this case, the terminal velocity of the particle must be determined by means of Newton's resistance law, and the electrical mobility cannot be obtained simply by multiplying n by the values in Table 12.1.

12.2.3 Terminal Velocity in the Non-Stokes Region (Re_p >1)

For particle motion beyond the Stokes region, $Re_p > 1$, the electrical force is balanced by Newton's drag force (Equation 4.14). Equating Equation 12.5 and Equation 4.14 gives

$$neE = \frac{\pi}{8} C_D \rho_g d_p^2 V_{TE}^2 \qquad (12.15)$$

Unlike the Stokes region, the drag coefficient C_D in Newton's region is a function of a particle's Reynolds number, Re_p. In order to solve the terminal velocity, V_{TE}, C_D must be obtained. Substituting Re_p (Equation 4.8) into Equation 12.15 and rearranging gives the following equation:

$$C_D(Re_p^2) = \frac{8neE\rho_g}{\pi\eta^2} = C \qquad (12.16)$$

where C is a constant. The quantity of $C_D(Re_p^2)$ can be determined because the right side of Equation 12.16 does not contain the terminal velocity. Once the quantity of $C_D(Re_p^2)$ is known, the terminal velocity of the particle can be determined using two methods. One of the methods is graphical. From Figure 4.3, there must be one point on the curve that satisfies Equation 12.16 and one of the following equations: Equation 4.16, Equation 4.17, or Equation 4.18. The two chosen equations represent two curves in the graph of C_D vs. Re_p. The intersecting point of the two curves in Figure 4.3 will be the Reynolds number of the particle. The terminal velocity of the particle can be calculated from its Reynolds number.

The second method (the preferred and simpler method in most cases) involves the determination of the terminal velocity, which is calculated directly from the value of $C_D(Re_p^2)$ using an empirical equation:[1]

$$V_{TE} = \left(\frac{\eta}{\rho_g d_p}\right) \exp(-3.070 + 0.9935\,Y - 0.0178\,Y^2) \qquad (12.17)$$

where

$$Y = \ln\left[C_D(Re_p^2)\right] = \ln\left[\frac{8neE\rho_g}{\pi\eta^2}\right] \qquad (12.18)$$

Equation 12.18 is accurate within 3% for particle motion with $1 < Re_p < 600$ and within 7% for particle motion with $0.5 < Re_p < 1000$. When $Re_p < 1$, particle motion is in the Stokes region and Equation 12.12 should be used. When $Re_p > 1000$, the drag coefficient is constant (Newton's region in Figure 4.3, $C_D = 0.44$) and the terminal velocity can be directly solved from Equation 12.15.

Example 12.2: *Determine the terminal velocity of a 10 μm particle between two parallel plates with a potential difference of 2000 V. The distance between the two plates is 0.02 m and the particle has 500 element charges.*

Solution: The electrical field intensity

$$E = \frac{\Delta W}{\Delta x} = \frac{2000}{0.02} = 100,000 \left(\frac{V}{m}\right)$$

From Table 12.1, the electrical mobility of the particle with 500 charges

$$B_E = n \times 9.7 \times 10^{-11} = 4.85 \times 10^{-8}\,\frac{m^2}{V \cdot s}$$

From Equation 12.14

$$V_{TE} = B_E E = 4.85 \times 10^{-8} \times 100,000 = 0.00485 \ (m/s)$$

12.3 PARTICLE CHARGING

The separation of a particle in air by electrical means has two distinct stages:

- First, the particle must acquire electrical charges (positive or negative ions, or electrons).
- Second, the particle must be accelerated in an electrical field to reach the collection media (a surface or a filter).

From Equation 12.5, the electrical force (or Coulomb force) exerted on a particle is directly proportional to the number of elemental charge n and the field intensity E. In the previous section, electrical field intensities for typical simple geometries were discussed. In this chapter, charging mechanisms for particles will be discussed.

Ions charged to a particle can be positive, negative, or a combination of both. If the charging ions contain both positive and negative ions, the charging is *bipolar*. If the charging ions contain only positive or only negative ions, the charging is *unipolar*. Although electrical precipitation can be realized under conditions of bipolar charging of particles, unipolar charging is far more effective and is inevitably employed in electrostatic precipitation (ESP) equipment. This is because the ions with opposite polarity on the same particle neutralize each other and reduce the electrical force. In unipolar charging, theoretically there is no difference between positive and negative charging, because both are equally effective for the same degree of charge. Practically, other factors must be considered to determine the charging polarity. Positive charging, which produces positive gas ions or strips off electrons from gases, is preferred in industrial gas cleaning processes because of its greater stability and high efficiency. The positive charging also generates ozone during the process of stripping off electrons from oxygen molecules. Excessive ozone concentration is prohibitive for indoor environments. Therefore, negative charging is preferred for residential air cleaning.

There are two distinct particle-charging mechanisms: field charging and diffusive charging. *Field charging* is done by ion attachment to particles in an electrical field driven by the electrical force. *Diffusive charging* depends on the thermal force (Brownian motion) of ions, not the electrical force. Practically, the field-charging process predominates for particles larger than 1 μm, and the diffusive-charging process predominates for particles smaller than 0.1 μm. Both processes are important for particles between 0.1 and 1 μm.

12.3.1 Field-Charging Process

Let us consider an initially uncharged spherical particle of diameter d_p, which is placed suddenly in a unipolar electrical field of intensity E_0 and ion density N_0. The particle will immediately be charged by the gas ions, and this charging process will continue until the repelling field set up by the accumulation of charge on the particle becomes sufficiently strong to prevent any further ions from reaching the particle. The mechanism of particle charging basically depends on the flow of gas ions to the particles. The corona ions follow the lines of electric force through the gas, except for small irregularities, which are in the order of magnitude of the mean-free path of the ions, or about 0.1 μm in standard air conditions. Hence, the ion-path irregularities are negligible for particles larger than about one micron but must be taken into account for submicron particles.

The presence of either a dielectric or a conducting particle in an electrostatic field causes lines of electric force to concentrate in the neighborhood of the particle with a consequent increase in the electric field at the particle's surface. This distortion of the electric field may readily be calculated for the case of a sphere. The distortion is greatest for a conducting particle, and, for a dielectric particle, it diminishes with the dielectric constant. The field and equipotential lines for the case of a conducting sphere in a uniform electric field are shown in Figure 12.3a. The lines of force are seen to be attracted to the sphere, and calculation shows that the total number of lines of force passing through the sphere is increased by a factor of three, compared with the undistorted field. The maximum field at the surface for the particle occurs along the axis of the sphere parallel to the direction of the field and is also three times greater than the undisturbed field.

The dashed lines represent the limits of the field passing through the sphere. All gas ions traveling along the lines of force within these limits will strike the sphere and impart charge to it. Any charge imparted to the sphere immediately changes the field configuration and hence changes the rate of charging. The field configuration for the gas where the particle has received half of its saturation charge is shown in Figure 12.3b. It is clear that the particle now receives charge from a smaller portion of the field's area, and the rate of charging is correspondingly reduced. As the particle continues to receive charge, the incoming field diminishes in both area and magnitude until charging finally ceases altogether and the charging process is complete (Figure 12.3c).

From Figure 12.3b, during the charging process, the gas ion current I coming to the sphere is

$$I = J_E A(t). \qquad (12.19)$$

where J_E is the ion flux toward the particle and $A(t)$ is the area perpendicular to the ion flow direction measured at a point several sphere diameters to the left of the particles where the field is essentially undisturbed by the presence of the particle. The drift or forward velocity V_{TE} of the gas ions in the undisturbed electric field is $B_{Ei}E_0$. Hence, the ion flux is given by

$$J_E = C_{i0} B_{Ei} E_0 \qquad (12.20)$$

where B_{Ei} is the electrical mobility of ions, not the particle, and C_{i0} is the undisturbed ion concentration. The charging current coming to the particle is given by

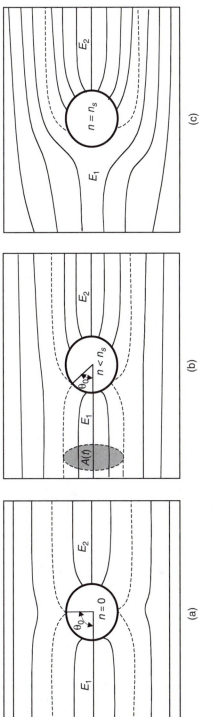

Figure 12.3 The charging process of a conducting spherical particle, where n is the number of element charge and n_s is the saturated element charge: (a) uncharged at the initial time, (b) partially charged, and (c) charge reaches the saturation and charging ceased.

$$I = C_{i0}B_{Ei}E_0A(t) \tag{12.21}$$

where it is assumed that the gas ions are singly charged, i.e., they carry one electric charge. The value of $A(t)$ is calculated from the total electric flux $\psi(t)$, which enters the particle at time t:

$$A(t) = \frac{\psi(t)}{E_0} \tag{12.22}$$

The calculation of $\psi(t)$ is a problem in electrostatics and can be determined using the following equations.[2] The electric field E_1, at any point on the surface of a conducting sphere that is placed in an initially uniform field, may be shown by the methods of electrostatics to be

$$E_1 = 3E_0\cos\theta \tag{12.23}$$

This field is attractive over the half of the sphere that faces the oncoming gas ions. Assume that at time t, a total number n of gas ions have accumulated on the particle. The electric charge of the ions on the particle produces a repelling field, which tends to prevent additional ions from reaching the particle. The magnitude of this repelling field E_2 at the surface of the sphere is

$$E_2 = \frac{4K_E ne}{d_p^2} \tag{12.24}$$

where K_E is defined by Equation 12.2. Hence, the net electric field E at any point on the sphere is given by the sum of E_1 and E_2, or

$$E = E_1 + E_2 = 3E_0\cos\theta - \frac{4K_E ne}{d_p^2} \tag{12.25}$$

The total electric flow entering the sphere is clearly given by the integral of E over the portion of the sphere for which E is positive. This positive portion is enveloped by the angle θ_0, as shown in Figure 12.3.

$$\psi(t) = \int_0^{\theta_0}\left(3E_0\cos\theta - \frac{4K_E ne}{d_p^2}\right)\pi d_p^2\sin\theta\,d\theta \tag{12.26}$$

Integrating Equation 12.26 is easily evaluated by conventional methods, but the details are not essential to the argument and are omitted. The result is

$$\psi(t) = \frac{3}{4}\pi d_p^2 E_0 \left(1 - \frac{4K_E ne}{3E_0 d_p^2}\right)^2 \qquad (12.27)$$

The limiting or saturation value of the particle charge occurs when $n = n_s$ and $\psi(t) = 0$. Denoting the saturation charge by $n_s e$ and letting $\psi(t) = 0$, we find the value of n_s to be

$$n_s = \frac{3E_0 d_p^2}{4K_E e} \qquad (12.28)$$

It is convenient to write Equation 12.27 in terms of n_s so that the expression for $\psi(t)$ becomes

$$\psi(t) = \pi n_s K_E e \left(1 - \frac{n}{n_s}\right)^2 \qquad (12.29)$$

Substituting Equation 12.29 into Equation 12.22 and then into Equation 12.21 yields the expression for the ion current I coming to the particle:

$$I = \frac{d\psi(t)}{dt} = C_{i0} eB_{Ei} \left[\pi n_s K_E e \left(1 - \frac{n}{n_s}\right)^2\right] \qquad (12.30)$$

or

$$\frac{d(n/n_s)}{dt} = \pi C_{i0} K_E eB_{Ei} \left(1 - \frac{n}{n_s}\right)^2 \qquad (12.31)$$

Integrating Equation 12.31, and noting that $n = 0$ at $t = 0$, we have

$$\frac{n}{n_s} = \frac{\pi C_{i0} K_E eB_{Ei} t}{\pi C_{i0} K_E eB_{Ei} t + 1} \qquad (12.32)$$

The factor $\pi N_0 K_E eB_E$ has the dimension of (1/time), so it is convenient to consider $1/\pi N_0 K_E eB_E$ as the charging time constant τ_E, which determines the rate or rapidity of charging. Rewrite Equation 12.32 in the form

$$\frac{n}{n_s} = \frac{t}{t + \tau_E} \qquad (12.33)$$

where

$$\tau_E = \frac{1}{\pi C_{i0} K_E e B_{Ei}} \qquad (12.34)$$

Clearly, the smaller the value of τ_E, the shorter the time of charging is, and vice versa. Equation 12.33 is plotted in Figure 12.4 with t/τ_E. The charging rate is relatively rapid for $t < 2\tau_E$ and relatively slow for $t > 4\tau_E$. Unlike other time constants that are exponentially related to the rates of change, the time constant for particle charging is inversely related to the rate of particle charging, which is a slower process than those of exponential functions. For example, when $t = \tau_E$, the particle charging reaches its 50% saturation charge; when $t = 3\tau_E$, the particle charging reaches its 75% saturation charge. This means that a relatively long time will be required for a particle to reach 99% of its saturation charge.

It is of interest to determine representative values for τ_E under practical conditions for electrical precipitation. The ion mobility B_E varies somewhat, but not markedly, for different gases. It is also different for positive and negative ions, although, again, the difference is not large. For air-cleaning purposes, the value of B_E for average air ions (positive and negative ions) with a single charge may be taken as the representative value, i.e., $B_{Ei} = 1.5 \times 10^{-4}$ m²/V·s. The typical air ion concentration C_{i0} is 5 $\times 10^{14}$ ion/m³ under standard atmospheric conditions. Therefore, for these typical values, the time constant of charging is

$$\tau_E = \frac{1}{\pi C_{i0} K_E e B_E} = \frac{1}{340} = 0.003 \ (\text{sec})$$

This is a very short time compared to the treatment time of air in a precipitator. For typical ESP devices, such as the cylindrical-wire precipitator, the ion concentrations (N_0) can be much higher than 5×10^{14} ion/m³. Thus, the charging time constant can

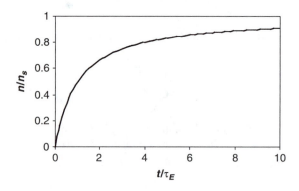

Figure 12.4 Normalized field particle charging rate (n/n_s) with the normalized charging time (t/τ_E).

be less than a millisecond. Therefore, the charging time of a suspended particle is so small that it may be neglected in comparison with the treatment time of the air in the precipitator. A particle in an electrical field can also be assumed to be charged to its maximum capacity instantly, similar to the relaxation time of a particle assumed to be at its terminal velocity instantly. This maximum or saturation charge $n_s e$, acquired by a *conducting particle,* is defined by Equation 12.28. Substituting $e = 1.6 \times 10^{-19}$ C and $K_E = 9 \times 10^9$ Nm²/C² into Equation. 12.28 gives the number of saturation charges for the conducting particle:

$$n_s = 0.00052 \, E_0 d_p^2 \tag{12.35}$$

where E_0 is the field intensity in V/m and d_p is the particle diameter in μm. The saturation charge n_s is proportional both to the electric field E_0 and to the square of particle size. Apparently, particle charge increases rapidly with particle size.

Modification for Dielectric Particles

The only modification introduced by considering a dielectric particle instead of a conducting particle is the introduction of the dielectric constant (permittivity) of the particle ε_p for the particle saturation charge n_s.

$$n_s = \left(\frac{3\varepsilon_p}{\varepsilon_p + 2} \right) \frac{E_0 d_p^2}{4 K_E e} \tag{12.36}$$

and the number of charges of the particle at time t:

$$n = \left(\frac{3\varepsilon_p}{\varepsilon_p + 2} \right) \frac{E_0 d_p^2}{4 K_E e} \left(\frac{t}{\tau_E + t} \right) \tag{12.37}$$

where ε_p is the dielectric constant, τ_E is the time constant for particle charging defined in Equation 12.34, and t is the time of particle charging. The modification is due to the fact that the dielectric particle distorts the electric field by a factor of $3\varepsilon_p/(\varepsilon_p + 2)$. The factor $3\varepsilon_p/(\varepsilon_p + 2)$ is a measure of the field distortion due to the particle. If $\varepsilon_p = 1$, the factor reduces to a value of 1. That is, there is no disturbance of the field, which is expected. For large values of ε_p, the factor approaches 3, the value for a conducting particle. Ordinary dielectric materials have values of ε_p between 1 and 10. For example, ε_p for transformer oil is about 2.0 and for marble about 8.0. The corresponding values of the distortion factor are 1.5 and 2.56, respectively. Thus, the saturation charge for an oil droplet would be only $1.5/3.0 = 0.5$, and the saturation charge for a marble particle would be $2.56/3 = 0.85$ that of a metallic particle. Values of ε_p for typical materials are listed in Appendix 7.

Modification for Nonspherical Particles

The primary effect of nonspherical particles on the charging process is that they distort the field in a somewhat different manner than spherical particles do. Such modified distortions of the electric field will not materially affect the time factor of the charging equation, but they will alter the maximum or saturation charge. In general, for particles of the same volume or of the same surface area, the maximum field at the surface of the nonspherical particle will be somewhat greater than that for the equivalent spherical particle and would thus tend to raise the saturation charge on the nonspherical particle. Offsetting this, however, is the tendency of the particle charge to produce a higher field in just those regions for which the distorted external field is highest. As a result, the charge on spherical particles applies to equivalent nonspherical particles and is accurate enough for many purposes. Note that the particle size must be in diameters of equivalent volume, not the aerodynamic diameter, because charging is related to the actual particle surface area rather than to aerodynamic behavior.

12.3.2 Diffusive-Charging Process

The charging of particles by an ion current in an electric field, which was discussed in the preceding section, becomes less effective as particle size decreases. It is necessary to determine the importance of ion diffusion in the charging of very small particles. An exact theory of particle charging of very small particles would take into account the simultaneous effects of the electric field and of ion diffusion, but it is mathematically difficult to obtain. A consideration of the exact theory combining both field- and ion-diffusion processes leads to significantly different results only in the particle size region from about 0.2 μm to 0.5 μm diameters, whereas results of the ion diffusion process are reasonably accurate for particles smaller than about 0.2 μm.

Ions present in a gas share the thermal energy of the gas molecules and, in general, obey the laws of kinetic theory. The Brownian motion of the ions causes them to diffuse through the gas, and in particular to collide with any particles that may be present. These collided ions will, in general, adhere to the particles because of the attractive electrical forces that come into play as the ions approach the particles. Ion diffusion, therefore, provides a particle-charging mechanism that does not depend on an externally applied electric field. An electric field will aid in charging the particle but is not necessary for the diffusion-charging process. During a charging process, the accumulation of electric charge on a particle gives rise to a repelling field, which tends to prevent additional ions from reaching the particle. The rate of charging, therefore, decreases as charge accumulates on a particle and ultimately will reach a saturation charge. Particle charge under these conditions depends on the thermal energies of the ions and ion concentration, on the particle size, and on the time of exposure.

The first significant study of particle charging by diffusion of unipolar ions seems to have been made by Arendt and Kallmann, who gave the rate of particle charging valid for the particle that has already taken an appreciable charge.[3] The derivation of the Arendt and Kallmann equation is rather complicated. Essentially, the same equation can be derived by the simpler method outlined by White.[2] In kinetic theory, it is shown that the concentration of a gas ion C in a potential field is not uniform but varies according to the equation[4]

$$C = C_0 \exp\left(\frac{\Delta E}{kT}\right) \qquad (12.38)$$

where C_0 is the undisturbed ion concentration, ΔE is the potential energy in J, k is Boltzmann's constant, and T is the absolute temperature.

Let us apply the equation to the case of gas ions in the neighborhood of a suspended particle. If the charge on the particle is ne, the potential energy of a gas at a distance r from the particle will be

$$\Delta E = -\frac{K_E ne^2}{r} \qquad (12.39)$$

where $K_E = 9 \times 10^9$ Nm²/C², as defined in Equation 12.2. Hence, the ion concentration distribution near the particle is

$$C_i = C_{i0}\left(-\frac{K_E ne^2}{rkT}\right) \qquad (12.41)$$

At the surface of the particle $r = d_p/2$, Equation 12.40 becomes

$$C_i = C_{i0}\left(-\frac{2K_E ne^2}{d_p kT}\right) \qquad (12.41)$$

The number of ions that strike the surface of the particle per second is, from kinetic theory

$$\frac{C_i u_{rms}}{4}\left(\pi d_p^2\right) = \frac{\pi}{4} d_p^2 C_i u_{rms} \qquad (12.42)$$

where $\bar{u}_i$ is the mean velocity of the ions. Typical $\bar{u}_i$ for air ions is 240 m/s under standard atmospheric conditions. Assuming that every ion that strikes the particle is captured, the charging rate is

$$\frac{dn}{dt} = \frac{\pi}{4} d_p^2 \bar{u}_i C_{i0} \exp\left(-\frac{2K_e ne^2}{d_p kT}\right) \qquad (12.43)$$

Table 12.2 Particle Charge (n) for Diffusion Charging Process under Standard Atmospheric Conditions[a]

Particle Diameter	Time (s)				
d_p (μm)	10^{-3}	10^{-2}	10^{-1}	1	10
0.01	0.00907	0.0646	0.217	0.412	0.613
0.1	0.646	2.17	4.12	6.13	8.15
0.3	3.82	9.25	15.2	21.3	27.3
0.5	8.17	17.6	27.6	37.7	47.8
1	21.7	41.2	61.3	81.5	102
3	92.5	152	213	273	334
5	176	276	377	478	579
10	412	613	815	1017	1219
30	1524	2129	2735	3340	3947
50	2763	3772	4782	5792	6802
100	6134	8153	10172	12192	14211

[a] Calculated from Equation 12.44: $C_{i0} = 5 \times 10^{14}$ ion/m³.

Integrating the previous equation yields the formula for the diffusion particle charge:

$$n(t) = \frac{kTd_p}{2K_E e^2} \ln\left(1 + \frac{\pi K_E C_{i0} \bar{u}_i e^2 d_p t}{2kT}\right) \tag{12.44}$$

Under standard atmospheric conditions, T = 293 K, $k = 1.38 \times 10^{-23}$ J/K, $e = 1.6 \times 10^{-19}$ C, $\bar{u}_i$ = 240 m/s, Equation 12.44 is reduced to

$$n(t) = 8.77 d_p \ln\left(1 + 2.18 \times 10^{-15} C_{i0} d_p t\right) \tag{12.45}$$

where d_p is in μm, C_{i0} in ion/m³, and t in s. Values of n for a typical ion concentration ($C_{i0} = 5 \times 10^{14}$ ion/m³) are given in Table 12.2 for a range of values of d_p and t.

12.3.3 Total Charge of Particles

Field-charging and diffusion-charging mechanisms are both active in the corona discharge. As previously discussed, the field-charging process generally predominates for particles larger than about 0.5 μm in diameter and for corona fields of a few hundred kilovolts per meter and higher, whereas the ion-diffusion process is predominant for very fine particles of about 0.2 μm in diameter and smaller. Both the field- and the diffusion-charging mechanisms may be regarded as limiting or asymptotic processes, which are sufficiently accurate in their dominant particle-size ranges. However, for the intermediate size range between about 0.2 μm and 0.5 μm, both charging effects must be taken into account.

The total charge of a particle, n_t, can be approximated as the sum of field charging and diffusion charging for most practical air-cleaning processes. Equation 12.37 and Equation 12.44 may be considered reasonably satisfactory for the entire range of particle size and electric-field strength encountered in electrical precipitation. Therefore, the total charge of a particle (n_t) with a permittivity ε_p, in an electrical field with an intensity E and an ion concentration C_{i0}, over a time period t, can be expressed as

$$n_t(t) = \left(\frac{3\varepsilon_p}{\varepsilon_p + 2}\right)\frac{Ed_p^2}{4K_E e}\left(\frac{t}{\tau_E + t}\right) + \frac{kTd_p}{2K_E e^2}\ln\left(1 + \frac{\pi K_E C_{i0}\bar{u}_i e^2 d_p t}{2kT}\right) \quad (12.46)$$

Under standard atmospheric or room air conditions, T = 293 K, $\bar{u}_i$ = 240 m/s, $k = 1.38 \times 10^{-23}$ J/K, $e = 1.6 \times 10^{-19}$ C, Equation 12.46 is reduced to

$$n_t(t) = 0.00052Ed_p^2\left(\frac{\varepsilon_p t}{(\varepsilon_p + t)(\tau_E + t)}\right) + 8.77d_p\ln\left(1 + 2.18 \times 10^{-11} C_{i0}d_p t\right) \quad (12.47)$$

where E is the field intensity in V/m, d_p is the particle diameter in μm, C_{i0} is the undisturbed ion concentration in ion/m³, τ_E is the charging time constant in s as defined in Equation 12.34, and t is the charging time in s. Note that d_p is in μm, while others are in SI units in Equation 12.47 to simplify the numerical magnitudes of variables.

Example 12.3: *Determine the total charge of a particle of 1 μm in diameter in 0.1 s under standard room air conditions. The particle has a permittivity of 5 and is in an electrical field of 500 kV/m. The ion concentration surrounding the particle is 7 × 10¹⁴ ion/m³.*

Solution: Note that the electrical mobility for air ions B_{Ei} = *1.5 × 10⁴ m²/V·s.* From Equation 12.34, the field charging time constant is

$$\tau_E = \frac{1}{\pi C_{i0}K_E e B_{Ei}} = \frac{1}{\pi \times 7 \times 10^{14} \times 9 \times 10^9 \times 1.6 \times 10^{-19} \times 1.5 \times 10^{-4}} = 0.0021 \ (s)$$

Substituting the values in Equation 12.47 gives

$$n_t(t) = 0.00052Ed_p^2\left(\frac{\varepsilon_p t}{(\varepsilon_p + t)(\tau_E + t)}\right) + 8.77d_p\ln\left(1 + 2.18 \times 10^{-11} C_{i0}d_p t\right)$$

$$= 0.00052 \times 5 \times 10^5 \times 1^2\left(\frac{5 \times 0.1}{(5 + 2)(0.0021 + 0.1)}\right)$$

$$+ 8.77 \times 1 \times \ln(1 + 2.18 \times 10^{-11} \times 7 \times 10^{14} \times 1 \times 0.1)$$

$$= 182 + 64 = 246 \ (ions)$$

Although Equation 12.46 is explicit in expressions, it is developed under restricted assumptions. One of the assumptions is treating the field charging and diffusion charging individually. In reality, a particle in an electrical field is subjected to both types of charging simultaneously. The differential equation for the dual charging mechanisms is too complicated to be solved by analytical methods. Thus, numerical solution is necessary.[5, 6]

Another error source results from neglecting the effects of the kinetic energy in the charging field. Armington derived a field-charging equation including the effects of the kinetic energy and the persistence of momentum of the ions approaching the particles exposed in an ion field.[7] This leads to a larger particle charge and better agreement with experiments. In summary, although the total particle charging calculated using Equation 12.46 is somewhat conservative and lower than experimental results, it is sufficiently accurate for most air-cleaning applications.[8]

12.3.4 Charging Limit

From Equation 12.46, if the electrical field intensity is infinitive (such as near the wire of a tube-wire precipitator), or if the diffusion charging time is very long (such as the particle's charging in an open atmosphere), the total charge of a particle could be very large theoretically. Practically, there is an upper limit to the maximum amount of charge that can be held by a particle at a given size. This maximum charge limit, n_{max}, is determined by the surface properties of the particle.

For negatively charged solid particles, the maximum charge is reached when the electrical field at the particle surface (as described by Equation 12.8) reaches the value required for spontaneous emission of electrons from the surface. When this limit is reached, the electrons at the surface of the particle get crowded, resulting in a surface electrical field so strong that it causes electrons to be ejected from the particle. At such a particle surface electrical field intensity, E_{max}, the maximum charge, becomes

$$n_{max} = \frac{E_{max}d_p^2}{4K_E e} \quad \text{(for solid particles)} \qquad (12.48)$$

where E_{max} is the particle surface field intensity for spontaneous emission of electrons. Note that Equation 12.48 can be derived directly from Equation 12.36 by assuming $\varepsilon_p = \infty$ (nonconductive) and replacing E_0 with E_{max}. At standard conditions

$$E_{max} = 9 \times 10^8 \text{ V/m} \quad \text{(for negatively charged solid particles)}$$

$$E_{max} = 2.1 \times 10^{10} \text{ V/m} \quad \text{(for positively charged solid particles)}$$

The difference in E_{max} values for positively and negatively charged particles is due to emission of electrons or positive ions from the particle surface. Because emission of ions is much more difficult than emission of electrons from the particle surface, the E_{max} for positively charged particles is much higher than the E_{max} for negatively charged particles.

For liquid particles, the limiting factor is the surface tension instead of surface electric field intensity. The maximum charging limit for liquid, n_{max}, is called the *Rayleigh limit* and is given as

$$n_{max} = \left(\frac{2\pi\gamma d_p^3}{K_E e} \right)^{\frac{1}{2}} \quad \text{(for liquid particles)} \qquad (12.49)$$

where γ is the surface tension in N/m. For water, $\gamma = 0.073$ N/m.

The Rayleigh limit is the control limit for liquid particles because it is smaller than the emission limit. The Rayleigh limit is useful in atmospheric studies. When mutual repulsion of electric charges within the droplet exceeds the bonding force of surface tension, the droplet will disintegrate and form smaller droplets. These smaller droplets, with the same total volume as the particle before disintegration, have higher charging capacity that may be below the Rayleigh limit. Otherwise, the disintegration will continue until a new balance between the electrical repulsion force and the surface tension force is achieved.

12.3.5 Corona Discharge

From the previous discussions, particle charging for meaningful air cleaning requires a high concentration of ions. Due to the high mobility of ions, the lifetime of free ions is short. Therefore, ions must be produced continuously. In principle, airborne particles may be charged in numerous ways by taking advantage of the varied electric activity associated with many physical and chemical phenomena. For example, natural fogs carry appreciable charges because of cosmic-radiation effects, and ordinary dust generally is charged by the mere act of grinding or even by dispersion from a bulk state. In the real world, uncharged airborne particles are rare.

On the other hand, despite the wealth of particle charging phenomena available (such as flame ionization, static electrification, electrolytic charging, and photoelectric emission of electrons), only a few of these phenomena can achieve the requirement for large-scale, high-efficiency air cleaning. Most natural charging processes can only produce bipolar charging at a low degree of charge. As a general principle, economic considerations require that the particle charge be made as large as possible, as the Coulombic force is proportional to the charge. Theory and long experience have shown that the high-voltage corona discharge is by far the most efficient and applicable means to produce a meaningful amount of unipolar ions for a very high particle charge for air-cleaning purposes.

To produce a corona discharge, a nonuniform electrostatic field must be established. Such a nonuniform electrostatic field can be formed between an active electrode, such as a wire, and a passive receiving electrode, such as a plate or a concentric tube. Air is a good insulator under normal conditions. However, under a

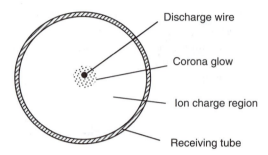

Figure 12.5 Schematic diagram of a corona discharge.

strong electrical field, air becomes conductive. As shown in Figure 12.5, the electrical field intensity near the wire can be extremely high, and ionization processes are confined to or near the glow region adjacent to the wire. To ensure a corona discharge, a minimum field intensity E_{min} near the wire must be maintained. This minimum field intensity was experimentally determined by White:[8]

$$E_{min} = 3000 + 127 d_w^{-\frac{1}{2}} \quad (\text{kV/m}) \tag{12.50}$$

where E_{min} is in kV/m and d_w is the diameter of the wire in m. For a wire 1 mm in diameter, the E_{min} is 7016 kV/m, and for a 2 mm wire the E_{min} is 5840 kV/m.

Positive Corona

In the positive corona, the wire is positive and the tube is negative, thus the electrons are drawn to the positive wire. Near the glow region, electrons are stripped from air molecules and form positive ions. While the electrons are drawn to the positive wire, negative air ions are driven to the passive receiving tube area. The most probable source of primary electrons required to maintain the positive corona is the release of electrons from gas molecules by ultraviolet light quanta radiated from the visible glow region. These quanta, in mixed gases such as air, have ample energy to ionize gas molecules with lower ionization potentials. The ionization must occur very close to the corona glow because most gases are quite opaque to the short wavelength radiation of the ultraviolet region. Release of electrons from the passive receiving electrode (tube) by impact of the positive ions is energetically impossible for the field with a low intensity near the tube.

The smooth, uniform, and visible glow that envelopes the corona wire in the positive corona is evidence of the diffuse nature of the ionization process. It is clear that the wire itself serves merely as a collecting electrode for the electrons and performs no ionization function. As long as the wire is smooth and round, it exerts no effect that could tend to concentrate the discharge into other nonuniform glow, such as brushes or streamers. Fundamentally, the positive corona is entirely a gas process through the photoelectric release of electrons from the gas molecules.

Negative Corona

In the negative corona, the wire is negative and the tube is positive, thus the electrons are emitted from the active electrode. Free electrons from the ionization zone travel to the ion charge region and collide with gas molecules to form negative ions. The ability to form negative ions is a fundamental property of gas molecules. Some gases, such as nitrogen, hydrogen, helium, neon, and argon, if sufficiently pure, have no affinity for electrons and hence do not form negative ions. Therefore, negative coronas do not occur in these gases. On the other hand, oxygen, chlorine, sulfur dioxide, and many other gases have strong electron affinities and are able to produce highly stable negative coronas. Consequently, negative corona characteristics are highly sensitive to gas composition.

The visual appearance of the negative corona is different from the positive corona. In the glow region of a negative corona, a series of localized glow points or brushes appear to be in a rapid, dancing motion over the wire surface. Existence of the localized brushes in a negative corona is evidence of long ionization paths of electrons. The higher the voltage, the longer the brushes within the sparkover are. A significant percentage of free electrons generated in the active region of the corona travels the entire distance to the outer tube, an event that may greatly increase the total corona current, and thus the ionization efficiency.

In summary, the unipolar negative corona may be visualized as consisting of two primary regions. The first region is the active glow region around the discharge wire, which consists of a copious amount of positive and negative ions, free electrons, and both excited and normal molecules, all interacting with one another and with the wire. The second region is the passive region between the glow and the passive electrode (tube). This region contains neutral molecules, plus a small amount of negative ions and electrons that have been created in the active region, and they move toward the passive electrode under the influence of the electric field. Under normal conditions, the active region is very small and is essentially a plasma with no net charge; it provides a copious source of electrons for the corona. The passive region, on the other hand, occupies practically the entire volume of the corona and possesses a relatively large electric space charge. The passive region is where the particle charging and collection take place.

Particles introduced into the corona charging field will have the same polarity as the active wire electrode. Thus, a positive corona will produce positively charged particles, and a negative corona will produce negatively charged particles. Ion concentration in a corona charging field ranges from 10^{12} to 10^{15} ion/m^3. Typically, a negative corona is about 10 times more efficient than a positive corona. This can be explained from a particle mechanics point of view: In a positive corona, the corona current carriers are positive gas ions with relatively low mobility; in a negative corona, the corona current carriers are electrons with a mobility about 1000 times smaller than gas ions. The high mobility of electrons makes it much easier for the electrons to collide with particles and makes the negative corona a more efficient ion generator. As a result, industrial precipitators exclusively employ negative coronas because of their higher efficiency and stability. However, a negative corona produces an excessive amount of ozone at the active electrode region. Therefore,

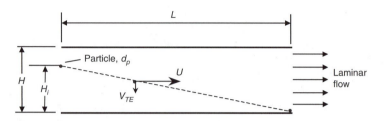

Figure 12.6 Particle trajectory between two parallel planes of a precipitator under laminar flow conditions.

for air cleaning in residential and animal buildings, a positive corona must be used to prevent ozone generation.

12.4 PARALLEL-PLATE ELECTROSTATIC PRECIPITATORS (ESPs)

ESPs use electrical force to collect charge particles for air cleaning. We will discuss two flow conditions: laminar flow and perfect mixing. The perfect mixing flow can be considered a very high turbulent flow condition. For turbulent flows in between laminar and complete mixing, turbulent intensity must be known, and numerical solutions are often necessary for the calculation of particle collection efficiencies.

12.4.1 Laminar Flow Model

Consider that a particle with a diameter d_p enters the gap between the parallel plates at a height of H_i from the collecting plate and is being attracted to the plate at a velocity V_{TE}. The particle-laden air has a mean velocity of U, and the plates have a length of L. The total distance between the two plates is H, as shown in Figure 12.6. If the particle reaches the bottom plate before it exits, the particle is assumed to be collected. Thus, the particle collection efficiency is the ratio of H_i to H:

$$\xi = \frac{H_i}{H} \tag{12.51}$$

Under certain conditions (such as when the plate is long, the retention time is long, the particle size is large, or there is combination of these factors), H_i can be larger than H, indicating that all particles traveled to a lateral distance larger than H and were collected on the plate. Practically, when $H_i > H$, the particle collection efficiency is 1.

Particles that enter the gap between the parallel plates with a smaller distance than H_i will be collected. Let us consider the general situation under which the particle is subjected to both field and diffusion charging. Because the particle charge n is time dependent, the terminal velocity (V_{TE}) is time dependent under diffusion charging, the distance H_i should be written as

$$H_i = \int_0^{L/U} V_{TE}(t)\,dt \tag{12.52}$$

where L/U is the retention time of a particle in the ESP. From Equation 12.12 and Equation 12.44

$$V_{TE}(t) = n(t)eEB$$
$$= \left(\frac{3\varepsilon_p}{\varepsilon_p + 2}\right)\frac{BE^2 d_p^2}{4K_E}\left(\frac{t}{\tau_E + t}\right) + \frac{BEkTd_p}{2K_E e}\ln\left(1 + \frac{\pi K_E C_{i0}\bar{u}_i e^2 d_p t}{2kT}\right) \tag{12.53}$$

Substituting Equation 12.53 into Equation 12.52 and integrating over the retention time period L/U gives

$$H_i = a_1\left[\frac{L}{U} - 2.3\tau_E \ln\left(\frac{U\tau_E + L}{U}\right)\right] + a_2\left[\left(\frac{L}{U} + \frac{1}{a_3}\right)\ln\left(\frac{a_3 L + U}{U}\right) - \frac{L}{U}\right] \tag{12.54}$$

where

$$a_1 = \left(\frac{3\varepsilon_p}{\varepsilon_p + 2}\right)\frac{BE^2 d_p^2}{4K_E} \tag{12.55}$$

$$a_2 = \frac{BEkTd_p}{2K_E e} = \frac{C_c EkT}{6\pi\eta} \tag{12.56}$$

$$a_3 = \frac{\pi K_E C_{i0}\bar{u}_i e^2 d_p}{2kT} \tag{12.57}$$

Apparently, the first term at the right side of Equation 12.54 is the distance traveled by the particle caused by the field charging, and the second term is the distance traveled by the particle caused by diffusion charging. As discussed in the section on particle charging, field charging is predominant for particles larger than 1 μm in diameter. Furthermore, the charging time constant for a typical ESP for air cleaning is much smaller than the retention time. Typically, if $\tau_E < 0.1\ L/U$, the particle can be considered charged to saturation charge instantly, and Equation 12.54 is reduced to

$$H_i = \left(\frac{3\varepsilon_p}{\varepsilon_p + 2}\right)\frac{BE^2 d_p^2 L}{4K_E U} \quad \text{(for } d_p > 1\ \mu m, \text{ and } \tau_E < 0.1 L/U\text{)} \tag{12.58}$$

For particles smaller than 1 μm and relatively short (typically $L/U < 10\tau_E$) retention time, Equation 12.54 should be used.

Example 12.4: *In a parallel-plate ESP, the distance between adjacent parallel plates is 10 mm. The plates are 0.4 m long, and the mean air velocity passing through the ESP is 2 m/s. The voltage between the plates is 3000 V. The particle has a permittivity of 5, and the ion concentration is 7×10^{14} ion/m³. Determine the particle collection efficiency for 1 μm particles. Evaluate the error in the instantaneous saturation charge of the particles when only the field charging is considered.*

Solution: From Example 12.3, the time constant for field charging is 0.0021 s. The retention time for the air in the ESP, $L/U = 0.4/2 = 0.2$ s, which is 95 times longer than the charging time constant. Therefore, the particle can be considered instantaneously charged to saturation charge. Applying Equation 12.58 gives

$$B = \frac{C_c}{3\pi\eta d_p} = \frac{1.17}{3\pi \times 1.81 \times 10^{-5} \times 10^{-6}} = 6.86 \times 10^9 \left(\frac{m}{N \cdot s}\right)$$

$$E = \frac{\Delta W}{\Delta x} = \frac{3,000}{0.01} = 300,000 \ \ V/m$$

$$H_i = \left(\frac{3\varepsilon_p}{\varepsilon_p + 2}\right)\frac{BE^2 d_p^2 L}{4K_E U} = \left(\frac{3 \times 5}{5+2}\right)\frac{6.86 \times 10^9 \times 300,000^2 \times 10^{-12} \times 0.4}{4 \times 9 \times 10^9 \times 2}$$

$$= 0.0073 \ (m)$$

From Equation 12.52

$$\xi = \frac{H_i}{H} = \frac{0.00735}{0.02} = 0.37$$

Now, let us take into account the effect of diffusion. From Equation 12.54 and Equation 12.55

$$a_1 = \left(\frac{3\varepsilon_p}{\varepsilon_p + 2}\right)\frac{BE^2 d_p^2}{4K_E} = 0.03675 \ (m \cdot s)$$

$$a_2 = \frac{BEkTd_p}{2K_E e} = \frac{6.86 \times 10^9 \times 300,000 \times 1.36 \times 10^{-23} \times 293 \times 10^{-6}}{2 \times 9 \times 10^9 \times 1.6 \times 10^{-19}} = 0.00285 \ (m/s)$$

$$a_3 = \frac{\pi K_E C_{i0} \bar{u}_i e^2 d_p}{2kT} = \frac{\pi \times 9 \times 10^9 \times 7 \times 10^{14} \times 240 \times (1.6 \times 10^{-19})^2 \times 10^{-6}}{2 \times 1.36 \times 10^{23} \times 293}$$

$$= 15,250 \ (s^{-1})$$

Substituting these values into Equation 12-54 gives

$$H_i = a_1 \left[\frac{L}{U} - 2.3\tau_E \ln\left(\frac{U\tau_E + L}{U} \right) \right] + a_2 \left[\left(\frac{L}{U} + \frac{1}{a_3} \right) \ln\left(\frac{a_3 L + U}{U} \right) - \frac{L}{U} \right]$$

$$= 0.03675 \times \left[0.2 - 2.3 \times 0.0021 \ln(\frac{0.0042 + 0.4}{2} \right]$$

$$+ 0.00285 \times \left[(0.2 + \frac{1}{15,250}) \ln\left(\frac{15,250 \times 0.4 + 2}{2} \right) - 0.2 \right]$$

$$= 0.00763 + 0.00400 = 0.01163 \quad (m)$$

The particle collection efficiency becomes

$$\xi = \frac{H_i}{H} = \frac{0.01163}{0.02} = 0.58$$

This represents a 21% lower particle collection efficiency when only the field charging is considered, or a 57% error. However, when the particle size increases, the first term is increasingly predominant and the effect of diffusion charging can be neglected.

12.4.2 Complete Mixing Flow Model

In air cleaning, the flow rate passing through an ESP is usually turbulent to some extent. Under turbulent conditions, the particles that were driven to the collecting plate keep reentering the airflow. In an extreme case, the turbulence is very high and the air passing through the ESP is completely mixed.

Let us consider a particle motion under the complete mixing situation. We also assume that when a particle reaches the surface of the receiving plate, the particle is collected. When the particles and the carrying air are in a complete mixing process, particles do not move unidirectionally toward the receiving tube. Rather, their movement is analogous to settling under complete mixing conditions. As shown in Figure 12.7, only the particles very close to the receiving tube surface can be collected when the air is completely mixed. Assume that near the tube boundary, there is a very thin layer of air within which the airflow is laminar and the particle is traveling toward the tube surface. Within an infinite time interval dt, particles will travel a small distance, $\delta = V_{TE}dt$, within this laminar flow layer.

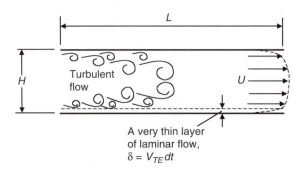

Figure 12.7 Illustration of a particle collection layer at the collecting plate of a precipitator under complete mixing flow conditions.

Let us consider a unit depth: The area $L \times 1$ can be defined as the particle collection zone, within which all particles with a velocity of V_{TE} are collected. The fraction of particles collected by the receiving tube is then the ratio of the area of $L\delta$ to the cross section area of LH.

On the other hand, the fraction of particles collected in a unit volume of air is the reduction in particle concentration, dC_{pn}/C_{pn}. Therefore

$$\frac{dC_{pn}}{C_{pn}} = -\frac{L\delta}{LH} = -\frac{V_{TE}}{H} dt \qquad (12.59)$$

The minus sign $(-)$ indicates that the particle concentration is decreasing with time. Note that V_{TE} in Equation 12.59 is defined by Equation 12.53. Let the particle concentration at the entrance of ESP be C_{pn0} and C_e at the exit, and the retention time be L/U. Substituting Equation 12.53 into Equation 12.59, and integrating over the concentration domains of $[C_{pn0}, C_e]$ with the corresponding retention time domain $[0, L/U]$, gives the concentration of a given size of particle at the exit of the ESP:

$$C_e = C_{pn0} \exp\left(-\frac{H_i}{H}\right) \qquad (12.60)$$

where H_i is defined by Equation 12.54. The collection efficiency for a given diameter of particles can then, according to the definition, be obtained from Equation 12.60:

$$\xi = 1 - \frac{C_e}{C_{pn0}} = 1 - \exp\left(-\frac{H_i}{H}\right) \qquad (12.61)$$

Clearly, the particle collection under a complete mixing condition never reaches 1 and is substantially lower than the collection efficiency under a laminar flow condition. For example, in Example 12.4, if the airflow is completely mixed, the collection efficiency is

$$\xi = 1 - \exp\left(-\frac{0.01163}{0.02}\right) = 0.44$$

Compared with 0.58 for laminar flow, complete mixing is 27% less efficient. From the previous example, the airflow through an ESP should be designed with less mixing condition so that particles can maintain a unilateral movement toward the collection plates and increase the particle collection efficiency.

12.5 TUBE-WIRE ELECTROSTATIC PRECIPITATORS (ESPS)

As discussed in the previous sections, only corona charging can effectively produce a meaningful amount of ions for air cleaning purposes. The following analysis of tube-wire ESPs, as shown in Figure 12.5, includes particle collection efficiency under laminar and turbulent flow conditions and design parameters affecting the ESP performance. As far as collection efficiency is concerned, we assume that a particle is collected when it reaches the surface of the receiving tube.

12.5.1 Particle Collection in Laminar Flow

Let us consider a particle with a diameter d_p, flowing through the ESP along the axial direction. Based on particle charging analysis, we can reasonably assume that the particle charging time constant is much smaller than the particle retention time in the precipitator and that the particle always reaches its saturation charge n_s in most practical precipitators. The electrical force exerted on the particle causes a motion perpendicular to the airflow direction with a terminal velocity V_{TE}, as defined by Equation 12.12. As shown in Figure 12.8, because the terminal velocity of the particle is the relative velocity of the particle to the airflow in the radial direction

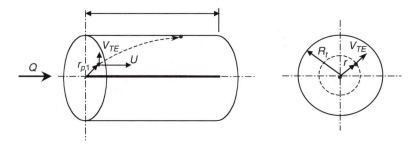

Figure 12.8 Particle trajectory in a tube-wire ESP with an initial radial position of r_{p1} in a laminar flow. The terminal velocity decreases as the particle approaches the collecting tube because the electrical intensity also decreases as r increases.

$$V_{TE} = \frac{dr}{dt} \tag{12.62}$$

Assume that the particle is at an initial radial location r and that gravitational force is negligible. Equation 12.12 can rewritten as

$$\frac{dr}{dt} = neBE = \frac{neC_c\Delta W}{3\pi\eta d_p \ln(R_t / R_w)} \frac{1}{r} \tag{12.63}$$

where n is the number of charges on the particle, e is the element charge, C_c is the particle slip correction coefficient, ΔW is the voltage between the wire and the tube, η is the viscosity of air, and R_t and R_w are the radii of the tube and the wire, respectively.

To solve Equation 12.63, it is necessary to determine the particle charging n first, because it is proportional to the terminal velocity of the particle. It is very complicated when the charging process and the separation process are combined in the analysis because both processes are functions of the radial position of the particle. It is useful, from a design point of view, to analyze the collection efficiency of a particle with a given charge n at the entry of the ESP. The charge n then remains constant in the ESP (Figure 12.8), even though the particle's radial position is changing constantly during the separation process. This assumption can be realized by charging the particles before they enter the separation section.

Assume that the particle reaches the tube after a time t; then Equation 12.63 can be integrated over the radius domain $[r_{p1}, R_t]$ corresponding to the time domain $[0, t]$

$$\int_{r_{p1}}^{R_t} rdr = \int_0^t \left[\frac{neC_c\Delta W_1}{3\pi\eta d_p \ln(R_t / R_w)} \right] dt \tag{12.64}$$

yields

$$\frac{1}{2}(R_t^2 - r_{p1}^2) = \frac{neC_c\Delta Wt}{3\pi\eta d_p \ln(R_t / R_w)} \tag{12.65}$$

The collection efficiency for a given size of particle d_p, as described in Equation 11.21, is the ratio of two circular areas:

$$\xi = \frac{R_t^2 - r_{p1}^2}{R_t^2 - R_w^2} \tag{12.66}$$

The radius of the wire is much smaller than the radius of the tube, thus R_w^2 in the denominator can be neglected. Substituting Equation 12.65 into Equation 12.66, and simplifying, gives

$$\xi = \frac{2neC_c \Delta Wt}{3\pi\eta d_p R_t^2 \ln(R_t / R_w)} \tag{12.67}$$

Note that the retention time of the particle in the precipitator, t, can be expressed in different forms: $t = L/U$ or $t = LA/Q$, where L is the length, U is the average air velocity, A is the cross-section area, and Q is the flow rate of the precipitator. Thus, Equation 12.67 can be written as

$$\xi = \frac{2neC_c \Delta WL}{3\pi\eta d_p R_t^2 \ln(R_t / R_w)U} \tag{12.68}$$

or, noting that $A = \pi R_t^2$

$$\xi = \frac{2neC_c \Delta WL}{3\eta d_p \ln(R_t / R_w)Q} \tag{12.69}$$

Equations 12.67 through 12.69 can be misleading, because it appears that the particle collection efficiency is inversely proportional to the particle diameter. In fact, the particle collection efficiency for an ESP is directly proportional to the particle diameter and the retention time. This is because Equation 12.67 contains the particle charge n, which is variable depending on the field intensity, charging time, and particle size. It may be extremely difficult to determine the charge analytically. However, for a given ESP, the charge is approximately proportional to the square of the particle diameter, i.e., $n \propto d_p^2$, as discussed in previous sections. Therefore, the d_p in the denominator is cancelled by the d_p^2 contained in n.

As a numerical example, Figure 12.9 shows the particle collection efficiency of a cylindrical ESP vs. particle diameter at different retention times. The ESP has the following parameters: $R_t = 0.1$ m, $R_t/R_w = 100$, $L = 1$ m, $\Delta W = 3000$ V. The particles are assumed to have a charge of $n = 1.5 \times 10^{14} d_p^2$. The retention time $t = V/Q$, where V and Q are the volume and flow rate of the ESP, respectively.

For a given design, collection efficiency will be 1.0 for particles larger than d_{100}, although the calculated value is greater than 1.0. Apparently, the particle collection efficiency increases rapidly as the particle size increases. Particle collection efficiency decreases rapidly as the retention time decreases. Equations 12.71 through 12.75 provide an analytical tool for precipitator design. As expected, to increase the particle collection efficiency, an ESP should have a small tube diameter, a long retention time, and a high potential difference between the wire and the tube.

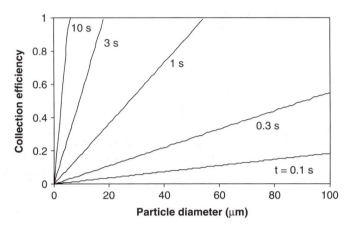

Figure 12.9 Particle collection efficiency vs. particle diameter at different retention times for an ESP under standard room air conditions with the following parameters: R_t = 0.05 m, R_t/R_w = 100, L = 1 m, ΔW = 3000 V. Particle charge n = 1.5 × 10¹⁴ d_p^2. Retention time $t = V/Q$.

Example 12.5*: Calculate the collection efficiency for particles of 10 µm in diameter in a cylindrical ESP with the following parameters: R_t = 0.1 m, R_t/R_w = 100, L = 1 m, ΔW = 5000 V, ε_p = 5, Q = 0.01 m³/s. Before entering the ESP separation cylinder, the particles have been charged to a level approximately n = 1.2 × 10¹⁴ d_p^2. Assume that the airflow in the ESP is laminar and under standard room air conditions (T = 293 K, P = 101.325 kPa).*

Solution: The retention time of particles is

$$t = \frac{LA}{Q} = \frac{1 \times \pi \times 0.1^2}{0.01} = 3.14 \ (s)$$

Substituting the data in Equation 12.67, note that the charge of the particles is proportional to the square of the particle diameter, *n = 1.2 × 10¹⁴ d_p^2*

$$\xi = \frac{2neC_c \Delta Wt}{3\pi\eta d_p R_t^2 \ln(R_t/R_w)}$$

$$= \frac{2 \times 1.2 \times 10^{14} \times (10 \times 10^{-6})^2 \times 1.6 \times 10^{-19} \times 5,000 \times 3.14}{3\pi \times 1.81 \times 10^{-5} \times 10 \times 10^{-6} \times 0.1^2 \times \ln(100)} = 0.77$$

12.5.2 Particle Collection in Complete Mixing Flow (High Turbulent Intensity)

When the particles and the carrying air are in a complete mixing process, particles do not move unidirectionally toward the receiving tube. Rather, particle movement

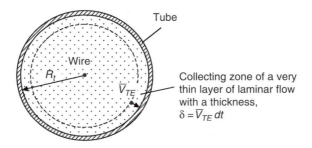

Figure 12.10 Particle collection of a tubing ESP under complete mixing conditions.

is analogous to settling under completely mixed conditions. As shown in Figure 12.10, only the particles very close to the receiving tube surface can be collected when air is completely mixed. Assume that near the tube is a very thin layer of air within which the airflow is laminar. Thus, the particle is traveling unidirectionally toward the tube surface. Within an infinite time interval dt, the particle will travel a small distance, $\delta = \overline{V}_{TE}\,dt$, within this laminar flow layer; $\overline{V}_{TE}$ is the average terminal velocity in the particle collection zone. The area $2\pi R_t \delta$ can be defined as the particle collection zone, within which all particles with a velocity of $\overline{V}_{TE}$ are collected. The fraction of particles collected by the receiving tube is then the ratio of the area $2\pi R_t \delta$ to the cross-section area of the receiving tube. On the other hand, the fraction of particles collected in a unit volume of air is the reduction in particle concentration, dC_{pn}/C_{pn}. Therefore

$$\frac{dC_{pn}}{C_{pn}} = -\frac{2\pi R_t}{\pi R_t^2}\,\overline{V}_{TE}\,dt = -\frac{2\overline{V}_{TE}}{R_t}\,dt \qquad (12.70)$$

The minus sign (−) indicates that the particle concentration decreases with time. Because the flow is completely mixed, the particle charging can be considered constant at an average value between the wire and the tube. Particle charging under a complete mixing condition is very complicated. We use the following assumptions for the following analysis: The effect of the diffusion charge on $\overline{V}_{TE}$ is small compared with the effect of mixing and the estimation of the average field charging. Therefore, the particle charge can be considered time independent and $\overline{V}_{TE}$ is constant within the collection zone, as shown in Figure 12.10.

Because $\overline{V}_{TE}$ is constant, Equation 12.70 can be integrated directly to obtain the particle concentration every time the ESP passes:

$$C_{pn}(t) = C_{pn0}\,\exp\left(-\frac{2\overline{V}_{TE}}{R_t}\,t\right) \qquad (12.71)$$

where C_{pn0} is the particle concentration in the supply air entering the ESP. Let the particle concentration leaving the ESP be C_e. Note that the retention time of the particle in the precipitator, t, equals L/U or LA/Q, where L is the length, U is the average air velocity, A is the cross-section area, and Q is the flow rate of the precipitator. So Equation 12.71 can be written as

$$\frac{C_e}{C_{pn0}} = \exp\left(-\frac{2\overline{V}_{TE}L}{R_t U}\right) = \exp\left(-\frac{2\pi\overline{V}_{TE}R_t L}{Q}\right) \tag{12.72}$$

The particle collection efficiency, according to the definition, can be obtained from Equation 12.72 and is called the Deutsch–Anderson equation:

$$\xi = 1 - \frac{C_e}{C_{pn0}} = 1 - \exp\left(-\frac{2\pi\overline{V}_{TE}R_t L}{Q}\right) \tag{12.73}$$

Note that $\overline{V}_{TE}$ in the complete mixing condition is a kind of "average" terminal velocity in the collection zone. This average terminal velocity is derived from the average field charging of the particles. The particles in the collection zone are subjected to electrical field intensity E_o, which is, from Equation 12.10, equivalent to

$$E_0 = \frac{\Delta W}{R_t \ln(R_t / R_w)} \tag{12.74}$$

Substituting Equation 12.74 into Equation 12.12 gives

$$\overline{V}_{TE} = \frac{C_c enE_0}{3\pi\eta d_p} = \frac{C_c en}{3\pi\eta d_p}\frac{\Delta W}{R_t \ln(R_t / R_w)}$$

Substituting Equation 12.75 into Equation 12.73 gives the particle collection efficiency in an ESP under complete mixing conditions:

$$\xi = 1 - \exp\left[-\frac{2C_c ne\Delta W\, L}{3\eta d_p \ln(R_t / R_w)Q}\right] \tag{12.75}$$

Similar to the laminar flow conditions (Equations 12.67 through 12.69), Equation 12.75 contains the particle charge n, which is approximately proportional to the square of the particle diameter. Figure 12.11 shows the particle collection efficiencies at various particle diameters of a tube-wire ESP under two extreme flow conditions: laminar flow and complete mixing.

From Figure 12.11, separation efficiency under a complete mixing condition is much lower than the laminar flow for large particles. Under a complete mixing condition, no particle size can achieve a 100% separation efficiency, a fact apparent in many ESPs. For a given ESP, particle collection efficiency depends on the airflow conditions. In most air-cleaning cases, airflow in an ESP is neither laminar nor

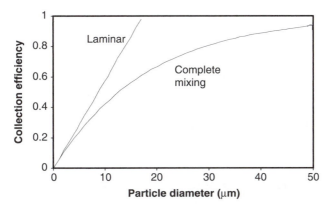

Figure 12.11 Comparison of particle collection efficiency under two flow conditions within a tube-wire ESP: laminar flow and complete mixing. $R_t = 0.05$ m, $R_t/R_w = 100$, $L = 1$ m, $\Delta W = 3000$ V, $Q = 10$ L/s. Particle charge $n = 1.5 \times 10^{14} \, d_p^2$. (Predicted using Equation 12.69 and Equation 12.75.)

complete mixing, but rather turbulent to some extent. Therefore, the two particle separation efficiency curves for laminar flow and complete turbulent flow in Figure 12.11 define the range of the particle separation efficiency for that given ESP. The particle separation efficiency for the ESP in Figure 12.11 should be higher than that of complete mixing flow but lower than that of laminar flow, provided that other factors, such as particle bouncing and detachment, are ignored. The complexity of the effect of turbulent flow on particle separation results in extreme difficulty in predicting particle collection efficiency under actual conditions. Nonetheless, Equation 12.69 and Equation 12.75 provide a guideline for designing ESP parameters. For a given design, minimizing turbulent intensity in the separation chamber is key to increasing the particle separation efficiency.

12.6 DESIGN OF ESP

12.6.1 Design Considerations

Basic engineering ESP design is derived from the fundamentals outlined in the preceding sections and from the large body of accumulated field experience. Early ESPs were designed principally by analogy with prior installations and by certain rule-of-thumb procedures. This approach, although necessary under the conditions existing at the time, is obviously limited in scope and provides little guidance for advancements or improvements except by slow and costly trial-and-error methods. However, since the late 1940s it has become possible to rely more and more on fundamentals. Field experience is still a major factor, but scientific principles and fundamental methods are being utilized to an increasing degree in interpreting and applying this experience.

The type and size of precipitator are determined by the basic properties of the gas and the particles handled, by the gas flow, and by the required collection efficiency. Generally, pipe-type precipitators are used for small gas flows, for collection of mists and fogs, and frequently for applications requiring water-flushed electrodes. Duct-type precipitators are used for larger gas flows, for dry collection, and sometimes also for water-flushed service. Collecting-electrode size for pipes is usually from about 0.15 m in diameter by 2 m long (for small units) up to 0.3 m in diameter by 5 m long (for large units). In duct precipitators, the collecting plates vary in size from 0.6 to 1.0 m wide by 2 m high up to 2 to 2.5 m wide by 6 to 8 m high. Many designs for collecting plates have been used. Fundamental requirements of collecting-plate design include the following:

- Good corona and high sparking-voltage characteristics
- Shielded or shadow zones for particle collection to keep reentrainment losses at a minimum
- Good rapping characteristics
- High mechanical strength coupled with light weight

Corona-electrode design is determined chiefly by application conditions, such as gas temperature, nature and concentration of the dispersed, and presence of corrosive gases or particles. Most industrial precipitators use steel or steel alloy corona wires of about 2.5 mm in diameter. Square wires of 3 mm to 6 mm have advantages in few applications because of their greater cross section. Lead-covered and Hastalloy wires are used for sulfuric acid mist precipitators. Very fine tungsten wires of 0.12 to 0.25 mm diameter are used for air-cleaning precipitators because of their much lower ozone-generation properties. Corona currents are usually in the range of 0.03 to 3 mA/m of discharge wire, with voltages of the order of 30 kV to 100 kV for single-stage precipitators, and 10 kV to 15kV for two-stage precipitators.[8]

The basic importance of methods and equipment for supplying electrical energy or corona power to precipitators can scarcely be overemphasized. It is perhaps obvious that, because a precipitator functions by electric forces acting on the particles, precipitator performance can be no better than the electrical energization. Nevertheless, this is undoubtedly one of the most misunderstood and frequently overlooked factors of precipitator design and operation.

Contrary to initial expectation, the best precipitator performance is obtained not with steady or so-called pure direct current, but rather with intermittent or pulsating waveforms. This is largely because of the higher voltages and currents that can be maintained with intermittent voltages under the sparking conditions that commonly exist in precipitators. For these reasons, most precipitators are powered by high-voltage unfiltered rectifier sets, although pulse methods, somewhat similar to those used in high-power radar equipment, may be used with advantage.

The corona electrodes in larger precipitators usually are subdivided into multiple groups, or sections, frequently referred to as *high-tension bus sections*. These sections are individually powered by separate rectifier sets to reduce the bad effects of precipitator sparking and equipment outages, and to provide better matching of corona voltages and currents to the electrical characteristics of the gas and dust. Ideally, the best precipitator performance would be obtained by energizing each corona wire individually from a separate rectifier set, obviously an uneconomic arrangement. In

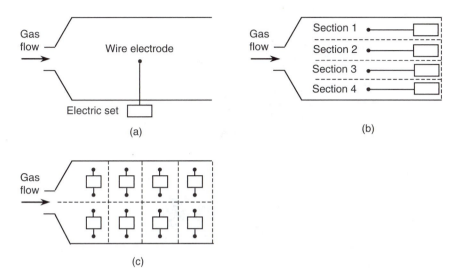

Figure 12.12   Illustration of good and poor sectionization arrangements for high ESP performance requirements. (a) Very poor — single electrode set and single section.
(b) Poor — multielectrodes but only longitudinal sections. (c) Good — both cross-
sectional and longitudinal sections.

practice, the goal is an economic balance among degree of sectionalization, precipitator size, and cost. The effect of high-tension sectionalization on precipitator efficiency is typically very large. For example, a fly-ash precipitator designed for a gas
flow of 50 m³/s with just one section typically might produce a collection efficiency
of only 60%, but the same precipitator with two corona sections might have an
efficiency of 90%. With four sections, it might be 99% efficient. The much greater
efficiency of the four-section arrangement basically results from the substantially
higher precipitator voltage that is possible, compared with the single large section.
Good and poor sectionalization arrangements are shown in Figure 12.12.

Mechanical design includes factors such as the precipitator shell, supporting
structure, suspension and alignment of the electrodes, high-tension electrode frame,
and hopper arrangement. A typical design for duct-type precipitators used for collection of fly ash from pulverized coal–fired power boilers is illustrated in Figure
12.13. In addition, heat insulation, gas pressure, gas temperature, and corrosion
resistance are important considerations for many applications. Finally, the particle
or dust removal system for emptying the hoppers plays a major role in the successful
operation of gas-cleaning installations.

Shell design and the supporting steel are essential structural problems solved by
the methods common to this branch of engineering. Material for the precipitator
shell is usually steel with appropriate heat insulation, or *gunniting*. For corrosive
gases and particles, the shell interior must be of corrosion-resistant materials, such
as lead for sulfuric acid mist and tile block for paper mill salt cake. Pressurized
shells are required in some applications.

Both the corona and the collecting electrodes must be mounted accurately and
aligned to maintain the corona gaps as near the design values as possible. Electrode

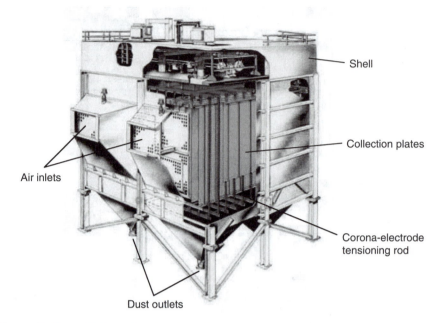

Figure 12.13   An ESP for collection of coal fly ash in a power plant. (Courtesy of Environmental Elements Corp.)

alignment should be maintained within 5%, which corresponds to about 5 mm for a 200 mm duct.

The high-tension frame should be rigid enough to maintain structural stability yet permit rapping or vibration of the corona wires. Suspension of the corona electrodes is of the greatest importance. In addition to the requirement for accurate alignment, the suspension must be designed to prevent corona-wire breakage. Breakage may occur either because of arc-burning at a support or guide point or because of fatigue failure under rapping stresses. Both of these causes of failure are fundamental in nature and can be prevented by proper design, e.g., by use of very low-resistance electrical contacts at wire supports to prevent arc formation and support arrangements that minimize rapping stresses.

High-tension electrical insulators are required to support the discharge electrode frame and also to provide a sufficient level of high-voltage insulation. Ceramic insulators generally are used, and porcelain, fused silica, and fused aluminum are most common. Provision must be made to protect the insulators from fouling by moisture, dust, or mists. Air sweepage, stream or electric heating, and oil seals are used for this purpose, according to the application.

For dry-tape precipitators, hoppers are usually placed under the collecting electrodes to catch and store collected material. Storage capacity, hopper angle, prevention of moisture condensation, and adequate means of dust removal are essential considerations in hopper design. For example, it is not uncommon for precipitator operation to be nullified because of overflowing hoppers or because of air leakage from the dust removal system, which reentrains the collected dust and causes it to be lost from the precipitator. In some applications, so-called wet-bottom receivers replace the ordinary hopper. This is exemplified particularly in paper mill precipitators, where the press requires that the collected material be returned in the form of slurry. Other arrangements that have found some use include flat-bottom receivers with scrapers or drags to remove the dust.

In the case of mists or fogs, removal of collected particles from the plates or pipes is relatively simple, because the resulting liquid usually drains off by gravity. However, for dry particles, some form of rapping or vibration is usually necessary to remove the collected material from the electrodes. As a rule, this must be done with some finesse in order to maintain a proper balance between adequately cleaning the plates and keeping reentrainment or rapping losses at a minimum.

In general, rappers may be classified into two categories:

- The impulse type, which produces an impact or hammer blow
- The vibrator type, which produces a continuous vibration of some duration

The impact type appears to be most useful for corona-electrode cleaning.

Impact rappers have been designed based on just about every known principle, from various cam or lifting mechanisms to dropping a weight against the electrode or the electrode-support frame — or even lifting and dropping the collecting plates themselves — to various types of rolling balls or moving weights actuated by pneumatic or magnetic forces. One fairly definite conclusion is that sound design and good quality control are essential and may be more important than the operating principle. Common difficulties experienced with rapper systems are excessive complication; fatigue failure of metal parts; inability to perform consistently in the moist, dirty, and corrosive atmospheres often encountered; lack of rapping flexibility and stability; and high initial and maintenance costs.

Since its introduction in 1949, the magnetic impulse rapper has found wide acceptance as a practical means of providing closely controlled continuous rapping to eliminate the highly objectionable rapping puffs characteristic of intermittent rapping systems. This rapper system, which provides precisely controlled impact blows over a wide intensity range and at a normal rapping frequency of one or two raps per minute per electrode, is essentially continuous. Vibrator-type rappers usually are powered by compressed air or by electricity. Vibration frequencies commonly are in the range of 50 to 100 per second. Vibrators usually are operated intermittently.

12.6.2 Integration into Plant Systems

Electrostatic precipitators for industrial applications usually are components of integrated systems or processes. For example, in chemical manufacturing, precipitators may serve to preclean process gases ahead of a catalyst operation, or they may be used for fractional precipitation of dusts and vapors produced in processes that are based on one or more steps for cooling and condensing vapors to fumes. Another system-type operation occurs in the cleaning of fly ash from pulverized coal–fired boilers in electric generating stations. In this case, air pollution control is the primary consideration, but the precipitator must nevertheless be fitted into an integrated operation for handling the boiler gases.

The system factors associated with these integrated operations play a major role in the design, operation, and performance of precipitators. Factors such as gas temperature; gas humidity; physical and chemical characteristics of the suspended particles; particle size, resistivity, corrosiveness, etc.; total gas flow; degree of cleaning; and structural features of the precipitator are all largely determined by the system. Raw materials, such as the coal burned in a power plant or the clay and lime fed to the kiln in a cement plant, frequently are dominant factors in the operation and performance of the precipitator.

For these reasons, rational design of precipitators should always include full knowledge of the plant process and raw materials used. In the case of new processes, pilot-scale precipitator tests generally are advisable, if not essential, to the design of full-scale units. The pilot step provides necessary information on basic and practical design factors that usually cannot be obtained by other means. However, it should be noted that pilot precipitator investigations require fundamental knowledge and skill both in operation and in interpretation of results. Unfortunately, there are many examples of pilot-precipitator tests where ignorance or neglect of fundamentals has led to large financial losses, waste of valuable time, and disappointing results. The natural tendency is to attribute these disappointments to the electrical-precipitation process rather than to the defects of the pilot investigations.

To insure achievement of these goals, pilot-precipitator programs must be based on the application of fundamental principles, rather than on trial-and-error or random-empirical procedures. The principal objectives usually are

- To measure the basic properties of the gas and particulates that govern precipitator operation
- To obtain the necessary information to specify the optimal physical design of a full-scale precipitator installation

12.6.3 Typical Configurations of ESPs

Based on the fundamentals of parallel-plate and tube-wire electrostatic precipitators, ESPs can be designed in many configurations. Some of these configurations were developed for special control purposes and collection efficiency; others evolved for economic reasons. Typical ESP configurations include plate-wire (the most common variety), flat-plate precipitator, and tubular precipitator. Wet precipitators and multistage precipitators can be made from any of the previous three configurations or a combination of them.

Plate-wire ESPs are composed of parallel plates of sheet metal and high-voltage electrodes. Dust-laden air flows through the channels formed by the plates. These parallel plates can be very tall and can be arranged in many airflow paths. As a result, this configuration can handle a high airflow rate. To dislodge the collected dust in the ESP, the plates are arranged in sections, often in three or four series. The power supplies are often sectionized in the same fashion to obtain higher operating voltages and better reliability. Dust also accumulates on the electrode wire and must be cleaned periodically. The power supply consists of a step-up transformer and high-voltage rectifiers to convert AC voltage to pulsing DC voltage in a range of 20,000 to 100,000 V, as needed. Auxiliary equipment and control systems are needed to allow the voltage to be adjusted to the highest level possible without excessive sparking and to protect the power supply and electrodes in the event a heavy arc or short-circuit occurs.

In practical design, there is an airflow path without the particle charging between the electrodes and hoppers. This portion of air, which does not get cleaned and is called *sneakage*, usually takes 5 to 10% of the total airflow. Antisneakage baffles are usually used to force the sneakage flow to mix with the main dust-laden airflow for later cleaning sections. However, the sneakage flow at the last section will not be cleaned. Losses due to sneakage can reduce the particle collection efficiency.

Another major factor that affects the particle collection efficiency is the resistivity (in ohm-m) of the collected particles. Because the particles accumulate continuously on the ESP plates, the ion current must pass through the dust layer to reach the ground plates. This ion current creates an electric field in the layer, and it can become large enough to cause local electrical breakdown, which is called *back-corona*. When back-corona occurs, new ions of the opposite polarity are injected into the wire-plate gap, where they reduce the charge on the particles and the collection efficiency and sometimes cause sparking. Back-corona is prevalent when the resistivity of the layer is high, typically higher than 2×10^{13} ohm-m. For lower resistivities, the operation of the ESP is not impaired by back coronas, but resistivities much higher than 2×10^{13} ohm-m considerably reduce the collection efficiency because of the severe back corona. At resistivities lower than 2×10^{10} ohm-m, the particles are held on the plates so loosely that rapping and nonrapping reentrainment becomes much more severe. Care should be exercised in measuring or estimating resistivity, because it is strongly affected by variables such as temperature, moisture, gas composition, particle composition, and plate surface characteristics.

Flat-plate precipitators use flat plates instead of wire for the high-voltage electrodes. The flat plates increase the average electric field and provide increased surface area for the particle collection. Corona cannot be generated on flat plates by themselves, so the corona-generating electrodes and particle charging are placed ahead of or behind the flat plate collection zones. These corona-generating electrodes may be sharp-pointed needles or wires. Unlike plate-wire or tubular configurations, the flat-plate design operates equally well with either negative or positive polarity; thus it is suitable for residential building air cleaning. Flat-plate ESPs have a wide application for high-resistivity particles with small mass median diameters (1 to 2 μm). These applications especially emphasize the strengths of the design, because the electrical dislodging forces are weaker for small particles than for large ones.

Tubular precipitators have the wire electrodes running along the axes of the tubes. Tubular ESPs have typical applications in sulfuric acid plants, coke-oven by-product gas cleaning (such as tar removing), and iron and steel sinter plants. The tube may be circular, square, or hexagonal honeycomb, with gas flow upward or downward. A tubular ESP is essentially a one-stage unit and is unique in having all gas pass through the electrode region. The high-voltage electrode operates at one voltage for the entire length of the tube, and the current varies along the length as the particles are removed from the gas stream. Tubular ESPs comprise only a small portion of the ESP population and are most commonly applied when the particles are either wet or sticky. These ESPs, usually cleaned with water, have particle reentrainment losses of a lower magnitude than those of dry ESPs.

Wet precipitators can be configurations of any of the ESPs discussed previously, with wet surfaces instead of dry ones. The wetting liquid is usually water. The liquid fluid may be applied intermittently or continuously to wash the collected particles into a sump for disposal. Wet ESPs can improve the particle collection efficiency substantially, because the wet collection surfaces prevent reentrainment, especially during rapping or back-corona. The disadvantage is the increased complexity of the wash and disposal of the slurry, which must be treated more carefully than dry dust.

Two-stage precipitators are composed of several ESPs, discussed previously, in series. The ESPs discussed earlier are generally in parallel mode, that is, the discharging and collecting electrodes are side by side. The two-stage precipitator is a series device with the discharge electrode, or ionizer, preceding the collector electrodes. Advantages of this configuration include more time for particle charging and less back-corona. The configuration is economical for small sites, especially for indoor air cleaning. This type of ESP is typically used for flow rates less than 25 m^3/s and for fine, sticky particles such as oil mists, fumes, or smokes, because there is little electrical force to hold the particles on the collecting plates. Cleaning may be done by water wash or in-place detergent spraying of the collector, followed by air-blow drying. Two-stage precipitators are considered different types of devices than large, high-volume, single-stage ESPs. The smaller units are usually sold as preengineered, packaged systems.

DISCUSSION TOPICS

1. Electric force between two charged particles is analogous to the gravitational attracting force between two masses. Compare the corresponding variables in these two phenomena.
2. Why are particle charging and separation the two major steps in electrostatic precipitation? How are these two stages related?
3. In particle-charging calculations, which particle diameter should be used: aerodynamic diameter or geometric diameter? Why?
4. In air-cleaning devices, the mechanical mobility of a particle is constant. However, the electric mobility may not be constant during the process. Why?
5. What is the difference between the saturation charge and the charging limit (or maximum charge) of a particle?
6. Charges on a particle depend on the particle size and the charging mechanisms. In what particle size range is field charging predominant and in what size range is diffusion charging predominant?
7. Can you design an experiment to determine the permittivity of a particle? Assume that you have a parallel-plate ESP, a particle counter, and a particle supply source.
8. Airflow conditions in an ESP affect the particle collection efficiency. Laminar flow is preferred over turbulent flow. Therefore, reducing turbulent intensity in an

ESP is a critical design parameter that must be considered. How can you reduce the turbulent intensity?

9. Give the rationale behind the assumption that the charging time constant is negligible in a typical air-cleaning ESP.

10. Is an ESP more effective for large particles than for small particles?

11. How would you design an ESP that can effectively remove particles smaller than 1 μm? What are the key variables for such a design?

12. What additional information is needed to quantify the particle collection efficiency in an ESP under turbulent flow conditions (the flow is neither laminar nor perfect mixing)?

In your opinion, under perfect mixing conditions, what is the best averaging method to calculate the average charge in the particle-collecting zone near the tube surface? Is it possible that all particles are charged to the saturation charge at the radial position of R_w (where R_w is the radius of the wire)?

PROBLEMS

1. A negative ion generator has a power rating of 200 W. The power supply is 24 V DC. Assume that the generator has an efficiency of 8% (i.e., 8% of the power consumed is used to generate ions). How many electrons are produced per second by this ion generator?

2. A tube-wire electrostatic precipitator (as shown in Figure 12.2c) has a wire diameter of 3 mm and a tube diameter of 30 cm. The potential between the wire and the cylinder is 5000 V. Plot the field intensity E vs. R_L. Use the x axis for R_L from 1.5 mm to 150 mm and the y axis for E in kV/m.

3. Determine the electrical mobility of a 3 μm particle in an electrical field of 300 kV intensity. Assume that the ion concentration in the atmosphere is 5×10^{14} ion/m³ and diffusion charging is neglected.

4. Determine the electrical mobility of a 0.5 μm particle in an electrical field of 300 kV intensity. The retention time of the particle in the electrical field is 0.2 s. What is the change in percentage of the terminal velocity of the particle between the entry and exit of the electrical field?

5. What is the terminal velocity of a 20 μm particle ($\varepsilon_p = 5$) with standard density in a 1000 kV/m electrical field? The particle motion may be beyond the Stokes region.

6. Plot the charges of particles due to the field and diffusion charging in an electric field with an intensity of 500 kV and an ambient ion concentration of 5×10^{14} ion/m³, respectively, as a function of time (from 0 to 10 s) for particles of 0.1, 0.3, 0.5, 1, 3, and 5 μm in diameter. Assume that the air conditions are standard.

7. For a uniform electric field of 500 kV/m and an ion concentration of 10^{14} ion/m³, how long does it take to charge a 1.5 μm particle ($\varepsilon_p = 5$) to 95% of its saturation charge?

8. What is the maximum charge that a 1 μm spherical solid particle can hold? To reach this charge, what is the equivalent electrical field intensity?

9. In an electrical field that has an ion concentration of 5×10^{14} ion/m³, determine the relationship between the charging time constant and the particle diameter under standard room conditions. Plot the relationship on a semilog graph, with the particle diameter on log scale.

10. A particle 2 μm in diameter is in an air flow passing through two parallel plates with a potential difference of 3000 V. The permittivity (ε_p) of the particle is 7. The air velocity is 1.2 m/s. The distance between the plates is 2 cm. The length of plates along the airflow is 20 cm. How many element charges has the particle acquired?

11. As shown below, a portable ESP designed for home air cleaning has a flow rate of 50 l/s. Particle-laden air was drawn into the electrostatic field, composed of a set of paired parallel plates. The plates are 600 mm wide and 300 mm long. The field intensity between the two plates can be treated similarly to that of two infinite plates. The voltage between each pair of plates is 2000 V. The average permittivity of the particles is 5. All indoor air properties are under standard conditions. Determine the particle collection efficiency for particles of 5 μm in diameter. If the particle collection efficiency must be 99%, how would you redesign the ESP?

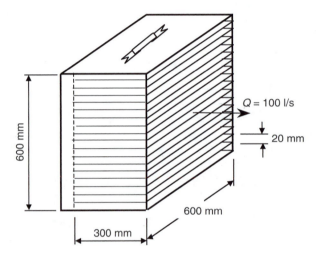

12. In a parallel-plate ESP, the distance between adjacent parallel plates is 100 mm. The plates are 3 m long, and the mean air velocity passing through the ESP is 1 m/s. The voltage between the plates is 3000 V. The particle has a permittivity of 5, and the ion concentration is 7×10^{14} ion/m³. Determine the particle collection efficiency for 2.5 and 10 μm particles. Evaluate the error when only the field charging for the 2.5 μm particles is considered.

13. Determine the particle collection efficiency vs. particle diameter at different retention times for a tube-wire ESP under standard room air conditions with the following parameters: $R_t = 0.1$ m, $R/R_w = 100$, $L = 3$ m, $\Delta W = 5000$ V, $\varepsilon_p = 5$. Assume that the particle charge at the entry of the ESP is zero. The retention time $t = V/Q$, where V is the volume of the tube.

14. Prove that the collection efficiency for laminar flow is higher than for perfect mixing in a parallel-plate ESP. Assume that all design configurations and particle size distributions are the same. (Hint: You may prove this mathematically from the collection efficiency equations for both models, show it graphically using different H_t/H values, or both.)

15. Determine the collection efficiency for particles of 10 μm in diameter in a tube-wire ESP with the following parameters: $R_t = 0.08$ m, $R_w = 3$ mm, $L = 3$ m, ΔW = 5000 V, $Q = 0.01$ m³/s. Particles have been charged to 70% of their saturation

charges under the same field intensity at the entry of the ESP. Assume that the airflow in the ESP is laminar and under standard room air conditions.

16. In problem 15, if the flow is complete mixing, what will the collection efficiency be for the same sized particles?

REFERENCES

1. Hinds, C.W., *Aerosol Technology — Properties, Behavior and Measurement of Airborne Particles,* John Wiley & Sons, New York, 1999.
2. White, H.J., Particle charging in electrical precipitation, *Trans. Am. Ins. Electr. Engineers,* 70:1198–1191, 1951.
3. Arendt, A. and Kallmann, H., The mechanism of charging mist particles, *Z. Physics,* 35: 421, 1926.
4. Loeb, L.B., *Kinetic Theory of Gases,* McGraw-Hill, New York, 1934, 95.
5. Liu, B.Y.H. and Kapadia, H., Combined field and diffusion charging of aerosol particles in the continuum regime, *J. Aerosol Sci.,* 9: 227–242, 1978.
6. Lawless, P.A., Particle charging bounds, symmetry relations, and an analytical charging model for the continuum regime, *J. Aerosol Sci.,* 27: 191–215, 1996.
7. Armington, R.E., unpublished Ph.D. thesis, University of Pittsburgh, Pittsburgh, PA, 1957.
8. White, H.J., *Industrial Electrical Precipitation,* Addison-Wesley, Reading, MA, 1963.

CHAPTER 13

Control of Gaseous Pollutants

Gaseous pollutants in air, such as carbon monoxide, the oxides of nitrogen, the oxides of sulfur, and unburned hydrocarbons, are by far the major pollutants in terms of mass. For outdoor air, approximately 90% by mass of the air pollutants are gases, with most being the result of combustion of fossil fuels and biomass. For many indoor environments, the ratio of gaseous pollutants to the total could be higher. Absolute concentrations of gaseous pollutants in indoor air, however, are relatively low. Four general categories of gaseous pollutant control technologies are commonly used to remove gaseous pollutants from air: adsorption, absorption, catalytic conversion, and source reduction. This chapter describes the principles of the first three categories.

After completing this chapter, the reader will be able to

- Apply the principles of adsorption, adsorption, and catalytic conversion to analyze an air-cleaning process
- Determine the adsorption efficiency (isotherm) and service life or regeneration cycle (adsorption wave) of an adsorption bed
- Design an adsorption bed (filter) using active carbon for air cleaning
- Design a counterflow absorption tower using the mass balance, Henry's law, or gas–liquid equilibrium data
- Determine the number of ideal stages for a cascade counterflow absorption tower
- Determine the ratio of gas to liquid for a counterflow absorption tower

13.1 GAS CONTROL MECHANISMS

In this chapter, we will discuss three basic mechanisms for the separation of gaseous pollutants from the air stream: adsorption, absorption, and catalysis.

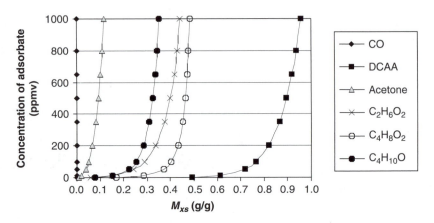

Figure 13.1 Adsorption isotherms of activated carbon for selected VOCs using Equation 13.1 and data in Appendix 8. DCAA = dichloroacetic acid ($C_2H_2Cl_2O_2$).

13.1.1 Adsorption

Adsorption is a process in which a gas or a liquid is attracted to the surface of a solid or a liquid. In the context of gas control, adsorption is the process in which a gas of concern is attracted to a surface and thus separated from the air stream. The adsorbing material is referred to as the *adsorbent*, and the gas or vapor being adsorbed is referred to as the *adsorbate*. Adsorption is a surface phenomenon, that is, the process only happens at the interacting surface of the gas and the surface.

Adsorption processes can be classified as either physical adsorption or chemical adsorption (also called *chemisorption*), depending on the nature of adsorption forces involved. Physical adsorption is caused by intermolecular attractive force, or the van der Waals force, thus it is also called *van der Waals* adsorption. Chemisorption, on the other hand, involves chemical reaction forces between the adsorbent and the adsorbate, which essentially forms a layer of compound between the adsorbate and the outmost layer of the adsorbent.

It is important to distinguish between physical adsorption and chemisorption because the adsorption type affects the selection of adsorbent and the design of the process. The distinction is usually clear. When the distinction is in doubt, one or more of the following criteria can be used:

Heat release rate — The driving force for physical adsorption is the intermolecular attraction between the adsorbent and the adsorbate at the adsorption surface (van der Waals force). There is no change in the molecular properties of the adsorbent and adsorbate. The bonding force at the adsorption interface is small, and the heat released during the physical adsorption is relatively low, approximately the same order as the energy for liquefaction at about 20 kJ/g·mole. In a chemisorption, there are electron or ion exchanges between the adsorbent and the adsorbate, thus a more vigorous reaction and larger heat release than the physical adsorption occurs. The heat release rate for a chemisorption is in the same order of the corresponding bulk chemical reaction, typically in the range of 20 to 400 kJ/g·mole.

Reversibility — A physical adsorption layer can be reversed by reducing the pressure at the temperature at which the adsorption took place. Condensation of water vapor on a car window in winter (a physical adsorption phenomenon) can be removed by blowing warm air on the window (a process to reduce the partial pressure of water vapor in the air). A chemisorption layer, on the other hand, requires much more deliberated conditions to remove, especially on many metal surfaces that requires very high temperature or positive ion bombardment to remove the adsorbate.

Equilibrium pressure — Because physical adsorption of a gas is related to liquefaction or condensation, it only occurs at pressures and temperatures close to those required for liquefaction. Therefore, if the gas partial pressure in the bulk air, P_g, at a given temperature is much lower than the gas partial pressure at the adsorption layer P_a (typically $P_g < 100P_a$), no significant adsorption will take place. There are, however, low-pressure adsorptions, notably in the case of fine porous adsorbents, because of capillary effect. Chemisorption, on the other hand, can take place at much lower pressures and much higher temperatures than physical adsorption.

Thickness of reaction layers — At equilibrium, a physical adsorption layer can be several molecules thick, whereas a chemisorption layer can only be one molecule thick, because the newly formed compound layer prevents the further reaction of the adsorbent and the adsorbate. However, physical adsorption can continue after the chemisorption layer is completed. Therefore, this criterion may be deceptive, because both physical adsorption and chemisorption occur at the same adsorption interface.

13.1.2 Absorption

Absorption is a process involving bulk penetration of gases into the structure of the solid or liquid by means of diffusion. From a theoretical or an experimental point of view, the difference between adsorption and absorption is sometimes confusing. Generally, one can distinguish between adsorption and absorption using one of the following two criteria:

Site of reaction — Absorption is a *volumetric* phenomenon, whereas adsorption is a *surface* phenomenon. For example, when ammonia gas passes through a water spray, absorption occurs. The gaseous ammonia molecules will be captured by the water droplets. Each captured ammonia molecule dissolves in the body of a water droplet, rather than simply clinging to its surface. Eventually, the entire water droplet volume will contain ammonia, until an equilibrium concentration in the water droplets is reached. Conversely, activated carbon may also be used to take up the ammonia in an air stream. In this case, the ammonia is held on the surface of the activated carbon by adsorption.

State changes — An absorption process may result in a state change (from solid to liquid) of the adsorbent, whereas an adsorption process will not change the state of the adsorbent. For example, when alumina is specially prepared, or when silica is produced in its gel form, these substances can be used to take up moisture from the air at room temperatures. The moisture is captured but may be removed by raising the temperature or by distillation. On the other hand, granular calcium chloride may be used as a desiccant. During the moisture-removal process, the calcium chloride is converted into a hydrate and reaches a saturation point at $CaCl_2 \cdot 2H_2O$. After this saturation, further moisture tends to cause the calcium chloride to lose its crystalline shape and dissolve into the water it adsorbed, that is, change the state from solid to liquid. Thus, moisture removal by alumina or silica gel will be adsorption, and by calcium chloride will be absorption.

13.1.3 Sorption

When both adsorption and absorption occur simultaneously, it is called *sorption*. In practice, sorption is a common phenomenon, although in some cases adsorption may be predominant, and in other cases absorption may be predominant. In the case of water condensation, when water vapor condenses on a dry board surface, adsorption occurs at the surface. At the same time, however, condensed water vapor diffuses into the board structure — an absorption process.

13.1.4 Catalytic Conversion

Sometimes gaseous pollutants can be changed into benign substances via chemical reactions. Catalytic conversion uses a *catalyst*, a substance that can increase the chemical reaction rate of other reactants but does not undergo permanent changes itself, to accelerate the reaction and remove pollutants from the air. Heterogeneous catalysts supported on high surface porous areas consist of several subprocesses, such as chemical reaction, pore diffusion, and mass transfer. Catalytic conversion has been widely used in gas emission control for combustion processes, as well as in automobiles.

13.2 ADSORPTION PRINCIPLES

At the surface of a solid or liquid, the van der Waals forces are unbalanced. The attractive force normal to the surface tends to grab adjacent molecules of gaseous or liquid state. Thus, when a gas is allowed to reach its equilibrium at a solid or liquid surface, the gas concentration is always found to be higher in the immediate vicinity of the surface than in the free gas phase.

Two primary variables affect the adsorption efficiency :

* The surface area of the adsorbent
* The affinity of the adsorbent for the adsorbate

Because adsorption occurs at the surface, a large surface area of the adsorbent is essential to increase the adsorption efficiency. For gas adsorption, the surface of the adsorbent's internal pores is more important than the apparent outside area of the adsorbent. When the pores' diameters are sufficiently small, say only several times the diameter of the adsorbent gas molecules, an enormous surface area will be available for adsorption. For example, charcoal contains an effective adsorption surface area of 10^5 to 10^6 m^2/kg, compared with the apparent surface area of 0.05 to 0.5 m^2/kg. In combustion air pollution control, molecular sieves have been developed to remove SO_2, NO_x, and H_g emissions. Such molecular sieves are made of metal aluminosilicates with very fine pore diameters of a similar order to the molecular diameters of the adsorbant gases, typically around 3 to 10 Å (10^{-10} m). These metal sieves have extremely high gas removal efficiency, higher than 99% for SO_2, for example. However, due to these very fine pore diameters, the sieve can become saturated with the adsorbate quickly and must be regenerated frequently using heated purging gas. The adsorption and regeneration processes are expensive and require frequent maintenance. Therefore, such fine-pore molecular sieves are not likely to be economical for large-volume air cleaning that has either a large flow rate or a high pollutant concentration in the airflow.

Adsorbent affinity for a given adsorbate — that is, the attraction between the adsorbent and adsorbate — is also required for adsorption. For example, charcoal (activated carbon) has an affinity for hydrocarbons. Thus, activated carbon is an ideal adsorbent for many hydrocarbon gases, such as CO, CO_2, and many organic compounds. Silica gel, on the other hand, has an affinity for water, and can thus can be used for adsorbing moisture from the air.

Activated carbon is made by the carbonization of coal, wood, fruit pits, or coconut shells, and is activated by treatment with hot air or steam. Activated carbons can be made in many forms, depending on the need: pellet and granular forms are common for adsorption beds; fibrous structures are common for air filters. Silica gel is a granular product made from gel precipitation by sulfuric acid treatment of a sodium silicate solution. It is commonly used in drying air or other gases below the temperature of 250°C. Both activated carbon and silica gel can be regenerated by heating between 175 and 350°C.

Affinities can also be designed. When pore diameters are changed, a molecular sieve can also change its adsorption affinities. When the pore diameter is 30 Å, some light gases, such as NH_3 and H_2O, can be adsorbed. When the pore diameter is increased to 40 Å, larger molecules, such as CO_2, SO_2, H_2S, C_2H_4, and C_2H_6, can be adsorbed. Further increasing the pore diameters can enable the adsorption of large organic molecules, such as benzene (C_6H_6), phenol (C_6H_6O), and Toluene (C_7H_8).

Although numerous theories have been developed about adsorption phenomena, none of these theories at the present can be generally applied to a wide range of experimental data.[1, 2] A great deal of experimental adsorption data, however, has been collected at equilibrium. For example, the amount of adsorbate adsorbed per unit of adsorbent at equilibrium is measured vs. the partial pressure of the adsorbate. In general, these studies have found that increasing the partial pressure of the adsorbate at a given temperature results in increased adsorption, whereas increasing the temperature of the adsorbate in the air at a given partial pressure results in decreased adsorption. This explains why many regeneration processes involve heating the adsorbent. It is also generally true that adsorption is more efficient for adsorbates with higher molar mass than for those with lower molar mass under the same partial pressure and temperature. This implies that heavier gases or vapors are easier to adsorb than lighter gases.

13.2.1 Adsorption Isotherm

An *adsorption isotherm* is a plot of adsorbent capability vs. the adsorbate concentration or partial pressure at a given temperature. Adsorption capability can be measured in terms of capacity (gram of adsorbate adsorbed per gram of adsorbent at saturation, g/g), or the total volume or mass of adsorbate adsorbed. Evidently, the efficiency of an adsorption process is closely related to its adsorbent's capacity and adsorbate's concentration.

Physical and thermodynamic properties of different gases are particularly useful for engineers and scientists who design adsorption systems, and analysis of isotherms can yield important data. Yaws et al. studied the interaction between activated charcoal (the most commonly used adsorbant) and volatile organic compounds (VOCs), which are primary pollutants in many situations.[3] Here, the adsorption capacity of activated charcoal for 243 VOCs and the adsorption capacity of activated carbon as a function of the VOC concentration was developed:

$$\log_{10} M_{xs} = a_1 + a_2 \log_{10} C + a_3 (\log_{10} C)^2 \tag{13.1}$$

where M_{xs} is the adsorption capacity at equilibrium, in grams, of adsorbate (VOC) per 100 g of adsorbent (carbon); C is the concentration of VOC in ppmv; and a_1, a_2, and a_3 are correlation constants for specific VOCs in Appendix 8. The correlation constants in Appendix 8 were determined from regression of available literature data.

In actual carbon air-cleaning devices, such as adsorption beds or adsorption filters, equilibrium absorption is rarely achieved. The real adsorption capacity for plant operating conditions is commonly 30 to 40% of equilibrium. Therefore, the total carbon requirement for an adsorption system is recommended to be 3 to 4 times of the value calculated from the equilibrium capacity (Equation 13.1). This design of adsorption capacity is sometimes called *working adsorption capacity*. Factors that can reduce the adsorption capacity include loss of adsorption zone, moisture in the air, heat waves, or residue moisture in the carbon. The adsorption isotherms of activated carbon for selected VOCs using Equation 13.1 and Appendix 8 are shown in Figure 13.1 on page 398.

Example 13.1: In an automobile assembling shop, the concentration of n-butanol ($C_4H_{10}O$) in the room air is 5 ppmv. The density of n-butanol is 3.06 kg/m³ under standard room air conditions. A carbon filter bed is used for air cleaning, and the airflow rate is 0.1 m³/s through the filter. Determine the following:

a. The adsorption capacity of an activated carbon filter
b. The total carbon mass needed for the bed, assuming the working adsorption capacity is 40% of the saturated adsorption and the bed service life is one year

Solution:

a. From Appendix 8, we have the adsorption constants of $C_4H_{10}O$:

$$a_1 = 0.89881$$

$$a_2 = 0.32534$$

$$a_3 = -0.03648$$

From Equation 13.1, we have

$$\log_{10} M_{xs} = a_1 + a_2 \log_{10} C + a_3 (\log_{10} C)^2$$

$$= 0.89881 + 0.32534 \log_{10}(5) - 0.03648[\log_{10}(5)]^2 = 1.1084$$

$$M_{xs} = 10^{1.1084} = 12.8 \left(\frac{g \ of \ C_4H_{10}O}{100 \ g \ of \ carbon} \right)$$

b. The total amount of n-butanol passing through the carbon bed in one year is

$$M = QC\rho_g t = 0.1 \frac{m^3 air}{s} \times 0.000005 \frac{m^3 n - bu}{m^3 air} \times 3.05 \frac{kg}{m^3 n - bu} \times (365 \times 24 \times 3600)s$$

$$= 48.1 \ (kg)$$

By taking into account the 40% actual adsorption factor, the actual carbon mass needed can be calculated as follows:

$$M_{carbon} = \frac{M}{M_{xs}} \times \frac{1}{40\%} = 48.1 \ (kg \ of \ n - bu) \div 0.128 \frac{kg \ of \ n - bu}{kg \ of \ carbon} \times 2.5$$

$$= 939.5 \ (kg \ of \ carbon)$$

13.3 ADSORPTION WAVE

To determine the rate at which an adsorbent will become saturated, we use a concept called the *adsorption wave*. Let us consider a polluted air stream passing through a stationary bed of adsorbent. The pollutant gas in the air has an initial concentration of C_0. As the polluted air stream passes through the bed and the time increases, the amount of pollutant gas adsorbed on the bed increases. However, this adsorption process does not occur uniformly throughout the bed. The layer near the air inlet adsorbs most of the pollutant gas initially and soon becomes saturated. The remaining portion of the bed adsorbs little pollutant gas and can be essentially considered gas free. As shown in Figure 13.2, the dark areas are the saturated adsorbent, and the dotted areas are the areas of active adsorption, or the *adsorption zone*. The pollutant concentrations are assumed to be C_0 at the beginning of the adsorption zone and 0 at the end of the adsorption zone. In the adsorption zone, the pollutant concentration is essentially reduced sharply from C_0 to zero in an S-shape, hence the term *adsorption wave*.

The character of the adsorption wave depends on the adsorption capacity of the adsorbent and the flow rate or retention time of the polluted air. As more polluted air enters the adsorption bed, a longer section of the initial layers of the bed become

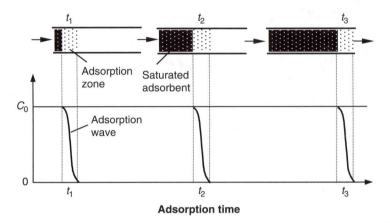

Figure 13.2 Progressive adsorption wave through a stationary adsorption bed. The dark areas are saturated adsorbent, and the dotted areas are adsorption zones.

saturated, and thus the adsorption zone moves toward the outlet. At some later time, the leading edge of the adsorption zone reaches the end of the adsorption bed length L, and concentration of the pollutant at the exit air stream increases rapidly. This increase of pollutant concentration at the exit is known as the *breakthrough point*, or *break point*. Arbitrarily, the break point can be determined based on the measurable concentration of the pollutant in the exit air stream. Such concentrations may vary, depending on the application and the threshold limits of allowable pollutant emission standards. For many toxic chemicals, a measurable concentration at the exit, even if it is less than 1% of the inlet concentration, can be defined as the break point.

The particular shape of the adsorption wave is extremely important. The pollutant concentration within the adsorption zone could be very steep or quite flat. As shown in Figure 13.3, the gas-phase pollutant concentration C (in mass of pollutant per volume of air), or the solid-phase pollutant concentration C_x (in mass of pollutant per mass of adsorbent), is plotted vs. the length of the adsorption bed, x, along the total bed length L. Concentration C varies from zero to C_0, and C_x varies from zero to C_{xs}, saturated concentration in the adsorbent. Assuming that the adsorbent behind the adsorption zone is saturated, the concentrations are C_0 and C_{xs} at the beginning of the adsorption zone (x_1) and are both essentially zero at the end of the zone (x_2). The polluted air moves through the bed with a velocity U relative to the adsorbent; the adsorption zone moves toward the end of the bed with a velocity V_{az}, relative to the bed. From a mathematical point of view, U is much greater than V_{az} (by several orders of magnitude), and we can let U represent the velocity of airflow relative to the fixed adsorption zone from left to right, and let V_{az} represent the velocity of the adsorption bed passing through the fixed zone from right to left (Figure 13.3).

To develop the governing equations for the adsorption wave, we look at the mass balance of the adsorbate (pollutant) for the adsorption zone between x_1 and x_2, as shown in Figure 13.3. The rate of pollutant transferred into the adsorption zone by the airflow must equal the rate of that transferred out and adsorbed by the adsorbent. In many air-quality problems, the pollutant concentration in the air is very low, and

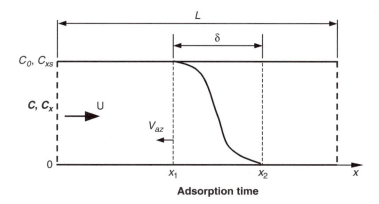

Adsorption time

Figure 13.3 A schematic description of an adsorption wave in a stationary adsorption bed.

the total mass flow rate is practically the air mass flow rate. In an adsorption zone, the pollutant concentration in gas phase at x_2 is essentially zero, and the pollutant concentration in solid phase at x_1 is essentially C_{xs}. Making those simplifications, the pollutant mass balance can be written as

$$QC_1 = \rho_x A V_{az} C_{xs}$$ (13.2)

where Q is the volumetric airflow rate in m³/s, C_1 is the pollutant concentration in the inlet airflow in kg/m³ of air, ρ_x is the density of the adsorbent in kg/m³, A is the cross-sectional area of the adsorption bed, C_{xs} is the saturated solid pollutant concentration in the adsorbent, and V_{az} is the velocity of the adsorption bed relative to the adsorption zone.

General solutions for Equation 13.2 are possible for a specific adsorption process, provided that the relationship between C_1 and C_{xs} is experimentally determined. In the following discussion, we use as an example the adsorption of VOCs by carbon, as described in Equation 13.1. In order to solve V_{az} explicitly in Equation 13.2, we must solve Equation 13.1 for the relationship between C_1 and C_{xs}. For many adsorption processes, the second-order term $(log_{10}C)^2$ in Equation 13.1 is negligible. For pollutant concentrations lower than 50 ppmv, this assumption would result in an error of less than 5%. For most indoor air quality problems, gaseous pollutants (except carbon dioxide) are generally much lower than 50 ppmv. Thus, Equation 13.1 can be rewritten as follows:

$$C = 10^{\left(-\frac{a_1}{a_2}\right)} M_{xs}^{\left(\frac{1}{a_2}\right)}$$ (13.3)

where a_1 and a_2 are adsorption correlation constants for a specific adsorption process and can be found in Appendix 8 or experimentally determined.

Note that in Equation 13.1 and Equation 13.3, C is in ppmv and M_{xs} is in gram of pollutant per 100 g of adsorbent. M_{xs} can also be interpreted as the saturated concentration of solid-phase adsorbate (pollutant) in the adsorbent. In Equation 13.2, C_l is in kilogram of pollutant per cubic meter of air, and C_{xs} in gram of pollutant per gram of adsorbent, thus $C_l = \rho_g C$ and $C_{xs} = 0.01 M_{xs}$, where ρ_g is the density of the pollutant gas. The pollutant gas is air in an air-quality problem. Substituting expressions of C_l and C_{xs} into Equation 13.3 gives the Freundlich equation:

$$C_1 = \kappa C_{xs}^{\frac{1}{\zeta}}$$ (13.4)

where κ must have the same units as C_1, and ζ is a pure number. For adsorption of VOCs in carbon and from Equation 13.1

$$\kappa = 10^{\frac{2-a_1}{a_2}} \rho_g \left(\text{in } \frac{\text{mass of adsorbate}}{\text{volume of air}} \right)$$ (13.5)

$$\zeta = \frac{1}{a_2}$$ (13.6)

Note that κ and ζ would have different values for different adsorbates and adsorbents. They could also vary even for the same adsorbate and adsorbent under different temperatures. Therefore, care must exercised when determining the κ and ζ in a specific adsorption process.

Because we have assumed that equilibrium of pollutant exists at position x_1, then $C_1 = C_0$. Substituting Equation 13.4 into Equation 13.2 gives

$$QC_1 = \rho_x A V_{az} \left(\frac{C_1}{\kappa} \right)^{\zeta}$$ (13.7)

Solving Equation 13.7 for the velocity of adsorption wave gives

$$V_{az} = \frac{Q}{\rho_x A} \kappa^{1/\zeta} C_0^{\left(\frac{\zeta-1}{\zeta} \right)}$$ (13.8)

or

$$V_{az} = \frac{U}{\rho_x} \kappa^{1/\zeta} C_0^{\left(\frac{\zeta-1}{\zeta} \right)}$$ (13.9)

The adsorption wave velocity therefore depends on the airflow velocity across the bed, the adsorbent density, and the adsorption wave shape characterized by κ and ζ.

From Figure 13.3, the thickness of the adsorption wave ($\delta = x_2 - x_1$) is also an important characteristic of the wave. In order to determine δ, we consider the mass transfer in the gas phase in a differential thickness dx of the adsorption zone. As the pollutant gas passes through a distance of dx, the pollutant gas concentration changes by dC. The rate of mass transfer of the pollutant gas $\dot{m}_g$, into a solid phase is the product of the volumetric airflow rate and the change of pollutant gas concentration, namely

$$\dot{m}_g = QdC \qquad (13.10)$$

where C should be in mass per volume of air.

On the other hand, the change of mass transfer rate $\dot{m}_g$ can also be derived from the general theory of mass transfer. The mass flow rate of pollutant is proportional to the difference in pollutant concentration between the actual value in the gas phase, C, and the equilibrium value on the surface of the solid phase, C_e. Further, the mass transfer rate also depends on the volume of the bed, Adx, and the mass transfer coefficient K_x. The mass transfer coefficient takes into account the film resistance and the effective interfacial area of the adsorbent. Values of K_x vary over a typical range of 5 to 50 s^{-1}. Thus, the mass transfer rate of the pollutants over the dx can be written as

$$dm_g = K_x A(C_e - C)dx \qquad (13.11)$$

Note that within the adsorption zone, the equilibrium concentration of pollutant gas is approximated by the Freundlich equation:

$$C_e = \kappa C_x^\zeta \qquad (13.4)$$

Substituting Equation 13.4 into Equation 13.11, and equating Equation 13.10 and Equation 13.11, gives

$$QdC = - K_x A(C - \kappa C_x^\zeta)dx \qquad (13.12)$$

From Equation 13.2, assuming equilibrium within the adsorption zone at a given moment

$$C_x = \frac{QC}{\rho_x A V_{az}} \qquad (13.13)$$

Substituting Equation 13.8 for V_{az} into Equation 13.13 gives

$$C_x = \frac{C_x^\zeta}{\kappa} C_0^{1-\zeta} \qquad (13.14)$$

Substituting Equation 13.14 into Equation 13.12 gives the relationship between C and x within the adsorption zone:

$$QdC = -K_x A(C - C^\zeta C_0^{1-\zeta})dx \tag{13.15}$$

or

$$\frac{K_x A}{Q} dx = -\left(\frac{dC}{C - C^\zeta C_0^{1-\zeta}}\right) \tag{13.16}$$

From Equation 13.16, the thickness of the adsorption zone approaches zero if the mass transfer coefficient at the surface approaches infinity. The zone thickness may become infinitely long when $\zeta = 1$. In this case, the pollutant concentration in the main air stream is the same as the concentration inside the film at the solid surface at a given position, and thus there is no driving force for mass transfer from the gas phase to the solid phase, and the equation gives an infinite depth of the adsorption zone. Because ζ normally is not zero for a usable adsorbent, Equation 13.16 predicts the adsorption zone thickness with a reasonable success if the integration limits are modified as follows.

The width of an adsorption zone is limited by x_1 and x_2, and its concentration by C_0 and 0, corresponding to x_1 and x_2. Theoretically, the concentration can never reach zero. A $C = 0$ value will result in a infinite zone depth. For mathematical convenience, we use the adsorption efficiency, $\xi_x = 1 - C/C_0$, as the variable varying from 0 to 1, corresponding to C varying from C_0 to 0. Then Equation 13.16 can be integrated as

$$\int_{x_1}^{x_2} \frac{K_x A}{Q} dx = -\int_0^1 \frac{d\xi_x}{\xi_x - \xi_x^\zeta} = -\int_0^1 \frac{d\xi_x}{\xi_x(1 - \xi_x^{\zeta-1})} \tag{13.17}$$

The value on the right side of the Equation 13.17 is undefined with the limits [0, 1]. However, if the limit values are close to 0 and 1, the integral is defined. For example, say that we take ξ_x to be within 1% of the limiting values (i.e., 0.01 and 0.99). This is equivalent to defining the adsorption zone depth such that C approaches within 1% of its limiting values of 0 and C_0. Based on this arbitrary selection of adsorption efficiency and integrating limits, Equation 13.17 becomes

$$\frac{K_x A}{Q} \delta = \ln\left(\frac{0.99}{0.01}\right) - \frac{1}{\zeta - 1} \ln\left(\frac{1 - 0.01^{\zeta-1}}{1 - 0.99^{\zeta-1}}\right) \tag{13.18}$$

If one wishes the adsorption efficiency ξ_x to be within 5% of limiting values, then the constant values 0.01 and 0.99 in Equation 13.18 can be replaced by 0.05 and 0.95, respectively.

The general shape of the gas concentration vs. distance can be integrated from Equation 13.16 between the limits of $[x_1, x]$ and $[0.99C_0, C]$:

$$x = x_1 + \frac{Q}{K_x A}\left(\ln \frac{0.99C_0}{C} + \frac{1}{\zeta - 1}\ln \frac{1-(C/C_0)^{\zeta-1}}{1-0.99^{\zeta-1}} \right) \qquad (13.19)$$

Because the adsorption zone width δ is defined by Equation 13.18, and the adsorption bed velocity V_{az} is defined by Equation 13.8, the time for breakthrough of the adsorption bed, t_x, can be calculated. If we assume that the time required to establish the adsorption zone to its full thickness at the inlet is zero, it gives

$$t_x = \frac{L-\delta}{V_{az}} \qquad (13.20)$$

Equation 13.20 gives a conservative estimation of the operating time of an adsorption bed before breakthrough occurs and regeneration or replacement of adsorbent is required. Note that the value of δ depends on the ξ_x value we choose. For example, Equation 13.18 is based on a ξ_x value of 0.01 at x_1, the leading edge of the adsorption zone. Other ξ_x values can be chosen to determine the breakthrough time.

Example 13.2: *In an automobile assembling shop, the concentration of n-butanol ($C_4H_{10}O$) in the room air is 0.000013 kg/m^3. An activated carbon filter is used in an air recirculation system to remove the n-butanol. The bulk density of the activated carbon is 400 kg/m^3, and the mass transfer coefficient of the adsorbent $K_x = 20$ 1/s. The airflow rate of the recirculation system is 100 l/s. The cross-sectional area of the bed is 2 m^2. The designed adsorption efficiency is $\xi_x = 0.9$. That is, the breakthrough point of the bed is considered to be when the outlet n-butanol concentration reaches 10% of the inlet concentration. Determine the following:*

    a. The speed of the adsorption zone
    b. The thickness of the adsorption zone
    c. The length of the bed if the bed is to be regenerated or replaced every two months

Solution:

    a. To determine the speed of adsorption zone, Equation 13.8 should be used. From Appendix 8, we have the adsorption constants of $C_4H_{10}O$:

$$a_1 = 0.89881$$

$$a_2 = 0.32534$$

$$a_3 = -0.03648$$

The density of polluted air at standard conditions, $\rho_g = 1.2$ kg/m³, is

$$\kappa = 10^{\frac{2-a_1}{a_2}} \rho_g = 10^{\frac{2-0.89881}{0.32534}} \times 1.2 = 2,910$$

$$\zeta = \frac{1}{a_2} = \frac{1}{0.32534} = 3.074$$

Substituting the values into Equation 13.8 gives the speed of the adsorption zone:

$$V_{az} = \frac{Q}{\rho_x A} \kappa^{\frac{1}{\zeta}} C_0^{\left(\frac{\zeta-1}{\zeta}\right)} = \frac{0.1}{400 \times 2} \times 2,910^{\frac{1}{3.074}} \times 0.0000128^{\frac{3.074-1}{3.074}} = 8.35 \times 10^{-7} \ (m/s)$$

b. The thickness of the adsorption zone can be calculated from Equation 13.18, noting that the lower and upper limits for ξ_x are 0.1 and 0.9, respectively, in this case.

$$\delta = \frac{Q}{K_x A}\left[\ln\left(\frac{0.9}{0.1}\right) - \frac{1}{\zeta-1}\ln\left(\frac{1-0.1^{\zeta-1}}{1-0.9^{\zeta-1}}\right)\right]$$

$$= \frac{0.1}{20 \times 2}\left[2.197 - \frac{1}{3.074-1}\ln\left(\frac{1-0.1^{3.074-1}}{1-0.9^{3.074-1}}\right)\right] = 0.00355 \ (m)$$

c. From Equation 13.20, the breakthrough time is designed to be 2 months, that is

$$t_x = 2 \times 30 \times 24 \times 3,600 = 5.184 \times 10^6 \ s$$

So, the needed length of the filter bed is

$$L = V_{az}t_x + \delta = 8.35 \times 10^{-7} \times 5.184 \times 10^6 + 0.00355 = 4.33 \ m$$

13.4 REGENERATION OF ADSORBENTS

Unlike many filters that must be replaced when they are clogged, adsorbents in an adsorption bed can be regenerated when they reach the breakthrough point. During regeneration, the pollutant-laden gas flow must be stopped or switched to another, unsaturated adsorption bed. A regeneration fluid is passed through the saturated bed, and the pollutant gases adsorbed on the adsorbent are desorbed into the regeneration fluid. At higher temperatures, gases tend to be volatilized and easier to detach from the adsorbent. Thus regeneration fluids are usually heated. Steam and hot air are common regeneration fluids. However, one disadvantage of hot fluid regeneration is that the adsorption bed is heated substantially during the process, and subsequent adsorption cannot be performed until the bed has cooled down, if there is to be no loss in adsorption efficiency.

The regeneration process of an adsorption bed is essentially adsorption reversed. During regeneration, the adsorption zone becomes a desorption zone, and the desorption speed is in the opposite direction of the adsorption speed. Because it a reverse process, the desorption speed can be written in the same format as the adsorption speed:

$$V_{rz} = \frac{Q_r}{\rho_x A} \kappa_r^{1/\zeta_r} C_r^{\left(\frac{\zeta_r - 1}{\zeta_r}\right)} \tag{13.21}$$

where Q_r is the flow rate of the regeneration fluid flow in m^3/s, κ_r is the concentration factor in kg/m^3 (the same units as C_r), C_r is the pollutant concentration at the exit of the bed during the desorption, and ζ_r is a pure number for regeneration of an adsorption bed. The subscript $_r$ refers to the corresponding properties during the regeneration process.

It is important to note that the values of κ and ζ can be substantially different for an adsorption and its corresponding desorption process. Such disparity can be caused by differences in the temperature, carrier fluid, and adsorbent properties of the absorbant between the desorption and adsorption conditions. The pollutant concentration at the exit of the bed, C_r, during desorption is usually different from the pollutant concentration at the inlet of an adsorption bed. However, the equilibrium saturation values for the adsorbate, C_x, are the same for both the adsorption and the desorption process. Because the bed is saturated at the beginning of the desorption process, and because desorption is essentially a reverse process of adsorption, from Equation 13.4, we can rewrite the pollutant concentrations at the outlet of a desorption bed and the inlet of the adsorption bed as

$$C_r = \kappa_r C_{xs}^{\zeta_r} \tag{13.22}$$

$$C_0 = \kappa C_{xs}^{\zeta} \tag{13.23}$$

Combining Equation 13.22 and Equation 13.23 gives the relationship between the concentrations at the outlet of the desorption and at the inlet of the adsorption:

$$C_r = \kappa_r \left(\frac{C_0}{\kappa}\right)^{\zeta_r/\zeta} \tag{13.24}$$

Substituting Equation 12.24 into Equation 12.21, gives

$$V_{rz} = \frac{Q_r \kappa_r}{\rho_x A} \left(\frac{C_0}{\kappa} \right)^{\left(\frac{\zeta_r - 1}{\zeta} \right)} \qquad (13.25)$$

Like the analysis of an adsorption zone, the width of a desorption zone, δ_r, can be established. For example, taking regeneration efficiency to be within 1% of the limiting values (i.e., 0.01 and 0.99) is equivalent to defining the desorption zone depth such that C approaches within 1% of its limiting values of 0 and C_r. Thus Equation 13.18 can be rewritten as

$$\frac{K_{xr} A}{Q_r} \delta_r = \ln\left(\frac{0.99}{0.01} \right) - \frac{1}{\zeta_r - 1} \ln\left(\frac{1 - 0.01^{\zeta_r - 1}}{1 - 0.99^{\zeta_r - 1}} \right) \qquad (13.26)$$

where K_{xr} is the mass transfer coefficient for the regeneration process (usually different from K_x during the adsorption process). As with Equation 13.18, if one wishes the regeneration efficiency to be within 5% of limiting values, then the constant values 0.01 and 0.99 in Equation 13.26 can be replaced by 0.05 and 0.95, respectively.

The regeneration time then becomes

$$t_r = \frac{L - \delta_r}{V_{rz}} \qquad (13.27)$$

The desorption zone width is much smaller than the bed length, so the regeneration time t_r can be approximated as

$$t_r = \frac{L}{V_{rz}} = \frac{\rho_x A L}{Q_r \kappa_r} \left(\frac{\kappa}{C_0} \right)^{\left(\frac{\zeta_r - 1}{\zeta} \right)} \qquad (13.28)$$

13.5 DESIGN CONSIDERATIONS FOR ADSORPTION PROCESSES

Adsorption process designs vary greatly with different applications. Technically, the type of adsorber to be used should be determined first, and then parameters that affect the adsorption efficiency should be calculated. There are four adsorber operation modes:

Regenerative adsorbent — The bed can be regenerated when it becomes saturated.
Nonregenerative — The adsorbent must be replaced when it becomes saturated.
Batch operation — The adsorption process can be interrupted, for example, to replace the adsorbent.
Continuous operation — The adsorption process cannot be interrupted.

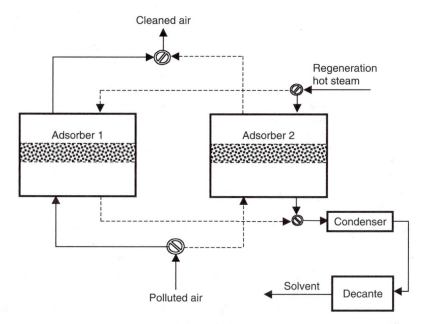

Figure 13.4 A two-pass regenerative adsorber.

Generally, for a gaseous pollutant less than 1 ppmv entering the bed, a nonregenerative adsorber may be used. The adsorbent and the adsorbate can be discarded if they are not volatile. When a gaseous pollutant is several ppmv or higher, a regenerative adsorber may more economical. Because the adsorbent can become saturated quickly and frequent replacement is not practical, regeneration is a better alternative. When an adsorption process cannot be interrupted and the pollutant must be removed continuously from the air stream, a continuous operation adsorber is needed. Such continuous operation usually requires a multipass adsorber. Figure 13.4 shows a sketch of a two-pass regenerative adsorber. To ensure that the adsorption process is uninterrupted, one adsorber is adsorbing and the other is desorbing. The solid lines show the areas where the process is active, and the dashed lines show the areas where the process is not active.

Many factors affect the efficiency of an adsorber. Often, capital and operating cost are the major limiting factor in achieving a desirable efficiency.[4] However, in the design of a high-efficiency adsorber, at a minimum the following design parameters should be considered:

A sufficient retention time for interaction between the pollutant gas and the adsorbent should be provided. It is detrimental to adsorption efficiency if the gas moves quickly through the bed. Typical retention time within the adsorber should be between 0.6 and 6 seconds.[5, 6] For this retention-time limit, the face velocity at the gas inlet of the adsorber is recommended between 0.1 and 0.5 m/s. Note that the long retention time and low face velocity often result in a large volume within the adsorber. If small volume is a requirement, other types of gas removal, such as catalytic converters, should be considered.

For multigas pollutants, it is sometimes necessary to remove competing gases prior to the adsorber, especially for high concentrations. For example, for an adsorber designed for odor removal in an animal facility, ammonia may need to be removed first, because the odor concentration is much lower than the ammonia concentration, and the ammonia may be highly competitive with the odor.

A uniform gas distribution will ensure that interaction occurs between the pollutant gas and the adsorbent and will improve the efficiency of the adsorber.

The capability of regenerating the adsorbent after it reaches saturation should be supplied. Unless the pollutant gas concentration is very low and replacement of the adsorbent is inexpensive, regeneration should be considered.

13.6 ABSORPTION PRINCIPLES

The absorption process involves a mass transfer by means of molecular and turbulent diffusion. This process includes a pollutant A, mixed with a stagnant carrying gas B and a stagnant absorbing liquid C. For example, in an absorption of ammonia, a carrying air stream (carrying gas B) with a concentration of ammonia gas (pollutant A) passes through a water stream (absorbing liquid C). The ammonia concentration in the air is absorbed into the water, and the air stream is cleaned. Ammonia in the water can be distilled and recovered.

Absorption is a basic chemical engineering unit operation. In air cleaning, it is also commonly referred to as *wet scrubbing*. When polluted air is brought into contact with the scrubbing liquid (absorber), the pollutant gases are absorbed by the liquid and the air is cleaned. Wet scrubbing of particulate matter using a spray tower and venturi scrubber is based on the principles of impaction and coagulation, as described in Chapter 11. Wet scrubbing of gaseous pollutants is based on diffusion. One should not use the impaction and coagulation principles to analyze the cleaning efficiency of gaseous pollutants, because for gases transportation and mass transfer are governed by gas kinetics, whereas for particles, transportation and mass transfer are governed by particle mechanics (except for very fine particles).

The following factors determine the rate of absorption:

Pollutant concentration gradients at the interfaces of liquid and gas — The concentration gradient provides a primary driving force for absorption. The larger the gradients, the faster the absorption will be.

Interface surface area between the scrubbing liquid and gas — The absorption rate is proportional to the interface area. Increasing the interface area between the liquid and the gas can be achieved by dispersing the scrubbing liquid into the gas flow. The dispersing mechanisms include the following:

Packed towers, which employ shapes such as rings and saddles for packing, so that the total wetted surface area is increased

Spray towers, which increase the liquid flow rate and/or spray small liquid droplets into the gas flow

Venturi absorbers, which force the gas flow to pass through a high-velocity liquid jet

Mass diffusion coefficient of the pollutant gas — This is a given property for a given application. If the diffusion coefficient of the concerned pollutant gas is too small, cleaning mechanisms other than absorption must be considered.

Solubility of the pollutant gas with respect to the scrubbing liquid — This is one of the most important parameters determining the absorption rate. For a particular pollutant gas, the scrubbing liquid must have a high solubility for it. For example, water is an excellent absorber for ammonia gas because the solubility of ammonia to water is very high.

Turbulence intensity of the gas flow — Turbulence intensity can increase substantially the interaction between the absorber and the polluted gas. For example, spraying liquid counter to the gas flow can disturb the gas flow pattern and increase the local turbulence intensity near the liquid droplets. Different shapes of packing material can also increase the turbulence intensity near the wetted surface.

The reverse of absorption is called *desorption* or *stripping*. In a stripping process, the pollutant in liquid state is transferred to surrounding gases, and the principles are the same as for absorption. For example, when hot steam passes through an organic oil, some volatile components will be vaporized from the liquid oil and carried away by the steam, leaving only nonvolatile oil.

13.7 MASS BALANCE FOR WET SCRUBBERS

13.7.1 Henry's Law and Gas–Liquid Equilibrium

It is important to understand the equilibrium distribution of the gas fraction and the liquid fraction of a pollutant in an absorption process. Consider an absorption process in which a pollutant A is mixed with a stagnant carrying gas B and makes contact with a stagnant absorbing liquid C. The mass transfer of pollutant A from a gas phase into a liquid phase, in mole fraction (or mass fraction), is described by Henry's law:

$$P_A = Hx_A \qquad (13.29)$$

where P_A is the gas phase partial pressure of the pollutant A in Pa, x_A is the mole fraction of the pollutant A in liquid phase, and H is the Henry's law constant in Pa/mole fraction for the given system. Henry's law constants for some typical gases in water are listed in Table 13.1 (from the National Research Council, 1929, with converted SI units).[7]

Table 13.1 Henry's Law Constants for Some Typical Gases in Water

Temperature (°C)	$H \times 10^{-6} \left(\dfrac{Pa}{\text{mole fraction of the system}} \right)$									
	CO_2	CO	C_2H_6	C_2H_4	He	H_2	H_2S	CH_4	N_2	O_2
0	72.8	3520	1260	552	12900	5790	26.8	2240	5290	2550
10	204	4420	1890	768	12600	6360	36.7	2970	6680	3270
20	142	5360	2630	1020	12500	6830	48.3	3760	8040	4010
30	186	6200	3420	1270	12400	7290	60.9	4490	9240	4750
40	233	6960	4230		12100	7510	74.5	5200	12400	5350

From Equation 13.29, x_A is proportional to the partial pressure of the pollutant A in gas phase and not related to the total pressure. For most air-quality problems, the pollutant concentrations are low (typically less than 5% of the system mass), thus the P_A is low, and Henry's law can be used as a first estimation. In Equation 13.29, the Henry's law constant H does not depend on the total pressure P of the pollutant and the carrying gas. In many design problems, it is often more convenient to use the relationship between the mass or mole fraction of the pollutant in gas phase and liquid phase, which is a direct description of mass transfer between the two phases. When both sides of Equation 13.29 are divided by the total pressure P, Henry's law can be rewritten as

$$y_A = \frac{P_A}{P} = H' x_A \qquad (13.30)$$

where y_A becomes the mole fraction of pollutant A in gas phase. H' is the Henry's law constant, depending on total pressure P, and equals

$$H' = \frac{H}{P} = \frac{mole\ fraction\ of\ A\ in\ gas\ phase}{mole\ fraction\ of\ A\ in\ liquid\ phase} \qquad (13.31)$$

Example 13.3: *What is the mole fraction of dissolved oxygen in water under standard atmospheric conditions (20°C, 101.325 kPa), and what is the concentration of O_2 in mg of O_2 per liter of water, when the system (air and water) is in equilibrium?*

Solution: The Henry's law constant for O_2 in water at 20°C is 4010×10^6 Pa/mass fraction of the system (Table 13.1). Because air is an ideal gas and O_2 takes 21% of the volume in air, the partial pressure of O_2 is 21% of the standard atmospheric pressure:

$$P_{O2} = 0.21 \times 101,325\ (Pa) = 21,278\ Pa$$

Substituting H and P_{O2} values into Equation 13.29 gives

$$x_{O2} = \frac{P_{O2}}{H} = \frac{21,278}{4,010 \times 10^6} = 5.306 \times 10^{-6}\ (mole\ fraction\ in\ liquid\ phase)$$

The liquid phase here is the water and the dissolved oxygen. Therefore, 5.306×10^{-6} mole of oxygen is dissolved in one mole of the total system, which includes water and oxygen. In this case, the mole fraction of dissolved O_2 is so small compared with the water that the water can be treated as the whole system. Because the molar weight of O_2 is 16,000 mg/mole, and water is 10,000 mg/mole, one liter of water is approximately 1,000 g and thus 100 mole, the mass of dissolved oxygen in one liter of water (i.e., the concentration) is

$$C_{O2} = 5.306 \times 10^{-6} \left(\frac{mole\ of\ O_2}{mole\ of\ system} \right) \times 100 \left(\frac{mole\ of\ system}{liter\ of\ water} \right) \times 16,000 \left(\frac{mg\ O_2}{mole\ O_2} \right)$$

$$= 8.49 \left(\frac{mg\ O_2}{liter\ of\ water} \right)$$

This result is very close to the actual free oxygen content in the water at 20°C. Generally, Henry's law is sufficiently accurate for pollutant gases at low partial pressures (i.e., low concentrations). However, when the partial pressure of a pollutant gas is high, Henry's law may overestimate the absorption rate. The following example shows the equilibrium distribution of a gas ammonia–water system at 1 atmosphere pressure and 20°C (Figure 13.5). The experimental molar fraction is nonlinear with respect to partial pressure. As a first estimation, the molar fraction in liquid can be considered linear when the ammonia partial pressure is lower than 5 kP (or 5% in volume of the carrying gas, or 5% in mole fraction).

Equilibrium distribution for many gases is expressed in mole fraction in gas phase vs. mole fraction in gas liquid phase. To establish an equilibrium distribution of a pollutant mass transfer, consider a constant pressure system immersed in an isothermal bath so that the system can maintain a constant temperature and pressure (Figure 13.6a). The actual equilibrium distribution is usually in a concave shape. In order for a liquid to absorb a pollutant from a carrying gas, the molar fraction of the pollutant in the gaseous state must be high enough, compared with its molar fraction in the liquid state. When the process is above the distribution curve, absorption occurs. When the process is below the distribution, desorption occurs. Therefore, an absorption-unit operating line should be located above the equilibrium line. An

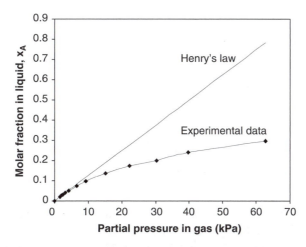

Figure 13.5 Equilibrium distribution for a gaseous ammonia–water system at 20°C (for Henry's law, $y_A = 1.2x_A$ for ammonia at low concentrations). Experimental data from Perry, J.H., ed., *Chemical Engineers' Handbook*, 5th ed., McGraw-Hill, New York, 1973.

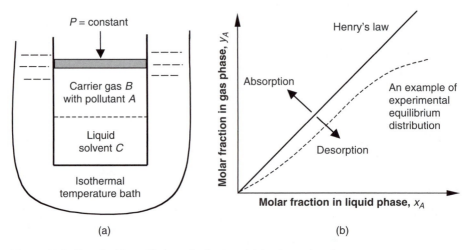

Figure 13.6 Gas–liquid equilibrium distribution. (a) Isothermal and constant-pressure setup. (b) Equilibrium distribution on molar/mass bases.

Table 13.2 Equilibrium Data for Ammonia–Water System

x_A, Ammonia in Liquid Phase	y_A, Ammonia in Gas Phase, $P = 101,325$ Pa	
	20°C	30°C
0	0	0
0.0126		0.0151
0.0167		0.0201
0.0208	0.0158	0.0254
0.0258	0.0197	0.0321
0.0309	0.0239	0.0390
0.0405	0.0328	0.0527
0.0503	0.0416	0.0671
0.0737	0.0657	0.105
0.0960	0.0915	0.145
0.137	0.150	0.235
0.175	0.218	0.342
0.210	0.298	0.463
0.241	0.392	0.597
0.297	0.618	0.945

Source: Perry, J.H., ed., *Chemical Engineers' Handbook*, 5th ed., McGraw-Hill, New York, 1973.

operating line is an important parameter in designing a multistage absorption scrubber, which will be further described in this section.

Equilibrium data can be found in several handbooks such as *Chemical Engineers' Handbook* (Perry, 1984) and *Handbook of Chemistry and Physics* (Lide, 2001).[8, 9] Table 13.2 lists equilibrium data, in mole fraction in gas phase vs. mole fraction in liquid phase, for an ammonia–water system at two temperatures.

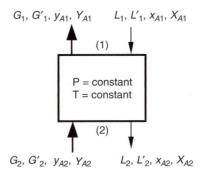

Figure 13.7 Mass flow diagram of a single-stage gas–liquid system.

13.8 SINGLE-STAGE COUNTERFLOW GAS–LIQUID SYSTEM

Let us again consider an absorption process including a pollutant A mixed with a stagnant carrying gas B and a stagnant absorbing liquid C. The word *stagnant* refers to that fact the carrying gas B is insoluble to the absorbing liquid C, and the absorbing liquid C does not vaporize into the carrying gas B. Thus, only the pollutant A can change its phase between gas and liquid. To determine the operating charac-teristics of a counterflow gas–liquid system, we start by analyzing the mass balance of the system, including the pollutant, the carrying gas, and the absorption liquid.

A single-stage counterflow gas–liquid system is sketched in Figure 13.7 and described using the following symbols: G is the total gas flow rate (including the pollutant gas and the carrying gas), in moles per unit time; G' is the stagnant carrying gas flow rate alone, in moles per unit time; y_A is the mole fraction of pollutant A in gas phase; Y_A is the mole ratio of pollutant A in gas phase. L is the total liquid flow rate (including the pollutant in liquid phase and the absorption liquid), in moles per unit time; L' is the stagnant absorption liquid flow rate alone, in moles per unit time; x_A is the mole fraction of pollutant A in liquid phase; X_A is the mole ratio of pollutant A in liquid phase. The subscripts $_1$ and $_2$ indicate the flow at which side of the system. In this system, the mole fractions and mole ratios for pollutant A in the gas and liquid phases have the following relationships:

$$X_A = \frac{x_A}{1 - x_A} \tag{13.31}$$

$$Y_A = \frac{y_A}{1 - y_A} \tag{13.32}$$

The mass balance for the pollutant A in this single-stage system can be expressed as the total moles of pollutant flow into the scrubber set equal to the total moles of the pollutant flow out the scrubber, both in gas and liquid phases:

$$L_1 x_{A1} + G_2 y_{A2} = L_2 x_{A2} + G_1 y_{A1} \qquad (13.33)$$

Or, the mass balance can be expressed in mole ratio as

$$L_1' X_{A1} + G_2' Y_{A2} = L_2' X_{A2} + G_1' Y_{A1} \qquad (13.34)$$

$$L_1' \frac{x_{A1}}{1 - x_{A1}} + G_2' \frac{y_{A2}}{1 - y_{A2}} = L_2' \frac{x_{A2}}{1 - x_{A2}} + G_1' \frac{y_{A1}}{1 - y_{A1}} \qquad (13.35)$$

To solve the preceding equations, the relationship of x_A and y_A at equilibrium must be available. For a pollutant at low concentration, Henry's law (Equation 13.29) can be used. For a pollutant at very high concentration, experimental data of equilibrium distribution between x_A and y_A at the given conditions (temperature and pressure) should be used.[10]

Example 13.4: *A mixture of air and hydrogen sulfide is forced to pass through a single-stage counterflow water absorption scrubber, as shown in Figure 13.7. The mole fraction of the H_2S in gas phase is 0.00005 (50 ppmv) at the inlet. The total pure airflow rate into the scrubber is 80 moles per second, and the pure water flow rate into the scrubber is 10 moles per second. Assuming that the gas–water system is at an equilibrium state, the temperature is 30°C, and the atmospheric pressure is 101,325 Pa, find the mole fraction of H_2S in gas phase at the exit of the absorption tower. Assume that the system within the tower is binary (for gas, it is air-H_2S, and for liquid, it is water-H_2S).*

Solution: From Figure 13.7 and the problem description, we know several variables. Pure water flow rate is the same as the total liquid flow rate at the inlet and outlet because the H_2S concentration is very low and water is considered a stagnant liquid in this process:

$$L_1 = L_1' = L_2' = 10 \left(\frac{mole \; H_2O}{s} \right)$$

The mole fraction of H_2S in gas phase at inlet is

$$y_{A1} = 0.00005 \left(\frac{mole \; H_2S}{mole \; air} \right)$$

The total H_2S flow rate is pure airflow rates at both inlet and exit of the scrubber:

$$G_1' = G_2' = 80 \left(\frac{mole \; air}{s} \right)$$

Substituting the preceding values into Equation 13.35 gives

$$10\frac{0}{1-0}+80\frac{y_{A2}}{1-y_{A2}}=10\frac{x_{A2}}{1-x_{A2}}+80\frac{0.00005}{1-0.00005} \qquad (13.a)$$

Because the concentration of H_2S is very low (mole fraction << 5%), we use Henry's law. At 30°C and one atmospheric pressure P = 101.325 Pa, the Henry's law constant is

$$H' = \frac{H}{P} = \frac{60.9\times10^6}{101,325} = 601\left(\frac{mole\ fraction\ in\ gas}{mole\ fraction\ in\ liquid}\right)$$

Substituting $x_{A2} = y_{A2} / H' = 0.001664 y_{A2}$ into Equation 13.a and rearranging gives

$$80\frac{y_{A2}}{1-y_{A2}}-10\frac{0.001664y_{A2}}{1-0.001664y_{A2}}-0.004=0$$

$$0.11646y_{A2}^2 - 79.98y_{A2}+0.004=0$$

$$y_{A2} = 0.000002774\left(\frac{mole\ H_2S}{mole\ air}\right)$$

This mole fraction of H_2S at the exit in gas phase, y_{A2}, is equivalent to 2.774 ppmv concentration of H_2S in the exit air. The mole fraction of H_2S in liquid phase is so small that essentially $L_2 = L'_2$.

13.9 MULTISTAGE COUNTERFLOW GAS–LIQUID SYSTEM

13.9.1 Mass Balance of the Multistage System

To reduce further the concentration of pollutant A in its gas phase, the single-stage gas–liquid system can be repeated in series, that is, the gas stream G_1 leaves the absorber and enters the next stage to make contact again with a fresh liquid stream L_1. However, this can result in the consumption of a large amount of absorbing liquid and is not practical economically. Ideally, the same absorbing liquid stream should be used to increase the mass transfer of pollutant A from gas phase into liquid phase. This can be done using a *multistage counterflow gas–liquid system*.

As shown in Figure 13.8, the liquid stream L leaves stage 1 and then makes contact with a gas stream G in stage 2, which has a higher concentration (mole

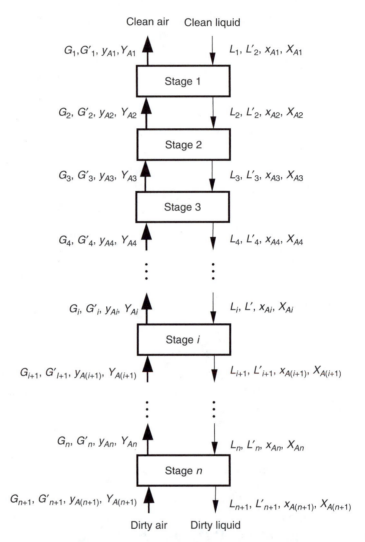

Figure 13.8 Schematic diagram of a multistage counterflow absorption process.

fraction) than G_1. The liquid stream L (with L_1 being the freshest and L_{n+1} the dirtiest) and the gas stream G (with G_1 being the cleanest air and G_{n+1} the dirtiest air) both flow through n stages in opposite directions. This system is very similar to a countercurrent heat exchanger, in which the outlet heat stream approaches to the temperature of the inlet hot stream. Here, the outlet mass flow rate of A in liquid phase more closely approaches the inlet mass flow rate of A in gas phase.

From Figure 13.8, the overall component mass balance for pollutant A in the system can be written as

$$L_1 x_{A1} + G_{n+1} y_{A(n+1)} = L_{n+1} x_{A(n+1)} + G_1 y_1 \qquad (13.36)$$

where x_A and y_A are mole fractions, and L and G are the mass flow rates, in mole/s or kg/s, for liquid flow and total gas flow. As shown in Figure 13.7, x_{A1} and y_{A1}, $x_{A(n+1)}$ and $y_{A(n+1)}$ are called *terminal fractions,* which are usually given for a specific design problem.

Writing the overall mass balance for first $(i\text{-}1)$ stages, we have

$$L_1 x_{A1} + G_i y_{Ai} = L_i x_{Ai} + G_1 y_1 \qquad (13.37)$$

Solving Equation 13.37 for y_{Ai}, we obtain the operating line for a multistage counterflow gas–liquid absorption scrubber:

$$y_{Ai} = \frac{L_i}{G_i} x_{Ai} + \frac{G_1 y_{A1} - L_1 x_{A1}}{G_i} \qquad (13.38)$$

In Equation 13.38, each pair of x_{Ai} and y_{Ai} are in equilibrium and L_1, G_1, x_{A1}, and y_{A1} are usually given in a practical design. The relationship between x_{Ai} and y_{Ai} is plotted in Figure 13.9. Assuming that the equilibrium line follows Henry's law and is thus a straight line, the operating line for an absorption process is a curve that is approximately parallel to the equilibrium line, but may bend upward or downward slightly, depending on the efficiency of each stage and the actual equilibrium between x_{Ai} and y_{Ai}. In a counterflow absorption process, as the dirty air passes through the absorption liquid, the total airflow rate and the total liquid flow rate both increase as the stage number i increases, thus the slop of the operating line (L_i/G_i in Equation 13.38) varies slightly among the different stages.

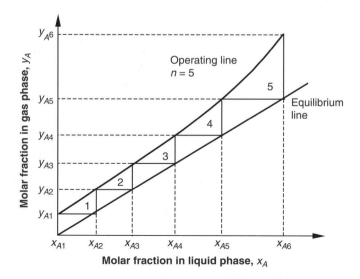

Figure 13.9 A general sketch of the operating line and the equilibrium line for a multistage gas–liquid absorption system ($n = 5$ in the diagram).

In many practical problems, the mass fractions of a particular pollutant in both the gas and the liquid streams are small. This results in the slope of the operating line being approximately constant, and thus a straight line approximately parallel to the equilibrium line. For example, a 3000 ppmm concentration of carbon dioxide in the air is only 3% of the total airflow, so assuming that the total mass flow rates for the air and the absorption liquid are the same, the slope (L_i/G_i) of the operating line would only vary a maximum of 6.2%. For very large mass fractions of a pollutant, the operating line can be highly nonlinear.

An important application of Equation 13.38 is to determine the number of ideal stages required for a particular absorption process to achieve the pollutant removal from $y_{(An+1)}$ to y_{A1}, for a total gas flow rate $G_{(An+1)}$ using an absorption fluid with an initial flow rate L_1. There two methods to accomplish this: a graphical method and an analytical method. The graphical method is simpler and generally more satisfactory for most two-component situations. The graphical method is applicable to nonlinear equilibrium line or operating line situations, whereas the analytical method applies strictly to situations in which both the equilibrium line and the operating line are straight lines.

13.9.2 Ideal Number of Stages — Graphical Method

To use the graphical method, one must know the equilibrium line and the operating line on the same coordinates. Figure 13.8 shows a typical n-stage ($n = 5$) cascade absorber. The two ends of the operating line have coordinates of (x_1, y_1) and (x_{n+1}, y_{n+1}). The ideal number of stages can be determined by changing the gas-phase pollutant concentration from y_{n+1} at the inlet to y_1 at the outlet and changing the liquid-phase pollutant concentration from x_{n+1} at the outlet and x_1, at the inlet (Figure 13.7).

The gas concentration leaving stage 1 is y_1, and by definition the process reaches equilibrium so the liquid concentration leaving stage 1 must be x_2. Thus, the point (x_2, y_1) must lie on the equilibrium line (Figure 13.8). The point (x_2, y_1) can be found by drawing a straight line horizontally from the point (x_1, y_1) on the operating line rightward and intersecting with the equilibrium line. The next step is to draw a vertical line from point (x_2, y_1) upward until it intersects with the operating line. The intercepting point on the operating line has a coordinate (x_2, y_2). This step, defined by three points (x_1, y_1), (x_2, y_1), and (x_2, y_2), represents one ideal stage — stage 1. By repeating this process, starting from the point (x_2, y_2) on the operating line, we can define the second stage. This process continues until the last vertical line reaches the other end (x_{n+1}, y_{n+1}). In Figure 13.8, the last stage is stage 5.

In reality, the point (x_{n+1}, y_{n+1}) may not meet the actual end of the operating line. In this case, a fractional stage results. From a design point of view, the number of stages can only be an integer, and that number should be the upper cut number. For example, if the solution is $n = 5.2$, the actual stages should be six, to ensure the outlet gas concentration is according to specification. If the number of stages is given, one can modify the operating line by changing the ratio of total liquid to total gas (L/G), because L/G defines the slope of the operating line.

13.9.3 Ideal Number of Stages — Analytical Method

In most rectification applications, the total mass flow rates of absorption liquid and gas (pollutant and carrying gas) do not change appreciably from the inlet to outlet. This is especially true in many air-cleaning applications, where pollutant concentration is typically lower than 5%. Therefore, the subscripts of L and G can be omitted, and the operating line expressed in Equation 13.38 is a straight line. When both the equilibrium and the operating lines are straight over a given range of concentration, the number of ideal stages can be calculated analytically.

Let the equation of the equilibrium line be

$$y_A = ax_A + b \tag{13.39}$$

where a and b are constant. This is a more generic form of Henry's law, with an intersecting point of b on the y_A axis. If the $(i\text{-}1)$ stage is ideal (refer to Figure 13.9)

$$y_{A(i-1)} = ax_{Ai} + b \tag{13.40}$$

Substituting Equation 13.40 for x_{Ai} into Equation 13.38, for ideal stages and constant mass flow rates of both liquid and gas phases

$$y_{Ai} = \frac{L}{aG}(y_{a(i-1)} - b) + y_{A1} - \frac{L}{G}x_{A1} \tag{13.41}$$

Let us define an absorption factor S as the ratio of the slope of the operating line to the slope of the equilibrium line:

$$S = \frac{L}{aG} \tag{13.42}$$

Because both the operating line and the equilibrium line are straight, the slopes of both lines are constant, thus the absorption factor S is constant, and Equation 13.41 can be expressed as

$$y_{Ai} = Sy_{A(i-1)} - S(ax_{A1} + b) + y_{A1} \tag{13.43}$$

Referring to Figure 13.7 and Figure 13.9, and by the definition of Equation 13.39, the quantity $ax_{A1}+b$ is the concentration of the gas phase that is in equilibrium with the inlet liquid phase concentration ax_{A1}. From Figure 13.10, if we use the symbol y^* to represent the gas-phase concentration in equilibrium with a special liquid phase, then

$$y^*_{A1} = ax_{A1} + b \tag{13.44}$$

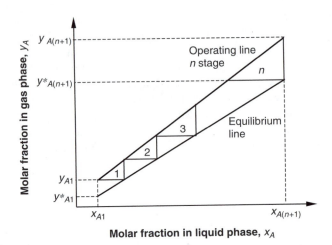

Figure 13.10 Sketch for an analytical method to determine the number of ideal stages.

Note in Equation 13.39, Equation 13.40, and Equation 13.44, b is the intercepting value at y_A at $x_A = 0$. For gases obeying Henry's law, $b = 0$ and $a = H'$. Substituting Equation 13.44 into Equation 13.43 gives

$$y_{Ai} = Sy_{A(i-1)} - Sy_{A1}^* + y_{A1} \tag{13.45}$$

Using Equation 13.45, the value of y_{Ai} can be calculated, step by step, starting from stage 1 (Figure 13.9). For stage 1, $i = 2$ and $y_{Ai = yA2}$ in Equation 13.45. Equation 13.45 becomes

$$y_{A2} = Sy_{A1} - Sy_{A1}^* + y_{A1} = y_{A1}(1 + S) - Sy_{A1}^* \tag{13.46}$$

For stage 2, $i = 2$, and $y_{Ai = yA3}$, and eliminating y_{A2} by using Equation 13.46, Equation 13.45 becomes

$$y_{A3} = y_{A1}(1 + S + S^2) - y_{A1}^*(S + S^2) \tag{13.47}$$

For the ith stage, Equation 13.45 becomes

$$y_{A(i+1)} = y_{A1}(1 + S + S^2 + \cdots + S^i) - y_{A1}^*(S + S^2 + \cdots + S^i) \tag{13.48}$$

Using the sums of geometric series

$$1 + S + S^2 + \cdots + S^i = \frac{1 - S^{i+1}}{1 - S} \tag{13.49}$$

$$S + S^2 + \cdots + S^i = S\frac{1-S^i}{1-S} \tag{13.50}$$

Substituting Equation 13.39 and Equation 13.50 into Equation 13.48 gives

$$y_{A(i+1)} = y_{A1}\frac{1-S^{i+1}}{1-S} - y_{A1}^{*}S\frac{1-S^i}{1-S} \tag{13.51}$$

For an absorption cascade with n stages, $i + 1 = n$ in Equation 13.51

$$y_{An} = y_{A1}\frac{1-S^n}{1-S} - y_{A1}^{*}S\frac{1-S^{n-1}}{1-S} \tag{13.52}$$

In Equation 13.52, the number of stages n is not explicitly solved for a given set of operating and equilibrium lines. However, it can be used in the form of a chart relate n, S and inlet and outlet gas concentrations. Such a chart can provide a useful tool for design because it presents ranges of design parameters from which a designer can choose. The application of Equation 13.52 and a chart for different absorption process designs can be found in several handbooks, such as the *Chemical Engineers' Handbook* edited by Perry.[8]

The number of ideal stages can also be solved explicitly from Equation 13.45. For stage n, substituting $i = n + 1$ into Equation 13.45 gives

$$y_{A(n+1)} = Sy_{An} - Sy_{A1}^{*} + y_{A1} \tag{13.53}$$

From Figure 13.9 we can see that

$$y_{An} = y_{A(n+1)}^{*} \tag{13.54}$$

Thus, Equation 13.53 becomes

$$y_{A1} = y_{A(n+1)} - S(y_{A(n+1)}^{*} - y_{A1}^{*}) \tag{13.55}$$

We can then rewrite Equation 13.52 in terms of S^{n+1}:

$$S^{n+1} = \frac{S(y_{A(n+1)} - y_{A1}^{*}) + (y_{A1} - y_{A(n+1)})}{y_{A1} - y_{A1}^{*}} \tag{13.56}$$

and substituting $y_{A1} - y_{A(n+1)}$ from Equation 13.55 into Equation 13.56 gives

$$S^n = \frac{y_{A(n+1)} - y^*_{A(n+1)}}{y_{A1} - y^*_{A1}} \tag{13.57}$$

Taking logarithms from Equation 13.57 and solving for n yields

$$n = \frac{\ln\left[\dfrac{y_{A(n+1)} - y^*_{A(n+1)}}{y_{A1} - y^*_{A1}}\right]}{\ln S} \tag{13.58}$$

Substituting S from Equation 13.55 into Equation 13.58 gives

$$n = \frac{\ln\left[(y_{A(n+1)} - y^*_{A(n+1)})/(y_{A1} - y^*_{A1})\right]}{\ln\left[(y_{A(n+1)} - y_{A1})/(y^*_{A(n+1)} - y^*_{A1})\right]} \tag{13.59}$$

One special case of Equation 13.59 is when the operating line and the equilibrium line are parallel. In this case, the absorption factor S is unity and Equation 13.58 and Equation 13.59 are indeterminate. When $S = 1$

$$y_{A(i+1)} = y_{A1}(1+n) - y^*_{A1}n \tag{13.60}$$

and the number of ideal stages is

$$n = \frac{y_{A(n+1)} - y_{A1}}{y_{A1} - y^*_{A1}} = \frac{y_{A(n+1)} - y_{A1}}{y_{A(n+1)} - y^*_{A(n+1)}} \tag{13.61}$$

Example 13.5: *The ammonia concentration in an air stream that enters a cascade counterflow water scrubber is 3000 ppmv. The total gas (air and ammonia) flow rate is 1 m³/s. The scrubbing water flow rate is 1 l/s. For low concentrations and standard room air conditions (20°C and 1 atm) ammonia approximately follows Henry's law with $H' = 1.2 \dfrac{\text{mole fraction in gas}}{\text{mole fraction in liquid}}$. The molar volume for ideal gases is 24 l/mole. The molar weight of ammonia and water are 17 and 10 g/mole, respectively. If 90% of ammonia in the air is to be absorbed by the water, find the following:*

a. The exit concentration of ammonia in mg of NH_3 per liter of water
b. The number of ideal stages required

Solution:

a. First we determine the terminal mole fractions. Because ammonia and air can be considered ideal gases, the volume fraction is the same as the mole fraction. As shown in Figure 13.8, the mole fraction of ammonia at the inlet of gas stream is

$$y_{A(n+1)} = \frac{0.003 \ (m^3 \ NH_3 \ / \ s)}{0.997 \ (m^3 \ air \ / \ s)} = 0.00301 \left(\frac{mole \ NH_3}{mole \ air} \right)$$

The total ammonia mass flow rate is based on molar volume of 24 l/mole:

$$\frac{0.00301 \ (mole \ NH_3 \ / \ s)}{0.024 \ (m^3 \ NH_3 \ / \ mole)} = 0.125 \left(\frac{mole \ HN_3}{s} \right)$$

Of the total ammonia flow in gas phase, 90% is absorbed in the water. Thus, at the exit of the water stream, the total ammonia in liquid phase is

$$0.125 \ (mole \ NH_3 \ / \ s) \times 0.9 = 0.1125 \ (mole \ HN_3 \ / \ s)$$

The scrubbing water flow rate is 1 l/s = 1000 g/s = 100 mole/s. Thus, the mole fraction of ammonia in liquid phase at the exit of water stream is

$$x_{A(n+1)} = \frac{0.1125 \ (mole \ NH_3 \ / \ s)}{100 \ (mole \ water \ / \ s)} = 0.001125 \left(\frac{mole \ NH_3}{mole \ water} \right)$$

The ammonia concentration C in the exit water stream in terms of milligram of ammonia per liter of water can be calculated as

$$C = \frac{0.1125 \ (mole \ NH_3 \ / \ s)}{1 \ (L \ water \ / \ s)} = \frac{0.1125 \ (mole \ NH_3 \ / \ s) \times 17,000 \ (mg \ / \ mole \ NH_3)}{1 \ (L \ water \ / \ s)}$$

$$= 1913 \left(\frac{mg \ NH_3}{L \ water} \right)$$

b. Other mole fractions are determined as follows. The ammonia concentration in the entering water flow is $x_{A1} = 0$, and from Equation 13.44, noting that $a = H' = 1.2$

$$y_{A1}^* = 1.2x_{A1} = 0$$

$$y_{A(n+1)}^* = 1.2x_{A(n+1)} = 1.2 \times 0.001125 = 0.00135 \left(\frac{mole \ HN_3}{mole \ water} \right)$$

Because for air 1 m³/s = 41.67 mole/s, then the mole fraction of ammonia at the exit gas stream is

$$y_{A1} = \frac{0.125 \ (mole \ NH_3 \ / \ s) \times 0.1}{41.67 \ (mole \ air \ / \ s)} = 0.0003 \left(\frac{mole \ NH_3}{mole \ air} \right)$$

Substituting y_{A1} y_{A1}^{*}, $y_{A(n+1)}$ and $y_{A(n+1)}$ into Equation 13.59 gives the number of ideal stages:

$$n = \frac{\ln\left[(y_{A(n+1)} - y_{A(n+1)}^{*}) / (y_{A1} - y_{A1}^{*})\right]}{\ln\left[(y_{A(n+1)} - y_{A1}) / (y_{A(n+1)}^{*} - y_{A1}^{*})\right]}$$

$$= \frac{\ln\left[(0.00301 - 0.00135) / (0.0003 - 0)\right]}{\ln\left[(0.003 - 0.0003) / (0.00135 - 0)\right]} = 2.45 \rightarrow 3$$

13.10 DESIGN OF LIQUID–GAS RATIO

A practical concern for the design of absorption systems is the determination of the minimum ratio of liquid flow rate to gas flow rate. We first consider the absorption liquid flow rate (L') and the carrying gas flow rate (G') because they are constant in a binary system. The ratio of L'/G' is therefore the same for all stages in a cascade absorption tower. With reference to Figure 13.7, noting that $L_1' = L_2'$ and $G_1' = G_2'$, the mass balance of Equation 13.34 can be rewritten as

$$L'(X_{A1} - X_{A2}) = G'(Y_{A1} - Y_{A2}) \tag{13.62}$$

The minimum liquid-to-gas flow ratio can therefore be written as

$$\left[\frac{L'}{G'}\right]_{min} = \frac{Y_{A1} - Y_{A2}}{X_{A1} - X_{A2}} \tag{13.63}$$

As shown in Figure 13.7, the molar ratios of the pollutant in gas phase, Y_{A1} and Y_{A2}, and the carrying gas flow rate G' are known by the design criteria. The molar ratio of the pollutant in liquid phase at the liquid exit, X_{A2}, is also known by the design criteria. Thus, if we can determine the molar ratio of the pollutant in liquid phase at the inlet, X_{A1}, we can determine the minimum absorption liquid flow rate L' or the liquid to gas flow ratio L'/G'.

In order to find this quantity, we can use one of two methods. For a liquid–gas system that follows Henry's law, one can find x_{A1} from the equilibrium line using the known y_{A1} and then calculate X_{A1} and Y_{A1} using Equation 13.31 and Equation 13.32. If the equilibrium line is not linear, a linearization technique may be used to

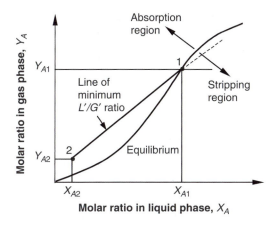

Figure 13.11 Graphical position of the equilibrium line and the operating line with a minimum molar ratio of absorption liquid flow to carrying gas flow.

determine X_{A1} from the X_A vs. Y_A graph (Figure 13.11). To begin, we define an initial point (point 2) by the two known ratios X_{A2} and Y_{A2}. The intersection of Y_{A1} and X_{A1} (point 1) can then be found by drawing a line from Y_{A1} parallel to the X_A axis and intersecting with the equilibrium line. The x-coordinate of point 1 is then X_{A1}. The slope of the line between points 1 and 2 is the minimum ratio of L'/G'. If L'/G' is smaller than this minimum value, point 1 will fall into the stripping region and desorption will occur.

After X_{A1} has been determined, the minimum liquid-to-gas flow ratio can be calculated using Equation 13.62. Substituting Equation 13.31 and Equation 13.32 into Equation 13.63, the minimum L'/G' can be expressed in terms of molar fractions x_A and y_A.

$$\left[\frac{L'}{G'}\right]_{min} = \frac{\left[y_{A1}(1-y_{A2})-y_{A2}(1-y_{A1})\right](1-x_{A1})(1-x_{A2})}{\left[x_{A1}(1-x_{A2})-x_{A2}(1-x_{A1})\right](1-y_{A1})(1-y_{A2})} \tag{13.64}$$

Example 13.6: *In Example 13.5, assume that the counterflow water scrubber is single stage. The ammonia mole fraction in liquid phase at the water exit reaches only 30% of its equilibrium value. What water flow rate should be used to achieve the same ammonia removal result (i.e., 90% of ammonia in gas phase is removed)?*

Solution: From the calculations in Example 13.5

$$y_{A1} = 0.0003 \left(\frac{mole\ NH_3}{mole\ air}\right)$$

$$y_{A2} = y_{A(n+1)} = 0.00301 \left(\frac{mole\ NH_3}{mole\ air}\right)$$

From Equation 13.30, because this is a single stage system, the ammonia mole fraction in liquid phase at the exit of water stream at 30% of its equilibrium value is

$$x_{A2} = 0.3 \times \frac{y_{A2}}{H'} = \frac{0.00301}{1.2} = 0.000753 \left(\frac{mole\ NH_3}{mole\ water} \right)$$

$$x_{A1} = 0$$

Substituting the preceding values into Equation 13.64 gives the water-to-gas mole ratio:

$$\left[\frac{L'}{G'} \right]_{min} = \frac{\left[y_{A1}(1 - y_{A2}) - y_{A2}(1 - y_{A1}) \right](1 - x_{A1})(1 - x_{A2})}{\left[x_{A1}(1 - x_{A2}) - x_{A2}(1 - x_{A1}) \right](1 - y_{A1})(1 - y_{A2})}$$

$$= \frac{\left[0.0003(1 - 0.00301) - 0.00301(1 - 0.0003) \right](1 - 0)(1 - 0.000753)}{\left[-0.000753 \right](1 - 0.0003)(1 - 0.00301)}$$

$$= 3.6 \left(\frac{mole\ water}{mole\ air} \right)$$

Because the airflow rate is 1 m³/s = 41.67 mole/s under standard conditions, one mole of water is 0.01 kg, and the water flow rate is

$$Q_l = \left[\frac{L'}{G'} \right] \times 41.67 \left(\frac{mole\ water}{s} \right) \times 0.01 \left(\frac{kg}{mole\ water} \right)$$

$$= 1.5 \left(\frac{kg\ water}{s} \right)$$

13.11 CATALYTIC CONVERSION

13.11.1   Principles of Catalytic Conversion

Catalytic conversion is another major method of gas control. Catalysts have been widely used in chemical and petrochemical processes, where fast reaction and low temperature are required. The adsorption of gases onto a catalyst provides a chemical shortcut, in which reactants are converted to products much more rapidly than normal. When a catalyst is involved in a chemical reaction, the process can be conducted at lower temperatures, resulting in energy and reaction material savings. In addition, because the reaction rate is greatly increased, smaller reactor volume can be used. In several known catalytic reaction mechanisms, the catalyst forms intermediate compounds with the reactant. However, many catalytic processes and principles have not been fully explained or understood. Much developmental work in the catalytic conversion area has been achieved through elaborate experimental design involving trial and error for many types of materials.

Catalytic conversion is especially useful for gas control applications, such as emission reduction from automobiles. For example, ethylene and oxygen adsorb onto specific sites on a platinum (Pt) catalyst and rapidly convert to carbon dioxide and water:

$$C_2H_4 + 3O_2 \longrightarrow 2CO_2 + 2H_2O \tag{13.65}$$

Given the same reactants, a different catalyst may produce different products. In contrast to the platinum catalyst just mentioned, a vanadium (V_2O_5) catalyst selectively produces, from ethylene and oxygen, mostly partially oxidized product such as aldehyde:

$$C_2H_4 + \frac{1}{2}O_2 \longrightarrow CH_3CH = O \tag{13.66}$$

The theoretical basis of catalysis is differences in the chemical energies required by different processes. Each chemical process requires a certain amount of energy, called an *activation energy* (E_A), to proceed. This energy includes such necessities as the energy required to break certain bonds or to move reactants into place. Figure 13.12 shows the net enthalpy change against reaction coordinates of a hypothetical reaction Catalysts effectively reduce the activation energy required for the reaction through various means, such as arranging the reactant molecules in favorable positions or creating intermediary compounds. Because less energy is required for each molecular reaction, the reaction proceeds at a much higher rate than usual.

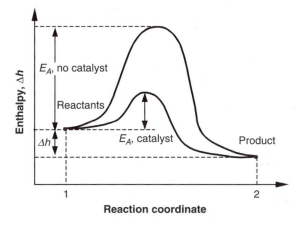

Figure 13.12 Reaction energy path. Points 1 and 2 are the initial and final states of reaction, respectively.

The rate of reaction r_c is inversely proportional to the exponential of the activation energy in mole/m^3·s. This relationship is expressed through the rate constant k, in 1/s, as in Equation 13.67 and Equation 13.68, where C is concentration at any given time of reactants, in mole/m^3; subscripts $_A$, $_B$, and $_C$ denote the different reactants involved, and superscripts ^{a}, ^{b}, and ^{c} denote their respective reaction orders. The rate constant is related to the preexponential function k_0, which is proportional to the number of active sites, or areas where the reaction can occur, on the catalyst. R and T are the universal gas constant and the absolute temperature, respectively.

$$r_c = kC_A^a C_B^b C_C^c \cdots \tag{13.67}$$

$$k = k_0 \exp\left(-\frac{E_A}{RT}\right) \tag{13.68}$$

The activation energy for the entire reaction represents the slowest of all steps involved in converting reactants to products, or *rate-limiting steps*, and the overall reaction rate can never be greater than this. The difference between the energy states of the reactants and products is the exothermic heat of reaction.

An instructive example of how the catalyst functions to decrease the activation energy is illustrated by the important environmental application of the oxidation of the pollutant CO to CO_2:

$$CO + \frac{1}{2}O_2 \longrightarrow CO_2 \tag{13.69}$$

This reaction requires about 700°C in the absence of a catalyst, because the rate-limiting step is the thermal dissociation of O_2 into O atoms. The activation energy has been determined experimentally to be about 40 kcal/mol (1 kcal = 4187 J). In the presence of Pt or Pd, however, the dissociation of O_2 occurs catalytically on the metal surface and the rate-limiting step becomes the reaction of adsorbed CO with adsorbed O atoms. The activation energy is reduced to less than 20 kcal and the reaction occurs at 100°C. Thus, the catalyst changes the rate-limiting step and provides a less energetic path for the formation of the product. The catalyst provides a chemical shortcut and lowers the energy necessary for the reaction so that CO can be abated from vehicles and power plants at modest temperatures.

It is of great importance to recognize that a catalyst changes neither the energetics nor the equilibrium of the initial and final states, but rather affects only the rate of approach to the final product state (i.e., the process between 1 and 2 of reaction coordinate in Figure 13.12). Thus, the reaction enthalpy ΔH and reaction free energy, ΔG (and consequently the equilibrium constant K_e) are not altered by the presence of a catalyst.

Catalysts influence selectivity by preferentially lowering the activation energy for a particular step in the reaction sequence, thereby increasing the rate at which this step proceeds. Selectivity is an issue for many reactions in which multiple

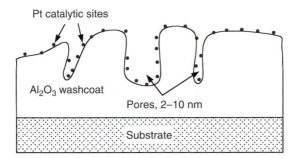

Figure 13.13 Pt catalytic sites dispersed on a high surface-area Al_2O_3 carrier bonded to a substrate.

products can occur in parallel. For example, with the V_2O_5 catalyst, the reaction path leading to the formation of the aldehyde product is favored because it has lower activation energy than the complete combustion to CO_2 and H_2O. For the reaction using Pt, the opposite is true. Heterogeneous catalysts are those that are of a different state than one or more of the reactants, while homogeneous ones are those that are of the same state. In catalytic air pollution control, all processes will utilize solid heterogeneous catalysts, through which gaseous reactants pass.

13.11.2   Dispersed Catalyst Components

In many industrial reactions, the number of reactant molecules converted to products in a given time is directly related to the number of catalytic sites available to the reactants. It is therefore common practice to maximize the number of active sites by dispersing the catalytic components, such as Pt, Fe, Ni, Rh, Pd, CuO, PdO, and CoO, onto a high surface-area carrier, such as Al_2O_3, SiO_2, TiO_2, or SiO_2-Al_2O_3, increasing the number of sites on which chemisorption and catalytic reaction can occur. The carriers themselves are seldom catalytically active, but they do play a major role in maintaining the overall stability and durability of the finished catalyst.

The most commonly used carrier in catalysis, especially for environmental applications, is Al_2O_3, which we will use as an example to illustrate how to develop a model of a heterogeneous catalyst. Figure 13.13 shows a few select 2- and 10-nm pores of a high surface-area Al_2O_3, into which Pt has been deposited by solution impregnation. The Pt particles, or crystallites, are represented as dots. When the Al_2O_3 is bonded to a monolithic honeycomb support, it is called a *washcoat*. The internal surface of the Al_2O_3 is rich in surface OH⁻ species, which cover the entire surface and are part of the walls of each pore. It is on these internal walls and at the OH⁻ sites that the catalytic components are bound. The catalytic surface area is the sum of all the areas of the active catalytic components — in this example, Pt. The smaller the individual size of the crystallites of the active catalytic material (higher catalytic surface area), the more sites are available for the reactants to interact. As a rough approximation, one assumes that, for a process controlled by kinetics, the higher the catalytic surface area, the higher the rate of reaction is.

The tiny Pt-containing particles shown in Figure 13.13 are dispersed throughout the porous Al_2O_3 carrier network and generate a high-Pt surface area. This procedure maximizes the catalytic areas but also introduces other possible rate-controlling physical processes, such as mass transfer of the reactants to the catalytic sites. Each of these processes has a rate influenced by the hydrodynamics of the fluid flow, the pore size, the structure of the carrier, and the molecular dimensions of the diffusing molecules.[11, 12]

13.11.3   Steps of Heterogeneous Catalysis

To maximize reaction rates, it is essential to ensure accessibility of all reactants to the active catalytic sites dispersed within the internal pore network of the carrier. As an example, but as a technique that can be applied to general cases of reactants and products, let us consider reaction Equation 13.69, in which CO and O_2 molecules are flowing through a bed of a heterogeneous catalyst. To be converted to CO_2, the following physical and chemical steps must occur:

1. **Bulk diffusion of reactants** — CO and O_2 must contact the outer surface of the carrier containing the catalytic sites. To do so, they must diffuse through a stagnant thin layer of gas in close contact with the catalyzed carrier. Bulk molecular diffusion rates vary approximately with $T^{3/2}$ and typically have activation energies of $E_1 = 2 - 4$ kcal/mol.

2. **Pore diffusion of reactants** — Because the bulk of the catalytic components are internally dispersed, the majority of CO and O_2 molecules must diffuse through the porous network toward the activation sites. The activation energy for pore diffusion E_2, is approximately half that of a chemical reaction, or about 6 to 10 kcal/mol.

3. **Adsorption of reactants** — Once molecules of CO and O_2 arrive at the catalytic site, O_2 dissociates quickly and chemisorption of both O and CO occurs on the adjacent catalytic sites. The kinetics generally follows exponential dependence on temperature, for example, $\exp\left(-E_3/RT\right)$, where E_3 is the activation energy, which for chemisorption is typically greater than 10 kcal/mol.

4. **Catalytic reaction** — An activated complex forms between adsorbed CO and adsorbed O with an energy equal to that of the total activation energy (because this is the rate-limiting step). At this point, the activated complex has sufficient energy to convert to adsorbed CO_2. Kinetics also follows exponential dependence on temperature, for instance $\left(-E_4/RT\right)$, with activation energies typically greater than 10 kcal/mol.

5. **Desorption of products** — CO_2 desorbs from the site following exponential kinetics, for example, $\exp\left(-E_5/RT\right)$, with activation energies typically greater than 10 kcal/mol.

6. **Pore diffusion of products** — The desorbed CO_2 diffuses through the porous network toward the outer surface, with an activation energy and kinetics similar to those in step 2.

7. **Bulk diffusion of products** — CO_2 must diffuse through the stagnant layer and, finally, into the bulk gas. Reaction rates follow $T^{3/2}$ dependence. Activation energies are also similar to step 1, less than 2 to 4 kcal/mol.

Steps 1 and 7 represent bulk mass transfer, which is a function of the specific molecules, the dynamics of the flow conditions, and the geometric surface area (outside or external area) of the catalyst of the carrier. Pore diffusion, illustrated in steps 2 and 6, depends primarily on the size and shape of both the pore and the diffusing reactants and products. Steps 3, 4, and 5 are related to the chemical interactions of reactants and products (CO, O_2, and CO_2 respectively) at the catalytic site(s).

Any of the seven steps listed previously can be rate limiting and control the overall rate of reaction in a heterogeneous catalyst. The rate-limiting mechanism can be determined based on the differences in activation energy, pore diffusion, and mass transfer. Figure 13.14 shows an example of the conversion of CO and O_2 to CO_2 as the temperature of the catalyst is increased. In the chemical kinetic control region, the reaction of adsorbed CO with adsorbed O is slow relative to the diffusion, and thus is rate limiting. As the temperature increases to 200°C, the reaction steps with high activation energies increase quickly. The control of the overall conversion rate shifts from chemical reaction to pore diffusion. Here, the rate of conversion of CO and O is faster than the rate at which they can be supplied, and a concentration gradient exists within the product. This is referred to as *intraparticle diffusion*, in which the catalytic metals deep within the carrier are not being completely utilized or have an effectiveness factor of less than 1. The effectiveness factor is the ratio of the actual rate to the theoretical maximum rate and can be thought of as a measure of catalyst utilization. When temperature increases to about 400°C, both chemical reaction and pore diffusion become faster than the bulk mass transfer can occur. At temperatures higher than 450°C, the conversion efficiency is relatively independent of the temperature, becoming rate limiting. In this regime, the CO and O_2 are converted to CO_2 as soon as they arrive at the outer or external surface of the catalyst/carrier. The concentration of reactant and product is essentially zero within the product, and the effectiveness factor is close to zero.

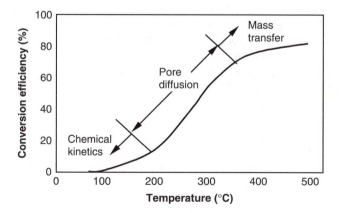

Figure 13.14 Catalytic conversion efficiency for CO → CO2 at different temperatures and under different conversion mechanisms.

The slope of the conversion vs. temperature curve can give a qualitative picture of the rate-controlling steps. The steeply rising lower part of the curve is indicative of chemical control (either step 3, 4, or 5). The relatively flat, temperature-insensitive portion reflects bulk mass transfer control (either step 1 or 7), whereas the intermediate portion is characteristic of pore diffusion control (either step 2 or 7).

13.11.4   The Arrhenius Equation

We can derive a very simple expression to aid in determining which step is rate-determining by using the Arrhenius equation. By taking the natural logarithm of Equation 13.68, we obtain a linear equation.

$$\ln k = \ln k_0 - \frac{E}{R}\left(\frac{1}{T}\right)$$ (13.70)

Equation 13.70 can generate a series of straight lines whose slopes are directly related to the activation energy of the rate-controlling step.

As shown in Figure 13.14, when the reaction of CO plus O_2 to CO_2 is controlled by one of the chemical steps, bulk and pore diffusion of reactants to the active sites is fast. The concentration of reactants within the catalyst/carrier is essentially uniform. When pore diffusion is the limiting step, the concentration of reactants decreases from the outer periphery of the catalytic surface toward the center of the pore or the bottom of the washcoat. Finally, with mass transfer control, the concentration of reactants approaches zero at the boundary layer near the outside surface of the catalyst.

The efficiency with which a catalyst functions in a process depends on what controls the overall reaction rate. If the kinetics of a process are measured and found to be in a regime in which chemical kinetics are rate controlling, the catalyst should be made with as high a catalytic surface as possible. This is accomplished by increasing the catalytic component loading and/or dispersion, so that every catalytic site is available to the reactants. Furthermore, the catalytic components should be dispersed uniformly throughout the interior of the carrier with an effectiveness factor approaching 1.

When it is known that a process will have significant pore diffusion limitations, the carrier should be selected with large pores and the catalytic components located as close to the surface as possible to improve the effectiveness factor. To enhance the transport rate, one can decrease the thickness of the carrier or washcoat to decrease the diffusion path of reactants and products.

The rate of mass transfer is enhanced by increasing turbulence in the bulk gas and increasing the geometric surface area (i.e., external area) of the support. Clearly, increasing the catalytic surface area, the loading of the catalytic components, or the size of the pores will have no effect on enhancing the rate of mass transfer, because these catalyst properties to not participate in the rate-limiting step.

DISCUSSION TOPICS

1. Why is adsorption referred to as a surface phenomenon and absorption referred to as a volumetric phenomenon?
2. What is the primary driving force during an adsorption process, and what is the driving force for absorption?
3. Explain why the isotherm and adsorption wave are important in an adsorption process. What are the practical implications of these two terms?
3. Discuss the most important factors that affect an absorption process. Can you single out the factors that are design factors, that is, the factors that can be manipulated by a designer to alter the outcome of the process?
4. Why can Henry's law be directly used in most cases of indoor air quality problems, without as much concern for experimental equilibrium data as in many other chemical processes?
5. Many catalytic conversion processes are neither well defined nor understood. If we treat these unknown catalytic processes as black boxes, can you speculate on what is happening at the interface of a catalyst and the reactants, from a physical, chemical, and biological point of view?

PROBLEMS

1. Plot the adsorption isotherms of activated carbon for the following VOCs using equation 13.1 and data in Appendix 8: carbon monoxide (CO), dichloroacetic acid ($C_2H_2Cl_2O_2$), acetone (C_3H_6O), propionic acid ($C_3H_6O_2$), n-butyric acid ($C_4H_8O_2$) and n-butanol ($C_4H_{10}O$). Use the x-axis for adsorbent adsorption capacity in (g/g) and the y-axis as pollutant gas concentration. How large will the error be if the third term on the right side of Equation 13.1 is neglected?
2. In order to determine the adsorption isotherm of activated carbon for ammonia, the following experiment was conducted under standard room air conditions. Three identical activated carbon filters A, B, and C, each containing 500 g of carbon, were used. Air with three levels of ammonia concentration, 10, 50, and 100 ppmv, was ventilated through each filter, and 5, 6, and 10 g of ammonia were collected on filters A, B, and C, respectively. Assume that Equation 13.1 applies, and determine the adsorption correlation constant a_1, a_2, and a_3 for ammonia.
3. A mixture of air and carbon monoxide is forced to pass through a single-stage counterflow water scrubber, as shown in Figure 13.7. The total gas (air and CO) flow rate entering the scrubber is 200 l/s, with a CO concentration of 200 ppmv. The total pure water flow rate into the scrubber is 500 g/s. Assuming that the gas–water system is at equilibrium state, that the temperature is 20°C, and the atmospheric pressure is 101,325 Pa, find the CO concentration (in ppmv) in the exiting gas stream. Assume that the system within the tower is binary (for gas it is air-CO, and for liquid it is water-CO).
4. The total dirty airflow rate entering a counterflow absorption is 1 m³/s with a scrubber is ammonia concentration of 150 ppmv. The entering water flow rate is 1 l/s with no ammonia. If the tower is designed to absorb 90% mass of ammonia in the air, find the terminal ammonia mole fractions of x_{A1}, y_{A1}, $x_{A(n+1)}$, and $y_{A(n+1)}$, and plot the operating line. The carrying air and scrubbing water is at standard room conditions: 20°C and 1 atmospheric pressure.

5. Imagine you are designing a counterflow ammonia cascade absorber. The ammonia mole fractions in gas phase are 0.07 and 0.018 at the gas inlet and outlet, respectively. The ammonia mole fractions in liquid phase are 0.04 and 0.012 at the water inlet and outlet, respectively. The operating line is assumed to be a straight line. The ammonia absorption experimental equilibrium data are shown in Table 13.2. Determine the number of ideal stages using graphical methods.
6. Derive the relationship between the liquid flow rate and the number of ideal stages for a counterflow absorber at a given gas flow rate. Assume that the operating line and equilibrium line are given (i.e., y_{AI}, y^*_{AI}, $y_{A(n+1)}$, and $y^*_{A(n+1)}$ are all given). Assume that all operating lines and equilibrium lines are straight.
7. The carbon monoxide concentration in an air stream is 5000 ppmv and enters a cascade counterflow water scrubber. The pure airflow rate is 500 l/s. The scrubbing water flow rate is 1 l/s. The concentration of CO in the water at the inlet is negligible. For low concentrations and standard room air conditions, 20°C and one atmospheric pressure, CO follows Henry's law (see Table 13.1), and the molar volume for ideal gases is 24 l/mole. The molar weight of CO and water are 28 and 10 g/mole, respectively. If 99% of carbon monoxide in the air is to be absorbed by the water, determine the number of ideal stages.
8. Use the equation derived in problem 6 and data obtained in problem 7 to plot the liquid flow rate vs. the number of ideal stages. If the number of stages were doubled, by what percentage would the absorbing liquid be reduced?

REFERENCES

1. Young, D.M. and Crowell, A.D., *Physical Adsorption of Gases*, Butterworth & Co., London, 1962.
2. Wark, K., Warner, C.F., and Davis, W.T., *Air Pollution*, Addison-Wesley Longman, Berkeley, CA, 1998, 220–277.
3. Yaw, C.L., Bu, L., and Nijhawan, S., Determining VOC adsorption capacity, *Pollut. Eng.*, Feb. 1995, 34–37.
4. Vatavuk, W.M., *QAQPS Cost Manual*, EPA 450/3-90-006, 4th ed., U.S. EPA QAQPS, Washington DC, Jan. 1990.
5. McCabe, W.L., Smith, J.C., and Harriott, P., *Unit Operations of Chemical Engineering*, 5th ed., McGraw-Hill, New York, 1976.
6. Buonicore, A. and Davis, W.T., ed., *Air Pollution Engineering Manual*, Van Norstrand, New York, 1992.
7. National Research Council, *International Critical Tables, Volume III*, McGraw-Hill, New York, 1929.
8. Perry, J.H., ed., *Chemical Engineers' Handbook*, 5th ed., McGraw-Hill, New York, NY, 1973.
9. Lide, D.R., *Handbook of Chemistry and Physics*, CRC Press, Boca Raton, FL, 2001.
10. Geankoplis, C.J., *Transport Processes and Unit Operations*, 2nd ed., Allyn and Bacon, Boston, MA, 1984.
11. Farrauto, R. and Bartholomew, C., *Fundamentals of Industrial Catalytic Processes*, Kluwer Scientific Publisher, Amsterdam, 1997.
12. Morbidelli, M., Garvriilidis, A., and Varma, A., *Catalyst Design: Optimal Distribution of Catalysts in Pellets, Reactors and Membranes*, Cambridge University Press, Cambridge, UK, 2001.

CHAPTER 14

Ventilation Requirements and Measurement

The purpose of ventilation is to supply fresh air to, and meet the heating/cooling requirements and air quality of, occupants within an indoor environment. Ventilation fulfils this purpose by bringing fresh air into the airspace to replace and dilute the heat, moisture, and gaseous and particulate pollutants that eventually build up indoors. Ventilation is effective in controlling temperature, relative humidity, and gaseous concentration. Particulate pollutant control may be different from the control of gaseous pollutants and temperature, because particulate matter (PM) has distinct properties, such as gravitational settling, inertia, and spatial distribution. Particulate matter may require a combination of abatement techniques. Techniques such as filtration and dust source control, together with ventilation, have been used to keep PM at an acceptable level. Ventilation control for temperature and for gaseous and particulate pollutants can be similar when the airspace is completely mixed. For typical indoor environments, the thermal settling velocity of particles of concern is much smaller than the air velocity within the airspace, and thus the airflow patterns prevail in the transportation of particles, and the particle settling-velocity is often negligible. When air velocity is low and the particle size is large, particle-settling velocity should be considered.

In this chapter, principles of ventilation are reviewed from the perspective of mass and energy balance. In the following analysis, whether we are dealing with the entire airspace or a zone within an airspace, perfect mixing is assumed. Specific ventilation strategies under incomplete mixing and ventilation effectiveness are discussed in Chapter 15. Ventilation rate is usually an unavoidable variable in solving indoor air quality problems. Typical methods of ventilation rate measurement, including direct airflow rate measurements and indirect measurements such as tracer gas and calorimetry methods, are included in this chapter. Other important factors, such as air leakage (including infiltration and exfiltration), have substantial effects on a ventilation system and ventilation effectiveness. *Air leakage* refers to the air enters or exits from undesigned passages, such as cracks in windows, doors, or structures. In extreme cases, excessive air leakage can cause the failure of ventilation. Therefore, determination of the air leakage rate and source is important in identifying ventilation effectiveness problems and will be included in Chapter 15.

By completing this chapter, the reader will be able to

- Determine the required minimum ventilation rate to maintain acceptable levels of indoor environmental parameters, such as temperature, relative humidity, and airborne pollutant concentrations
- Select the air-cleaning capacity (including the cleaning efficiency and flow rate) of a recirculation or filtration device to meet the room air-cleaning requirement
- Determine the room ventilation rate using different tracer methods
- Determine the local and room mean air ages using tracer injection methods, including the following:
  - Pulse tracer method
  - Step-up method
  - Step-down method
  - Calorimetry method
- Evaluate the effectiveness of air mixing using mean air ages for a ventilated room of which the ventilation rate and volume are known
- Measure ventilation rate using available commercial instrumentation and procedures for applications such as troubleshooting and pollutant emission evaluation

14.1 VENTILATION REQUIREMENTS FOR COMPLETELY MIXED AIRSPACES

In general, three types of variables are of concern and can be controlled by ventilation in an indoor environment:

- Temperature (sensible heat)
- Relative humidity
- Particular airborne pollutant

Ventilation requirement refers to as the minimum ventilation rate that must be maintained to keep all three variables within acceptable levels. Ventilation requirements for a ventilated airspace differ from its functions and atmospheric conditions. For many buildings minimum ventilation rates are based on temperature control. In other specialty buildings, such as an animal holding room in a cold climate, ventilation may be based on moisture balance control; ventilation for a welding workshop may be based on fume control.

14.1.1 Mass and Energy Conservation of a Ventilated Airspace

We start by considering a room as a single control volume with a volume V, within which air is completely mixed. As shown in Figure 14.1, supply air for the room has a volumetric flow rate Q_s, enthalpy h_s (including sensible heat h_{ss} and latent heat h_{ws}), moisture content w_s, and pollutant concentration C_s. Air being exhausted from the room has a volumetric flow rate Q_e, enthalpy h_e (including sensible heat h_{se}

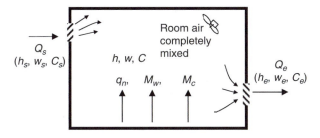

Figure 14.1 Heat and mass transfer of a ventilated airspace.

and latent heat h_{we}), moisture content w_e, and pollutant concentration C_e. Other psychrometric properties, such as specific volume v and relative humidity RH, can be found from a psychrometric chart or table (Appendix 3). Within the room airspace, the total sensible heat transfer rate is q_t, the moisture production rate is $\dot{W}_p$, and the pollutant production rate is $\dot{M}_p$. Note that the mass flow rate of supply air and exhaust air can be expressed by the volumetric flow rate and the specific volume; the energy and mass balances for the room as a single control volume can be described in terms of total air mass flow rate, enthalpy, mass of water, and mass of a specific pollutant.

$$\frac{Q_s}{v_s} - \frac{Q_e}{v} = \frac{d}{dt}\left(\frac{V}{v}\right) \tag{14.1}$$

$$h_s\frac{Q_s}{v_s} - h\frac{Q_e}{v} + q_t = \frac{d}{dt}\left(\frac{hV}{v}\right) \tag{14.2}$$

$$w_s\frac{Q_s}{v_s} - w\frac{Q_e}{v} + \dot{W}_p = \frac{d}{dt}\left(\frac{wV}{v}\right) \tag{14.3}$$

$$C_s\frac{Q_s}{v_s} - C\frac{Q_e}{v} + \dot{M}_p = \frac{d}{dt}\left(\frac{CV}{v}\right) \tag{14.4}$$

where v_s and v are specific volume of the supply air and the room air, respectively, and V is the volume of the airspace. Because air in the room is completely mixed, the properties in the exhaust air are the same as in the room air. That is, $v_e = v$, $h_e = h$, and $C_e = C$.

Equation 14.1 is the mass balance for the total air ventilation. The mass flow rate of the supply air minus the mass flow rate of exhaust air equals the change of mass of air within the airspace. Equation 14.2 is the energy balance for the sensible heat. The sensible heat in the air is the air mass flow rate multiplied by the sensible heat content of the air. Thus, the sensible heat in the supply air, minus the sensible heat in the exhaust air and plus the total sensible heat production, equals the change of sensible heat in the room air. Similarly, partial mass balances for moisture and a specific pollutant are described by Equation 14.3 and Equation 14.4, respectively.

The total sensible heat transfer rate q_t of the ventilated airspace is the sum of all heat loss or gain through and within the airspace:

$$q_t = q_{oc} + q_b + q_e + q_r \tag{14.5}$$

where q_{oc} is the sensible heat production rate by occupants; q_b is the heat transfer through the airspace shelter (typically through walls, ceiling, floor, and perimeter), including radiant heat transfer. Heat transfer though the shelter can be a heat loss or heat gain. If it is a heat loss, q_b is negative (−), and if it is a heat gain, q_b is positive (+) in Equation 14.5; q_e is the heat production rate from equipment such as lights, ovens, and motors. For an unventilated airspace, when q_t is negative, it is called *heating load* because supplemental heat is needed to maintain a desirable temperature. When q_t is positive, it is called *cooling load* because cooling is needed. For a ventilated airspace, the heat transfer rate caused by enthalpy differences in the supply and the exhaust air must be considered. In reality, there is always some exchange in sensible heat and latent heat, but the total heat remains essentially constant.

The total moisture production rate $\dot{W}_p$ and the total airborne pollutant production rate $\dot{M}_p$ within an enclosed airspace can be determined as follows:

$$\dot{W}_p = \dot{W}_{oc} + \dot{W}_b + \dot{W}_e \tag{14.6}$$

$$\dot{M}_p = \dot{M}_{oc} + \dot{M}_b + \dot{M}_e \tag{14.7}$$

where the subscripts *oc, b,* and *e* represent occupants, building, and equipment, respectively. The moisture production rate for humans at different activity levels, laboratory animals, and production animals can be found in books such as *ASHRAE Fundamentals* (2001) and *ASAE Standards* (2000).[1, 2]

The pollutant production rate $\dot{M}_p$ varies widely, depending on the types of pollutants and sources. Indirect methods for determining the particle production rate are discussed in Chapter 9. There is apparently no easy way to quantify the total pollutant production rate unless the sources are well defined. In many cases, there are multiple sources of pollutant production, which is usually in more than one phase (gaseous, liquid, or solid). Although it is difficult to characterize the production rate in a simple formula for most types of pollutants, one of common indoor pollutants, carbon dioxide, can often be readily calculated based on the load of the occupants. Because carbon dioxide is a metabolic product, on average every 24.6 kJ of total heat production (*THP*) will produce one liter of carbon dioxide for humans and animals under standard indoor conditions. The mass production rate of carbon dioxide in kg/s is known as

$$\dot{M}_{p-co2} = \rho_{co2} \frac{THP}{24600} \tag{14.8}$$

where ρ_{co2} is the density of carbon dioxide, which is 1.83 kg/m³ at 20°C and one atmospheric pressure; *THP* is the total heat production of occupants in kJ/s. Note that the total heat includes both sensible heat and latent heat, whereas the heat production in Equation 14.5 only includes the sensible heat. Carbon dioxide can be used as an indoor air quality indicator for ventilation control. A high carbon dioxide concentration usually indicates there is not enough ventilation for the airspace.

Analytical solutions of transient mass and heat transfer described in Equations 14.1 through 14.4 can be very complicated, because each variable in those equations could be time dependent. Many analytical solutions can be obtained by assuming that one or several variables (such as the total heat and moisture production or ventilation rate) are constants. When transient heat transfer must be considered, numerical algorithms and computer simulations are useful tools for solutions. Although ventilation requirements may vary substantially between the transient state and steady state, design criteria for ventilation requirements are usually based on steady-state calculations. For specific ventilation facilities, such as a fan hood used only occasionally, the ventilation requirement may be based on the transient state.

14.1.2 Sensible Heat Balance Ventilation Requirement

Applying mass conservation to supply and exhausted air, the rate of mass of air entering the room (Q_s/v_s) must be equal to the mass of air that is exhausted from the room (Q_e/v_e). Noting that all d/dt terms are zeros at steady state.

$$\frac{Q_s}{v_s} = \frac{Q_e}{v_e} \qquad (14.9)$$

Applying energy conservation to sensible heat, the total sensible heat contained in the supply air plus the net sensible heat transfer rate, must be equal to the total sensible heat removed by the exhaust air. Substituting Equation 14.9 into Equation 14.2 for steady state yields

$$h_{ss}\frac{Q_s}{v_s} + q_t = h_{es}\frac{Q_e}{v_e} \qquad (14.10)$$

where h_{ss} and h_{es} are sensible heat for supply air and exhaust air, respectively, in kJ/kg of dry air. Substituting Equation 14.9 into Equation 14.10 gives the volumetric ventilation rates for *sensible heat balance (or temperature balance) control* for exhaust air Q_e and supply air Q_s, respectively, in m³/s:

$$Q_e = \frac{q_t v_e}{(h_{es} - h_{ss})} \qquad (14.11)$$

$$Q_s = \frac{q_t v_s}{(h_{es} - h_{ss})} \qquad (14.12)$$

14.1.3 Moisture Balance Ventilation Requirement

Substituting Equation 14.9 into Equation 14.3 for the steady state gives the volumetric ventilation rates for *moisture (or relative humidity) balance control* for exhaust air Q_e and supply air Q_s, respectively, in m³/s:

$$Q_e = \frac{\dot{W}_p v_e}{(w_e - w_s)} \tag{14.13}$$

$$Q_s = \frac{\dot{W}_p v_s}{(w_e - w_s)} \tag{14.14}$$

where w is the water content of air in kilogram of water vapor per kilogram of dry air (kg/kg). Subscripts s and e represent supply and exhaust air, respectively. $\dot{W}_p$ is the water vapor production rate in kg/s within the airspace.

14.1.4 Pollutant Balance Ventilation Requirement

When mass balance is applied to pollutant balance ventilation control, care should be taken in regard to the units of pollutant concentration. Pollutant concentrations are commonly expressed in milligram of pollutant per cubic meter of air (mg/m³) for solid and liquid pollutants, and in ppm (in volume or mass) for gaseous pollutants. Care must be exercised to convert these common pollutant concentrations into consistent SI units. From Equation 14.4, particulate pollutant mass balance at the steady state can be written as

$$Q_e = \frac{\dot{M}_p v_e}{C_e v_e - C_s v_s} \tag{14.15}$$

$$Q_s = \frac{\dot{M}_p v_s}{C_e v_e - C_s v_s} \tag{14.16}$$

where C_e and C_s are the pollutant concentrations of exhaust air and supply air, respectively, in kilograms of pollutant per cubic meter of air. $\dot{M}_p$ is the mass production rate of the particulate pollutant in kilograms per second.

For gaseous pollutants, concentrations are commonly expressed in parts per million in volume (ppmv). Because 1 ppmv = 10^{-6} m³ of gas per cubic meter of air, Equation 14.15 and Equation 14.16 can be written as

$$Q_e = \frac{\dot{M}_p v_c v_e}{C_{e-ppmv} v_e - C_{s-ppmv} v_s} \times 10^6 = \frac{Q_{cp} v_e}{C_{e-ppmv} v_e - C_{s-ppmv} v_s} \times 10^6 \tag{14.17}$$

$$Q_s = \frac{\dot{M}_p v_c v_s}{C_{e-ppmv} v_e - C_{s-ppmv} v_s} \times 10^6 = \frac{Q_{cp} v_s}{C_{e-ppmv} v_e - C_{s-ppmv} v_s} \times 10^6 \quad (14.18)$$

where the gas concentrations C_e and C_s are in ppmv; $\dot{M}_p$ is the mass production rate for the gas in kg/s, and Q_{cp} is the volumetric gas production rate in m³/s; v_c is the specific volume of the gaseous pollutant in m³/kg. The conversion factor 10^6 is to ensure that the ventilation rate is in m³/s when the gas concentration is in ppmv.

14.1.5 Ventilation Graph

The preceding analysis defines the minimum ventilation rates for the control of temperature, moisture, and a specific pollutant. The minimum ventilation rate must be the highest of the three ventilation rates to balance sensible heat, moisture, and the given pollutant. Because sensible heat and moisture content are functions of air temperature, the ventilation requirements can be plotted vs. the outside temperature at given indoor conditions, including the required temperature, relative humidity, and an acceptable concentration of a pollutant, such as carbon dioxide or particles. Such a plot is called a *ventilation graph* (Figure 14.2), in which the minimum ventilation rates Q_{min} are required to maintain acceptable indoor environmental conditions (temperature, humidity, and a specific pollutant concentration) with respect to supply air temperatures. The minimum ventilation for an indoor environment should follow the uppermost portion of the three curves in a ventilation graph. Therefore, to determine the minimum ventilation rate, one must calculate all ventilation requirements for balances of temperature, moisture, and specific pollutant

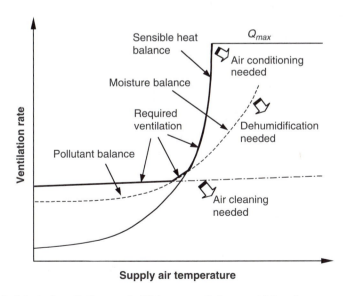

Figure 14.2 A typical ventilation graph. Minimum ventilation must follow the uppermost portion of the three curves to maintain an acceptable indoor environment. T_d is the heat deficit temperature.

concentration as described previously, and select the highest of the three as the minimum ventilation. If the minimum ventilation rate is selected based on humidity balance or pollutant concentration balance, supplemental heat is needed to maintain a desirable room temperature.

In practice, a ventilated airspace has a maximum ventilation capacity Q_{max}. At maximum ventilation, if the indoor temperature exceeds the required (usually the setpoint) temperature, air conditioning must be used to maintain the desirable temperature. If the relative humidity is too high at the maximum ventilation, dehumidification should be applied. If an air pollutant of concern exceeds the acceptable concentration, air-cleaning equipment should be added to ventilation system.

Example 14.1: *There are 30 students in a classroom. Each student produces 120 W of sensible heat and 100 g of water vapor per hour (or 70 W latent heat). The room air temperature is 20°C, and the specific volume is 0.84 m3/kg. The carbon dioxide concentration of the supply air is 400 ppmv. The supply air has a temperature of 0°C, a specific volume of 0.775 m3/kg, and relative humidity of 80%. Heat loss through the building shelter is 2 kW, and heat gain from equipment and lights are 1.5 kW. If we want to maintain an indoor environment in the classroom with a carbon dioxide concentration of less than 800 ppmv, relative humidity of less than 50%, and room temperature of 20°C, what should the minimum ventilation be for the supply air? At the chosen minimum ventilation rate, what are the actual temperature, relative humidity, and carbon dioxide concentration in the room, if there is no heating or humidification? Atmospheric conditions are assumed to be at sea level and standard, and the room air is completely mixed.*

Solution: From Appendix 7, we can find the psychrometric properties of supply air (0°C and 80% RH) and exhaust air (20°C and 50% RH) as follows:

$$w_s = 0.003, \ v_s = 0.773 \text{ m}^3/\text{kg}, \ h_{ss} = 0 \text{ kJ/kg}$$

$$w_e = 0.0073, \ v_s = 0.84 \text{ m}^3/\text{kg}, \ h_{es} = 20 \text{ kJ/kg}$$

From Appendix 7, the specific volume of CO_2 at 20°C, $v_c = 0.546$ m³/kg.

To determine the ventilation requirement, all three ventilation rates for sensible heat balance, moisture balance, and CO_2 balance must be calculated.

Ventilation for sensible heat balance, using the supply air rate, from Equation 14.12, is

$$q_t = q_{oc} + q_b + q_e + q_r = 0.12 \times 30 - 2 + 1.5 = 3.1 \ (kW)$$

$$Q_s = \frac{q_t v_s}{(h_{es} - h_{ss})} = \frac{3.1 \times 0.773}{20 - 0} = 0.12 \left(\frac{m^3}{s} \right)$$

Ventilation for moisture balance, using the supply air rate, from Equation 14.14, is

$$\dot{W}_p = 0.1 \left(\frac{kg}{person \cdot h} \right) \times 30 \ person \times \frac{1 \ h}{3600 \ s} = 0.00083 \ \frac{kg}{s}$$

$$Q_s = \frac{\dot{W}_p v_s}{(w_e - w_s)} = \frac{0.00083 \times 0.773}{0.0073 - 0.003} = 0.149 \ \left(\frac{m^3}{s} \right)$$

Ventilation for moisture balance, using the supply air rate, from Equation 14.18 and Equation 14.17, is

$$Q_{cp-CO_2} = M_c v_c = \frac{THP}{24600} = \frac{(0.12 + 0.07) \times 30}{24600} = 0.000232 \ \left(\frac{m^3 \ of \ CO_2}{s} \right)$$

$$Q_s = \frac{Q_{cp} v_s}{C_{e-ppmv} v_e - C_{s-ppmv} v_s} \times 10^6$$

$$= \frac{0.000232 \times 0.773}{800 \times 0.84 - 400 \times 0.773} \times 10^6 = 0.494 \ \left(\frac{m^3}{s} \right)$$

Because the highest ventilation requirement is for CO_2 balance control, the minimum supply airflow rate should be 0.494 m³/s. At this ventilation rate, the actual room temperature and relative humidity, assuming there is neither heating nor humidification, can be determined by solving the new sensible heat and moisture content in the room when $Q_s = 0.494$ m³/s. From Equation 14.12 and Equation 14.14

$$h_{es} = \frac{q_t v_s}{Q_s} + h_{ss} = \frac{3.1 \times 0.773}{0.494} + 0 = 4.6 \ \left(\frac{kJ}{kg} \right)$$

$$w_e = \frac{\dot{W}_p v_s}{Q_s} + w_s = \frac{0.00083 \times 0.773}{0.494} + 0.003 = 0.0032 \ \left(\frac{kg \ of \ water}{kg \ of \ air} \right)$$

From the psychrometric chart in Appendix 3, we can find that the temperature is 4.6°C and the relative humidity is 61%.

14.2 AIR CLEANING EFFICIENCY WITH RECIRCULATION AND FILTRATION

Consider an enclosed airspace with an air filter (Figure 9.7). The air filter has an overall filtration efficiency ξ_f, with a flow rate Q_f passing through the filter. If the room air is completely mixed, the pollutant concentrations in the exhaust air and the supply air to the filter are the same as the room pollutant concentration C. Under a general nonisothermal condition, $v_s \neq v$, and at steady state, the mass balance of a specific pollutant for the room can be rewritten from Equation 9.28 as

$$\dot{M}_p = (Q + \xi_f Q_f + V_d A) C_f - QC_s \frac{v_s}{v} \tag{14.19}$$

where $\dot{M}_p$ is the mass production rate of the pollutant in kg/s; Q is the exhaust volumetric ventilation rate in m³/s; ξ_f is the overall filter efficiency; Q_f is the flow rate passing through the filter in m³/s; V_d is the deposition velocity in m/s; C_f is the pollutant concentration of the room with filter system in kg/m³; C_s is the pollutant concentration of supply air in kg/m³; and v_s and v are specific volumes of supply air and room air, respectively, in m³/kg.

When the room has no filter, the pollutant production remains the same. The deposition velocity V_d and pollutant concentration of the supply air, C_s, also remain the same as that in the room with a filter. The pollutant concentration within the airspace changes from C_f to C when there is no filter. The new mass balance is deduced from Equation 14.19 and becomes

$$\dot{M}_p = (Q + V_d A) C - QC_s \frac{v_s}{v} \tag{14.20}$$

Because the pollutant production rates remain the same regardless of the filter, Equation 14.19 and Equation 14.20 are equal. The only change will be in the pollutant concentration in the room. Equating Equation 14.19 and Equation 14.20, and rearranging, gives

$$\frac{C_f}{C} = \frac{Q + V_d A}{Q + \xi_f Q_f + V_d A} \tag{14.21}$$

If we define the overall air cleaning efficiency of the room, that is, the fraction of pollutant concentration reduction resulting from air filtration within the room, as ξ_r, then

$$\xi_r = \frac{C - C_f}{C} = \frac{\xi_f Q_f}{Q + \xi_f Q_f + V_d A} \tag{14.22}$$

Note that the overall room air cleaning efficiency is always smaller than the filter efficiency. When the flow rate through the filter, Q_f, is substantially higher than the room ventilation rate Q, the room cleaning efficiency is similar to the filter efficiency. In Equation 14.22, the deposition velocity V_d is usually an unknown variable and very difficult to measure. If the particle size distribution and the room ventilation

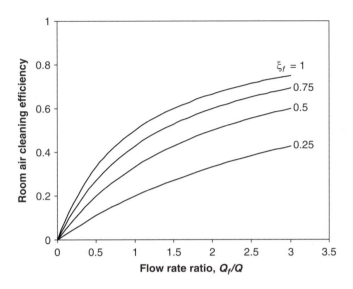

Figure 14.3 The overall efficiency for room air cleaning using filters at different filtration efficiencies.

systems are defined, V_d can be calculated as described in Chapter 9. For a gaseous pollutant, the deposition is negligible compared with the removal by ventilation. Thus, for gases

$$\xi_r = \frac{C - C_f}{C} = \frac{\xi_f Q_f}{Q + \xi_f Q_f} \qquad (14.23)$$

From Equation 14.23, the overall room air cleaning efficiency is a function of filter efficiency, flow rate through the filter, and room ventilation rate. If the number of particles deposited on surfaces is much less than the number of particles removed by ventilation and filtration, Equation 14.23 can also be used to estimate the maximum particle removal efficiency for a room with a filter under complete-mixing conditions. Figure 14.3 shows the relationship between the room air cleaning efficiency and the ratio of flow rate through the filter to the room ventilation rate (Q_f/Q). Clearly, at a given filter efficiency ξ_f, the overall room air cleaning efficiency increases as the ratio Q_f/Q increases. On the other hand, increasing the flow rate through the filter means more frequent filter maintenance and higher operating cost.

Example 14.2: Determine the maximum room air cleaning efficiency for a room with a ventilation rate of 500 l/s. The room has an air-cleaning device with a filtration efficiency of 85% for particles, and the flow rate passing through the filter is 200 l/s. Assume that deposition is negligible compared to particle removal through ventilation. To reduce the pollutant emission from the room by at least 50%, which is equivalent to achieving a 50% or higher room air cleaning efficiency, how many such air cleaning devices are needed?

Solution: Neglecting the deposition for maximum possible cleaning efficiency, from Equation 14.22,

$$\xi_r = \frac{\xi_f Q_f}{Q + \xi_f Q_f} = \frac{0.85 \times 0.18}{0.5 + 0.85 \times 0.18} = 0.234 = 23.4\%$$

For $\xi_r = 50\%$,

$$Q_f = \frac{Q\xi_r}{\xi_f(1-\xi_r)} = \frac{0.5 \times 0.5}{0.85 \times (1-0.5)} = 0.588 \, (m^3 \, / \, s) = 588 \, (l \, / \, s)$$

Thus, three air-cleaning devices are needed, each with a 200 l/s flow rate.

14.3 TERMINOLOGY OF AIR AGE

Before we discuss different ventilation rate measurement methods, it is necessary to define different terms used in room ventilation studies. As shown in Figure 14.4, a room with a volume of V is ventilated at a total airflow rate of Q. Under complete mixing conditions, the fates of air and a specific pollutant may be indistinguishable, because that pollutant is following the path of the airflow.

Let us consider that an air molecule or a pollutant particle, n_i, enters the room airspace. At a specific location within the room airspace, the following time periods for the air molecule can be defined:

- **The internal age τ_i** — The time that has elapsed since n_i entered the room
- **The residual time τ_r** — The remaining time that n_i will spend in the room
- **The residence/retention time τ** — The total time (age) between when n_i entered the room and left it

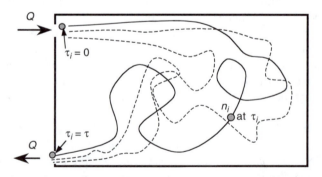

Figure 14.4 The path of an air molecule (n_i) in a ventilated room at different ages. The dotted lines show the paths either of the same particle entering the room at different times or of different particles entering the room at the same time.

The three time periods have a relationship of

$$\tau_i + \tau_r = \tau \qquad (14.24)$$

Practically, it is less meaningful to consider just the age of one air molecule or a single particle. Rather, the mean age of a population of air or particles of a control volume or the entire room is considered. Thus, the mean age of air refers to the mean age of air of the concerned volume.

The other frequently used term for in ventilation studies is *air turnover rate*. The air turnover rate is defined as

$$air\ turnover\ rate = \frac{Q}{V} \qquad (14.25)$$

The reciprocal of the air turnover rate is the *air turnover time*, sometimes called the *nominal time* of the ventilation system. If the ventilation rate Q is in room volume per hour, the turnover rate is in air changes per hour (ACH). Apparently, the air turnover time is the same as the time constant of a ventilation system for pollutant removal in completely-mixed room. It should be pointed out that the air turnover time is different from the mean age of room air. The calculation of mean age of room air will be discussed in later sections of this chapter.

14.4 TRACER GAS METHODS FOR VENTILATION RATE MEASUREMENT

Some of the most commonly used methods to measure room ventilation rate involve tracer gas. The measurement of the total flow rate with a tracer gas is based on the assumption that the room air is completely mixed, and therefore artificial mixing during measurement is often necessary. Depending on the type of tracer injection method, the tracer gas techniques are classified by the following categories:

Rate of decay — A short duration of gas is injected at a constant rate into the airspace to establish a uniform concentration within the room, and the gas concentration decay is recorded.

Constant injection rate — Tracer gas is injected into the room at a constant rate, and the gas concentration is recorded during the time period of the gas injection.

Constant concentration method — Tracer gas is injected into the airspace at a controlled rate so as to maintain a constant gas concentration in the room. Meanwhile, the injection rate of the tracer gas is recorded.

Injection of tracer gas can be done by either active or passive techniques. *Active techniques* involve using pressurized systems, such as pumps, to release and sample the tracer gas. *Passive techniques* involve utilizing molecular diffusion to release and sample the tracer gas. Active methods are suitable for unoccupied buildings and can be done in short period of time. Passive methods are more suitable for occupied buildings and obtain long-term averages. For all tracer gas methods, time-dependent ventilation rates pose special problems, because the actual ventilation rate is systematically different from the average rate. Furthermore, incomplete mixing within the room airspace can introduce another large error source. In reality, most ventilation systems' ventilation rates fluctuate due to the change of heating or cooling loads, or even to power fluctuations, and a certain level of incomplete mixing occurs. In other cases, incomplete mixing ventilation, such as displacement ventilation, is desirable. It is often a challenging task to determine the ventilation rate for a room or a building.

14.4.1 Theoretical Analysis

The basis for all three tracer gas methods mentioned previously is the mass balance of the tracer gas and air. To simplify the following analysis, isothermal conditions are assumed. Because the temperatures of air and the tracer gas are constant, the mass flow rate is proportional to the volumetric flow rate. Under complete mixing, the mass balance of tracer gas within the room is

$$V\frac{dC}{dt} = Q_g - QC \qquad (14.26)$$

where V is the volume of room, C is the tracer gas concentration, Q is the room ventilation rate, and Q_g is the tracer gas flow rate. Solving Equation 14.26 for room ventilation gives

$$Q = \frac{1}{C}\left(Q_g - V\frac{dC}{dt}\right) \qquad (14.27)$$

Equation 14.27 gives the instantaneous mass balance of the tracer gas. In practice, the instantaneous mass balance equation cannot be used to obtain the ventilation rate, because dC/dt cannot be measured instantaneously. Further, an average ventilation rate during a period of time is more useful, especially for time-dependent ventilation. Integrating Equation 14.27 between a time interval $[t_1, t_2]$ gives

$$\int_{t_1}^{t_2} Q\,dt = \int_{t_1}^{t_2} \frac{Q_g}{C}dt + V\ln\left(\frac{C(t_1)}{C(t_2)}\right) \qquad (14.28)$$

Because the ventilation rate and tracer injection rate may be time dependent, we use their time average rate during the time period $[t_1, t_2]$ and obtain

$$\overline{Q} = \frac{\overline{Q_g}}{C} + \frac{V}{(t_2 - t_1)}\ln\left(\frac{C(t_1)}{C(t_2)}\right) \qquad (14.29)$$

where the over bar denotes the time average during the time period of $[t_1, t_2]$:

$$\overline{Q} = \frac{1}{t_2 - t_1} \int_{t_1}^{t_2} Q(t)\,dt \tag{14.30}$$

$$\frac{\overline{Q_g}}{C} = \frac{1}{t_2 - t_1} \int_{t_1}^{t_2} \frac{Q_g(t)}{C(t)}\,dt \tag{14.31}$$

At a constant tracer gas injection rate, Equation 14.29 becomes

$$\overline{Q} = Q_g\left(\overline{\frac{1}{C}}\right) + \frac{V}{(t_2 - t_1)} \ln\left(\frac{C(t_1)}{C(t_2)}\right) \tag{14.32}$$

Note that for a varying gas concentration, the inverse of the average concentration $(1/\overline{C})$ is different from the average of inverse concentration $\overline{(1/C)}$. The inverse of an average is generally smaller than the average of an inverse:

$$\overline{\frac{1}{C}} \le \overline{\frac{1}{C}} \tag{14.33}$$

14.4.2 The Rate of Decay Method

When the rate of decay technique is used, a small amount of tracer gas is initially injected into the airspace. The measurement of tracer gas concentration is started after the gas injection has stopped. In Figure 14.5, note that the measurement start time t_1 can be at any time after the tracer injection has stopped. Such elapsed time after the tracer has stopped should be minimal, in order to have a sufficient concentration difference during the measurement period t_1 and t_2. Because the tracer gas flow rate $Q_g = 0$ during the measurement time period $[t_1, t_2]$, Equation 14.32 becomes

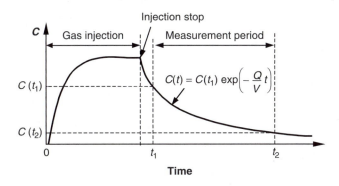

Figure 14.5 Time sequence of a rate of decay method.

$$\overline{Q} = \frac{V}{(t_2 - t_1)} \ln\left(\frac{C(t_2)}{C(t_1)}\right) \tag{14.34}$$

The rate of decay method is a two-point estimate of the ventilation rate based on two gas concentration readings. Using the average values, the decay is not necessarily exponential, and the ventilation rate may not be constant during the measurement period.

If the ventilation rate is constant, the decay becomes exponential, and a direct integration of Equation 14.27 gives

$$Q = \frac{V}{(t_2 - t_1)} \ln\left(\frac{C(t_2)}{C(t_1)}\right) \tag{14.35}$$

The tracer gas concentration is

$$C(t) = C(t_1)\exp\left(-\frac{Q}{V}t\right) \tag{14.36}$$

Errors due to incomplete mixing in rate of decay methods have been analyzed by Etheridge and Sandberg.[3] To minimize the uncertainties, the measurement time period is suggested to be approximately the same as the room air turnover time. That is, the suggested measurement time is

$$t_2 - t_1 \approx \frac{V}{Q} \tag{14.37}$$

Suggested Measurement Procedures

The total volume of the room must be measured, and a good perfect-mixing condition should be achieved. A mixing fan is often necessary to achieve good mixing. The tracer gas then can be introduced into the room either by someone walking around the room with a gas cylinder or by automatic injection through tubes. Multipoint gas injection is suggested to ensure complete mixing. After the gas injection, allow approximately 5 minutes for a room the size of a typical before recording the gas concentration. To obtain sufficient readings, the gas concentration at each measurement point should be taken at least once every 10 minutes for a typical-sized room. As a rule of thumb, the total measurement time should be equal to the time constant of the room ventilation system (V/Q), if this can be assessed in advance. Otherwise, the measurement period should be more than an hour.

Example 14.3: A rate of decay tracer method is applied to a room with a volume of 120 m³. After the tracer injection has stopped, tracer concentration at the exhaust

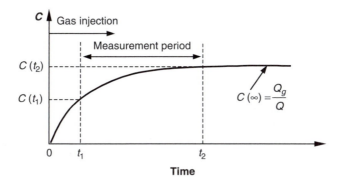

Figure 14.6 Time sequence of a rate of constant injection method.

air is immediately measured at 30 parts per billion in volume. After 5 minutes, the trace concentration was reduced to 10 ppb. What is the room ventilation rate? Assume that the room air is completely mixed and the ventilation rate is constant.

Solution: From the problem description, we know that

$$t_1 = 0 \text{ and } t_2 = 300 \text{ s}, \; C(t_1) = 300 \text{ ppb and } C(t_2) = 10 \text{ ppb}$$

From Equation 14.35

$$Q = \frac{V}{(t_2 - t_1)} \ln\left(\frac{C(t_2)}{C(t_1)}\right)$$

$$= \frac{120 \, m^3}{300 \, s - 0} \ln\left(\frac{30}{10}\right) = 0.439 \left(\frac{m^3}{s}\right)$$

14.4.3 The Constant Injection Method

When a constant injection rate method technique is used, the tracer gas is injected into the airspace at a constant rate. Integrating Equation 14.27 over a measurement time period $[t_1, t_2]$, as shown in Figure 14.6, and noting that the gas concentration may be time dependent, gives

$$\int_{t_1}^{t_2} Q(t)C(t)dt = Q_g(t_2 - t_1) - V\left[C(t_2) - c(t_1)\right] \tag{14.38}$$

The left side of Equation 14.38 has two time dependent variables, $Q(t)$ and $C(t)$. Applying the integral mean value theorem of Axley[4] to the left side, gives the ventilation rate

458 INDOOR AIR QUALITY ENGINEERING

$$Q(t) = \frac{Q_g(t_2 - t_1) - V[C(t_2) - C(t_1)]}{\int_{t_1}^{t_2} C(t)dt} \quad (t_1 \le t \le t_2) \tag{14.39}$$

Equation 14.39 shows that the constant injection method determines that the ventilation rate has occurred at some time during the measurement period. If the ventilation rate is constant during that period, then $Q(t)$ is a constant value and does not depend on time. In terms of average tracer gas concentration during the measurement period $\overline{C}$, Equation 14.39 can be rewritten as

$$Q(t) = \frac{Q_g}{\overline{C}} - \frac{V[C(t_2) - C(t_1)]}{\overline{C}(t_2 - t_1)} \quad (t_1 \le t \le t_2) \tag{14.40}$$

When the measurement takes hours or days, the second term at the right side of Equation 14.40 becomes negligible, thus

$$Q(t) = \frac{Q_g}{\overline{C}} \quad (t_1 \le t \le t_2) \tag{14.41}$$

Note that even if an average tracer gas concentration is used, the ventilation rate estimated may not be the same as the average ventilation rate during that measurement period, because the relationship between gas concentration and ventilation rate is nonlinear. A common practice is to inject the tracer gas over a long enough period of time to reach a gas concentration equilibrium $C(\infty)$. At the state of tracer gas equilibrium, the ventilation rate becomes

$$Q = \frac{Q_g}{C(\infty)} \tag{14.42}$$

In many practical situations, ventilation rate is not constant. Aside from on/off events, wind around the buildings can cause pressure differences and change the ventilation rate. Thus, the tracer gas concentration may never achieve an equilibrium state. Axley summarized the following general considerations to minimize errors caused by such variations:[4]

For ventilation rate variation having periods of variation $\tilde{\tau}$ smaller than the ventilation time constant [i.e., $\tilde{\tau} \le V/Q(t)$], the constant tracer gas injection method will provide accurate estimation of the mean ventilation rate, regardless of the magnitude of ventilation rate variation.

When $\tilde{\tau} \ge V/Q(t)$, and if $(t_2 - t_1) < \tilde{\tau}$, the constant tracer gas injection method will provide accurate estimates of near instantaneous ventilation rate.

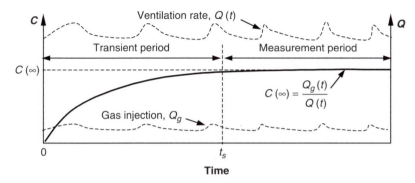

Figure 14.7 Schematic of time response of tracer gas concentration, injection rate, and room ventilation rate with a constant concentration method.

If $\tilde{\tau} \geq V/Q(t)$ and $(t_2 - t_1) \approx \tilde{\tau}$, the constant tracer gas injection method may underestimate or overestimate the ventilation rate, depending on the timing of the integration interval relative to the ventilation fluctuation.

If $\tilde{\tau} \geq V/Q(t)$ and $(t_2 - t_1) \gg \tilde{\tau}$, the constant tracer gas injection method will underestimate the average ventilation rate.

Suggested Measurement Procedures

Similar to the rate of decay method, the room volume must be measured, and a mixing fan is often necessary to ensure complete mixing within the room. The constant injection rate method requires the identification of the inlet of the most supply air, where the tracer gas should be released and dispersed into the room. For mechanically ventilated rooms, this will not be a problem, because the air inlets are easy to identify. For naturally ventilated airspaces or leaky rooms, it may be challenging to identify where the supply air comes from. A visible smoke tracer may be used to identify the incoming airflow before the measurement. The other measurement procedures are the same as for the rate of decay method.

14.4.4 The Constant Concentration Method

When the constant concentration method is used, the tracer gas injection rate is controlled to maintain a constant concentration [i.e., the target concentration $C(\infty)$] within the measurement airspace. As shown in Figure 14.7, because the ventilation rate may be time dependent, the tracer gas injection rate must be adjusted accordingly to maintain a constant concentration in the room. The ventilation rate can be expressed by rewriting Equation 14.42 as

$$Q(t) = \frac{Q_g(t)}{C} \tag{14.43}$$

Although we will not have a detailed discussion, a qualitative analysis of dynamic response will assist the reader in understanding the limitations of the constant concentration method. As shown in Figure 14.7, from the beginning of the gas injection until the targeted concentration is reached, there is a time delay t_s, during which constant concentration method cannot be used. Depending on the error allowance and the room size, this time delay can be hours for a typical room. Therefore, this method can be very slow to reach the targeted tracer gas concentration. Furthermore, if the ventilation varies with the time, as would occur with a ventilation system with a fan on/off control system, there is another time delay between the ventilation rate change and the subsequent tracer injection rate change. Assume that this time delay is t_{si}. Then, the $Q(t)$ calculated using Equation 14.43 is actually the ventilation t_{si} from long before. For example, in a room of 40 m^3 air volume with a standard desktop circulation fan, t_{si} is approximately four minutes.

In summary, the constant tracer-gas concentration method is not suitable for predicting ventilation rate with a frequent variation. It is a slow measurement for large airspaces, because it takes a long time (typically several hours) to achieve a target concentration. In practice, this length of time is prohibitive for using the constant concentration method in many practical situations.

14.5 TRACER GAS METHOD FOR AIR AGE DISTRIBUTION MEASUREMENT

Air age is the residence time of an air population within a concerned airspace. An air population could be a small volume of the local air or the entire volume of room air. Unlike the measurement of ventilation rate, measurement of room air age using a tracer gas does not require complete mixing of the room air. Thus, air age can be used as a measure of pollutant removal efficiency.

14.5.1 Three Main Procedures to Determine Air Age Distribution

Air age and corresponding statistical distributions may be obtained by the injection of tracer gas or a pollutant at a point located either in the room or in the supply air duct. The population and the age are determined by the positions of the injection and the sampling points. Injecting gas into the supply air duct and sampling in the exhaust duct allows the residence time of the supply air to be determined. If the sampling point is in the room, the internal age of the local air (and tracer) is observed. Measurement of the residence time of the total internal population of pollutant requires injection into the room and sampling of the exhaust air. If the sampling point is also in the room, the local air age of the local pollutant concentration is obtained (Figure 14.8). The tracer-gas injection point in the room should be in close vicinity to the pollutant source.

The type of room air age distribution and air age is determined by one of the following the *injection* procedures: pulse method, step-up method, and step-down method.[3] As shown in Figure 14.8, the injection tracer gas rate is Q_g, and the room ventilation rate is Q. Tracer gas concentrations at the exhaust point and at an arbitrary

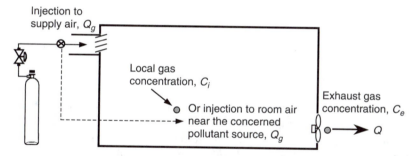

Figure 14.8 Schematic of tracer gas injection system.

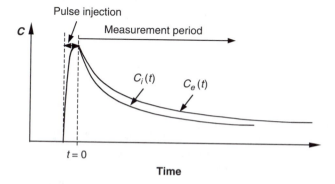

Figure 14.9 Schematic of time responses of tracer gas concentrations at the ith point in the room and the exhaust air, using the pulse method. The room air is completely mixed by a mixing fan during the gas injection. For a perfectly mixed airspace, the two concentration decay curves would overlap.

ith point within the room are C_e and C_i, respectively. The mean air ages for a local point and for the entire room are calculated as described in the following sections.

The Pulse Method

A small amount of tracer gas or pollutant is introduced to the supply air or at a point within the room, giving the room gas concentration a step change. The recorded gas concentration vs. time at the ith point within the room or at the exhaust gives the statistical density function of the corresponding air age distribution (Figure 14.9). The tracer gas concentrations at ith point and at the exhaust may not be the same if the room air is not completely mixed, thus the mean air ages for the local ith point and for the entire room may not be the same.

The mean local air age $\bar{\tau}_i$ and the mean room air age $\bar{\tau}_e$ are calculated as follows:

$$\bar{\tau}_i = \frac{\int_0^\infty tC_i(t)dt}{\int_0^\infty C_i(t)dt} \tag{14.44a}$$

$$\bar{\tau}_e = \frac{\int_0^\infty t^2 C_e(t)dt}{\int_0^\infty tC_e(t)dt} \qquad (14.44b)$$

From a practical point of view, the upper limits of the integrals must be a t_2, when the measurement stopped. The t_2 may not be the same at a local point and at the exhaust. If the measurement periods are the same for both the local point and the exhaust, the mean air ages will be different, because the gas concentrations are different. Detailed derivation of air mean ages for a local point and the overall room using an approach of air age frequency distribution has been described by Etheridge and Sandberg.[3]

The Step-up (Source) Method

A continuous and constant flow Q_g of tracer gas or pollutant is introduced to the supply air duct or at a point within the room. The recorded growth of gas concentration vs. time gives the cumulative age distribution.

When a step-up injection method is used, the initial gas concentration is zero (Figure 14.10). At a constant gas injection rate Q_g, the gas concentrations at the ith point and the exhaust eventually reach their target values $C_e(\infty)$ and $C_i(\infty)$, respectively. The mean local air age $\bar{\tau}_i$ and the mean room air age $\bar{\tau}_e$ are calculated as follows:

$$\bar{\tau}_i = \frac{\int_0^\infty t[C_i(\infty) - C_i(t)]dt}{\int_0^\infty [C_i(\infty) - C_i(t)]dt} \qquad (14.45a)$$

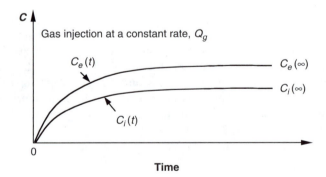

Figure 14.10 Schematic of time responses of tracer gas concentrations at the ith point in the room and in the exhaust air using the step-up method. For a perfectly mixed airspace, the two concentration curves would overlap.

$$\bar{\tau}_e = \frac{\displaystyle\int_0^{\infty} t[C_e(\infty) - C_e(t)]dt}{\displaystyle\int_0^{\infty} [C_e(\infty) - C_e(t)]dt} \tag{14.45b}$$

The Step-down (Decay, Washout) Method

After the concentrations have reached their equilibrium (target) values in a step-up procedure, the tracer gas or pollutant injection is stopped. The recorded decay of tracer concentration vs. time gives the complementary cumulative age distribution function (Figure 14.11).

When a step-down injection method is used, the step-up gas injection is first applied at a constant gas injection rate Q_g, until the gas concentrations at the ith point and the exhaust eventually reach their target values $C_e(\infty)$ and $C_i(\infty)$, respectively. At the target values, gas injection is stopped, and measurement of gas concentrations begins. The initial gas concentrations for the step-down stage, $C_e(0)$ and $C_i(0)$, are the target concentrations for the step-up stage, $C_e(\infty)$ and $C_i(\infty)$), for the exhaust and i^{th} point in the room, respectively. The mean local air age $\bar{\tau}_i$ and the mean room air age $\bar{\tau}_e$ are calculated as follows:

$$\bar{\tau}_i = \frac{\displaystyle\int_0^{\infty} C_i(t)dt}{C_i(0)} \tag{14.46a}$$

$$\bar{\tau}_e = \frac{\displaystyle\int_0^{\infty} tC_e(t)dt}{\displaystyle\int_0^{\infty} C_e(t)dt} \tag{14.46b}$$

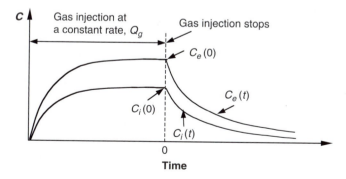

Figure 14.11 Schematic of time responses of tracer gas concentrations at the ith point in the room and the exhaust using the step-down method. For a perfectly mixed airspace, the two concentration curves would overlap.

From a theoretical point of view, the three injection methods are equivalent in the sense that they provide the same information. The pulse technique gives frequency distributions, the step-up method gives the cumulative distribution, and the step-down method gives the complementary cumulative distribution. All three injection methods are based on the assumption that the room airspace has no leakage (including infiltration and exfiltration), that is, all air comes in from the supply inlet and leaves via the exhaust outlet. However, in a real building, there is likely to be leakage through other openings than air supply inlets and exhaust outlets. This may introduce two sources of errors. Not all incoming air will be sampled in the exhaust duct. If not all incoming air is marked by the tracer, the final equilibrium concentration in the room will be unknown.

To overcome the problem that not all incoming air is marked, a modified step-down method is often applied. In this method, the gas is injected directly into the room and artificial mixing is used to create a uniform initial concentration $C(0)$. When the concentrations have reached a suitable initial level, the mixing fans are turned off and the decays of the concentrations are recorded. In this way, all air entering the room is marked by the tracer gas. However, the modified step-down procedure does not solve the problem of exfiltration. It introduces another disadvantage, in that the natural flow pattern in the room is at first destroyed, and some time is taken before it is restored.

The modified step-down method is the most widely used method in smaller rooms and ordinary homes. Its popularity stems from two circumstances. First, it is simpler than other methods. One does not have to control and measure the flow rate or the amount of gas injected. Second, in practice there is always an inflow of air through undesigned openings.

Similarly, a modified step-up method allows one to mark all air present in the airspace. The major drawback is that at the beginning, the natural airflow pattern within the room will be disturbed by the air motions set up by the artificial mixing.

Example 14.4: *During a step-down tracer experiment, the initial tracer concentration at a point i, $C_i(0)$, is measured as 50 parts per billion. The concentration then decreases at the following rate: $C_i(t) = C_i(0)\exp(-0.002t)$, where t is in seconds. Determine the local mean air age.*

Solution: From Equation 14.46a, the mean air age for the location i is

$$\overline{\tau}_i = \frac{\int_0^\infty C_i(t)dt}{C_i(0)} = \frac{\int_0^\infty 50\exp(-0.003t)dt}{50}$$

$$= \frac{\exp(-0.003 \times t)\big|_0^\infty}{-0.003} = 333(s)$$

14.6 CALORIMETRY METHOD TO ESTIMATE VENTILATION RATE

When precise information is needed about the ventilation rate in a room, two methods are used. One method is to attach a flow-measuring duct either upstream or downstream of each exhaust or supply duct. Potential methods for measuring airflow in such ducts have been outlined by Ower and Pankhurst and by Replogle and Birth.[5, 6] A second method for taking measurements of building air exchange rates is to conduct tracer gas experiments described in previous sections.

One major limitation of using tracer gas methods to measure ventilation rate is that sophisticated instrumentation is needed. When less precise estimates of the ventilation rate are needed, two other methods have been reported. In one of these methods, the ventilation rate is monitored by measuring the amount by which the carbon dioxide concentration of the ventilating air stream is increased by a supposedly known and steady rate of production of metabolic carbon dioxide.[7] In essence, this is the same as the constant tracer gas injection method, where the tracer gas is carbon dioxide and the constant injection is the production of CO_2 within the room.

In the same way that the tracer gas method of determining ventilation rate depends on monitoring changes within the airspace following a step change in tracer gas supply, it may also be possible to estimate ventilation rates by monitoring temperature changes following a step change in heat input. In this case, temperature is the "tracer." Because a steady-state temperature in a room involves many unknown variables, such as heat transfer rate and thermal capacitance of the building envelope, a transient calorimetry technique for airflow determination is described. In the following section, this theory is developed and the method is illustrated using experimental data.

14.6.1 Theory of Transient Calorimetry

The idealized temperature-time response of a completely mixed airspace is subject to a cyclical step change in sensible heat input. As shown in Figure 14.12, the room temperature increases when a supplemental heater with a capacity q_s is activated. The temperature will decrease exponentially after the heater is turned off. If the time interval for a complete cycle is short, say, less than 30 minutes, then it is reasonable to assume that the external climatic conditions are constant. Given this condition, the differential equations describing the energy balance of the airspace during the heating and cooling periods differ in only one term: the rate of sensible heat input Q_t.

Consider the entire ventilated airspace as a control volume, physically bounded by the inside surfaces of the building envelope. If there are no heat sinks within the airspace, the transient sensible heat balance for the control volume is described as

$$\frac{d(McT)}{dt} = \dot{m}_s c_s T_s - \dot{m}_e c_e T_e + q_s - q_b \qquad (14.47)$$

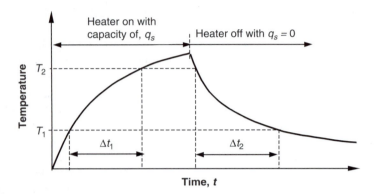

Figure 14.12 Temperature response of a ventilated airspace when it subjected to a heater-on and heater-off event.

where M is the total mass of air in the airspace, $\dot{m}$ is the mass flow rate of air, T is air temperature, c is the specific heat of air in kJ/kg°C, q_s is the supplemental heat, and q_b is the heat transfer rate through building shelter in kJ/s. The subscripts s and e refer to the properties of supply air and exhaust air, respectively.

In this estimation method, the temperature change from peak to peak (Figure 14.12) within the airspace is low, typically lower than 5°C. Therefore, the specific heat c can be treated as constant (i.e., $c = c_s = c_e$). For small temperature change and low pressure across the building ventilation systems, the mass flow rate of ventilated air can be considered constant: $\dot{m}_s = \dot{m}_e = \dot{m}$. Further, the airspace is completely mixed, and thus $T_e = T$. The heat transfer through the building can be expressed as

$$q_b = hA(T - T_b) \tag{14.48}$$

where h is the convective heat transfer coefficient, A is the inside surface area of airspace, and T_b is the temperature of the inside airspace surface, which is assumed constant with small-room air temperature fluctuation.

By invoking the preceding assumptions and the identities given in Equation 14.48, Equation 14.47 can be rewritten as follows:

$$\frac{dT}{dt} + \frac{\dot{m}c - hA}{Mc}T - \frac{\dot{m}cT_s + q_s - hAT_b}{Mc} = 0 \tag{14.49}$$

In Equation 14.49, all variables except $\dot{m}$ and h may be measured directly. During the heating period, the heat supply is constant at the rate q_s. During the cooling period, $q_s = 0$. Given the initial conditions depicted in Figure 14.12, the solutions to Equation 14.49 for the heating period Δt_1 and the cooling period Δt_2 are given by

$$\frac{B_1 - B_3T_2}{B_1 - B_3T_1} = \exp(-B_3\Delta t_1) \tag{14.50}$$

$$\frac{B_2 - B_3 T_1}{B_2 - B_3 T_2} = \exp\left(-B_3 \Delta t_2\right) \tag{14.51}$$

where

$$B_1 = \frac{\dot{m}cT_s + q_s - hAT_b}{Mc} \tag{14.52}$$

$$B_2 = \frac{\dot{m}cT_s - hAT_b}{Mc} \tag{14.53}$$

$$B_3 = \frac{\dot{m}cT_s - hT_b}{Mc} \tag{14.54}$$

14.6.2 Application of Calorimetry Method

The transient calorimetry method may be used to estimate the ventilation rate of an airspace, especially those difficult to measure directly. In Equation 14.50 and Equation 14.51, there are actually two unknowns: flow rate $\dot{m}$ and the convective heat transfer coefficient h. Other variables can be measured. However, the two unknowns were not expressed in explicit formulas; therefore, the reiteration method may be used to solve the flow rate.

An experiment may be conducted in an airspace in which a heater of known energy output q_s is cycled on and off. The temperature of the airspace can be measured immediately upstream of the exhaust outlet and plotted vs. time, as in Figure 14.12. From the temperature–time plot, the values for $T_1, T_2, \Delta t_1$, and Δt_2 can be selected. The supply air temperature T_s, and the building's physical parameters, total interior surfaces enclosing the airspace A, the volume of the airspace V, and the building surface temperature T_b, must be measured. It is ideal that the airspace internal surface temperature be uniform; otherwise, the heat transfer through the building shelter has to be the sum of heat transfer from different sections with different interior surface temperatures:

$$q_b = h \sum A_i (T - T_{bi)} \tag{14.55}$$

where the subscript i indicates sections of interior surface with different temperatures. Evidently, interior surface temperature will fluctuate slightly as the room air fluctuates, thus introducing an error source.

The transient calorimetry method for predicting ventilation rates appears to have potential applicability as part of a trouble-shooting procedure for buildings. Accuracy of the method depends on accuracy in temperature and time measurements. To this end, it is necessary to use a heat source that will provide a significant temperature lift within the room, typically 3 to 5°C. The elapsed time for the room temperature to first increase and then decrease between two temperature limits should be sufficiently long, say, 5 to 10 minutes in each of the heating and cooling segments of the temperature cycle, or ideally about the same as the ventilation constant V/Q if Q has a reasonable estimation. If the elapsed time is too long, however, conditions within the room or the supply air conditions could change, which would violate the basic assumptions on which this transient calorimetry method was based.

The transient calorimetry method requires accurate estimates of the heat transfer occurring at the boundaries of the airspace. The effect of errors in estimating the heat transfer coefficient or the building surface temperatures will be less significant in larger rooms with relatively still air.

In using the transient calorimetry method, it may also be necessary to account for a difference in the volumetric ventilation rate during a rapid heating period and a cooling period, especially for small airspaces and high ventilation rates. During rapid heating, the static pressure within the airspace increases, and the air delivery rate increases, especially for low-pressure propeller fans.

14.7 EFFECTIVENESS COEFFICIENT OF MIXING

Most of the analyses in this chapter have been based on the assumption that the room air is completely mixed. Ventilation strategies and air-handling systems vary widely among different types of buildings and applications. Although complete mixing is desirable for some applications, incomplete mixing is desirable for others. For example, in a displacement ventilation (also called *plug flow*) strategy, deliberate steps are taken to produce an air velocity that is uniform across the enclosure, such as a vertical-laminar flow clean room. Almost all ventilated airspaces have a certain degree of incompleteness of mixing, where there are regions of relatively stagnant air and regions in which air "short circuits" the majority of the enclosure.

If the circulation of air in an airspace is known to produce nonuniform pollutant concentrations, that is, partially mixed conditions, it would be useful to know which parts of the enclosure receive too little of the incoming air and which points receive too much. For example, point 1 in Figure 14.13 is located in a high-velocity region in which the concentration changes rapidly, whereas point 2 is in a low-velocity eddy region in which the concentration changes slowly. However, just because the concentration changes slowly does not mean that the concentration is large or small — only that it changes slowly. It would be useful to characterize mixing points 1 and 2 in a quantitative way.

Etheridge and Sandberg have developed a comprehensive theory to analyze nonuniform conditions within enclosed airspaces based on measuring the pollutant concentration over period of time t and computing a *dose* (or zero moment):[3]

$$Dose: \int_0^t C(t)dt \qquad (14.56)$$

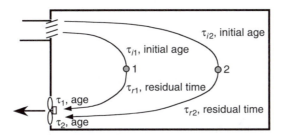

Figure 14.13 Different ages of different particles in a room airspace.

and a first moment of the concentration:

$$\text{The (first) moment:} \quad \int_0^t C(t)tdt \tag{14.57}$$

where t is the time period of interest. Partially mixed conditions can be characterized by an *effectiveness coefficient* E_v, defined as the ratio of residence time of the room air τ to the local mean air age $\overline{\tau}_i$.[8]

$$E_v = \frac{\tau}{\overline{\tau}_i} \tag{14.58}$$

where $\overline{\tau}_i$ is the local mean air age defined in Section 14.5.

The local mean air age can be calculated using different tracer injection methods, commonly step-down or step-up methods, and τ is the average room air residence time:

$$\tau = \frac{V}{Q} \tag{14.59}$$

where V is the volume and Q is the volumetric ventilation rate of the airspace.

For well mixed conditions, one can show mathematically that the local mean age $\overline{\tau}_i$ is equal to τ. Thus, if at a point in an enclosure, the effectiveness coefficient is unity ($E_v=1$), mixing at that point is similar to what would be predicted for the entire airspace. If $\overline{\tau}_i > \tau$, the effectiveness coefficient E_v is less than unity, and this means that the point in space is located in a stagnant region where mixing is poor. If the effectiveness coefficient E_v is greater than unity, mixing at that point in space is very vigorous, and one may imagine that makeup air passes through the point and "short circuits" the enclosure as a whole.

In summary:

$E_v > 1$ requires $\overline{\tau}_i < \tau$, and represents good mixing.
$E_v < 1$ requires $\overline{\tau}_i > \tau$, and represents poor mixing.

The effectiveness coefficient of mixing is a quantitative scalar to evaluate the level of mixing at a concerned location within the airspace. The greater the variation of E_v throughout the airspace, the more unevenly the supply air is distributed. Using tracer experiments, the effectiveness of mixing can be determined for a ventilated airspace. The following example illustrates the application of the effectiveness coefficient of mixing.

Example 14.5: *Let point 1 and point 2 be two locations of concern in a ventilated room with a volume of 200 m³ and a volumetric ventilation rate of 1.3 m³/s. A step-up tracer experiment was conducted. Within the measurement time period, tracer concentrations at point 1 and point 2 varied in the following forms:*

At point 1: $C_1(t) = C_1(\infty)\left[1 - \exp(-0.005t)\right]$ *when $0 < t < 500$ s*
At Point 2: $C_2(t) = C_2(\infty)\left[1 - \exp(-0.008t)\right]$ *when $0 < t < 500$ s*

Target tracer concentration at both points were also measured as $C_1(\infty)=50$ ppb and $C_2(\infty)=35$ ppb. Determine the effectiveness coefficient of mixing for points 1 and 2. Which point is better mixed?

Solution: For the room, the average air residence time is

$$\tau = \frac{V}{Q} = \frac{200}{1.3} = 154 \ (s)$$

For point 1: from Equation 14.45a

$$\int_0^{500} t\left[C_1(\infty) - C_1(t)\right]dt = \int_0^{500} C_1(\infty)t\left[1 - \exp(-0.005t)\right]dt$$

$$= C_1(\infty)\left[\frac{t^2}{2} + (1+0.005t)\frac{\exp(-0.005t)}{(10.005)^2}\right]\Bigg|_0^{500}$$

$$= 96,492\,C_1(\infty)$$

$$\int_0^{500} \left[C_1(\infty) - C_1(t)\right]dt = \int_0^{500} C_1(\infty)\left[1 - \exp(-0.005t)\right]dt$$

$$= C_1(\infty)\left[t - \frac{\exp(-0.005t)}{-0.005}\right]\Bigg|_0^{500} = 684\,C_1(\infty)$$

The local mean air age at point 1 is

$$\bar{\tau}_1 = \frac{\int_0^{500} [C_1(\infty) - C_1(t)]dt}{\int_0^{500} [C_1(\infty) - C_1(t)]dt} = \frac{96,492 C_1(\infty)}{684 C_1(\infty)} = 141(s)$$

$$E_{v1} = \frac{\tau}{\bar{\tau}_1} = \frac{153}{141} = 1.08$$

Similarly, for point 2, the local mean air age is

$$\bar{\tau}_2 = \frac{\int_0^{500} t[C_2(\infty) - C_2(t)]dt}{\int_0^{500} [C_2(\infty) - C_2(t)]dt} = \frac{\int_0^{500} C_2(\infty)t[1 - \exp(-0.008t)]}{\int_0^{500} C_2(\infty)[1 - \exp(-0.008t)]} = 178(s)$$

$$E_{v2} = \frac{\tau}{\bar{\tau}_2} = \frac{153}{178} = 0.86$$

Because $E_{v1} > 1$ and $E_{v2} < 1$, point 1 is better mixed than point 2.

14.8 FLOW RATE MEASUREMENT INSTRUMENTATION AND PROCEDURES

Various flow rate measurement instrumentation and procedures are summarized in Table 14.1. The values for volumetric or mass flow rate measurement are often determined by measuring pressure difference across an orifice, nozzle, or venturi tube.[9, 10] The various meters have different advantages and disadvantages. For example, the orifice plate is more easily changed than the complete nozzle or venturi tube assembly. However, the nozzle is often preferred to the orifice because its discharge coefficient is more precise. The venturi tube is a nozzle followed by an expanding recovery section to reduce net pressure loss. To assure and validate the accuracy of flow rate measurement instruments, appropriate calibration procedures should include documentation of traceability to the calibration facility. The calibration facility should provide documentation of traceability to respective standards.

14.8.1 Flow Measurement Instrumentation

Direct Measurement

Both gas and liquid flow can be measured quite accurately gravimetrically or volumetrically. While the direct method is commonly used for calibrating other metering devices, it is particularly useful when the flow rate is low or intermittent

Table 14.1 Volumetric or Mass Flow Rate Measurement

Measurement Means	Application	Range	Precision	Limitations
Orifice and differential pressure measurement system	Flow through pipes, ducts, and plenums for all fluids	Above Reynolds number of 5000	1 to 5%	Discharge coefficient and accuracy influenced by installation conditions
Nozzle and differential pressure measurement system	Flow through pipes, ducts, and plenums for all fluids	Above Reynolds number of 5000	0.5 to 2.0%	Discharge coefficient and accuracy influenced by installation conditions
Venturi tube and differential pressure measurement system	Flow through pipes, ducts, and plenums for all fluids	Above Reynolds number of 5000	0.5 to 2.0%	Discharge coefficient and accuracy influenced by installation conditions
Timing given mass or volumetric flow	Liquids or gases; used to calibrate other flowmeters	Any	0.1 to 0.5%	System is bulky and slow
Rotameters	Liquids or gases	Any	0.5 to 5.0%	Should be calibrated for fluid being metered
Displacement meter	Relatively small volumetric flow with high pressure loss	As high as 500 l/s, depending on type	0.1 to 2.0%, depending on type	Most types require calibration with fluid being metered
Gasometer or volume displacement	Short-duration tests; used to calibrate other flowmeters	Total flow limited by available volume of containers	0.5 to 1.0%	—
Thomas meter (temperatures of stream rise due to electrical heating)	Elaborate setup justified by need for good accuracy	Any	1%	Uniform velocity; usually used with gases
Element of resistance to flow and differential pressure measurement system	Used to check where system has calibrated resistance element	Lower limit set by readable pressure drop	1 to 5%	Secondary reading depends on accuracy of calibration
Turbine flowmeters	Liquids or gases	Any	0.25 to 2.0%	Uses electronic readout
Instrument for measuring velocity at point in flow	Primarily for installed system with no special provision for flow measurement	Lower limit set by accuracy of velocity measurement	2 to 4%	Accuracy depends on uniformity of flow and completeness of traverse
Heat input and temperature changes with steam and water coil	Checks value in heater or cooler tests	Any	1 to 3%	—
Laminar flow element and differential pressure measurement system	Measures liquid or gas volumetric flow rate; nearly linear relationship with pressure drop; simple and easy to use	50 ml/s 1 m³/s	1%	Fluid must be free of dirt, oil, and other impurities that could plug meter or affect its calibration
Magnetohydrodynamic flowmeter (electromagnetic)	Measures electronically conductive fluids, slurries; meter does not obstruct flow; no moving parts	0.006 to 600 L/s	1%	At present state-of-the-art, conductivity of fluid must be greater than 5:mho/cm
Swirl flowmeter and vortex shedding meter	Measure liquid or gas flow in pipe; no moving parts	Above Reynolds number of 10⁴	1%	—

Source: Adapted from ASHRAE, *Handbook of Fundamentals*, American Society of Heating, Refrigeration and Air Conditioning Engineers, Atlanta, 2001.

and when a high degree of accuracy is required. These systems are generally large and slow, but in their simplicity they can be considered primary devices.

The variable area meter or rotameter is a convenient direct-reading flowmeter for liquids and gases. This is a vertical, tapered tube in which the flow rate is indicated by the position of a float suspended in the upward flow. The position of the float is determined by its buoyancy and the upwardly directed fluid drag.

Displacement meters measure total liquid or gas flow over time. The two major types of displacement meters used for gases are the conventional gas meter, which uses a set of bellows, and the wet test meter, which uses a water displacement principle.

Indirect Measurement

The Thomas meter is used in laboratories to measure high gas flow rates with low pressure losses. The gas is heated by electric heaters, and the temperature rise is measured by two resistance thermometer grids. When the heat input and the temperature rise are known, the mass flow of gas is calculated as the quantity of gas that will remove the equivalent heat at the same temperature rise.

A velocity traverse (using a pitot tube or other single-point velocity-measuring instrument) measures airflow velocity or flow rate at multiple points. The total flow rate then is calculated by multiplying the single-point velocity of flow rate with the weighted cross-sectional area of the flow field. A common procedure is similar to the sampling in a duct described in Chapter 8, in which the number of points and locations were determined for rectangular or circular ducts. This method can be imprecise at low velocities and impracticable where many test runs are in progress.

Another field-estimating method measures the pressure drop across elements with known pressure drop characteristics, such as heating and cooling coils. If the pressure drop/flow rate relationship has been calibrated, the results can be precise. For fans, the entire fan curve (including flow rate vs. pressure drop at different fan speeds) should be obtained. If the method depends on rating data, it should be used for checking purposes only.

14.8.2 Venturi, Nozzle, and Orifice

Flow rate in a duct can be measured by a venturi meter, flow nozzle, or orifice plate (Figure 14.14). If the pressure drop/flow rate relationship has been calibrated, the results can be precise. The American Society of Mechanical Engineers (ASME) Standard MFC-3M describes the measurement of fluid flow in pipes using the orifice, nozzle, and venturi; ASME Standard PTC 19.5 specifies their flow rate.[11,9]

$$\dot{m} = \rho U_1 A_1 = \rho U_2 A_2 = A_2 \sqrt{\frac{2\rho(P_1 - P_2)}{1 - \beta^4}} \qquad (14.60)$$

where $\dot{m}$ is the mass flow rate in kg/s; U is the velocity of the stream in m/s; A is the flow area in m²; ρ is the density of the fluid in kg/m³; P is the absolute pressure in Pa; and $\beta = (D_2/D_1)$, where D_1 is the diameter of the duct and D_2 is the diameter

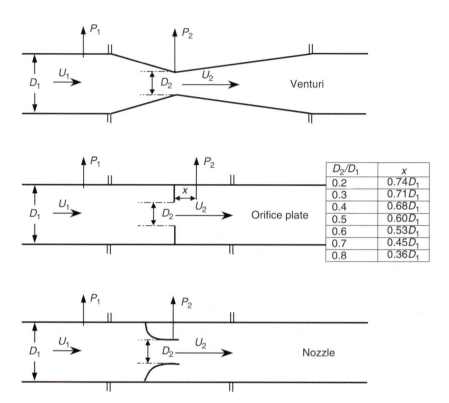

Figure 14.14 Schematic of measurement of flow rate using a venturi, nozzle, or orifice in a duct line.

of the throat of the measurement section. The subscript 1 refers to the entering conditions, and the subscript 2 refers to the throat conditions.

Note that in the case of an orifice plate, the location of P_2 measurement is not at the orifice cross section, but rather downstream from the orifice with a distance of x. The x values vary with the ratio of the duct diameter and the orifice diameter, as shown in Table 14.2. In Table 14.2 R_D is the Reynolds number of flow in the upstream duct, at $D = D$, as shown in Figure 14.14.

Because the flow through the meter is not frictionless, a coefficient, or correction factor C_f, is defined to account for friction losses. Equation 14.60 then becomes

$$\dot{m} = C_f A_2 \sqrt{\frac{2\rho(P_1 - P_2)}{1 - \beta^4}} \qquad (14.61)$$

The factor C_f is a function of geometry and Reynolds number. Values of C_f are given in ASME Standard PTC 19.5. The jet passing through an orifice plate contracts to a minimum area at the vena contracta located a short distance downstream from the orifice plate. The contraction coefficient, friction loss coefficient C_f, and approach factor $1/(1-\beta^4)^{0.5}$ can be combined into a single constant K_f, which is a

Table 14.2 Factors of C_f and K_f for Flow Rate Calculations

β (D_2/D_1)	Nozzle C_f	Nozzle K_f	Orifice C_f	Orifice K_f	Venturi C_f	Venturi K_f
0.1	$0.9975-2.0650 \times R_D^{-0.5}$	$0.9976-2.0651 \times R_D^{-0.5}$	$0.5961+0.2900 \times R_D^{-0.75}$	$0.5962+0.2900 \times R_D^{-0.75}$	0.984	0.9840
0.2	$0.9975\ -2.9203 \times R_D^{-0.5}$	$0.9983-2.9226 \times R_D^{-0.5}$	$0.5970+1.6406 \times R_D^{-0.75}$	$0.5974+1.6419\times R_D^{-0.75}$	0.984	0.9848
0.3	$0.9975-3.5766 \times R_D^{-0.5}$	$1.0016-3.5912 \times R_D^{-0.5}$	$0.5984+4.5208 \times R_D^{-0.75}$	$0.6008+4.5393 \times R_D^{-0.75}$	0.984	0.9880
0.4	$0.9975-4.1299 \times R_D^{-0.5}$	$1.0105-4.1838 \times R_D^{-0.5}$	$0.6003+9.2804 \times R_D^{-0.75}$	$0.6082+9.4015 \times R_D^{-0.75}$	0.984	0.9968
0.5	$0.9975-4.6174 \times R_D^{-0.5}$	$1.0302-4.7688 \times R_D^{-0.5}$	$0.6025+16.2122 \times R_D^{-0.75}$	$0.6222+16.7439 \times R_D^{-0.75}$	0.984	1.0163
0.6	$0.9975-5.0581 \times R_D^{-0.5}$	$1.0692-5.4216 \times R_D^{-0.5}$	$0.6035+25.5738 \times R_D^{-0.75}$	$0.6469+27.4117 \times R_D^{-0.75}$	0.984	1.0547
0.7	$0.9975-5.4634 \times R_D^{-0.5}$	$1.1443-6.2674 \times R_D^{-0.5}$	$0.6000+37.5977 \times R_D^{-0.75}$	$0.6883+43.1304 \times R_D^{-0.75}$	0.984	1.1288
0.8	$0.9975-5.8406 \times R_D^{-0.5}$	$1.2982-7.6013 \times R_D^{-0.5}$	$0.5846+52.4979 \times R_D^{-0.75}$	$0.7608+68.3232 \times R_D^{-0.75}$	0.984	1.2806

R_D = Reynolds number of fluid flow in the duct upstream of the orifice, or at D = D.

Source: ASME, Measurement of Fluid Flow in Pipes Using Orifice, Nozzle, and Venturi. ASME Standard MFC-3M-1989,1989.

function of geometry and Reynolds number. The orifice volumetric flow rate equations then become

$$Q = K_f A_2 \sqrt{\frac{2(P_1 - P_2)}{\rho}} \qquad (14.62)$$

where Q is the discharge flow rate in m³/s; A_2 is the orifice area in m²; and $P_1 - P_2$ is the pressure drop as obtained by pressure taps Pa and

$$K_f = C_f \sqrt{\frac{1}{1 - \beta^4}} \qquad (14.63)$$

Valves, bends, and fittings upstream from the flowmeter can cause errors. Long, straight pipes should be installed upstream and downstream from the flow devices to assure fully developed flow for proper measurement. ASHRAE Standard 41.8 specifies upstream and downstream pipe lengths for measuring the flow of liquids with an orifice plate. ASME Standard PTC 19.5 gives the piping requirements between various fittings and valves and the venturi, nozzle, and orifice. If these conditions cannot be met, flow conditioners or straightening vanes can be used.[9, 12, 13]

Nozzles are sometimes arranged in parallel pipes from a common manifold. Thus, the capacity of the testing equipment can be changed by shutting off the flow through one or more nozzles. Figure 14.15 shows a wind tunnel with a bank of flow

Figure 14.15 A wind tunnel in the Bioenvironmental Structures and Systems Laboratory, University of Illinois, has a bank of nozzles capable of measuring airflow rate from 0.03 to 14 m³/s with the appropriate number of nozzles open.

rate measurement nozzles. With an appropriate number of nozzles, the flow measurement capacity can vary from 0.03 to 14 m³/s for this apparatus.

14.8.3 Turbine (or Vane) Flowmeters

Turbine flowmeters are volumetric flow rate-sensing meters with a magnetic stainless steel turbine rotor suspended in the airflow. The air stream exerts a force on the blades of the turbine rotor, setting it in motion and converting the fluid's linear velocity to an angular velocity.

The design motivation for turbine meters is to have the rotational speed of the turbine proportional to the average fluid velocity and thus to the volume rate of fluid flow. The rotational speed of the rotor is monitored by an externally mounted pickoff assembly. Magnetic and radio frequency are the most commonly used pickoffs. The magnetic pickoff contains a permanent magnet and coil. As the turbine rotor blades pass through the field produced by the permanent magnet, a shunting action induces AC voltage in the winding of the coil wrapped around the magnet. A sine wave, with a frequency proportional to the flow rate, develops. With the radio frequency pickoff, an oscillator applies a high-frequency carrier signal to a coil in the pickoff assembly. The rotor blades pass through the field generated by the coil and modulate the carrier signal by shunting action on the field shape. The carrier signal is modulated as a rate corresponding to the rotor speed, which is proportional to the flow rate. With both pickoffs, frequency of the pulses generated becomes a measure of flow rate, and the total number of pulses measures total volume.[14]

Because output frequency of the turbine flowmeter is proportional to flow rate, every pulse from the turbine meter is equivalent to a known volume of fluid that has passed through the meter; the sum of these pulses yields total volumetric flow.

Turbine flowmeters should be installed with straight lengths of pipe upstream and downstream from the meter. The length of the inlet and outlet pipes should be according to manufacturers' recommendations or pertinent standards. Where recommendations of standards cannot be accommodated, the meter installation should be calibrated.

The lubricity of the process fluid and the type and quality of rotor bearings determine whether the meter is satisfactory for a particular application. In metering liquid fluorocarbon refrigerants, the liquid must not flash to a vapor (cavitate). This would cause a tremendous increase in flow volume. Flashing results in erroneous measurements and rotor speeds that can damage the bearings or cause a failure. Flashing can be avoided by maintaining adequate back pressure downstream of the meter.[15]

14.8.4 Measurement of Large Flow Rates Using a Single Vane Anemometer

In many ventilation applications, especially troubleshooting, measurement of flow rate from a large fan is needed. In such applications, direct measurement using flow meters is impractical because the cross-sectional area is very large, and the flowmeter covering the entire flow stream would be excessively large. In addition, the velocity across the flow area is not uniform as shown in Figure 14.17. A small

Figure 14.16 A small vane anemometer is used to measure the total flow rate of a large exhaust fan. The location of the vane anemometer must be calibrated to represent the total airflow rate of the fan.

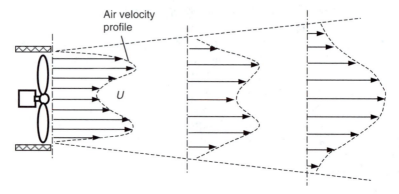

Figure 14.17 Velocity profile of a flow stream. (a) Upstream and downstream of a propeller fan mounted on a wall. (b) At the outlet of a circular duct.

turbine flow meter or a *vane anemometer* can be used for estimating the total flow rate. As shown in Figure 14.16, a small vane anemometer is mounted at the upstream of a fan of 0.6 m in diameter. The total flow rate of the large fan is proportional to the readout of the small vane anemometer. Such application requires calibration of the anemometer with the fan, or the measurement of the velocity profile across the flow field. If the velocity profile is known, one can find a vane anemometer location at which the flow velocity is the mean of the total flow field.

14.8.5 Airflow-Measurement Hood

Flow-measurement hoods are portable instruments designed to measure supply or exhaust airflow through diffusers and grilles in HVAC systems. A flow-measuring hood assembly typically consists of a fabric hood section, a plastic or metal base, an airflow-measuring manifold, a meter, and handles for carrying and holding the hood in place (Figure 14.18).

Figure 14.18 A TSI flow-measurement hood is used to measure airflow rate through an air diffuser in a room.

For volumetric airflow measurements, the flow-measuring hood is placed over a diffuser or grille. The fabric hood captures and directs airflow from the outlet or inlet across the flow-sensing manifold in the base of the instrument. The manifold consists of a number of tubes containing upstream and downstream holes in a grid pattern, designed to sense and average simultaneously multiple velocity points across the base of the flow-measuring hood. Air from the upstream holes flows through the tubes past a sensor and then exits through the downstream holes. Sensors employed by different manufacturers include swinging vane anemometers, electronic micro-manometers, and thermal anemometers. In the case of the electronic micromanometer sensor, air does not actually flow through the manifold, but the airtight sensor senses the pressure differential from the upstream-to-downstream series of holes. The meter on the base of the flow-measuring hood interprets the signal from the sensor and provides a direct reading of volumetric flow, in either an analog or a digital display format.

As a performance check in the field, the indicated flow of a measuring hood can be compared to a duct traverse flow measurement (using a pitot tube or a thermal anemometer). All flow-measuring hoods induce some back pressure on the air-handling system, because the hood restricts the flow coming out of the diffuser. This added resistance alters the true amount of air coming out of the diffuser. In most cases, this error is negligible and is less than the accuracy of the instrument. For proportional balancing, this error need not be taken into account, because all similar diffusers will have about the same amount of back pressure. To determine whether back pressure is significant, a velocity traverse can be made in the duct ahead of the diffuser with and without the flow-measuring hood in place. The difference in the average velocity of the traverse indicates the degree of back-pressure compensation required on similar diffusers in the system. For example, if the average flow rate is 0.5 m³/s with the hood in place and 0.52 m³/s without the hood, the indicated flow reading can be multiplied by 1.04 on similar diffusers in the system (0.52/0.5 = 1.04). As an alternative, the designer of the air-handling system can predict the reduction in airflow due to the additional pressure of the hood by using a curve supplied by the flow-measuring hood manufacturer.

DISCUSSION TOPICS

1. What does a minimum ventilation requirement mean? Would the minimum ventilation rate change for the same building with different types of pollutants?
2. If an air cleaner has a 90% particle removal efficiency, can the particle concentration in a ventilated room with such an air cleaner be practically reduced by 99%?
3. Can one determine the ventilation rate in a ventilated room by simply measuring the carbon dioxide concentration ?
4. What is the rationale behind the recommendation that when the rate of decay method is used to measure ventilation rate, the suggested measurement time period is approximately the same as the room air turnover time, that is, the suggested measurement time $t_2 - t_1 \approx V/Q$?
5. What are the differences between a local mean air age and the room mean air age in terms their practical implications to occupants?
6. Under what conditions that one can measure the room ventilation rate using tracer gas methods?
7. The calorimetry method is essentially a thermal tracer method. Why it is not as widely used as the tracer gas method? Name a few of its limitations.
8. What is the relationship between the mixing effectiveness coefficient (MEC) and the room air quality? Does a high value of MEC indicate better air quality?

PROBLEMS

1. A dining room has a capacity of 20 people; each person produces 200 W total heat. The carbon dioxide concentration in the supply air is 500 ppmv. Assume a supply air temperature of 16°C and 50% relative humidity. The room air is 22°C and 60% relative humidity. If the maximum CO_2 concentration in the room should be less than 1000 ppmv, what should the minimum ventilation be for the exhaust fan? (Assume complete mixing in the room.)
2. At the steady state, an animal holding room has an exhaust fan with a flow rate of 2000 l/s, maintaining the ammonia in the room at 5 ppmv. The specific volume of room air is 0.8 m³/kg. Ammonic concentration of the supply air is negligible. What will the ammonia concentration be if the ventilation rate is reduced to 1000 l/s?
3. In an auditorium with a ventilation rate of 6 m³/s, the carbon dioxide concentrations for the exhaust air and supply air are 800 ppmv and 400 ppmv, respectively. The specific volumes for exhaust and supply air are 0.82 and 0.78 m³/kg, respectively. The auditorium has an air volume of 8000 m³, and the air inside is assumed to be completely mixed. In a power failure, all exhaust fans are deactivated. Assume that there is no air exchange between the inside and outside of the auditorium. How long it will take for the CO_2 to reach 5000 ppmv after the power failure?
4. A residential house for sale has been on the market for a period of time, during which there was no ventilation and no human activity in the house. A house inspector measured the radon concentration of 10 pCi/l in the basement of the house. The basement has an air volume of 300 m³. There is an exhaust fan with a 40 l/s capacity in the basement. How long it will take to reduce the radon concentration below the safety level of 4 pCi/l? Assume that the radon concentration in the supply air is zero and the production rate during the ventilation is negligible.
5. An electronic processing room must meet the ISO cleanroom Class 4 standard, which allows a maximum particle concentration 1020 particles/m³ for particles 0.3

μm and larger. The air supply and distribution system for the room is shown in the figure below. Assume that the room air is completely mixed. The supply airflow is at a constant rate of 600 l/s. With a HEPA filter, the particle concentration in the supply air is zero. During a working shift, the particle concentration was measured at 1500 particles/m³ for particles larger than 0.3 μm. Analyze the following three approaches to making the room compliant with the Class 4 standard. Assume that particle production remains constant and deposition is negligible.

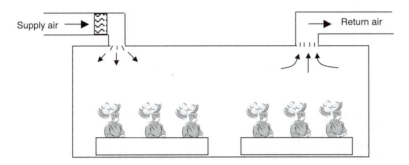

a. An electrostatic precipitation (ESP) air-cleaning system can be installed. The ESP has a flow rate of 200 l/s and a filter efficiency of 99% for particles larger than 0.3 μm. Determine whether the room meets the Class 4 criterion by installing the ESP system alone.

b. Assume that the particles produced in the room are primarily from clothing and human skin. A new type of work suit can reduce the particle production by 30%. Can changing the clothing alone meet the Class 4 criterion?

c. How much should ventilation be increased, without additional air cleaning or upgrading the work suits?

6. Assume that the emission rate of formaldehyde from building materials reduces exponentially and follows this relationship: $\dot{m}(t) = \dot{m}_0 \exp(-\lambda t)$, where $\dot{m}_0$ is the initial production rate and λ is a constant. During a building commissioning process, an inspector measures a formaldehyde concentration 80% higher than the allowed level. Seven days later, the inspector visits the building again and measures the formaldehyde concentration, which is now 30% higher than the allowed level. How many more days are needed for the formaldehyde concentration to be reduced to the required level? Assume that all building operating conditions, including ventilation, have been maintained the same throughout the entire time.

7. Determine the ventilation rate of a room with a volume of 120 m³ using a constant injection tracer method. From the start, the tracer concentration increases exponentially with time: $C(t) = C(\infty)\left[1 - \exp\left(-\frac{Q}{V}t\right)\right]$, where $C(\infty)$ is the tracer concentration when $t \to \infty$, Q is the ventilation rate, and V is the room volume. The following table shows the data recorded.

t, minutes	C(t), ppb in volume
0	0
5	200
10	250

8. In a constant injection tracer method, the tracer injection rate is 5 ml/min. The room volume is 100 m³. Determine the room ventilation rate using the recorded data shown in the following table.

t, minutes	C(t), ppb in volume	t, minutes	C(t), ppb in volume
0	140	8	140
1	150	9	160
2	200	10	200
3	220	11	220
4	210	12	210
5	170	13	170
6	140	14	190
7	160	15	180

9. During a step-down tracer experiment, the initial tracer concentration at the exhaust, $C_i(0)$, is measured as 100 parts per billion in volume. The concentration then decreases at the following rate: $C_i(t) = C_i(0)\exp(-\lambda t)$. The concentration is measured as 50 ppbv when $t = 5$ minutes. Determine the room mean air age.

10. In a step-up tracer experiment, the data shown in the following table were recorded at one point in the room. Determine the mean air age at the measurement point.

t, minutes	C(t), ppb in volume	t, minutes	C(t), ppb in volume
0	0	8	259
1	100	9	260
2	180	10	261
3	220	11	260
4	240	12	259
5	250	13	260
6	255	14	260
7	258	15	260

11. In a step-down tracer experiment for an office, tracer concentration data were recorded at a breathing-level point $C_i(t)$ and one point at the room air exhaust $C_e(t)$. Assume that at both measurement points, the tracer concentrations decrease exponentially with time: $C(t) = C_0(0)\exp(-\lambda t)$. The room has a volume of 40 m³ and a ventilation rate of 50 l/s. Determine whether the breathing-level air is better mixed than the exhaust air.

t, minutes	$C_i(t)$, ppb in volume	$C_e(t)$, ppb in volume
0	250	285
5	155	160
10	103	105

12. In a step-down tracer experiment for an office, tracer concentration data were recorded one point $C_i(t)$, as shown in the following table. The room has a volume of 60 m³ and a ventilation rate of 80 l/s. Determine the effectiveness coefficient of mixing at the measurement point.

t, minutes	$C_i(t)$, ppb in volume
0	350
1	250
2	210
3	180
4	160
5	150
6	140
7	130
8	125

REFERENCES

1. ASHRAE, *Handbook of Fundamentals*, American Society of Heating, Refrigeration and Air Conditioning Engineers, Atlanta, 2001, ch. 8.

2. ASAE Standards, *Engineering Practices and Data: Structure, Livestock and Environment*. American Society of Agricultural Engineers, St. Joseph, MI, 2000.

3. Etheridge, D. and Sandberg, M., *Building Ventilation — Theory and Application*, John Wiley & Sons, New York, 1996.

4. Axley, J. and Persily, A., *Integral Mass Balances and Pulse Injection Tracer Techniques*, Report NISTIR 88-3855, National Institute of Standards and Technology, U.S. Department of Commerce, 1988.

5. Ower, E. and Pankhurst, R.C., *The Measurement of Airflow*, 5th ed., Pergamon Press, Toronto, 1977.

6. Replogle, J.A. and Birth, G.S., *Instrumentation and Measurement for Environmental Sciences*, 2nd ed., Mitchell, B.W., ed., Publ. 13-82, American Society of Agricultural Engineers, St. Joseph, MI, 1983, ch. 5.

7. Penman, J.M. and Rashid, A., Experimental determination of airflow in a naturally ventilated room using metabolic carbon dioxide, *Building and Environment*, 17(4): 253–256, 1982.

8. Skaret, E. and Mathisen, H.M., Ventilation efficiency — a guide to efficient ventilation, *Transactions of ASHRAE*, 89(2):480–495, 1983.

9. ASME, *Part I: Measurement Uncertainty Instruments and Apparatus*, ANSI/ASME Standard PTC 19.1-85, 1985.

10. Benedict, R.P., *Fundamentals of Temperature, Pressure and Flow Measurements*, 3rd ed., John Wiley & Sons, New York, 1984.

11. ASME, *Measurement of Fluid Flow in Pipes Using Orifice, Nozzle, and Venturi*, Standard MFC-3M-85, 1989.

12. ASME, *Method for Establishing Installation Effects on Flowmeters*, ANSI/ASME Standard MFC-10M-94, 1994.

13. Miller, R.W., *Measurement Engineering Handbook*, McGraw-Hill, New York, 1983.

14. Woodring, E.D., Magnetic turbine flowmeters, *Instruments and Control Systems*, 6:133, 1969.

15. Liptak, B.G., ed., *Instrument Engineers Handbook, Vol. 1*, Chilton Book Company, Philadelphia, 1972.

CHAPTER 15

Ventilation Effectiveness and Air Distribution

The term *ventilation effectiveness* has been widely used to evaluate ventilation systems for air quality control. The application of ventilation effectiveness is often qualitative rather than quantitative, or else it is difficult to use in a field application. Air ages, for example, can be used to evaluate ventilation effectiveness qualitatively. Indeed, ventilation effectiveness is a very complicated matter that depends on a variety of variables, such as room airflow patterns, boundary conditions, obstructions within the airspace, pollutant sources, air leakage rate through the envelope, and ventilation systems. This chapter includes a method to quantify ventilation effectiveness for specific ventilation systems by using a ventilation effectiveness factor (VEF) and a ventilation effectiveness map (VEM). One of the advantages of a VEF is that a ventilation system can be its own control in a comparison of ventilation effectiveness, instead of requiring a control and a treatment. The self-comparison feature is particularly useful in system evaluation or in troubleshooting, because having an identical system to compare often proves very difficult and expensive.

One of the most important factors affecting ventilation effectiveness is room air distribution. This chapter discusses different mathematical models used in computational fluid dynamic (CFD) simulation and different methods and instrumentation to measure room air distribution. Air leakage measurement and analysis and the effect of air leakage on ventilation effectiveness are also discussed in this chapter. At the end of the chapter, a case study is presented to illustrate the effect of alternative ventilation schemes on particulate matter (PM) spatial distribution in rooms. Given the multitude and variety of factors affecting ventilation effectiveness, this chapter intends to provide the reader with a broad perspective on ventilation effectiveness. In that regard, much of the analysis in this chapter is qualitative rather than quantitative.

By completing this chapter, the reader will be able to

- Measure and characterize particle concentration spatial distribution in ventilated rooms
- Calculate ventilation effectiveness using a ventilation effectiveness factor (VEF) method that does not need a control room for comparison
- Understand the variability of using different computational fluid dynamic (CFD) models to study room airflow
- Measure room air distribution using different instruments and understand their advantages as well as their limitations
- Measure air leakage through mechanically ventilated rooms and its effect on ventilation effectiveness
- Improve ventilation effectiveness and indoor air quality using different ventilation schemes and ventilation systems

15.1 PARTICLE SPATIAL DISTRIBUTION

To appreciate the effort involved in quantifying ventilation effectiveness, we start with a case study of indoor PM spatial distribution with different ventilation schemes. A clear understanding of airborne particulate spatial distribution can provide important information for the improvement of ventilation system design and control strategies. In this case study, the particle mass spatial distribution was measured using a multipoint sampler under controlled conditions in a full-scale, mechanically ventilated laboratory room with three different ventilation systems. The experimental results show that particle mass spatial concentrations vary widely as a result of different ventilation systems and ventilation rates. Increasing the ventilation rate reduced the overall mean particle concentration with the same ventilation system. At the same ventilation rate, the ventilation effectiveness varied widely with different ventilation flow patterns.

15.1.1 An Experimental Case Study

In this case study, a test room of $5.5 \times 3.7 \times 2.4$ m ($L \times W \times H$) was set up within a room ventilation simulator, as shown in Figure 15.1. Two plenum rooms were attached to the test room to keep supply air and exhaust air uniform in temperature and relative humidity. Three ventilation schemes, with different inlet and outlet locations, and six ventilation rates were tested.

For easy discussion, the three ventilation systems are coded A, AB, and B. These ventilation systems will be used in Section 15.7, again to discuss the effect of ventilation scheme on ventilation effectiveness. Ventilation system A had a continuous slot inlet along one wall and a slot outlet on the opposite wall (Figure 15.2a). The air inlet was 50 mm wide, and the air outlet was 200 mm wide. The slot air inlet and slot outlet provided an approximate two-dimensional airflow pattern in the room. Airflow patterns were measured using particle image velocimetry (PIV). Inlet temperature and room temperature were measured using T-type thermocouples. Ventilation system AB was modified from ventilation system A. System AB had a continuous slot inlet and a slot outlet on the same side wall (Figure 15.2b). The air

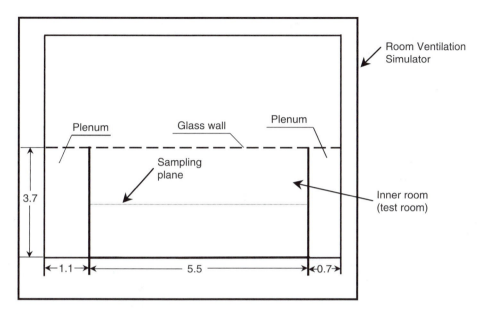

Figure 15.1 Floor plan of the test room for the case study (all dimensions are in meters).

inlet was 50 mm wide and the air outlet was 200 mm wide. The only difference between system AB and system A was the location of the outlet.

Ventilation system B was modified from ventilation system AB. System B had a continuous ceiling slot inlet with a hinged baffle and a slot outlet on the same sidewall (Figure 15.2c). Even with the minor change of air inlet from system A to system B, the inlet air jets, and thus the air distribution, were expected to be very different. The air jet in system A was a free jet, while the air jet in system B was a confined ceiling jet. The throw of these two jets can be substantially different. Therefore, systems A and B could be considered as two different air inlet systems.[1] The air inlet opening could be adjusted based on required inlet velocity in system B. The air outlet was kept at 200 mm wide. The only difference between system B and system AB was the location of the inlet.

15.1.2 Dust Generation and Measurement

A dust generation and emission system was developed to generate dust uniformly along the floor. A rotating-table dust generator was used to feed the dust into the dust emission system at a precise rate. Dust was uniformly emitted from the floor using a dust-emission system with 25 emitting ports evenly distributed over the entire floor area. The ports were on the bottom of the tube, and some dust could be collected in the tube. The tube was 12 mm above the floor. Each emission port had the same diameter of 1.6 mm. In order to maintain the same compressed air pressure at each port, the total opening area of the five ports in each branch tube was one-fifth of the cross-sectional area of the tube. The compressed air pressures were the same at both ends of the tube (7.1 kPa). Arizona dust (Power Technology, Inc.) was

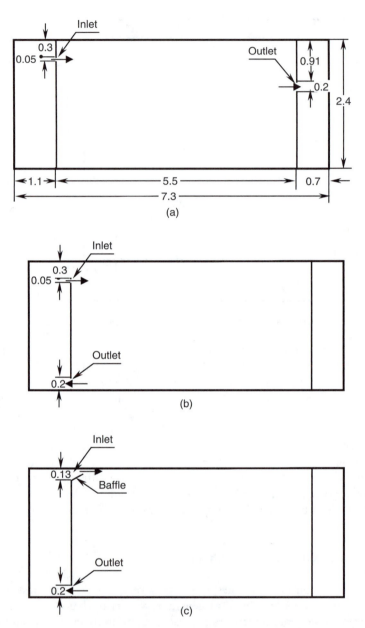

Figure 15.2 Three ventilation systems. (a) System A: slot air inlet and outlet at two side walls. (b) System AB: slot air inlet and outlet on the same side wall. (c) System B: slot air inlet attached to ceiling and outlet on the same side wall.

injected into the room airspace from the floor at a constant rate, using the dust-generation system. The airborne dust size distribution was measured using an aero-dynamic particle sizer (APS, Model 3320, TSI). The inlet sampling velocity was 0.3 m/s. Fifteen dust samples were recorded with the APS for system A at ventilation

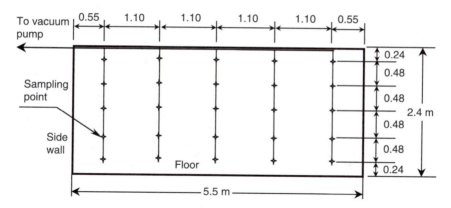

Figure 15.3 Twenty-five sampling points were evenly distributed across the cross section along the centerline of the room (all dimensions in meters) (side view).

rate 19.5 air changes per hour (ACH). The count mean aerodynamic diameter of the airborne dust was 1.63 μm, with geometric standard deviation σ_g of 1.59. The estimated mass median diameter was 3.1 μm. The density of the test dust was measured in the factory as 2.65 g/cm³.

Dust concentrations were measured at 25 points within the testing room. Measurement points were uniformly distributed in the central cross section (sampling plane in Figure 15.1) in the test room, as shown in Figure 15.3. The dust collector located downstream of each critical venturi orifice was a 37 mm diameter (0.8 μm porosity) filter housed in a holding cassette. As the surrounding air velocity in most of the sampling points was less than 0.5 m/s, the dust sampling inlet was placed perpendicular to the airflow to maintain all sampling close to isokinetic conditions according to the criteria of still air sampling. Filters were dried in a desiccant dryer for 24 hours and weighed on a precision electronic balance before and after the dust collection.

The dust spatial distributions in a ventilated airspace at six different ventilation rates, keeping all other parameters (dust mass production rate, M_p, outside dust concentration, C_o, and dust deposition rate, M_d) the same, are shown in Figure 15.4. The dust supply rate from the turning table was 3 mg/s. The net dust generation rate $(M_p–M_c)$ in the airspace, however, was only about 0.5 mg/s. The majority of the supplied dust from the turning table dust generator was lost in the transportation pipes and on the floor. All measured results show that there was a high variation in the dust spatial distribution within the mechanically ventilated airspace. At a low ventilation rate (8.6 ACH), the spatial dust concentration decreased gradually from the floor to the ceiling as expected (Figure 15.4a). The highest measured dust concentration near the floor was 30 times higher than the lowest concentration near the ceiling. The dust was not well mixed in the ventilated airspace at such a ventilation rate. The dust spatial distribution also shows that the highest dust concentration was at the right lower corner below the air outlet.

As the ventilation rate increased, the dust concentration peak gradually shifted from the outlet side to the inlet side (Figure 15.4.b through Figure 15.4f). The dust

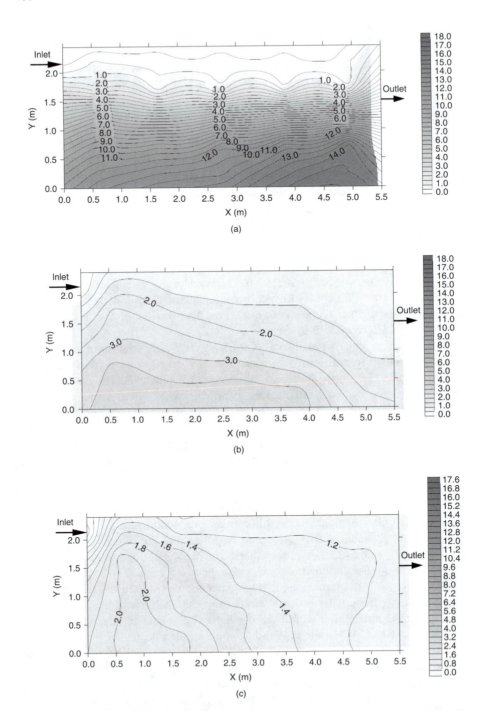

Figure 15.4 Effect of ventilation rate Q on dust spatial distribution (mg/m³). (a) $Q = 0.118$ m³/s.
(b) $Q = 0.264$ m³/s. (c) $Q = 0.378$ m³/s. (d) $Q = 0.566$ m³/s. (e) $Q = 0.755$ m³/s.
(f) $Q = 0.897$ m³/s.

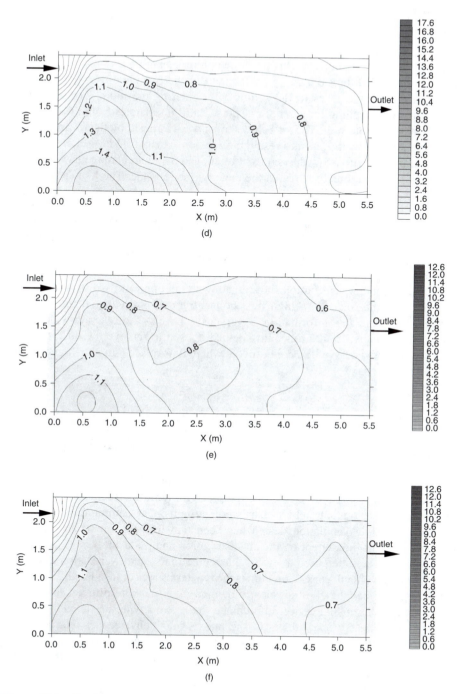

Figure 15.4 *Continued.*

concentration in the left half of the room was relatively higher than in the right half of the room. This dust spatial distribution can be explained by examining the flow pattern. When the free wall jet enters the room, it will attach to the boundary, if it approaches a solid boundary. This phenomenon of keeping a jet attached to a solid boundary is called the *Coanda effect*. Because of the Coanda effect, the incoming air from the inlet attaches to the ceiling after entering the room. The air stream travels along the ceiling for a certain distance, until it separates from the ceiling or reaches the opposite wall. The air below the jet was entrained by the jet and formed a reverse flow below the jet. As a result, the zone on the right side was more exposed to the air stream with a low dust concentration, and the recirculating air with a high dust concentration directly affected the zone on the left side. The higher the ventilation rate, the stronger the reverse recirculating air stream. As the reverse air stream reached the left wall, it turned upward and was entrained by the inlet jet again. As a result, a very dusty zone was formed below the air inlet jet. The spatial gradients of dust concentration became relatively lower as the ventilation rate increased. This phenomenon indicates that the airflow pattern has a direct effect on the dust spatial distribution and dust removal under the conditions tested.

The dust spatial distribution shows that dust is not uniformly distributed in the airspace and that there is a high dust concentration zone with all the different ventilation rates. This information is very useful when a dust control device is used. For example, if an air-cleaning device is used to remove the dust, the dust removal effectiveness will be improved when it is placed in the highest dust concentration zone.

15.2 CALCULATION OF VENTILATION EFFECTIVENESS

Ventilation effectiveness has been increasingly used in studies of ventilation and indoor air quality. Pollutant concentrations, especially particulate pollutants, vary widely within a ventilated airspace.[2, 3] Because the pollutant spatial distribution is not uniform, it is possible to improve the pollutant removal efficiency by properly designing the ventilation and air distribution systems. One common assumption in livestock building ventilation design has been that the location of air outlets or exhaust fans has little effect on the air distribution and pollutant removal.

Two terms are often used to evaluate a ventilation system: ventilation efficiency and ventilation effectiveness. Sometimes the two terms are confused with each other, and at other times they are not clearly defined; thus they are difficult to evaluate quantitatively. In this book, we differentiate these terms with the following definitions. *Ventilation efficiency* refers to the mass of air delivered per unit of power consumed by the ventilation system, in kilogram of air per watt of power (kg/W), at a given pressure. The more air is delivered per watt, the more efficient the ventilation system. Ventilation efficiency is also referred to as the ventilation efficiency ratio (VER) or energy efficiency ratio (EER).[4] The ventilation efficiency ratio is not a nondimensional term and has a value greater than zero. Ventilation efficiency is a criterion for energy and fan performance, not directly related to ventilation effectiveness and air quality control.

Ventilation effectiveness is the efficiency criterion for pollutant removal in the airspace of concern. Several qualitative methods, including air exchange efficiency, purging flow rate, and purging time, have been used to study ventilation effectiveness.[5] These qualitative methods can be analyzed further and represented by the mean age of air concept (as described in Chapter 14), which describes the relationship between pollutant concentrations in the airspaces of concern and the ventilation rate. Limitation of the mean age of air method includes ventilation rate dependence, tracer gas utilization, and inappropriateness for particulate pollutants. Roos (1999) analyzed pollutant removal efficiency, which quantifies the turnover time and pollutant emissions.[5] Liddament (1987) proposed a relative ventilation effectiveness to evaluate the ventilation effectiveness for a location of concern or an overall room airspace.[6]

Many existing methods to evaluate a ventilation system for air quality control need to compare two systems: control and treatment.[7,8] This multisystem comparison method is expensive and often not practical. The bases of comparison (with the control) vary widely and depend on many factors, including system configuration and ventilation rate. A method is needed that can be conveniently used to quantify ventilation effectiveness without the need of multisystem comparison and that is practical for field application.

15.2.1 Ventilation Effectiveness Factor (VEF)

To describe quantitatively the ventilation effectiveness for controlling indoor air quality, a *ventilation effectiveness factor* (VEF) can be used. The VEF is defined as

$$VEF = \frac{C_x - C_s}{C_m - C_s} \quad (C_m > C_s \text{ and } C_x > C_s) \tag{15.1}$$

where C_x is the pollutant concentration under complete mixing conditions in the airspace of concern and can be determined based on the mass balance of the pollutant of concern.

For a ventilated airspace at a steady state, the pollutant removed by the exhaust air must be equal to the sum of the pollutant brought in by the supply air and the pollutant produced within the airspace, regardless of complete or incomplete mixing conditions. The pollutant concentration of the exhaust air, C_e, can be measured. Under complete mixing conditions

$$C_x = C_e \tag{15.2}$$

C_s is the pollutant concentration of the supply air, C_m is the mean pollutant concentration of the airspace at N measured locations (if multiple point measurements are taken) or the point pollutant concentration at the location of concern. N can be any integer greater than 1, depending on how much detail the VEF in the airspace requires.

$$C_m = \frac{1}{V} \sum_{i=1}^{N} V_i C_i \qquad (15.3)$$

where V_i is the volume within which C_i is the representative concentration, V is the volume of the airspace, and

$$V = \sum_{i=1}^{N} V_i \qquad (15.4)$$

It is evident that the ratio V_i/V is the weighing factor for the mean concentration. This weighing factor is important for obtaining a reasonably accurate mean concentration, especially for airspaces with a large concentration gradient. If the measurement points are equally spaced in the airspace

$$C_m = \frac{1}{N} \sum_{i=1}^{N} C_i \qquad (15.5)$$

Substituting Equation 15.2 and Equation 15.3 into Equation 15.1 gives

$$VEF = \frac{V(C_e - C_s)}{\sum_{i=1}^{N} V_i C_i - VC_s} \qquad (15.6)$$

The VEF is nondimensional and independent of the ventilation rate. The higher the value of the factor, the more effective the ventilation system is for pollutant removal. When $VEF = 1$, the ventilation system is as effective as complete mixing. We will use $VEF = 1$ as the basis of comparison for the ventilation effectiveness. When $VEF > 1$, the ventilation system is said to be more effective, and when VEF < 1, the ventilation system is less effective. When $C_e < C_s$, the airspace becomes a settling chamber or an air cleaner, and the air cleaning efficiency, $\xi_r = 1 - C_e/C_s$, should be used to evaluate the system. When the pollutant concentration of supply air is negligible (i.e., $C_s \approx 0$), the VEF for a room becomes

$$VEF = \frac{C_e}{C_m} \qquad (15.7)$$

One useful feature of VEF is that the system being evaluated can be its own control. This feature is particularly valuable for field system evaluation or trouble hooting. From Equation 15.1 and Equation 15.7, the VEF compares the actual pollutant concentration with its own concentration under complete-mixing condi-

tions within the same airspace. Existing methods of evaluating ventilation effectiveness often require the comparison of two systems. In practice, it is usually difficult
to find an identical system for comparison.

Under complete-mixing conditions, the pollutant concentrations at all locations
within the airspace are the same (i.e., $C_m = C_e$), and thus, the VEF is always equal
to unity.

15.2.2 Ventilation Effectiveness under Incomplete-Mixing Conditions

In reality, most ventilated airspaces are incompletely mixed. For highly diffusive
pollutants, such as ammonia, the concentration within an airspace may be close to
the concentration under complete-mixing conditions. For less diffusive pollutants,
such as airborne dust, concentrations may vary widely within the airspace. As a
result, ventilation effectiveness factors can vary widely for different pollutants.

The VEF is affected by many factors, including the location of air inlets and air
outlets, air diffuser types, and obstructions (e.g., furniture, partitions, feeders, etc.)
within an airspace. The following case study serves as an example of calculating
the ventilation effectiveness factor. In livestock building ventilation system design,
it has been assumed that the location of the outlets (often the exhaust fans) has little
effect on ventilation effectiveness. Although this assumption is true for a completely
mixed airspace, it is often misleading in typical livestock buildings.

To obtain the mean concentration within an airspace, the pollutant concentration spatial distribution C_{mi} should be measured. One major limitation of quantifying ventilation effectiveness has been quantifying the pollutant concentration
spatial distribution.

As an example, Figures 15.5a and 15.5b show the airborne dust concentration
distributions of a central section in a ventilated room under two ventilation systems:
one with the outlet at the opposite wall (case A), and the other with the outlet at the
same wall as the inlet (case B). At a ventilation rate of 0.265 m³/s, the dust spatial
distribution and mean concentration for the two ventilation systems were substantially different. The mean dust concentrations within the room for case A and case
B were 2.82 mg/m³ and 0.8 mg/m³, respectively. This represents a 72% airborne
dust reduction in case B compared with case A. The highest dust concentration in
case A reached 4.5 mg/m³ over a large floor area; the highest dust concentration for
case B was lower than 2 mg/m³ and was confined to the corner near the outlet.
Because the air near the outlet was exhausted immediately, this location of high dust
concentration had little effect on the indoor air quality. Most locations within the
airspace had dust concentrations lower than 1 mg/m³ for case B.

Dust spatial distributions for cases A and B were measured at six different
ventilation rates between 0.11 and 0.9 m³/s, using a multipoint aerosol sampler
developed by Wang et al.[9] At each ventilation rate, the dust concentration measurements were repeated three times. The mean dust concentrations vs. ventilation rate
are presented in Figure 15.5. The vertical bars across each mean value represent the
standard deviation. The mean dust concentration for complete mixing was equal to
the dust concentration of the exhaust air under steady-state conditions. As a general
trend, the mean dust concentration decreased as the ventilation rate increased. Com-

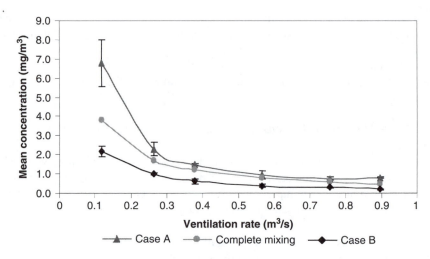

Figure 15.5 Dust concentration at different ventilation rates for case A and case B, compared with complete-mixing ventilation conditions.

pared with complete-mixing conditions, dust concentrations were consistently lower in case B and higher in case A.

Using Equation 15.6, the VEF can be calculated at each ventilation rate. Figure 15.6 shows the VEF vs. ventilation rate for cases A and B and complete-mixing ventilation. At six different ventilation rates, the mean VEF for case A is 0.69, that is, 69% as effective in dust removal compared with complete-mixing ventilation. The mean VEF for case B is 1.96, that is, 196% as effective in dust removal compared with complete-mixing ventilation. In terms of dust removal at locations of concern, the VEF for case B is 3 times more effective than case A, indicating that ventilation in case B is much more effective than in case A.

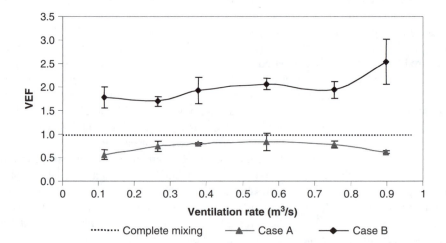

Figure 15.6 Ventilation effectiveness factors of cases A and case B vs. the ventilation rate.

From Figure 15.6, the VEFs were relatively consistent with respect to the changes in ventilation rates, especially considering the complexity of flow patterns and velocity field within the room airspace. This implies that unlike the pollutant concentration within the airspace, the VEF was not highly dependent on the ventilation rate. This is an important feature of the VEF, because it may be applied to a ventilation system with little concern for variation caused by ventilation rate. More measurements of VEFs for different types of buildings are needed to have a more definite conclusion on this finding.

From the preceding example, it is evident that incomplete mixing could be an effective method in air quality control if designed correctly. Individual ventilation systems require specific considerations to improve the ventilation effectiveness.

15.2.3 Ventilation Effectiveness under Zonal Ventilation Conditions

Zonal ventilation is the ventilation of only a portion of the airspace where thermal comfort or air quality is the primary concern. Fume hoods, for example, are a typical zonal ventilation system, in which the pollutant is exhausted at or near the source before it is mixed with the room air. A fresh air supply vent in an airplane is another zonal ventilation system, in which the air is directed to the person. In real buildings, such as livestock buildings, zones of concern (in terms of thermal comfort and air quality) are usually human- and animal-occupied zones. It may be more effective to use zonal ventilation to provide needed thermal comfort and air quality than complete mixing ventilation. To evaluate the effectiveness of such zonal ventilation, the ventilation effectiveness factor for a zone of concern, or ith location, can be rewritten from Equation 15.4 as

$$VEF_i = \frac{C_e - C_s}{C_i - C_s} \qquad (15.8)$$

where C_i is the pollutant concentration at the ith location — the zone of concern. C_e and C_s are pollutant concentrations of building exhaust air and supply air, respectively. VEF_i can be obtained by a point measurement.

15.2.4 Ventilation Effectiveness Map (VEM)

In ventilation system design and air quality control, a *ventilation effectiveness map* (VEM) may be useful to provide an overview of the ventilation effectiveness within the entire airspace, so that problematic zones can be identified and the system can be improved. A ventilation effectiveness map is a contour plot of the ventilation effectiveness factors, calculated using Equation 15.7, for the entire airspace. For example, using Equation 15.6, Figure 15.4b is converted into a VEM, as shown in Figure 15.8. The higher the VEM value, the more effective the ventilation system is in that area.

Because the VEM gives an overall picture of the ventilation effectiveness across the room airspace, it can be a useful tool for design, analysis, and modification of

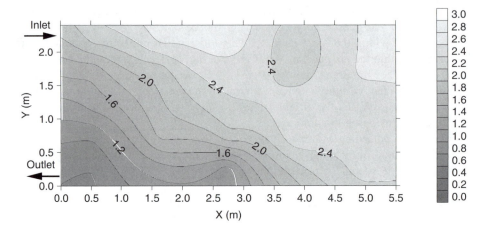

Figure 15.7 A ventilation effectiveness map for Case B: a full-sized room with a slot air inlet at the top of one side wall and an air outlet near the bottom of the same side wall.

a ventilation system. Each ventilation system has its specific VEM. Using a VEM, one can quickly quantify the ventilation effectiveness across the entire airspace. For example, all the VEF values for case B (Figure 15.7) are higher than 1, indicating that the system is very effective. Further, the VEF is much higher at the right side of the room, indicating a much cleaner zone for occupants.

Along with a ventilation graph, a VEM can be used as a tool for design, analysis, and research of ventilation systems. A valuable feature of the VEF is that a ventilation system can be its own control for comparison of ventilation effectiveness, instead of requiring a control and a treatment. This feature is particularly useful in system evaluation or troubleshooting, because it is often very difficult and expensive to have an identical system to compare.

15.3 GOVERNING EQUATIONS FOR ROOM AIR DISTRIBUTION MODELING

The transport and fate of indoor airborne pollutants, especially gases and small particulate matter, are primarily affected by the air distribution within the room. Therefore, characterization of airflow is particularly important for the implementation of air quality control strategies. For a ventilated airspace, even at very low air velocities, the airflow can be highly turbulent. Analytically solving turbulent room airflow is practically prohibitive. Numerical methods are widely used. Two alternative methods are currently available for turbulence modeling without directly simulating the small-scale turbulent fluctuations: *Reynolds-averaged Navier–Stokes* (RANS) and *large eddy simulation* (LES).

The RANS equations only represent transport equations for the mean flow quantities, with all the scales of the turbulence being modeled. The approach of permitting a solution for the mean flow variables greatly reduces the computational effort. The Reynolds-averaged approach is generally adopted for practical engineering calculations, and uses models such as Spalart–Allmaras, k-ε and its variants, k-ω.

LES provides an alternative approach, in which the large eddies are computed in a time-dependent simulation that uses a set of "filtered" equations. Filtering is essentially a manipulation of the exact Navier–Stokes equations to remove only the eddies that are smaller than the size of the filter, which is usually taken as the mesh size. The following turbulence models have been used in indoor air distribution modeling:

- Spalart–Allmaras model
- k-ε models:
  - Standard k-ε model
  - Renormalization-group (RNG) k-ε model
  - Realizable k-ε model
- k-ω models:
  - Standard k-ω model
  - Shear-stress transport (SST) k-ω model
- Reynolds stress model (RSM)
- Large eddy simulation (LES) model

In the following discussion, several examples are given to demonstrate the application of modeling room airflow using different models. The room geometry and ventilation scheme are shown in Figure 15.2c. The supply air temperature is approximately 24°C and is the same as the room air temperature. All simulated air velocity profiles are projected two-dimensional components at the middle of the experimental room. It is not the author's intention to compare the advantages or disadvantages of different models in the following discussion. Rather, the purpose is to show the magnitude of variations in the simulated results between models, and among the different parameter settings of the same model. Given the magnitude of variation and sensitivity to parameter settings of CFD models, readers are strongly recommended to use validated models to predict room air distributions.

15.3.1 The Spalart–Allmaras Model

The Spalart–Allmaras model is a relatively simple one-equation model that solves a modeled transport equation for the kinematic eddy (turbulent) viscosity. It was designed specifically for aerospace applications involving wall-bounded flows and has been shown to give good results for boundary layers subjected to adverse pressure gradients. It is also gaining popularity for turbomachinery applications.

On a cautionary note, the Spalart–Allmaras model is relatively new, and no claim is made regarding its suitability to all types of complex engineering flows. For instance, it may not be reliable yet to predict the decay of homogeneous, isotropic turbulence. Furthermore, one-equation models are often criticized for their inability to accommodate changes in length scale rapidly, and such a scale change might be necessary when the flow changes abruptly from a wall-bounded to a free shear flow.

In terms of computation, the Spalart–Allmaras model is the least expensive turbulence model of all those described here.

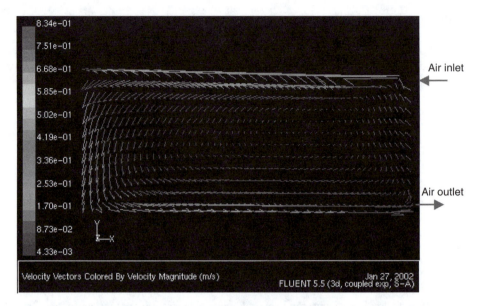

Figure 15.8 An example of calculated air distribution using the Spalart–Allmaras model, coupled, explicit, with $C_{b2} = 0.622$, Prandtl number = 0.667. The room geometry and ventilation scheme are shown in Figure 15.2c.

$$\frac{\partial}{\partial t}\left(p\tilde{v}\right) + \frac{\partial}{\partial x_i}\left(p\tilde{v}u_i\right) = G_v + \frac{1}{\sigma_{\tilde{v}}}\left[\frac{\partial}{\partial x_i}\left\{(\mu + p\tilde{v})\frac{\partial \tilde{v}}{\partial x_i}\right\} + C_{b2}p\left(\frac{\partial \tilde{v}}{\partial x_i}\right)^2\right] - Y_v \quad (15.9)$$

where G_v is the production of turbulent viscosity and Y_v is the destruction of turbulent viscosity that occurs in the near-wall region due to wall blocking and viscous damping; σ_v and C_{b2} are constants, and $\tilde{v}$ is the molecular kinematic viscosity.

An example of calculated room airflow distribution using the Spalart–Allmaras model is shown in Figure 15.8. As explained previously, this relatively simple one-equation model gave a one-rotation flow pattern without any secondary rotational flow. Even for a simple rectangular room geometry without obstructions, this simulation result may oversimplify actual flow patterns and velocity distribution. This will be evidenced in the next section, which presents measured air velocities for the same room.

15.3.2   The Standard k-ε Model

The standard k-ε model is a semiempirical model based on model transport equations for the turbulent kinetic energy k and its dissipation rate ε. The model transport equation for k is derived from the exact mathematical equation; the model transport equation for ε was obtained using physical reasoning and bears little resemblance to its mathematically exact counterpart.

In the derivation of the k-ε model, it was assumed that the flow is fully turbulent and that the effects of molecular viscosity are negligible. The standard k-ε model is therefore valid only for fully turbulent flows. The standard k-ε model clearly requires more computational effort than the Spalart–Allmaras model.

$$\frac{\partial}{\partial t}(\rho k) + \frac{\partial}{\partial x_i}(\rho k u_i) = \frac{\partial}{\partial x_i}\left[\left(\mu + \frac{\mu_t}{\sigma_k}\right)\frac{\partial k}{\partial x_i}\right] + G_k + G_b - \rho\varepsilon - Y_M \qquad (15.10)$$

$$\left[\left(\mu + \frac{\mu_t}{\sigma_\varepsilon}\right)\frac{\partial \varepsilon}{\partial x_i}\right] + C_{1\varepsilon}\frac{\varepsilon}{k}(G_k + C_{3\varepsilon}G_b) - C_{2\varepsilon}\rho\frac{\varepsilon^2}{k} \qquad (15.11)$$

where G_k represents the generation of turbulent kinetic energy due to the mean velocity gradients; G_b is the generation of turbulent kinetic energy due to buoyancy; Y_M represents the contribution of the fluctuating dilatation in compressible turbulence to the overall dissipation rate; $C_{1\varepsilon}$, $C_{2\varepsilon}$, and $C_{3\varepsilon}$ are constants; and σ_k and σ_ε are the turbulent Prandtl numbers for k and, respectively.

The turbulent (or eddy) viscosity μ_t is computed by combining k and ε as follows:

$$\mu_t = \rho C_\mu \frac{k^2}{\varepsilon} \qquad (15.12)$$

where C_μ is a constant.

An example of calculated room airflow distribution using the standard k-ε model is shown in Figure 15.9. It appears that, with the standard wall function, the air jet from the air inlet dissipates into the room somewhat frictionlessly. Without the standard wall function, the flow rotates near the air inlet. In both cases, the standard k-ε model may still oversimplify actual flow patterns and velocity distribution.

15.3.3 The RNG k-ε Model

The RNG-based k-ε turbulence model is derived from the instantaneous Navier–Stokes equations, using a mathematical technique called *renormalization group* (RNG) methods. The analytical derivation results in a model with constants different from those in the standard k-ε model, and additional terms and functions in the transport equations for k and ε. The RNG model is more responsive to the effects of rapid strain and streamline curvature than the standard k-ε model, which explains the superior performance of the RNG model for certain classes of flows.

Computations with the RNG k-ε model tend to take 10 to 15% more CPU time than with the standard k-ε model.

$$\frac{\partial}{\partial t}(\rho k) + \frac{\partial}{\partial x_i}(\rho k u_i) = \frac{\partial}{\partial x_i}\left(\alpha_k \mu_{eff}\frac{\partial k}{\partial x_j}\right) + G_k + G_b - \rho\varepsilon - Y_M \qquad (15.13)$$

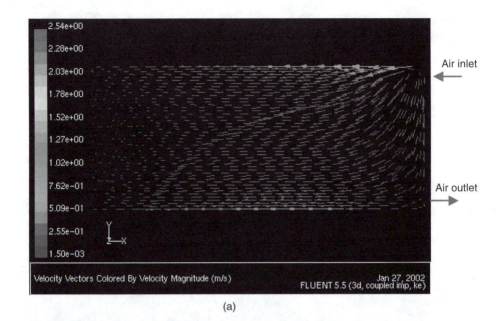

(a)

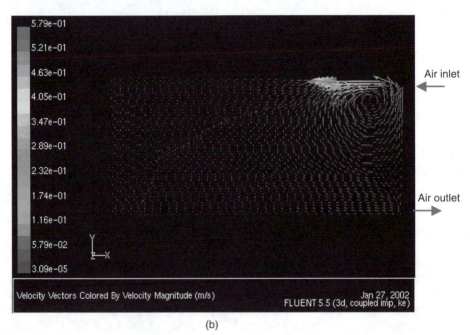

(b)

Figure 15.9 CFD results using the standard k-ε model, coupled, implicit, T = 25°C, relative humidity = 20%, inlet velocity = 0.74 m/s. (a) With standard wall function. (b) Without standard wall function.

$$\frac{\partial}{\partial t}(\rho\varepsilon) + \frac{\partial}{\partial x_i}(\rho\varepsilon u_i) = \frac{\partial}{\partial x_i}\left(\alpha_\varepsilon\mu_{eff}\frac{\partial\varepsilon}{\partial x_i}\right) + C_{1\varepsilon}\frac{\varepsilon}{k}(G_k + C_{3\varepsilon}G_b) - C_{2\varepsilon}\rho\frac{\varepsilon^2}{k} - R_\varepsilon \quad (15.14)$$

where G_k represents the generation of turbulent kinetic energy due to the mean velocity gradients; G_b is the generation of turbulent kinetic energy due to buoyancy; Y_M represents the contribution of the fluctuating dilatation in compressible turbulence to the overall dissipation rate; and the quantities α_k and α_ε are the inverse effective Prandtl numbers for k and ε, respectively.

An example of calculated room airflow distribution using the RNG k-ε model is shown in Figure 15.10. With the standard wall function, the air jet forms a large rotation airflow along the room boundaries with several large eddies within the room (Figure 15.10a). This flow pattern is quite stable and similar to the measurement. With standard wall function and a wall roughness constant of 0.35, a strong vortex formed near the air inlet, showing that the air jet has less effect on the room air distribution than the standard wall with zero wall roughness (Figure 15.10b). Apparently, this model is very sensitive to the parameter of roughness constant.

15.3.4 The Realizable *k*-ε Model

An immediate benefit of the realizable k-ε model is that it predicts more accurately the spreading rate of both planar and round jets. It is also likely to provide superior performance for flows involving rotation, boundary layers under strong adverse pressure gradients, separation, and recirculation. Some studies have shown that the realizable model provides the best performance of all the k-ε model versions for several validations of separated flows and flows with complex secondary flow features.

One limitation of the realizable k-ε model is that it produces nonphysical turbulent viscosities in situations where the computational domain contains both rotating and stationary fluid zones (e.g., multiple reference frames, rotating sliding meshes).

The realizable k-ε model requires only slightly more computational effort than the standard k-ε model.

$$\frac{\partial}{\partial t}(\rho k) + \frac{\partial}{\partial x_i}(\rho k u_i) = \frac{\partial}{\partial x_i}\left[\left(\mu + \frac{\mu_t}{\sigma_k}\right)\frac{\partial k}{\partial x_i}\right] + G_k + G_b - \rho\varepsilon - Y_M \quad (15.15)$$

$$\frac{\partial}{\partial t}(\rho\varepsilon) + \frac{\partial}{\partial x_i}(\rho\varepsilon u_i) = \frac{\partial}{\partial x_i}\left[\left(\mu + \frac{\mu_t}{\sigma_\varepsilon}\right)\frac{\partial\varepsilon}{\partial x_i}\right]$$

$$+ \rho C_1 S_\varepsilon - \rho C_2 \frac{\varepsilon^2}{k + \sqrt{v\varepsilon}} + C_{1\varepsilon}\frac{\varepsilon}{k}C_{3\varepsilon}G_b \quad (15.16)$$

where G_k represents the generation of turbulent kinetic energy due to the mean velocity gradients; G_b is the generation of turbulent kinetic energy due to buoyancy;

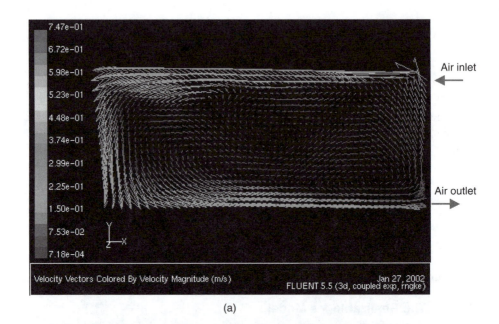

(a)

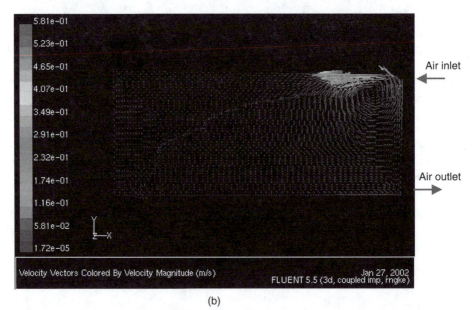

(b)

Figure 15.10 CFD results using the renormalization group (RNG) k-ε model, coupled, implicit, two outlets, $C_{mu} = 0.0845$, $C_{1\varepsilon} = 1.42$, $C_{2\varepsilon} = 1.68$. (a) Standard wall function and swirl dominated flow, with swirl factor = 0.07. (b) Standard wall function with a wall roughness constant = 0.35.

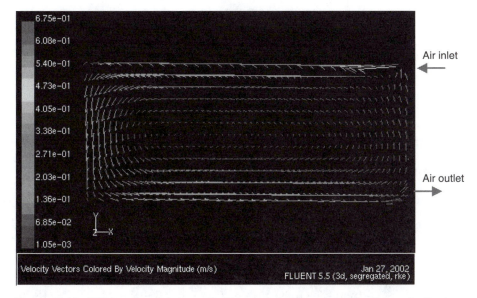

Figure 15.11 CFD results using the realizable k-ε model, segregated, implicit, and with standard wall function.

Y_M represents the contribution of the fluctuating dilatation in compressible turbulence to the overall dissipation rate; C_2 and $C_{1\varepsilon}$ are constants and σ_k and σ_ε are the turbulent Prandtl numbers for k and ε, respectively.

An example of calculated room airflow distribution using the realizable k-ε model is shown in Figure 15.11. Although this model is more complicated than the Spalart–Allmaras model, it gives a one-rotation flow pattern without any secondary rotational flow within the room.

15.3.5 The Standard k-ω Model

The standard k-ω model is an empirical model based on model transport equations for the turbulent kinetic energy k and the specific dissipation rate ω, which can also be thought of as the ratio of ω to k. The model predicts free shear flow spreading rates that are in close agreement with measurements for far wakes, mixing layers, and plane, round, and radial jets, and is thus applicable to wall-bounded flows and free shear flows.

The k-ω models require about the same computational effort as the k-ε models.

$$\frac{\partial}{\partial t}(\rho k) + \frac{\partial}{\partial x_i}(\rho k u_i) = \frac{\partial}{\partial x_i}\left(\Gamma_k \frac{\partial k}{\partial x_i}\right) + G_k - Y_k \tag{15.17}$$

$$\frac{\partial}{\partial t}(\rho\omega) + \frac{\partial}{\partial x_i}(\rho\omega u_i) = \frac{\partial}{\partial x_i}\left(\Gamma_\omega \frac{\partial \omega}{\partial x_i}\right) + G_\omega - Y_\omega \tag{15.18}$$

where G_k represents the generation of turbulent kinetic energy due to mean velocity gradients; G_ω represents the generation of ω. Γ_k and Γ_ω represent the effective diffusivity of k and ω, respectively; and Y_k and Y_ω represent the dissipation of k and ω due to turbulence.

15.3.6 The Shear-Stress Transport (SST) k-ω Model

The SST k-ω model is more accurate and reliable for a wider class of flows (e.g., adverse pressure gradient flows, airfoils, transonic shock waves) than the standard k-ω model because it includes the cross-diffusion term.

$$\frac{\partial}{\partial t}(\rho k) + \frac{\partial}{\partial x_i}(\rho k u_i) = \frac{\partial}{\partial x_i}\left(\Gamma_k \frac{\partial k}{\partial x_i}\right) + G_k - Y_k \tag{15.19}$$

$$\frac{\partial}{\partial t}(\rho \omega) + \frac{\partial}{\partial x_i}(\rho \omega u_i) = \frac{\partial}{\partial x_i}\left(\Gamma_\omega \frac{\partial \omega}{\partial x_i}\right) + G_\omega - Y_\omega + D_\omega \tag{15.20}$$

where G_k represents the generation of turbulent kinetic energy due to mean velocity gradients; G_ω represents the generation of ω. Γ_k and Γ_ω represent the effective diffusivity of k and ω, respectively (which are calculated as described later); Y_k and Y_ω represent the dissipation of k and ω due to turbulence; and D_ω represents the cross-diffusion term.

15.3.7 The Reynolds Stress Model (RSM)

Because the RSM accounts for the effects of streamline curvature, swirl, rotation, and rapid changes in strain rate in a more rigorous manner than one-equation or two-equation models, it has greater potential to give accurate predictions for complex flows. Use of the RSM is a must when the flow features of interest are the result of anisotropy in the Reynolds stresses.

However, the fidelity of RSM predictions is still limited by the closure assumptions employed to model various terms in the exact transport equations for Reynolds stresses. The modeling of the pressure-strain and dissipation-rate terms is particularly challenging and is often considered to be responsible for compromising the accuracy of RSM predictions.

On average, the RSM requires 50 to 60% more CPU time per iteration than the k-ε and k-ω models. Furthermore, 15 to 20% more memory is needed. The RSM may take more iterations to converge than the k-ε and k-ω models, due to the strong coupling between the Reynolds stresses and the mean flow.

$$\frac{\partial}{\partial t}(\rho \overline{u_i u_j}) + \frac{\partial}{\partial x_k}(\rho U_k \overline{u_i u_j})$$

$$= -\frac{\partial}{\partial x_k}\left[\rho \overline{u_i u_j u_k} + \overline{p(\delta_{kj} u_i + \delta_{ik} u_j)}\right] + \frac{\partial}{\partial x_k}\left[u\frac{\partial}{\partial x_k}\left(\overline{u_i u_j}\right)\right]$$

$$-\rho\left(\overline{u_i u_k}\frac{\partial U_j}{\partial x_k} + \overline{u_j u_k}\frac{\partial U_i}{\partial x_k}\right) - \rho\beta(g_i \overline{u_j \theta} + g_j \overline{u_i \theta})$$

$$+ \overline{p\left(\frac{\partial u_i}{\partial x_j} + \frac{\partial u_j}{\partial x_i}\right)} - 2\mu\overline{\frac{\partial u_i}{\partial x_k}\frac{\partial u_j}{\partial x_k}} - 2\rho\Omega_k\left(\overline{u_j u_m}\varepsilon_{ikm} + \overline{u_i u_m}\varepsilon_{jkm}\right)$$

(15.21)

Overall, the RSM gives good rotational airflow (Figure 15.12). When segregated, the flow pattern shows a vigorous circulation, with two large vortexes in the room even at a very low inlet air velocity. This air velocity profile may exaggerate the actual flow, especially in the vicinity of the boundaries. Without segregation, there is only one large rotational flow within the room and no secondary vortexes across the entire room. The difference in room airflow patterns between segregated and nonsegregated is so great that the results have no similarity. This again shows that the model is very sensitive to the parameters. It also shows that the model must be validated before the simulated data can have a meaningful interpretation.

15.3.8 The Large Eddy Simulation (LES) Model

In this model, large eddies are resolved directly, and small eddies are modeled by their averages. An attracting feature of LES is that, by modeling less of the turbulence (and solving more), the error induced by the turbulence model is reduced. Statistics of the mean flow quantities, which are generally of most engineering interest, are gathered during the time-dependent simulation.

The application of LES to room air simulation is still in its infancy. Typical applications to date have been for simple geometries. For complicated geometries, large computing capacity is required to resolve the complicated energy-containing turbulent eddies.

$$\frac{\partial \rho}{\partial t} + \frac{\partial}{\partial x_i}\left(\rho \overline{u}_i\right) = 0$$

(15.22)

$$\frac{\partial}{\partial t}\left(\rho \overline{u}_i\right) + \frac{\partial}{\partial x_j}\left(\rho \overline{u}_i \overline{u}_j\right) = \frac{\partial}{\partial x_j}\left(\mu\frac{\partial \overline{u}_i}{\partial x_j}\right) - \frac{\partial \overline{p}}{\partial x_i} - \frac{\partial \tau_{ij}}{\partial x_j}$$

(15.23)

where τ_{ij} is the subgrid-scale stress defined by

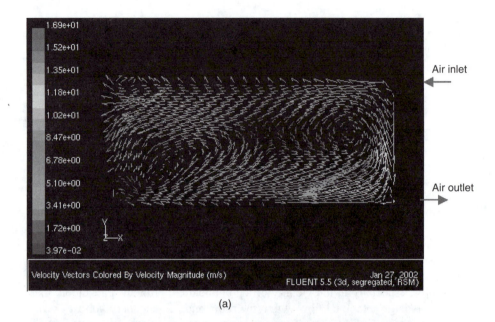

(a)

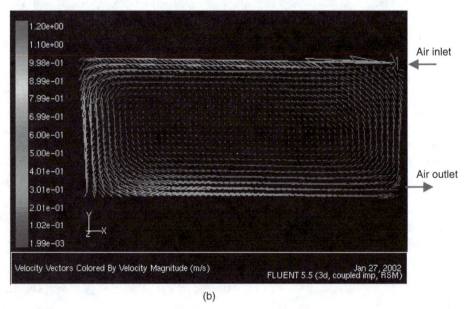

(b)

Figure 15.12 CFD results using the Reynolds stress model (RSM), temperature = 25°C, relative humidity = 20%, inlet velocity = 0.74 m/s, implicit, wall boundary condition from *k* equations with standard wall function. (a) Segregated. (b) Not segregated.

$$\tau_{ij} \equiv \rho \overline{u_i u_j} - \rho \overline{u}_i \overline{u}_j \tag{15.24}$$

Of all the models discussed previously, the LES model produces calculated airflow patterns best fitted to the measured data with good stability. Figure 15.13 shows three simulations with different simulation parameters. Apparently, the LES model has a stable output, because all three show secondary vortexes within the room. This may be explained as follows. In a room airflow field, the large eddies with low turbulence intensity carry the most kinetic energy. Those high-frequency turbulence flows only contribute a small fraction of the total energy. Typically, in a kinetic energy spectrum, turbulence flow with a frequency higher than 10 Hz contains less than 1% of the total kinetic energy. Therefore, the energy balance of the room airflow is primarily determined by the large eddies. Thus, even with the relatively simple energy balance equation in LES, the model captures the major energy contributors, and therefore predicts the room airflow to a closer proximity than many other, complicated models.

Unless the fundamentals of turbulence flow are fully explained, complicated turbulence models will likely continue to have difficulty in predicting room air with acceptable stability. The microscopic theory of turbulence is one of the few classical physics problems yet to be solved.

15.3.9 Measurement Results

Air velocity profiles for the same room as shown in Figure 15.2c under isothermal and nonisothermal conditions were measured using a particle imaging velocimetry method. In the measurement, the inlet air velocity was higher than the inlet velocity used in the preceding CFD simulations. Therefore, the flow patterns and air velocity vectors can only be compared qualitatively rather than quantitatively. Various air distribution measurement techniques and instrumentations will be discussed in the next section.

In Figure 15.14a, aside from the major rotational flow around the room boundary, there are many secondary vortexes across the entire room airspace under isothermal conditions. Under nonisothermal conditions, the large rotational flow was interrupted. In fact, the cold air jet dropped immediately after it entered the room, causing a reversed rotational flow in the rest of the room. The nonisothermal flow is much more vigorous than the isothermal flow for the same ventilation rate. This is because the buoyancy effect introduces additional kinetic energy to the total airflow.

In regard to the preceding simulations and measurements, it is noteworthy that simulation results are very different among models with similar room configurations and boundary and initial conditions. Even with the same model, the simulation results vary substantially with simulation parameter settings, such as varied coefficients and constants required by the model. These parameter settings can be very subjective, causing large variations in the simulated results.

Care must be taken when using a CFD model. It is highly recommended that a CFD model be validated using experimental data before its simulation results are interpreted or used.

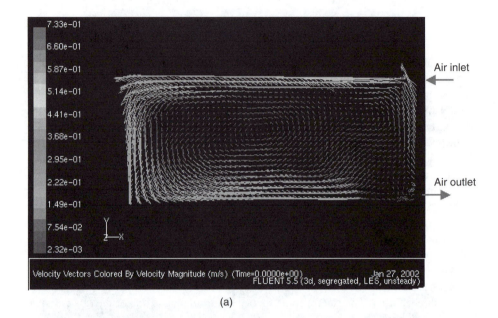

(a)

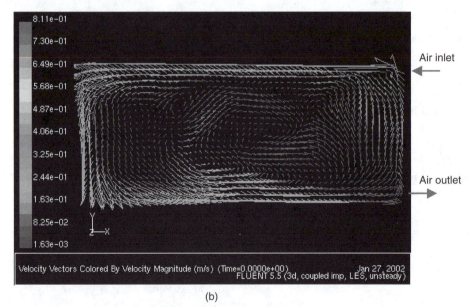

(b)

Figure 15.13 CFD results using large eddy simulation (LES), temperature = 25°C, relative
humidity = 20%, inlet velocity = 0.74 m/s. (a) Subgrid-scale model, segregated,
implicit. (b) Same as (a), except coupled implicit. (c) Subgrid-scale model,
segregated, implicit.

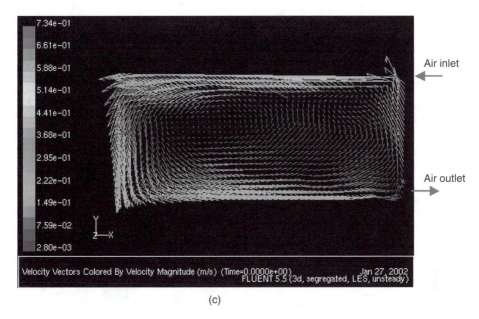

(c)

Figure 15.13 *Continued.*

On the other hand, a CFD model with simulated air distribution that is very different from the measurement does not necessarily indicate that the model is insufficient or inaccurate. As discussed earlier, many models are highly sensitive to simulation parameters, such as boundary roughness, swirling factors, wall functions, and segregation. Many of these factors are subjective and can cause a large variation in simulation results. These parameters can be adjusted through the validation process.

15.4 ROOM AIR DISTRIBUTION MEASUREMENT

Measurement techniques are especially important in room air distribution studies. Proper selection of the measurement instrumentation often determines the success of a study. In the existing literature, thermal anemometers, laser Doppler velocimetry, ultrasonic anemometers, flow visualization, and particle imaging velocimetry have all been used to measure airflow. Ideal instrumentation for the study of low-speed and high–turbulence intensity room airflow should possess the following characteristics:

- Fast time response
- High accuracy
- High spatial resolution
- Least disturbance to the flow field
- Multipoint or whole flow-field detection
- Flow direction detection
- Low cost
- Ease of use

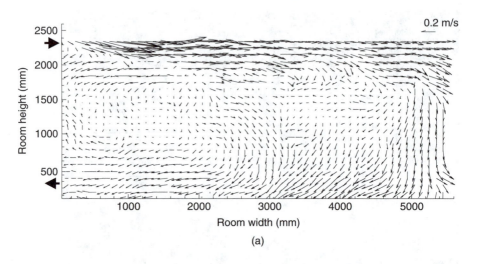

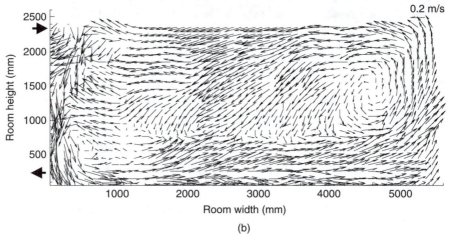

Figure 15.14   Air distribution at the center section of the room shown in Figure15.2c measured using PIV method. (a) Isothermal airflow patterns: Supply air temperature was the same as the room air temperature. (b) Nonisothermal airflow patterns of ΔT = T_r-T_s = 26°C. (From Zhao, L.Y. et al., Development of a 2-D particle imaging velocimetry system to study full-scale room airflows, *Trans. Am. Soc. Heat. Referig. Air Cond. Engr.*, 107(2): 434–444, 2001.)

15.4.1 Thermal Anemometers

Thermal anemometers are used to measure the fluid velocity by sensing the heat transfer changes from a small, electrically heated sensing element, such as hot wire, film, or bead sensors, exposed to the fluid motion. Hot film sensors are widely used in liquid or gas flows with particulates. For room air studies, hot-wire anemometers have been commonly used.

Because hot-wire anemometers can measure the velocity at a very high sampling frequency, turbulence intensity and turbulent kinetic energy can be measured effectively. The difficulty with hot wires is low-velocity measurements. Usually, hot wires are run at very high temperatures when compared with the fluid temperature, to ensure measurement stability. However, hot wires with high temperatures cause significant free convection near sensors and create flow velocity measurement error, especially in low-velocity flow fields such as indoor airflow, which is typically in the range of 0 to 0.25 m/s. Zhang evaluated the hot-wire anemometer measurement error for indoor airflow and determined that the uncertainty is typically 25% of mean velocities measured.[10] For air velocities lower than 0.15 m/s, hot-wire anemometers become unreliable because the thermal buoyancy of the hot wire element alone may create as high as 0.1 m/s convective air velocity.

Another difficulty with hot-wire anemometer measurement is velocity direction. Hot-wire anemometers cannot detect the exact flow direction or reverse flows (airflow blowing parallel to the probe support toward the probe). All of the velocity components perpendicular to the hot wire, except the reverse flow, contribute to the heat exchange. In addition, it is difficult to evaluate the effect of the three-dimensional velocity components. Three-dimensional anemometer probes are available, which consist of three perpendicular thin wires closely grouped. These are more expensive and less suited to low-velocity measurements, because thermal convection effects of the probes are greater than with omnidirectional anemometry.

Velocity measurements using hot-wire anemometers need accurate temperature compensation. Under isothermal conditions, hot-wire anemometers are convenient. For nonisothermal case studies, the measurement accuracy highly depends on the temperature measurements and temperature compensation. The temperature compensation can cause large measurement errors or require additional data processing equipment.

15.4.2 Laser Doppler Velocimetry

Laser Doppler velocimetry (LDV) is a method of measuring fluid velocities by detecting the Doppler frequency shift of a laser light that has been scattered by small particles moving with the fluid; thus it is an optical measurement of flow velocities. Two identical laser lights with different frequencies cross at one small volume of the flow field to form interference light fringes. Small particles, which are in the micrometer size range, are seeded into the flow field. When the particles pass through the measurement volume, the laser light is scattered. By detecting the frequency shift of the laser light, the time interval for particle travel through the light fringe can be calculated. Because the light fringe pattern is precisely determined by the laser light source, the particle velocity can be calculated. The velocities of particles represent the flow velocities.

Compared with hot-wire anemometers, LDV has the following advantages: no disturbance to the flow field, high measurement accuracy (0.1% error of the full range), flow direction detection, no temperature limitation, no low-velocity limitation, and flow reversal detection. The disadvantages are high cost, difficulty of use, and ability to measure only one point at a time.

LDV is widely used in small-scale mechanical pipe flow. In indoor airflow studies, LDV has been used to measure the airflow velocity distribution in reduced-scale room models. Because LDV is still a point measurement method, full-scale flow-field measurement requires an array of measurements throughout the room, which is very expensive, time consuming, and inaccurate in spatial velocity variation.

15.4.3 Flow Visualization

Flow visualization is a classic method in aerodynamic and hydrodynamic studies. Because water and air are transparent, flow patterns can be visualized by seeding visible particles or tracer gases, which are illuminated by a light source. Flow visualization has been used to verify existing physical principles and led to the discovery of numerous flow phenomena. Three tracers for flow visualization have been commonly used in aerodynamic studies: smoke-tube, smoke-wire, and helium bubbles. The smoke-tube method was the earliest method used to visualize flow. Its history goes back to 1893. In this method, smoke (i.e., a smokelike material such as vapors, fumes, or mist) is injected from a smoke generator directly into the flow field. This method has been widely used to study low turbulence flow.

Because some flow studies require small-scale details to verify fundamental flow phenomena, the smoke-wire method was developed by Raspet and Moore in the early 1950s. This method can provide very fine smoke filaments by vaporizing oil from a fine wire (with a wire diameter of about 0.1 mm) that is electrically heated. This method has been widely used to reveal fundamental flow structures.

To visualize flow at high turbulence intensity, the helium-bubble method was developed in 1936 to observe complicated flow structures by tracking the path of individual soap bubbles. Since then, the helium-bubble method has been widely used to study turbulent airflow. Many flow patterns can be visualized clearly by applying the preceding three methods. Many pictures of flow patterns can be found in Van Dyke.[11]

In indoor airflow studies, the flow visualization technique has been widely used to visualize airflow patterns in reduced-scale ventilated rooms. The original purpose of flow visualization was to record flow fields qualitatively. The possibility of acquiring quantitative measurement results motivated the development of new flow visualization techniques. Recent developments in digital image processing techniques and software makes it possible to obtain quantitative data from appropriate flow visualization.

15.4.4 Particle Image Velocimetry

Existing measurement instrumentation and technologies, such as thermal-based anemometers, LDV, and ultrasonic anemometry, are single-point measurements and do not provide complete information about the entire airflow field. In addition, these existing measurement technologies often require that the instrumentation be placed within the flow field of concern, thus disturbing the actual air distribution. Nonintrusive, full-scale, and instantaneous measurement techniques for airflow in entire room airspaces (vs. single-point measurements) are needed, especially in the area of developing CFD models.

A technique that uses particles and their images to measure flow velocities is called *particle image velocimetry* (PIV). The invention and development of PIV were originally for the study of experimental fluid mechanics.[12] PIV measures a 2D velocity vector map of a flow field at an instant of time by acquiring and processing images of particles seeded into the flow field. It is based on the principle that speed is equal to the displacement divided by the time interval. Using the PIV technique, the time interval during which the particles travel is recorded and the distance that particles travel is measured. Thus, particle speed can be calculated as simply distance traveled divided by time interval. Particle speed is the representation for the flow-field velocity.

Particle streak photography in combination with manual analysis of the image gave the semiquantitative measurement results of a flow field. Modern developments in digital image processing and optical instruments make image acquisition, processing, and analysis more automatic and quantitative. Particle tracking velocimetry (PTV) or particle streak velocimetry (PSV) are extensions of flow visualization. In PTV or PSV, the concentration of seeding particles should be dilute enough to form individual particle streaks, which can be analyzed automatically. In this operating mode, the spatial resolution of the method is not good enough to investigate complicated fine flow structures.

Low-image density PIV, high-image density PIV, and laser speckle velocimetry (LSV) are inventions to investigate complicated flow structures. LSV is based on the solid mechanics fact that coherent light scattered from solid surfaces forms speckle patterns. In LSV, dense particles are seeded into flow and illuminated by a laser light sheet to form laser speckle images. By analyzing the double-exposure laser speckle image using Yang's fringe method of interrogation, the speckle displacement information is extracted, and then the velocity data are obtained. High-image density PIV was developed as an improvement of LSV to overcome the high-density particle seeding difficulty of LSV in practical situations. PIV analysis images of a group of particles or of one particle falling into interrogation spots by autocorrelation or cross-correlation analysis methods, determined by the image acquisition mode: one frame, two exposures; or two frames, one exposure. Particle displacements are extracted to form velocity data information.

PIV techniques have been applied to a wide range of flow phenomena studies, including impinging jet flow, internal combustion engine flow, turbulent channel flow, etc. In different practical applications, according to the measurement requirement, the appropriate operating mode must be selected and specific system configurations are needed.

In general, a PIV system consists of illumination, image acquisition, particle seeding, and image processing and data analysis subsystems. Laser light is commonly used as the illumination source in PIV systems. Ideal tracer particles should be very small and follow the flow field. Different flow fields need different seeding particles. According to spatial resolution requirements, either photographic or couple-charged device (CCD) video cameras can be used to record particle images. If the resolution is high enough, CCD video cameras are preferred. To resolve velocity direction ambiguity during the image acquisition process, an image-shifting technique is needed. Image interrogation techniques include autocorrelation and cross-correlation image processing methods. Commercial software and hardware products for PIV image interrogation, such as INSIGHT from TSI, are available.

PIV started as a two-dimensional velocity measurement method and is being developed into a three-dimensional velocity measurement method. Stereoscopic photograph techniques have been employed to detect the three components of a velocity vector. Recently, holographic techniques have been researched to capture three-dimensional flow field information. Even though many PIV systems have been developed and some are becoming commercial products, most PIV experiments have been conducted on flows at very small scales — typically about a 100×100 mm field of view — and fairly low turbulent flow. Enlarging the study scale is limited by the camera resolution and flow-field illumination.

Scholzen and Moser developed a three-dimensional particle streak velocimetry system.[13] In their study, three cameras, a 120 mm thick white light sheet, and a digital image processing program were employed to acquire the particle streak image and extract particle displacement information. One camera with a relatively short exposure time setting was used to recognize the particle streak direction. The other two cameras, which were in the same setting but placed at different locations, formed stereoscopic photographs to obtain the three components (x, y, and z) of a velocity vector. The method was tested in a $2.4 \times 1.7 \times 1.2$ m ventilated space. The effective image area was 1.0×0.8 m. Good representative results were obtained, showing that the method is promising for indoor airflow study. However, the study seems to have been discontinued and there is no successive report of the study. From the investigators' experience, the technical difficulty in combining images from three cameras could be overwhelming, and the error derived from the process could be excessive using this three-camera technique.

In summary, although some PIV systems have been developed, such as INSIGHT from TSI, almost all PIV experiments have only studied flows in small enclosures, such as combustion chambers, with typically about 100×100 mm field of view, and with a fairly low turbulent flow.[12] Moreover, the illumination volume is very small, because the light sheet is usually very thin (in magnitude of several millimeters). With such a thin lighting sheet, it is difficult to study the 3D flow field, especially under transient conditions. Using PIV to study a large flow field, such as a full-scale room airspace, presents major challenges, such as camera resolution, flow-field illumination, particle seeding, and detailed boundary velocity conditions.[14] On the other hand, the PIV technique has advantageous features, including nonintrusiveness, whole flow-field measurement, and a low-velocity detecting threshold. These features are particularly suitable for room airflow studies, and the measurement data are extremely useful for CFD model development.

15.5 ROOM AIR DISTRIBUTION MEASUREMENT USING SPIV

This section discusses a stereoscopic particle imaging velocimetry (SPIV) system suitable for measurement of full-scale room 3D airflow. One challenge in indoor air quality studies is the measurement of three-dimensional air velocity profiles in an airspace so that the nature of airflow can be better understood and appropriate ventilation systems can be designed. There is much dispute over a variety of CFD models, primarily due to a lack of credible data to validate such models. The SPIV method is based on the principle of parallax to extract a third (z-direction) velocity component using two cameras. Sun and Zhang have developed a 3D algorithm in the particle streak mode (PSM) that requires only two cameras to acquire three velocity components (x, y, and z) and flow directions in a 3D volume, rather than in a thin layer of a plane.[15] The two-camera approach can greatly simplify image acquisition and data processing, and improve accuracy either by eliminating the error caused by the third camera image or by utilizing some special techniques employed to resolve directional ambiguity.

The 3D image volume can contain a full range of 3D velocities. In this 3D SPIV setup, two cameras are placed at different angles to view the illuminated field and to capture particle displacement images that contain the influence of the third velocity component. The parallax effect allows us to obtain different two-velocity component vector maps from each camera. The differences between the two images arise from the third velocity component and the geometric configuration of the two cameras. After image calibration, this third velocity component can be calculated. The two cameras are set at different exposure times, thus acquiring different streak lengths for the same particle path. The differences in the exposures then can be used to distinguish the flow directions.

15.5.1 SPIV System

The hardware of a SPIV system in the particle streak mode consists of a particle seeding system, an illumination system, an image recording system, and an image and data processing system. To visualize a flow for PIV, the flow must be seeded with small, neutrally buoyant tracer particles. The seeding particles should have the physical density of the host flow and should be as small as possible to guarantee good follower behavior, but large enough to reflect sufficient light in suitable illumination and to be captured by the image recording system. In PSM, the concentration of seeding particles should be low enough that individual particle streak images can be identified.

Illumination is a key component in a traditional PIV measurement system. Unlike an illumination system for traditional PIV, the light for this algorithm is not required to be transferred into a thin sheet, because it is used for volume measurement. The light source also should not be pulsed for this operating mode. The light intensity, however, should be high enough and its distribution be uniform enough in the whole observation volume to ensure clear images of streaks in the area of concern.

Two high-resolution cameras are required for the system. The shutters of the two cameras can be synchronously opened with a controller, and their exposure times can be set at high accuracy. If the airflow in a room of 5.5 m width is measured using the system and the diameters of the seeding particles are about 2 mm, a minimum resolution of about 2750 pixels horizontally is needed for the cameras to cover the entire room width. Ideally, one seeding particle should have two or more linear pixels (in the vertical or horizontal direction) to ensure a sharp image. Intensified cameras and some digital cameras can operate at low light levels, decreasing the requirement for the illumination system. Digital cameras have their own computer interface, and the captured images can easily be transferred to computers. Frame grabbing by computers in real time generally requires large computer capacity and high speed because of the high resolution of the photos. The purpose of image processing is to obtain information about the position, orientation, and size of every streak on the photos. Some commercial software for this kind of image processing is available. However, image processing is usually very time consuming if the illumination is not sufficient or the concentration of seeding particles is too high. Therefore, a high-speed personal computer with a large memory capacity is needed for data processing based on the algorithm presented in the following section, especially when the numbers of streaks on a pair of photos are high.

15.5.2 SPIV Algorithm

The most important step in a stereoscopic PIV study is to establish the relationship between a photo coordinate and the room coordinate that contains the particle of concern. Consider a generic case in which camera i (i = 1, 2, ..., n) views a particle with coordinates (x, y, z) in a three-dimensional room airspace, as shown in Figure 15.15. Assume that the local coordinates for camera i are (x_i', y_i', z_i'). The z_i' axis coincides with the optical axis of the camera lens, and the lens center is located at o_i (x_{0i}, y_{0i}, z_{0i}).

In Figure 15.15, the distance between the camera lens center and the film is d_i. In the photo image, the particle is the negative image of that on the film. Thus, the particle in the 2D photo image (X_i, Y_i), represents the actual relative location of the particle (x, y, z) in the 3D fluid field. The magnification from the film to photo image (including the transformation from ordinary-length unit to pixels, in which the photo image is measured) is M_i. The relationship between the room coordinate system, o-xyz, and camera i's coordinate system $o_i - x_i'y_i'z_i'$, can be written as

$$\begin{Bmatrix} x \\ y \\ z \\ 1 \end{Bmatrix} = \begin{bmatrix} a_{11i} & a_{12i} & a_{13i} & x_{0i} \\ a_{21i} & a_{22i} & a_{23i} & y_{0i} \\ a_{31i} & a_{32i} & a_{33i} & z_{0i} \\ 0 & 0 & 0 & 1 \end{bmatrix} \begin{Bmatrix} x_i' \\ y_i' \\ z_i' \\ 1 \end{Bmatrix} = \begin{bmatrix} A_i & O_i \\ 0 & 1 \end{bmatrix} \begin{Bmatrix} x_i' \\ y_i' \\ z_i' \\ 1 \end{Bmatrix} \quad (15.25)$$

where

$$O_i = \begin{bmatrix} x_{0i} & y_{0i} & z_{0i} \end{bmatrix}^T$$

and A_i denotes the 3 × 3 matrix of the factors a_{lmi} (l, m = 1, 2, 3), the direction cosines of the x_i', y_i', z_i' axes in the o-xyz coordinate system, and is an orthogonal matrix, for

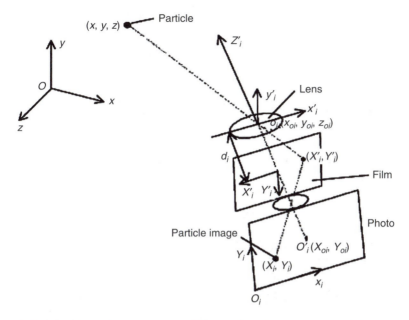

Figure 15.15 Coordinated systems for a SPIV algorithm: the room, one particle, ith camera, and the image associated with camera i. (From Sun, Y. and Zhang, Y., An algorithm of stereoscopic particle image velocimetry for full-scale room airflow studies, *Trans. Amer. Soc. Heat. Refrig. Air Cond. Engr.*, 110(1), 2004. With permission.)

$$A_i^{-1} = A_i^{T} \tag{15.26}$$

From Figure 15.15, the relationship between the particle position (x_i', y_i', z_i') in the camera i coordinates and its image position (X_i, Y_i) on the photo can be written as

$$\left. \begin{aligned} X_i &= \frac{M_i d_i}{z_i'} x_i' + X_{0i} \\ Y_i &= \frac{M_i d_i}{z_i'} y_i' + Y_{0i} \end{aligned} \right\} \tag{15.27}$$

Equation 15.27 can also be rewritten in the following form

$$\begin{Bmatrix} x_i' \\ y_i' \\ z_i' \\ 1 \end{Bmatrix} = \begin{bmatrix} 1 & 0 & -X_{0i} & 0 \\ 0 & 1 & -Y_{0i} & 0 \\ 0 & 0 & M_i d_i & 0 \\ 0 & 0 & 0 & 1 \end{bmatrix} \begin{Bmatrix} \dfrac{z_i'}{M_i d_i} X_i \\ \dfrac{z_i'}{M_i d_i} Y_i \\ \dfrac{z_i'}{M_i d_i} \\ 1 \end{Bmatrix} \tag{15.28}$$

Substituting Equation 5.28 into Equation 15.25 gives

$$
\begin{Bmatrix} x \\ y \\ z \\ 1 \end{Bmatrix} = \begin{bmatrix} c_{11i} & c_{12i} & c_{13i} & c_{14i} \\ c_{21i} & c_{22i} & c_{23i} & c_{24i} \\ c_{31i} & c_{32i} & c_{33i} & c_{34i} \\ 0 & 0 & 0 & 1 \end{bmatrix} \begin{Bmatrix} \dfrac{z_i'}{M_i d_i} X_i \\ \dfrac{z_i'}{M_i d_i} Y_i \\ \dfrac{z_i'}{M_i d_i} \\ 1 \end{Bmatrix} = C \begin{Bmatrix} b_i X_i \\ b_i Y_i \\ b_i \\ 1 \end{Bmatrix}
\qquad (15.29)
$$

where

$$
c_{kji} = a_{kji}, \ (k = 1, 2, 3; j = 1, 2)
$$

$$
c_{k3i} = -a_{k1i} X_{0i} - a_{k2i} Y_{0i} + a_{k3i} M_i d_i, \ (k = 1, 2, 3)
$$

$$
c_{14i} = x_{0i}
$$

$$
c_{24i} = y_{0i}
$$

$$
c_{34i} = z_{0i}
$$

$$
b_i = \frac{z_i'}{M_i d_i}
$$

Thus, from Equation 15.29 we have

$$
\begin{Bmatrix} b_i X_i \\ b_i Y_i \\ b_i \\ 1 \end{Bmatrix} = C^{-1} \begin{Bmatrix} x \\ y \\ z \\ 1 \end{Bmatrix} = \begin{bmatrix} d_{11i} & d_{12i} & d_{13i} & d_{14i} \\ d_{21i} & d_{22i} & d_{23i} & d_{24i} \\ d_{31i} & d_{32i} & d_{33i} & d_{34i} \\ 0 & 0 & 0 & 1 \end{bmatrix} \begin{Bmatrix} x \\ y \\ z \\ 1 \end{Bmatrix}
\qquad (15.30)
$$

After eliminating b_i, X_i and Y_i can derived from Equation 15.30:

$$X_i = \frac{d_{11i}x + d_{12i}y + d_{13i}z + d_{14i}}{d_{31i}x + d_{32i}y + d_{33i}z + d_{34i}}$$

$$Y_i = \frac{d_{21i}x + d_{22i}y + d_{23i}z + d_{24i}}{d_{31i}x + d_{32i}y + d_{33i}z + d_{34i}}$$

(15.31)

Because the photo image is two-dimensional, the z_i' cannot be determined based on the photo image taken by camera i only. In order to determine the third dimension, another photo image taken from an alternative location is required. Now, we consider two photos for the same particle at the point (x, y, z) taken by two different cameras (i.e., $i = 1, 2$). According to Equation 15.29

$$x = b_1(c_{111}X_1 + c_{121}Y_1 + c_{131}) + c_{141} = b_2(c_{112}X_2 + c_{122}Y_2 + c_{132}) + c_{142}$$

$$y = b_1(c_{211}X_1 + c_{221}Y_1 + c_{231}) + c_{241} = b_2(c_{212}X_2 + c_{222}Y_2 + c_{232}) + c_{242}$$

$$z = b_1(c_{311}X_1 + c_{321}Y_1 + c_{331}) + c_{341} = b_2(c_{312}X_2 + c_{322}Y_2 + c_{332}) + c_{342}$$

(15.32)

We define S_{11}, S_{21}, S_{31}, S_{12}, S_{22}, and S_{32} as follows:

$$S_{11} = c_{111}X_1 + c_{121}Y_1 + c_{131}$$
$$S_{21} = c_{211}X_1 + c_{221}Y_1 + c_{231}$$

$$S_{31} = c_{311}X_1 + c_{321}Y_1 + c_{331}$$
$$S_{12} = c_{112}X_2 + c_{122}Y_2 + c_{132}$$
$$S_{22} = c_{212}X_2 + c_{222}Y_2 + c_{232}$$

$$S_{32} = c_{312}X_2 + c_{322}Y_2 + c_{332}$$

(15.33)

Solving Equation 15.32 yields the solutions for x, y, and z:

$$\begin{pmatrix} x \\ y \\ z \end{pmatrix} = \frac{S_{11}(c_{242} - c_{241}) - S_{21}(c_{142} - c_{141})}{S_{12}S_{21} - S_{11}S_{22}} \begin{pmatrix} S_{12} \\ S_{22} \\ S_{32} \end{pmatrix} + \begin{pmatrix} c_{142} \\ c_{242} \\ c_{342} \end{pmatrix}$$

(15.33)

Equation 15.31 and Equation 15.34 are the relationships between the coordinates x, y, and z of the particle in the room coordinate system and its corresponding imaging coordinates X_i and Y_i on the photos taken by camera 1 and camera 2. There is no special condition for the setup of the cameras, such as distance between all cameras and the viewing plane. Theoretically, the two cameras with the same optical axis will not be able to measure the velocity component of the particles that are locating on the optical axis line when the cameras' shutters are triggered. In reality, it is impossible to position the two cameras with the same optical axis. Rather, the two cameras will be placed at different positions and orientations for obtaining all velocity components.

It is possible to obtain the cameras' parameters in Equation 15.31 or Equation 15.32 by direct measurements. However, it will be difficult to do so with satisfactory accuracy. Calibration is an alternative method to obtain these parameters. In Equation 15.31, we can assign d_{34i} to be 1. In that case, the values of c_{kji} ($k = 1, 2, 3; j = 1, 2$) and b_i in Equation 15.31 and Equation 15.32 are different from those in Equation 15.29 and Equation 15.30, but the relationships represented by Equations 15.29 and 15.30 with new c_{kji} and b_i are still valid. There are 11 unknown coefficients in Equation 15.31 for each camera, and these coefficients are indirectly restrained by Equation 15.26. Because of the nonlinearity of Equation 5.27, it is more convenient to calibrate the imaging system only using Equation 15.31 without considering Equation 15.26. Thus, we need to calibrate the imaging system with at least six different spatial points. After acquiring the data of all the calibrating points — the spatial point (x_j, y_j, z_j) and its image on the corresponding photo taken by camera i (X_{ij}, Y_{ij}) ($j = 1, 2, ..., n, n \geq 6$), we can get the values of the parameters by solving Equation 15.31 using the least-squares approach of linear regression problems.

15.5.3 Determination of Velocity Magnitude and Direction

To determine the three-dimensional location of every streak on photos and the velocity they represent, two cameras must be synchronized and the streaks in two images must be matched. First, we assume that the two camera shutters open at the same time and that the exposure time of camera 1 is t_1, and that of camera 2 is t_2. The difference between t_1 and t_2 must be large enough to be detected in image processing. From Equation 15.32, we have

$$\delta \equiv (S_{21}S_{32} - S_{31}S_{22})(c_{142} - c_{141}) + (S_{31}S_{12} - S_{11}S_{32})(c_{242} - c_{241})$$
$$+ (S_{11}S_{22} - S_{21}S_{12})(c_{342} - c_{341}) = 0 \tag{15.35}$$

Equation 15.35 is the relationship between the two images (X_i, Y_i) ($i = 1, 2$) of the same point (x, y, z) in the room airspace.

As shown in Figure 15.16, we select streak j on photo 1 taken by camera 1, and substitute its endpoint coordinates X_{1Aj} and Y_{1Aj} (or X_{1Bj} and Y_{1Bj}) into Equation 15.35. Then, by substituting the X_{2k} and Y_{2k} of the endpoints of every streak on photo 2 taken by camera 2, we can calculate δ_{Ak} or δ_{Bk} ($k = 1, 2, 2$ multiplied by the total number of streaks on photo 2, which is the calculated value of the part between the two equal marks in Equation 15.35.

If $|\delta_k|$ is less than a threshold, it implies that the velocity components can be calculated using the two images. The threshold limit is a preset value determined via iteration. We should substitute the coordinates of point C of streak j on photo 1 and the coordinates of the other endpoint of the streaks on photo 2 into Equation

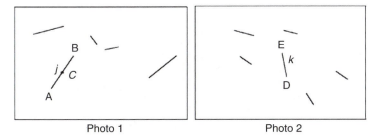

Figure 15.16 The match of streaks on two synchronous photos taken by two cameras from different view angles, respectively.

15.35. We select the streak whose second endpoint has minimum $|\delta_k|$ as the matching pair of streak j on photo 1. The direction of the streak on photo 2 is pointing from the first endpoint to the second endpoint, and the direction of the streak on photo 1 is in the direction from the endpoint that matches the first endpoint of the streak on photo 2 to its other endpoint.

For example, if endpoint A of streak j and endpoint D of streak k have a $|\delta_k|$ which is less than the threshold, then we substitute the coordinates of point C:

$$\left.\begin{aligned}
X_{1Cj} &= X_{1Aj} + \frac{t_2(X_{1Bj} - X_{1Aj})}{t_1} \\[2mm]
Y_{1Cj} &= Y_{1Aj} + \frac{t_2(Y_{1Bj} - Y_{1Aj})}{t_1}
\end{aligned}\right\} \tag{15.36}$$

and the coordinates of endpoint E of streak k into Equation 15.35.

If point E has the minimum $|\delta_k|$ with C coordinates of streak j among all streaks on photo 2, we will pair AB and DE, and denote that the streak direction on photo 1 is $A\,B$ and on photo 2 is $D\,E$. The position of the particle displacement in the room coordinates is determined by the position of point C on photo 1 and the position of point E on photo 2.

Differentiating Equation 15.31 in terms of x, y, and z leads to particle image displacements (X_i, Y_i) in the photo taken by camera i:

$$\left.\begin{aligned}
\Delta X_i &= \frac{\partial X_i}{\partial x}\Delta x + \frac{\partial X_i}{\partial y}\Delta y + \frac{\partial X_i}{\partial z}\Delta z \\[2mm]
\Delta Y_i &= \frac{\partial Y_i}{\partial x}\Delta x + \frac{\partial Y_i}{\partial y}\Delta y + \frac{\partial Y_i}{\partial z}\Delta z
\end{aligned}\right\} \tag{15.37}$$

Combining two synchronous photos taken by camera 1 and 2, respectively, we get the following equations from Equation 15.37 by dividing them by t_i:

$$
\begin{bmatrix}
\dfrac{\partial X_1}{\partial x} & \dfrac{\partial X_1}{\partial y} & \dfrac{\partial X_1}{\partial z} \\[2mm]
\dfrac{\partial Y_1}{\partial x} & \dfrac{\partial Y_1}{\partial y} & \dfrac{\partial Y_1}{\partial z} \\[2mm]
\dfrac{\partial X_2}{\partial x} & \dfrac{\partial X_2}{\partial y} & \dfrac{\partial X_2}{\partial z} \\[2mm]
\dfrac{\partial Y_2}{\partial x} & \dfrac{\partial Y_2}{\partial y} & \dfrac{\partial Y_2}{\partial z}
\end{bmatrix}
\begin{Bmatrix} V_x \\ V_y \\ V_z \end{Bmatrix}
=
\begin{bmatrix}
\Delta X_1/t_1 \\[1mm]
\Delta Y_1/t_1 \\[1mm]
\Delta X_2/t_2 \\[1mm]
\Delta Y_2/t_2
\end{bmatrix}
\tag{15.38}
$$

where

$$
\Delta X_1 = X_{\text{endpoint1}} - X_{\text{startpoint1}}
$$

$$
\Delta Y_1 = Y_{\text{endpoint1}} - Y_{\text{startpoint1}}
$$

$$
\Delta X_2 = X_{\text{endpoint2}} - X_{\text{startpoint2}}
$$

$$
\Delta Y_2 = Y_{\text{endpoint2}} - Y_{\text{startpoint2}}
$$

The elements in the matrix of Equation 15.38 can be readily derived from Equation 15.31. Solving Equation 15.38 with the least-squares approach gives the three-dimensional velocity components.

The SPIV system can be calibrated statically using rulers with known length and orientations in the SPIV coordinate systems (Figure 15.17). Figure 15.17a is the photo taken by camera 1, and Figure 15.17b is the photo taken by camera 2. In this example, rulers 1 and 2 were stuck on the back wall, and the positions of rulers 3, 4, 5, and 6 were between the back wall and the middle crossplane of the test room. During image processing, each ruler image on the second photo was cut into two parts of equal length. When only the upper part of a streak in the second photo was used in the data processing, the direction of the "stationary streak" should be downward, and vice versa. The measurement results of these ruler lengths are shown in Table 15.1. The average error between the SPIV measurement and the actual stationary streak is 3%, with a maximum of 5.4%. The maximum measurement error occurred on the ruler that was positioned almost in the z-direction (depth direction). All detected directions are correct.

After calibration, room air distribution can be measured. Figure 15.18 shows the two 2D images obtained using two digital cameras, and Figure 15.19 shows the 3D air distribution generated from the two 2D images in Figure 15.18.[23] This SPIV method is capable of obtaining 3D velocities in the entire room (not in just a slice of room airspace) with proper illumination and calibration.

(a)   (b)

Figure 15.17 A pair of photos of six rulers appear as stationary streaks in this static calibration process: (a) photo 1 (b) photo 2. (From Sun, Y. and Zhang, Y., Validation of a stereoscopic particle image velocimetry system for full-scale room airflow studies, *Trans. of Amer. Soc. Heat. Refrig. Air Cond. Engr.*, 109(2), 2003.)

Table 15.1 The Effects of Experimental Cases of Ventilation Systems on Dust Spatial Distribution

Test Cases	Ventilation System	Inlet Width (mm)	Inlet Temp (°C)	Room Temp (°C)	Air Exchange Rate (ACH[a])	Inlet Velocity (m/s)
Case 1	System A	50	24	24	19.5	1.78
Case 2	System AB	50	24	24	19.5	1.78
Case 3	System B	38	24	24	19.5	1.78

[a] ACH = air changes per hour

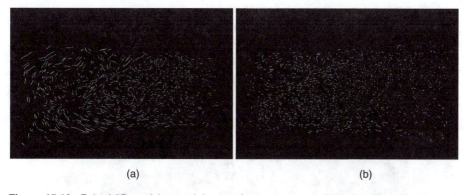

(a)   (b)

Figure 15.18 Paired 2D particle streak images from a test room 5.5 m long, 3.7 m wide, and 2.4 m high. (a) Camera 1 photo with 0.5 s exposure time. (b) Camera 2 photo with 0.25 s exposure time. Note that the length of a streak in photo (a) is twice as long as the corresponding streak in photo (b).

15.5.4 Examples of Room Air Distribution Measurements

To further illustrate the applications of SPIV for room air distribution measurement, a full scale, five-row section of a Boeing 767 aircraft cabin was developed in the Bioenvironmental Structures and Systems (BESS) Laboratory at the University

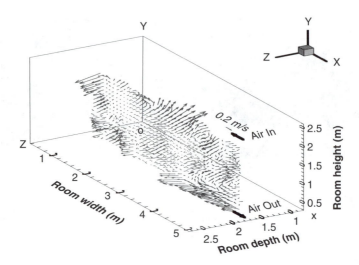

Figure 15.19 The 3D air velocity vectors calculated from the two 2D images in Figure 15.18.

of Illinois at Urbana-Champaign (Figure 15.20) to characterize experimentally the airflow in aircraft cabins. The mockup is a wood structure based on the actual dimensions of the Boeing 767 aircraft cabin. Inside the passenger cabin, necessary aircraft equipment, such as seats, internal panels, windows, and diffusers, forms the boundary of the airspace.

Figure 15.21 shows a sample front view of the measured vector maps without obstructions. This no-obstruction setup is to validate the SPIV measurement system and collect data for CFD-based model validation.

Figure 15.20 The full scale, five-row section of a Boeing 767 aircraft cabin used for characterization of airflows within the cabin.

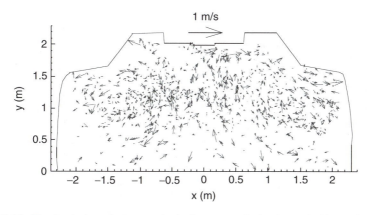

Figure 15.21 The front view of measured velocity vectors for isothermal without obstruction.

The vector profiles for specific z locations (perpendicular to the front view profile) allowed us to observe the change of flow pattern along the z-direction. Uniform grid data, generated from all the raw velocity data with interpolation, was used to obtain the vector profiles shown in Figure 15.22. The sequenced vector profiles of the air velocity are the velocity vectors at the elapsed time 0:00, corresponding to those raw velocity data in Figure 15.21. They are actually the projections of the air velocity on the cross section (x-y plane) at different locations along the z-direction. The maps of three cross sections were chosen to show the flow pattern profiles along the z-direction, which are respectively located in the middle three rows of the seats in the air cabin ($z = 1.3$ m, $z = 2.2$ m, and $z = 3.0$ m) from the front edge of the cabin, respectively.

As with many other measurement techniques, SPIV has its limitations. Velocity behind obstructions cannot be measured. Figure 15.23 shows the front view of all measured vectors of the cabin with seats and passengers. Evidently, the front row seats blocked velocity vectors, and only a partial velocity field could be measured.

15.6 AIR LEAKAGE

Air leakage (infiltration and/or exfiltration) through undesirable paths reduces ventilation effectiveness and interferes with energy conservation and ventilation control, especially in cold climate regions where the ventilation rate is at the minimum required. Air leakage is also an important indicator of the quality of construction and of building performance.

A considerable amount of research has been done on infiltration rates for residential and office buildings.[16–18] In this section, we will analyze the mechanism of air leakage and evaluate the data available in the literature so that guidelines and standards can be established for mechanically ventilated buildings. Although the following discussions are primarily focused on negatively ventilated buildings, the same principles can be applied to positively ventilated buildings.

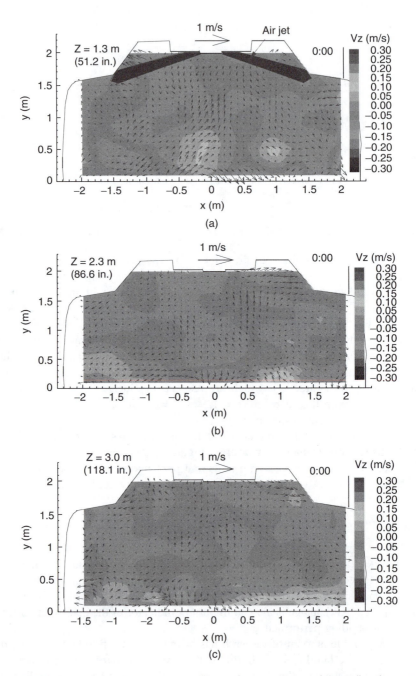

Figure 15.22 Sequenced velocity vector profiles and contour maps of the z-direction component at different depths. (a) z =1.3 m. (b) z = 2.3 m. (c) z =3.0 m.

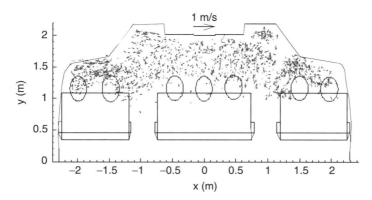

Figure 15.23 The front view of measured velocity vectors with obstructions of seats and passengers.

In a negative-pressure ventilation system, an exhaust fan forces the room air to move outside and creates a negative pressure across the building envelope. This negative pressure places a vacuum on the room airspace and allows outside air to enter the room through designated air inlets. The exhaust fan may be installed in the ceiling, exterior wall, or any other place to pull air out of the room. The air inlets may be located in the ceiling, walls, or floor. High air infiltration is an indication of poor building structure and performance. Infiltration generally occurs in the following forms:

Interflow — "Contaminated" air from adjacent interior zones leaks through common walls and other pathways, such as an interior door. In this case, cross-contamination occurs and ventilation effectiveness is reduced.

Inflow — Fresh air leaks through the building shell and mixes with the inside air. This form of infiltration is a common source of drafts. In cold climates, frost forms where the infiltrating cold air enters the room airspace and can cause water damage or be a nuisance.

Short-circuiting — Fresh outside air leaks through building shell openings adjacent to exhaust fans or return air outlets and is exhausted without mixing with the room air. This amount of infiltration cannot be felt by the occupants.

In these cases, ventilation effectiveness is adversely affected. Infiltration also adversely affects performance and controllability by reducing the actual amount of air that enters through the planned inlets. Zhang and Barber measured and modeled the infiltration rates through five rooms in a large, newly constructed building.[18] The data from that experiment are used as the basis for the following comparison and analysis of infiltration rates from different sources, such as structure, doors, and windows. All rooms had approximately equal volume/area ratios. Air infiltration through different components was analyzed in terms of a normalized infiltration rate, in l/s per m² of envelope surface area. Although the normalized terms are more accurate expressions for infiltration, sometimes they may not be consistent in dimensions and inconvenient for comparison from source to source. For example, the infiltration rate could be normalized in l/s per meter of perimeter for single-floor

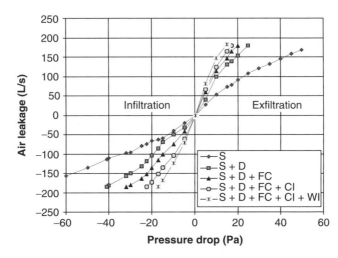

Figure 15.24 Air infiltration rates through different leaking sources of room envelopes. Each data point is an average of five data points from five identical rooms. S = structure, D = doors, CI = ceiling air inlet perimeters, WI = wall air inlet perimeters, and FC = fan covers.

buildings. Expressing infiltration rates as air change per hour (ACH) sometimes facilitates comparison of infiltration rates as a portion of the total ventilation rate. When ACH is used to compare different rooms, the volume/area ratio of the rooms should be consistent. Otherwise, the results could be misleading.

Infiltration rates through different leaking source components are plotted in Figure 15.24. Air leakage sources include structure (S), doors (D), ceiling air inlet perimeters (CI), wall air inlets (WI), and fan covers (FC). Total infiltration was 1.4 ACH at 20 Pa pressure across the room envelope. This infiltration rate was a substantial amount of the minimum ventilation rate for many types of rooms. Of the total ventilation air exchange in this case (defined as the air being exhausted), 53% entered through controlled inlets, 17% infiltrated through the structure, 13% leaked through the interior entry door, 9% leaked through fan covers, and 8% leaked through other sources, such as unsealed air inlet perimeters and fan covers. Because new building materials and construction technology were employed, these rooms were considered to be airtight. However, infiltration still contributed a large portion of the minimum ventilation rate and could significantly reduce the ventilation effectiveness, or even cause failure to control the required indoor environment.

15.6.1 Comparison of Literature Data on Air Leakage

Infiltration estimates from three sources are compared in Figure 15.25. The American Society of Agricultural Engineers presented recommendations for *tight* and *very tight* animal building constructions, based on data from the late 1950s.[17] Using the ASAE recommendation for very tight construction, the predicted infiltration for the same geometry of buildings as the tested rooms in Zhang and Barber's study at 20 Pa would be 3.5 l/s·m² envelope area, compared to the measured infil-

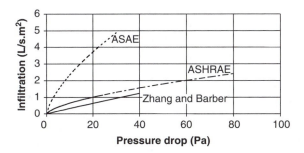

Figure 15.25 Normalized air infiltration rate, in liters per second per square meter of room envelope area, from three different sources showing substantial variation in predicted infiltration rates. (From ASAE, *ASAE Standards*, ASAE – The Society for Engineering in Agricultural, Food, and Biological Systems, St. Joseph, MI, 1994.; ASHRAE, *Fundamentals*, American Society of Heating, Refrigeration and Air Conditioning Engineers, Atlanta, 1993; and Zhang, Y. and Barber, E.M., Air leakage and ventilation effectiveness for confinement livestock housing, *Trans. Am. Soc. Agr. Engrs.*, 38(5):1501–1504, 1995.)

tration, which was 0.6 l/s·m² envelope area.[18] At 20 Pa, infiltration rates for office buildings were 1 l/s·m².[19] The total infiltration rate of rooms with more openings, such as doors, windows, and ducts, is expected to be higher. The infiltration rates for different types of buildings may increase with the pressure at a different rate because of differences in building rigidness. For example, when ΔP increases, gaps in a plastic fan cover tend to be wider and hence lead to larger infiltration.

15.6.2 Analysis of Air Infiltration

Infiltration through a building envelope occurs through different components, such as walls, ceilings, and fan covers. The infiltration rate though each component depends on the pressure drop across the component. Due to wind effect and restrictions of airflow, pressure drops across different components of the building envelope vary. For example, the pressure difference across the ceiling (ΔP_c) is likely to be smaller than the pressure difference across the fan (ΔP_f) because of the airflow restriction at the eave openings to the attic. The pressure drop across the leeward wall (ΔP_w) is likely to be lower than the pressure drop across the fan (ΔP_f windward wall) due to wind effect. Assuming that a building envelope has N components, infiltration through the ith component is expected to conform to Equation 15.39:[1]

$$Q_i = C_i (\Delta P_i)^{n_i} \tag{15.39}$$

where C_i is a discharging coefficient and n is an exponential coefficient and for pressure difference across that component ΔP_i.

Total infiltration through the building envelope is

$$Q = \sum_{i=1}^{N} C_i (\Delta P)^{n_i} \tag{15.40}$$

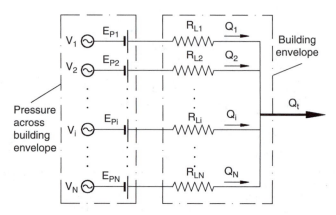

Figure 15.26 Analogue diagram of infiltration through a mechanically ventilated building envelope to an electrical circuit.

Infiltration through a building envelope can be expressed in an electric analogue diagram, as shown in Figure 15.26. In such an analogy, paired variables are

- Infiltration Q and current
- Infiltration resistance (R_L) and electrical resistance
- Pressure drop (ΔP^n) and voltage

Unlike an electric circuit, the infiltration rate is not linearly proportional to the pressure drop. For the ith component, E_{Pi} is the analogue of steady-state pressure drops created by the ventilation system, and V_i is the analogue of transient pressure drops caused by external disturbances such as wind. From Figure 15.26, Equation 15.40 can be written as

$$Q_t = \sum_{i=1}^{N} \frac{V_i + E_i}{R_{Li}} \tag{15.41}$$

Comparing Equation 15.40 and Equation 15.41 gives

$$\Delta P_i = \frac{V_i + E_i}{R_{Li}} \tag{15.42}$$

$$R_{Li} = \frac{1}{C_i} \tag{15.43}$$

Infiltration resistance R_L is an important parameter. Similar to the insulation value (RSI) of building materials, R_L can be a reference standard for the quality of building materials and construction. In order to provide precise control of ventilation rates, infiltration resistance should not be smaller than some standard value. More research is needed, however, to define that value.

In the case where only total infiltration (all components) is of concern, the overall infiltration resistance of the building envelope, R_{Lt}, and the pressure difference between the building airspace and outside, E_{Pt}, are used, and the infiltration rate can be expressed as

$$Q_t = C_t(\Delta P_t)^n = \frac{E_{Pt}}{R_{Lt}}$$

(15.44)

Existing infiltration data are all presented in the form of Equation 15.44.[17-20] In many applications, such as building commissioning, Equation 15.44 could be used. However, pressure variation among building components should be minimal when using Equation 15.44. Typically, wind speed should not exceed 6 km/h (i.e., pressure variation is less than 2 Pa) when measuring total infiltration.

The accuracy of infiltration measurement depends on the accuracy of measuring the pressure drop. Differentiating Equation 15.39 with respect to ΔP_i yields

$$dQ_i = C_i n_i \frac{\Delta P_i^{n_i}}{\Delta P_i} d(\Delta P_i)$$

(15.45)

The error ε_i of infiltration with respect to pressure drop can be calculated as the change of infiltration rate, dQ_i/Q_i. Combining Equation 15.39 and Equation 15.45 gives

$$\varepsilon_i = \frac{dQ_i}{Q_i} = n_i \frac{d(\Delta P_i)}{\Delta P_i}$$

(15.46)

Equation 15.46 shows that the error in infiltration ε_i is independent of infiltration resistance but proportional to the error in pressure drop [$d(\Delta P_i)$]. When ΔP_i is small, a small $d\Delta P_i$ can introduce a large error in the calculation of infiltration. This small $d\Delta P_i$ can be introduced by wind or by instrumentation and human error. Therefore, an infiltration rate measurement at low pressure drop (typically smaller than 5 Pa) could be quite inaccurate.

15.7 ALTERNATIVE VENTILATION AND AIR DISTRIBUTION SYSTEMS

The ventilation system and the resulting airflow patterns have a substantial effect on room air quality. For example, displacement ventilation can reduce the pollutant concentration at a local point to a level that is significantly lower than that of exhaust. Changing locations of air inlets or outlets can change the flow patterns and thus the pollutant spatial distribution. Again, we use a case study described in Section 15.1 to illustrate the effects of alternative ventilation systems on indoor air quality.

To study the effect of the locations of the air outlet and the air inlet on the dust spatial distribution in a mechanically ventilated airspace, three cases were tested in this study (Table 15.1). For each case, three replication measurements were taken for statistical comparison and analysis. The inlet air velocity was measured using an air velocity meter (Model 8330, TSI Inc). The air velocity meter was calibrated in the factory. The accuracy was ±5.0% of the reading or ±0.025 m/s, whichever is greater in the range of 0.13 to 20 m/s.

15.7.1 Effect of Outlet Location on Dust Spatial Distribution

Case 1 was conducted in ventilation system A, and case 2 was conducted in ventilation system AB (Table 15.1). The ventilation rates in case 1 and case 2 were the same at 19.5 ACH (0.264 m³/s). All other conditions were also the same in case 1 and case 2, except for the outlet location.

Comparing the dust spatial distributions between case 2 and case 1, as shown in Figure 15.27, reveals a significant difference in dust spatial distribution. The overall dust concentration in case 2 is substantially lower than in case 1. The dust concentration in most of the room is below 0.8 mg/m³ in Case 2, whereas the dust concentration in most of room in case 1 is above 2.0 mg/m³. The overall average dust concentration in case 2 is only 0.90 mg/m³, which is much lower than 2.30 mg/m³ in case 1. One reason is that the outlet of case 2 is located next to the high dust-concentration area so the exhaust air can remove more dust from the room than in case 1. This implies that outlet location has a significant effect on the dust spatial distribution and overall dust levels in a ventilated airspace. This information is important to improving ventilation effectiveness. The quantitative comparison of the dust spatial concentrations between case 1 and case 2 is shown in Figure 15.27c. This map shows the dust reduction percentage from case 1 to case 2. The dust concentration was reduced by 45 to 75% in most of the area within the test room.

15.7.2 Effect of Inlet Location on Dust Spatial Distribution

The effect of inlet location on the dust spatial distribution was studied in case 3 by keeping the same outlet as in case 2 but moving the inlet up to the ceiling. The ventilation rate was maintained at 19.5 ACH (0.264 m³/s). All other experimental conditions were also the same as case 2. Compared to case 2, the only difference was the inlet location. The comparison of dust spatial distributions between case 3 and case 2 is shown in Figure 15.28.

Comparing the dust spatial distributions between case 3 with case 2 shows that there was a small difference in dust spatial distributions. The overall average dust concentration in case 3 was 0.83 mg/m³, which is slightly lower than the 0.90 mg/m³ in case 2. In these two case studies, the ventilation rate, the inlet velocity, and the calculated inlet jet momentum were the same. The possible reason is the different jet momentum loss after the free inlet jet enters the airspace. The inlet air jet of

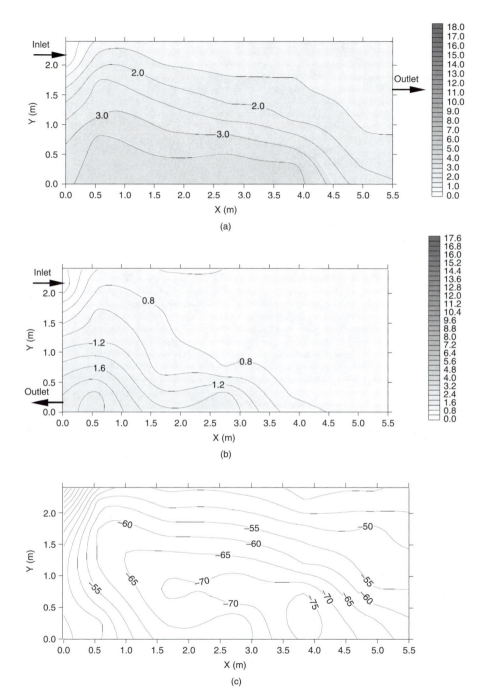

Figure 15.27 Effect of outlet location on dust spatial distribution (mg/m³). (a) Case 1: The outlet is 1.3 m from floor on the right side wall. (b) Case 2: The outlet is next to the floor on the left side wall. (c) Dust spatial concentration change of case 2 compared with case 1 (numbers are percentages).

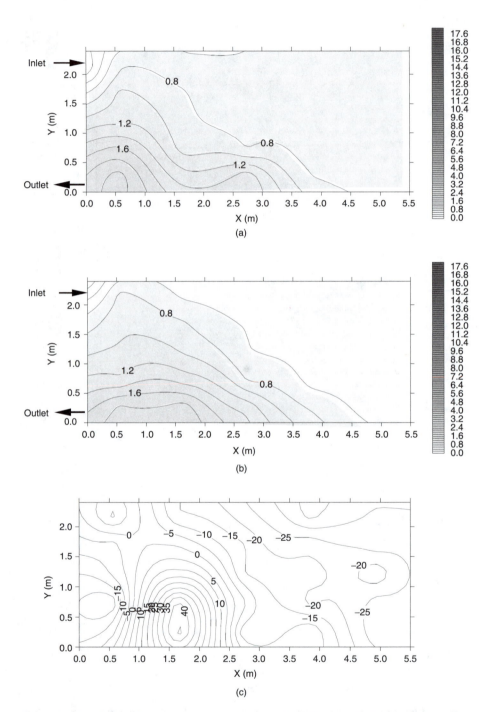

Figure 15.28 Effect of the inlet location on dust spatial distribution (mg/m³). (a) Case 2: The inlet is 0.3 m from the ceiling on the left side wall. (b) Case 3: The inlet is next to the ceiling on the left side wall. (c) Dust spatial concentration change of case 3 compared with case 2 (numbers are percentages).

case 3 was directed along the ceiling immediately after it entered the airspace, which caused minimum jet momentum loss. But in Case 2, the air jet entered the airspace not directly attached to the ceiling. It traveled a short distance before it was directed to the ceiling, because of the Coanda effect. Consequently, this difference in jet momentum may cause a difference in the flow pattern and the dust spatial distribution. Figure 15.27c shows the quantitative comparison of dust spatial concentration change from case 2 to case 3. The dust concentrations in most of the area in case 3 are lower than those in case 2, but in one area in the left side of the room, the dust concentration in case 3 is higher than that in case 2. When compared with case 2, the overall average dust concentration was 10% lower in case 3. This indicates that the inlet location had a minor effect on the dust spatial distribution and overall dust level of a ventilated airspace in these two cases.

It is widely believed that the location of air outlet has little effect on the airflow pattern, the heat, and the gas removal, and that only air inlet locations matters. However, the measurement in this study showed that for different locations of the outlet, dust concentration of the exhaust air, C_e, can be substantially different. As a result, the dust spatial distribution and the overall dust level can vary considerably. Comparison of ventilation effectiveness between two different ventilation systems showed that ventilation system design was important to indoor particulate matter control. It was observed that there was a high dust concentration zone in the ventilated room. Positioning the air outlet at the dustiest location can substantially improve dust removal effectiveness.

15.7.3 Effect of Air Cleaning on Dust Spatial Distribution

Internal air cleaning can effectively improve indoor air quality, especially when the cleaned air is properly distributed. The following example shows the effect of internal air cleaning on the room dust spatial distribution and removal efficiency.[21] In the study, two identical rooms were used. One was a control room and the other was a treatment room. In the treatment room, an air filter was installed to remove dust in an air recirculation system. The air inlet duct was placed where the highest dust concentration was expected, and the outlet duct was placed above the center alley (Figure 15.29b). The total airflow rate through the filter was 560 cfm (0.264 m³/s). The ratio of the airflow rate through the filter to the average room ventilation rate was 32%. The dust spatial concentrations at 27 points across the central section in the control room and the treatment room were measured using a multipoint sampler. The measured spatial dust concentrations with air filter showed that the overall average dust concentration was approximately 20% lower than the control room (Figure 15.29c). This agrees well with the predicted dust reduction efficiency by Equation 14.23. Apparently, a large flow rate for the air filter is required to improve the room air cleaning efficiency. The high dust concentration zone near the air inlet side disappeared in the treatment room. This indicates that some dust was removed from the dusty air by the air-cleaning recirculation system. On the other hand, as the air recirculation system affected the airflow pattern, so the dust spatial distribution was different from the control room's.

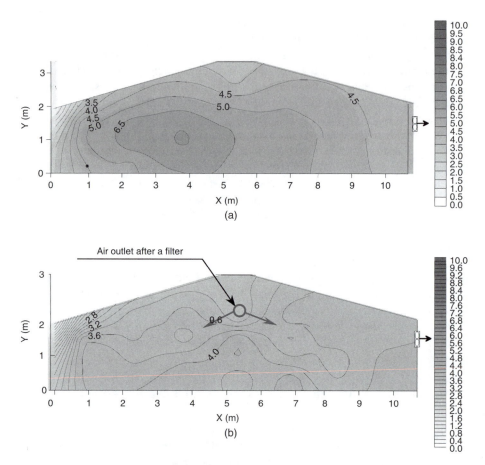

Figure 15.29 Comparison of average dust concentration spatial distribution between control room and treatment room with air cleaning (mg/m³). (a) Control room: Without air cleaning, C_{ave} = 5.02 mg/m³. (b) Treatment room: Air cleaning with a filter, C_{ave} = 3.82 mg/m³. (From Wang, X., Zhang, Y., and Riskowski, G.L., Dust spatial distribution in a typical animal building, in *Proc. Int. Symp. Dust Control for Animal Production Facilities*, Jutland, Denmark, May 30–June 2, 1999. With permission.)

In summary, indoor air quality, especially the spatial distribution of particulate matter, can be improved through appropriate design of ventilation schemes, location of air inlets and outlets, and use of air-cleaning technologies to clean the recirculation air. In many cases, displacement ventilation or zonal ventilation can substantially improve ventilation effectiveness.[22] Ventilation effectiveness for air quality control is an ongoing research area, and much has yet to be studied. It appears that standards for ventilation systems are difficult to establish or to apply in practice, because of the variation and complexity of room geometries and building operation conditions. However, performance criteria for a ventilation system may be implemented. Such ventilation system performance criteria include pollutant concentration threshold limits, temperature distribution, and controllability.

DISCUSSION TOPICS

1. What is the difference between ventilation efficiency and ventilation effectiveness?
2. Tracer gas methods have been used widely to measure ventilation rate and room air age. Is the room air age measurement a technique to study ventilation effectiveness rather than ventilation rate?
3. What criteria would you use to evaluate a computational fluid dynamics model in terms of its accuracy and the acceptability of the simulated results?
4. For an individual type of airflow slow measurement instrumentation, discuss its measurement principles, advantages, and limitations.
5. How does air leakage affect the ventilation effectiveness in a confinement room airspace? How could you minimize air leakage?
6. How do displacement ventilation systems work? Does a displacement ventilation system always give better ventilation effectiveness than complete mixing? What are the pros and cons of displacement ventilation?
7. If you are a design engineer, what kind of design tools would you like to assist in designing ventilation systems to improve ventilation effectiveness?

PROBLEMS

1. In order to determine the ventilation effectiveness of a building, particle concentrations at 12 locations are measured. Each measurement location is in the center of an equal-volume subzone, as shown in the two-dimensional diagram below. The dots are measurement points, and the numbers are particle concentrations in mg/m^3. The particle concentrations in the supply air and at the exhaust fan are also measured. Determine whether this ventilation system is more effective than a completely mixing ventilation system.

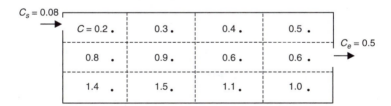

2. Particle concentrations and the weighted area at a cross section of a room were measured as shown in the following figure. Determine whether this ventilation system is more effective than a completely mixing ventilation system. The number in each subzone is the particle concentration, and the number in parentheses is the weighted area of that subzone.

3. It is known that the infiltration rate of a room varies with the pressure drop across the room in the following relation:

$$Q_i = 10(\Delta P)^{2/3}$$

If the pressure sensor has a range of 0 to 2 kPa, and measured a pressure drop across the room at 20 Pa, what will the maximum error (in %) of infiltration be?

4. Derive equations to prove that the displacement ventilation system (a) is more effective than the complete mixing ventilation (b), as shown below. Assume that the pollutants are uniformly produced along the floor and the room ventilation rates Q are the same. The room has a width W, height H, and length L.

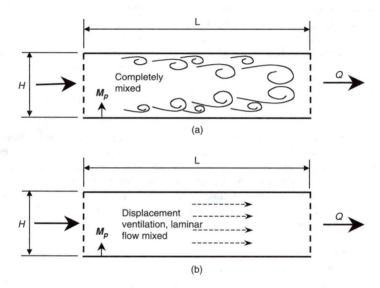

(a)

(b)

REFERENCES

1. ASHRAE, *Handbook of Fundamentals*, American Society of Heating, Refrigeration and Air Conditioning Engineers, Atlanta, 2001.
2. Kuehn, J.H., Predicting airflow patterns and particle contamination in clean rooms, *J. Aerosol Sci.*, 19(7):1405–1408, 1988.
3. Wang, X., *Measurement and Modeling of Concentration Spatial Distribution of Particulate Matter in Indoor Environments,* unpublished Ph.D. dissertation, University of Illinois at Urbana-Champaign, Urbana, IL, 2000.
4. Pratt, G.L., Ventilation equipment and controls, in *Ventilation of Agricultural Structures,* Hellickson, M.A. and Walker, J.N., eds., American Society of Agricultural Engineers, St. Joseph, MI, 1983, 54.
5. Roos, A., *On The Effectiveness of Ventilation,* unpublished Ph.D. dissertation, Technical University of Eindhoven, Eindhoven, The Netherlands, 1999.

6. Liddament, M.W., A review and bibliography of ventilation effectiveness — definitions, measurement, design and calculation, *Technical Note AIVC21*, Air infiltration and Ventilation Center, Berkshire, UK, 1987.

7. Persily A.K., Dols, W.S., and Nabinger, S.J., Ventilation effectiveness measurements in two modern office buildings, in *Proc. of Indoor Air '93*, 1993, 195–200.

8. Skaaret, E. and Mathisen, H.M., Test procedures for ventilation effectiveness field measurements, in *Proc. CIB Symp. on Recent Advances in Control and Operation of Building HVAC Systems,* Trondheim, Norway, 1985, 64–75.

9. Wang, X. et al., Development of a multipoint aerosol sampler using critical flow control devices, *Trans. Am. Soc. Heating, Refrig., Air Cond.*, 105(1):1108–1113, 1999.

10. Zhang, J. S., *A Fundamental Study of Two-Dimensional Room Ventilation Flows under Isothermal and Nonisothermal Conditions*, Ph.D. thesis, University of Illinois at Urbana-Champaign, Urbana, IL, 1991.

11. Van Dyke, M., *An Album of Fluid Motion*, Parabolic Press, Stanford, CA, 1982.

12. Adrian, R.J., Particle imaging techniques for experimental fluid mechanics, *Annu. Rev. Fluid Mech.*, 23:261–304, 1991.

13. Scholzen F. and Moser, A., Three-dimensional particle streak velocimetry for room airflow with automatic stereo-photogrammetric image processing, in *Proc. 5th Int. Conf. Air Distribution in Rooms*, ROOMVENT '96, Yokohama, Japan, July 17–19, 1996, 555–562.

14. Zhao, L.Y, *Measurement and Analysis of Full-Scale Room Airflow using Particle Image Velocimetry (PIV) Techniques*, Ph.D. dissertation, Library, University of Illinois at Urbana-Champaign, Urbana, IL, 2000.

15. Sun, Y. and Zhang, Y., An algorithm of stereoscopic particle image velocimetry for full-scale room airflow studies, *Trans. Am. Soc. Heat. Refrig. Air Cond. Engr.*, 110(1), 2004.

16. ASHRAE, *Fundamentals*, American Society of Heating, Refrigeration and Air Conditioning Engineers, Atlanta, 1993.

17. ASAE, *ASAE Standards,* ASAE — The Society for Engineering in Agricultural, Food, and Biological Systems, St. Joseph, MI, 1994.

18. Zhang, Y. and Barber, E.M., Air leakage and ventilation effectiveness for confinement livestock housing, *Trans. of Am. Soc. Agr. Engrs.*, 38(5):1501–1504, 1995.

19. Shaw, C.Y., Reardon, J.T., and Cheung, M.S., Changes in air leakage levels of six Canadian office buildings, *Am. Soc. Heat. Refrig. Air Cond. Eng. J.*, February 1993: 34–36.

20. ASHRAE, *Fundamentals*, American Society of Heating, Refrigeration and Air Conditioning Engineers, Atlanta, 2003.

21. Wang, X., Zhang, Y., and Riskowski, G.L., Dust spatial distribution in a typical swine building, in *Proc. Int. Symp. on Dust Control for Animal Production Facilities*, Jutland, Denmark, May 30–June 2, 1999.

22. Breum, N., Air exchange efficiency of displacement ventilation in a printing plant, *Ann. Occup. Hyg.*, 32(4):481–488, 1988.

23. Sun, Y. and Zhang, Y., Validation of a stereoscopic particle image velocimetry system for full-scale room airflow studies, *Trans. Am. Soc. Heat. Refrig. Air Cond. Engr.*, 109(2): 540–548, 2003.

Conversion Factors and Constants

Fundamental units:

1 micrometer (μm) = 1 micron (μ) = 10^{-6} m = 10^{-4} cm = 10^{-3} mm = 10^3 nm = 10^4 Å

1 in. = 2.54 cm

1 ft = 30.5 cm

1 yd = 91.44 cm

1 mile = 5280 ft = 1610 m

1 mile (nautical) = 1850 m

1 newton (N) = 1 kg·m/s^2 = 10^5 dyn

1 dyn = 1 g·cm/s^2

1 lb = 7000 grains = 454 g

1 g = 15.4 grains

1 ounce (mass) = 28.4 g

1 ton, long = 2240 lb = 1016 kg

1 ton, short = 2000 lb = 907 kg

Temperature

K = °C + 273

°F = 1.8(°C) + 32

Volume

1 ft^3 = 28,300 cm^3 = 28.3 l = 0.0283 m^3

1 m^3 = 35.3 ft^3

1 in.^{3} = 16.4 cm^3

1 gallon (U.S.) = 3.785 l

1 gallon (imperial) = 4.546 l

1 barrel = 42 gallons (U.S.) = 159 l

1 once (liquid, U.S.) = 29.6 cm^3

1 pint (liquid, U.S.) = 0.473 cm^3

1 quart (liquid, U.S.) = 0.946 cm^3

1 quart (liquid, U.S.) = 1.1365 cm^3

Flow rate

1 m^3/s = 2120 ft^3/min

1 ft^3/min = 28.3 l/min

1 ft^3/h = 0.47 l/min

Concentration
    1 grain/ft^3 = 2.29 g/m^3
    1 mg/m^3 = 1 μg/l = 1 ng/cm^3
    1 g/m^3 = 1 mg/l = 1 μg/cm^3
    1 mppcf = 35.3 particles/cm^3
       mppcf = million particles per cubic foot
Velocity
    1 ft/min = 0.508 cm/s
    1 ft/s = 30.5 cm/s
    1 mile/h = 1.61 km/h
    Velocity of sound in air at 20°C = 340 m/s = 34,000 cm/s
    Velocity of light in a vacuum = 3.00×10^8 m/s = 3.00×10^{10} cm/s
Acceleration
    Acceleration of gravity at sea level = 981 cm/s^2 = 32.2 ft/s^2
Pressure
    1 atm = 14.7 psia = 76 cm Hg = 760 mm Hg = 408 in. H$_2$O = 1040 cm H$_2$O
    1 atm = 1.01×10^6 dyn/cm^2 = 101 kPa (Pa = N/m^2)
    1 psi = 6.9 kPa
    1 in. of water = 250 Pa
    Vapor pressure of water at 20°C = 17.6 mm Hg = 2.34 kPa
Viscosity
    1 poise (P) = 1 g/cm·s = 1 dyn·s/cm^2 = 0.10 Pa·s = 0.10 N·s/m^2
    Viscosity of air at 20°C = 1.81×10^{-4} P = 1.81×10^{-5} Pa·s
Power
    1 Btu/h = 0.293 W
    1 ft·lb$_f$/min = 0.0226 W
    1 ton, refrigeration = 12000 Btu/h = 3520 W
    1 hp (boiler) = 981 W
    1 hp (horse power) = 746 W
Energy
    1 cal = 4.19 J = 0.00397 Btu
    1 ft·lb$_f$ = 1.36 J
    1 kW·h = 3600 kJ = 3412 Btu
Ideal gas law $PV = n_m RT$
    R = 8.314 J/K·mole for P in pascal
    R = 82.1 atm·cm^3/K·mole for P in atm and v in cm^3
    R = 8.31×10^7 dyn·cm/K·mole for P in dyn/cm^2 and v in cm^3
    R = 62400 mm Hg·cm^3/K · mole for P in mm Hg and v in cm^3
Volume of 1 mole of ideal gas at 20°C = 24.0 l
    Avogadro's number = N_a = 6.02×10^{23} molecules/mole
    Boltzmann's constant $k = R/N_a$ = 1.38×10^{-16} dyn·cm/K
Air at 20°C and 1 atm
    Density = 1.205 g/l = 1.205×10^{-3} g/cm^3 = 0.075 lb/ft^3
    Viscosity = 1.81×10^{-4} P = 1.81×10^{-5} Pa·s
    Mean free path = 0.666 μm
    Molecular weight = 28.9 g/mole
    Specific heat ratio = c_p/c_v = 1.40
    Diffusion coefficient = 0.19 cm^2/s
    Gas constant for dry air, R = 287.055 J/kg·K, when pressure P is in Pascal, mass
       in kg

Molecular weight for dry air = 28.9645

Composition by volume (dry air)

    N_2 — 78.1%

    O_2 — 20.9%

    Ar — 0.93%

    CO_2 — 0.031%

    Other — < 0.003%

Water at 20°C and 1 atm

    Viscosity = 0.0100 dyn·s/cm²

    Surface tension = 72.7 dyn/cm

    Vapor pressure = 17.6 mm Hg = 2.34 kPa

Water vapor at 20°C and 1 atm

    Diffusion coefficient = 0.24 cm²/s

    Density = 0.75×10^{-3} g/cm³

    Gas constant for water vapor, R = 461.52 J/kg·K, when pressure is P in Pascal, mass in kg

    Molecular weight for water vapor = 18.01534

Airborne Particle Properties at Standard Conditions

Particle Diameter dp (μm)	Slip Correction Factor Cc	Settling Velocity VTS (m/s)	Relaxation Time τ (s)	Mechanical Mobility B (m/N·s)	Diffusivity D (m²/s)	Coagulation Coefficient K (m³/s)	Correction Factor for Coagulation Coefficient β (m³/s)
0.001	224.33	6.7547E-09	1.0235E-08	1.3150E+15	5.3211E-06	5.9593E-16	0.00891
0.002	112.46	1.3545E-08	5.1585E-09	3.2963E+14	1.3338E-06	6.7205E-16	0.02005
0.003	75.17	2.0372E-08	3.4664E-09	1.4689E+14	5.9437E-07	7.2153E-16	0.03220
0.004	56.53	2.7234E-08	2.6206E-09	8.2846E+13	3.3522E-07	7.5908E-16	0.04505
0.005	45.34	3.4134E-08	2.1132E-09	5.3162E+13	2.1511E-07	7.8963E-16	0.05842
0.006	37.89	4.1069E-08	1.7751E-09	3.7017E+13	1.4978E-07	8.1544E-16	0.07221
0.007	32.56	4.8042E-08	1.5337E-09	2.7268E+13	1.1034E-07	8.3776E-16	0.08632
0.008	28.57	5.5052E-08	1.3528E-09	2.0933E+13	8.4702E-08	8.5737E-16	0.10069
0.009	25.46	6.2099E-08	1.2122E-09	1.6584E+13	6.7104E-08	8.7474E-16	0.11526
0.01	22.98	6.9183E-08	1.0997E-09	1.3469E+13	5.4499E-08	8.9024E-16	0.12999
0.02	11.80	1.4213E-07	5.9597E-10	3.4588E+12	1.3996E-08	9.7638E-16	0.27758
0.03	8.08	2.1904E-07	4.3088E-10	1.5794E+12	6.3907E-09	9.8577E-16	0.40916
0.04	6.23	3.0008E-07	3.5056E-10	9.1284E+11	3.6936E-09	9.5623E-16	0.51504
0.05	5.12	3.8543E-07	3.0426E-10	6.0030E+11	2.4290E-09	9.1057E-16	0.59662
0.06	4.38	4.7524E-07	2.7505E-10	4.2834E+11	1.7332E-09	8.6118E-16	0.65899
0.07	3.86	5.6964E-07	2.5568E-10	3.2333E+11	1.3083E-09	8.1373E-16	0.70708
0.08	3.47	6.6878E-07	2.4253E-10	2.5430E+11	1.0290E-09	7.7040E-16	0.74475
0.09	3.17	7.7276E-07	2.3358E-10	2.0637E+11	8.3504E-10	7.3170E-16	0.77476
0.1	2.93	8.8170E-07	2.2763E-10	1.7165E+11	6.9456E-10	6.9745E-16	0.79909
0.2	1.88	2.2625E-06	2.4039E-10	5.5059E+10	2.2279E-10	5.0855E-16	0.90826
0.3	1.55	4.2114E-06	3.0624E-10	3.0366E+10	1.2287E-10	4.3666E-16	0.94267
0.4	1.40	6.7563E-06	4.0010E-10	2.0552E+10	8.3161E-11	4.0088E-16	0.95900
0.5	1.32	9.9071E-06	5.1671E-10	1.5430E+10	6.2435E-11	3.7990E-16	0.96840
0.6	1.26	1.3666E-05	6.5404E-10	1.2317E+10	4.9840E-11	3.6619E-16	0.97446
0.7	1.22	1.8033E-05	8.1109E-10	1.0235E+10	4.1415E-11	3.5654E-16	0.97867
0.8	1.19	2.3006E-05	9.8725E-10	8.7478E+09	3.5396E-11	3.4935E-16	0.98176
0.9	1.17	2.8584E-05	1.1822E-09	7.6335E+09	3.0888E-11	3.4378E-16	0.98410

Particle Diameter dp (μm)	Slip Correction Factor Cc	Settling Velocity VTS (m/s)	Relaxation Time τ (s)	Mechanical Mobility B (m/N·s)	Diffusivity D (m²/s)	Coagulation Coefficient K (m³/s)	Correction Factor for Coagulation Coefficient β (m³/s)
1	1.15	3.4766E-05	1.3956E-09	6.7685E+09	2.7388E-11	3.3933E-16	0.98595
2	1.08	1.2974E-04	4.5332E-09	3.1574E+09	1.2776E-11	3.1908E-16	0.99375
3	1.05	2.8495E-04	9.4860E-09	2.0546E+09	8.3136E-12	3.1220E-16	0.99611
4	1.04	5.0037E-04	1.6252E-08	1.5221E+09	6.1589E-12	3.0872E-16	0.99722
5	1.03	7.7601E-04	2.4832E-08	1.2086E+09	4.8905E-12	3.0662E-16	0.99786
6	1.03	1.1119E-03	3.5225E-08	1.0022E+09	4.0551E-12	3.0522E-16	0.99827
7	1.02	1.5080E-03	4.7431E-08	8.5591E+08	3.4633E-12	3.0421E-16	0.99856
8	1.02	1.9643E-03	6.1451E-08	7.4690E+08	3.0222E-12	3.0345E-16	0.99876
9	1.02	2.4808E-03	7.7284E-08	6.6252E+08	2.6808E-12	3.0286E-16	0.99892
10	1.02	3.0576E-03	9.4930E-08	5.9526E+08	2.4086E-12	3.0239E-16	0.99905
20	1.01	1.2137E-02	3.7112E-07	2.9537E+08	1.1951E-12	3.0025E-16	0.99958
30	1.01	2.7239E-02	8.2864E-07	1.9641E+08	7.9473E-13	2.9953E-16	0.99973
40	1.00	4.8363E-02	1.4675E-06	1.4712E+08	5.9528E-13	2.9917E-16	0.99981
50	1.00	7.5509E-02	2.2877E-06	1.1760E+08	4.7586E-13	2.9895E-16	0.99985
60	1.00	1.0868E-01	3.2892E-06	9.7952E+07	3.9635E-13	2.9880E-16	0.99988
70	1.00	1.4787E-01	4.4720E-06	8.3928E+07	3.3960E-13	2.9870E-16	0.99990
80	1.00	1.9308E-01	5.8362E-06	7.3417E+07	2.9707E-13	2.9862E-16	0.99991
90	1.00	2.4431E-01	7.3817E-06	6.5246E+07	2.6401E-13	2.9856E-16	0.99992
100	1.00	3.0157E-01	9.1085E-06	5.8711E+07	2.3756E-13	2.9851E-16	0.99993
200	1.00	1.2053E+00	3.6350E-05	2.9333E+07	1.1869E-13	2.9829E-16	0.99997
300	1.00	2.7113E+00	8.1724E-05	1.9550E+07	7.9107E-14	2.9822E-16	0.99998
400	1.00	4.8195E+00	1.4523E-04	1.4661E+07	5.9322E-14	2.9818E-16	0.99998
500	1.00	7.5299E+00	2.2687E-04	1.1728E+07	4.7454E-14	2.9816E-16	0.99999

Relative Density of Common Aerosol Materials

Material	Density in 1000 kg/m³	Material	Density in 1000 kg/m³
Solids			
Aluminum	2.7	Natural fibers	1–1.6
Aluminum oxide	4.0	Paraffin	0.9
Ammonium sulfate	1.8	Plastics	1–1.6
Asbestos	2.0–2.8	Pollens	0.45–1.05
Asbestos, chrysotile	2.4–2.6	Polystyrene	1.05
Carnauba wax	1.0	Polyvinyl toluene	1.03
Coal	1.2–1.8	Portland cement	3.2
Fly ash	0.7–2.6	Quartz	2.6
Fly ash cenospheres	0.7–1.0	Sodium chloride	2.2
Glass, common	2.4–2.8	Sulfur	2.1
Granite	2.6–2.8	Starch	1.5
Ice	0.92	Talc	2.6–2.8
Iron	7.9	Titanium dioxide	4.3
Iron oxide	5.2	Uranine dye	1.53
Limestone	2.7	Wood (dry)	0.4–1.0
Lead	11.3	Zinc	6.9
Marble	2.6–2.8	Zinc oxide	5.6
Methylene blue dye	1.26		

Relative Density of Common Aerosol Materials *(Continued)*

Material	Density in 1000 kg/m³	Material	Density in 1000 kg/m³
		Liquids	
Alcohol	0.79	Mercury	13.6
Diobutyl phthalate	1.045	Oils	0.88–0.94
Dioctyl phthalate [DOP; di-2 (ethylhexyl) phthalate]	0.984	Oleic acid	0.894
Dioctyl sebacate	0.915	Polyethylene glycol	1.13
Hydrochloric acid	1.19	Sulfuric acid	1.84
		Water	1.0

Source: From Hinds, W.C., *Aerosol Technology*, John Wiley & Sons, New York, 1999.

K_p Values for Equation 3.41:

$$f(d_p) = \int_{-\infty}^{K_p} \frac{1}{\sqrt{2\pi}} \exp\left(-\frac{K_p^2}{2}\right) dK_p$$

K_p	0.00	0.01	0.02	0.03	0.04	0.05	0.06	0.07	0.08	0.09	K_p
-4.9	$.0^64792$	$.0^64554$	$.0^64327$	$.0^64111$	$.0^63900$	$.0^63711$	$.0^63525$	$.0^63348$	$.0^63179$	$.0^63019$	-4.9
-4.8	$.0^67933$	$.0^67547$	$.0^67178$	$.0^66827$	$.0^66492$	$.0^66173$	$.0^65869$	$.0^65580$	$.0^65304$	$.0^65042$	-4.8
-4.7	$.0^51301$	$.0^51239$	$.0^51179$	$.0^51123$	$.0^51069$	$.0^51017$	$.0^69680$	$.0^69211$	$.0^68765$	$.0^68339$	-4.7
-4.6	$.0^52112$	$.0^52013$	$.0^51919$	$.0^51828$	$.0^51742$	$.0^51660$	$.0^51581$	$.0^51506$	$.0^51434$	$.0^51366$	-4.6
-4.5	$.0^53398$	$.0^53241$	$.0^53092$	$.0^52949$	$.0^52813$	$.0^52682$	$.0^52558$	$.0^52489$	$.0^52325$	$.0^52216$	-4.5
-4.4	$.0^55413$	$.0^55169$	$.0^54935$	$.0^54712$	$.0^54498$	$.0^54294$	$.0^54098$	$.0^53911$	$.0^53732$	$.0^53561$	-4.4
-4.3	$.0^58540$	$.0^58163$	$.0^57801$	$.0^57455$	$.0^57124$	$.0^56807$	$.0^56503$	$.0^56212$	$.0^55934$	$.0^55668$	-4.3
-4.2	$.0^41335$	$.0^41277$	$.0^41222$	$.0^41168$	$.0^41118$	$.0^41069$	$.0^41022$	$.0^59774$	$.0^59345$	$.0^58934$	-4.2
-4.1	$.0^42066$	$.0^41978$	$.0^41894$	$.0^41814$	$.0^41737$	$.0^41662$	$.0^41591$	$.0^41523$	$.0^41458$	$.0^41395$	-4.1
-4.0	$.0^43167$	$.0^43036$	$.0^42910$	$.0^42789$	$.0^42673$	$.0^42561$	$.0^42454$	$.0^42351$	$.0^42252$	$.0^42157$	-4.0
-3.9	$.0^44810$	$.0^44615$	$.0^44427$	$.0^44247$	$.0^44074$	$.0^43908$	$.0^43747$	$.0^46594$	$.0^43446$	$.0^43304$	-3.9
-3.8	$.0^47235$	$.0^46948$	$.0^46673$	$.0^46407$	$.0^46152$	$.0^45906$	$.0^45669$	$.0^45442$	$.0^45223$	$.0^45012$	-3.8
-3.7	$.0^31078$	$.0^31036$	$.0^49961$	$.0^49574$	$.0^49201$	$.0^48842$	$.0^48496$	$.0^48162$	$.0^47841$	$.0^47532$	-3.7
-3.6	$.0^31591$	$.0^31531$	$.0^31473$	$.0^31417$	$.0^31363$	$.0^31311$	$.0^31261$	$.0^31213$	$.0^31166$	$.0^31121$	-3.6
-3.5	$.0^32326$	$.0^32241$	$.0^32158$	$.0^32078$	$.0^32001$	$.0^31926$	$.0^31854$	$.0^31785$	$.0^31718$	$.0^31653$	-3.5
-3.4	$.0^33369$	$.0^33248$	$.0^33131$	$.0^33018$	$.0^32909$	$.0^32803$	$.0^32701$	$.0^32602$	$.0^32507$	$.0^32415$	-3.4
-3.3	$.0^34834$	$.0^34665$	$.0^34501$	$.0^34342$	$.0^34189$	$.0^34041$	$.0^33897$	$.0^33758$	$.0^33624$	$.0^33495$	-3.3
-3.2	$.0^36871$	$.0^36637$	$.0^36410$	$.0^36190$	$.0^35976$	$.0^35770$	$.0^35571$	$.0^35377$	$.0^35190$	$.0^35009$	-3.2
-3.1	$.0^39376$	$.0^39354$	$.0^39043$	$.0^38740$	$.0^38447$	$.0^38464$	$.0^37888$	$.0^37622$	$.0^37364$	$.0^37114$	-3.1
-3.0	$.0^21350$	$.0^21306$	$.0^21264$	$.0^21223$	$.0^21183$	$.0^21144$	$.0^21107$	$.0^21070$	$.0^21035$	$.0^21001$	-3.0
-2.9	$.0^21866$	$.0^21807$	$.0^21750$	$.0^21695$	$.0^21641$	$.0^21589$	$.0^21538$	$.0^21489$	$.0^21441$	$.0^21395$	-2.9
-2.8	$.0^22555$	$.0^22477$	$.0^22401$	$.0^22327$	$.0^22256$	$.0^22186$	$.0^22118$	$.0^22052$	$.0^21988$	$.0^21926$	-2.8
-2.7	$.0^23467$	$.0^23364$	$.0^23264$	$.0^23167$	$.0^23072$	$.0^22980$	$.0^22890$	$.0^22803$	$.0^22718$	$.0^22635$	-2.7
-2.6	$.0^24661$	$.0^24527$	$.0^24396$	$.0^24269$	$.0^24145$	$.0^24025$	$.0^23907$	$.0^23793$	$.0^23681$	$.0^23573$	-2.6
-2.5	$.0^26210$	$.0^26037$	$.0^25868$	$.0^25703$	$.0^25543$	$.0^25386$	$.0^25234$	$.0^25085$	$.0^24940$	$.0^24799$	-2.5
-2.4	$.0^28198$	$.0^27976$	$.0^27760$	$.0^27549$	$.0^27344$	$.0^27143$	$.0^26947$	$.0^26756$	$.0^26569$	$.0^26387$	-2.4
-2.3	.01072	.01044	.01017	$.0^29903$	$.0^29642$	$.0^29387$	$.0^29137$	$.0^28894$	$.0^28656$	$.0^28424$	-2.3
-2.2	.01390	.01355	.01321	.01287	.01255	.01222	.01191	.01160	.01130	.01101	-2.2
-2.1	.01786	.01743	.01700	.01659	.01618	.01578	.01539	.01500	.01463	.01426	-2.1
-2.0	.02275	.02222	.02169	.02118	.02068	.02018	.01970	.01923	.01876	.01831	-2.0
-1.9	.02872	.02807	.02743	.02680	.02619	.02559	.02500	.02442	.02385	.02330	-1.9

K_p	0.00	0.01	0.02	0.03	0.04	0.05	0.06	0.07	0.08	0.09	K_p
-1.8	.03593	.03515	.03438	.03362	.03288	.03216	.03144	.03074	.03005	.02938	-1.8
-1.7	.04457	.04363	.04272	.04182	.04093	.04006	.03920	.03836	.03754	.03673	-1.7
-1.6	.05480	.05370	.05262	.05155	.05050	.04947	.04846	.04746	.04648	.04551	-1.6
-1.5	.06681	.06552	.06426	.06301	.06178	.06057	.05938	.05821	.05705	.05592	-1.5
-1.4	.08076	.07927	.07780	.07636	.07493	.07353	.07215	.07078	.06944	.06811	-1.4
-1.3	.09680	.09510	.09342	.09176	.09012	.08851	.08691	.08534	.08379	.08226	-1.3
-1.2	.1151	.1131	.1112	.1093	.1075	.1056	.1038	.1020	.1003	.09853	-1.2
-1.1	.1357	.1335	.1314	.1292	.1271	.1251	.1230	.1210	.1190	.1170	-1.1
-1.0	.1587	.1562	.1539	.1515	.1492	.1469	.1446	.1423	.1401	.1379	-1.0
-0.9	.1841	.1814	.1788	.1762	.1736	.1711	.1685	.1660	.1635	.1611	-0.9
-0.8	.2119	.2090	.2061	.2033	.2005	.0977	.1949	.1922	.1894	.1867	-0.8
-0.7	.2420	.2889	.2358	.2327	.2297	.2266	.2236	.2206	.2177	.2148	-0.7
-0.6	.2743	.2709	.2676	.2643	.2611	.2578	.2546	.2514	.2483	.2451	-0.6
-0.5	.3085	.3050	.3015	.2981	.29746	.2912	.2877	.2843	.2810	.2776	-0.5
-0.4	.3446	.3409	.3372	.3336	.3300	.3264	.3228	.3192	.3156	.3121	-0.4
-0.3	.3821	.3783	.3745	.3707	.3669	.3632	.3594	.3557	.3520	.3483	-0.3
-0.2	.4207	.4168	.4129	.4090	.4052	.4013	.3974	.3936	.3897	.3859	-0.2
-0.1	.4602	.4562	.4522	.4483	.4443	.4404	.4364	.4325	.4286	.4247	-0.1
K_p	0.00	0.01	0.02	0.03	0.04	0.05	0.06	0.07	0.08	0.09	K_p
0.0	.5000	.5040	.5080	.5120	.5160	.5199	.5239	.5279	.5319	.5359	0.0
0.1	.5398	.5438	.5478	.5517	.5557	.5596	.5636	.5675	.5714	.5753	0.1
0.2	.5793	.5832	.5871	.5910	.5948	.5987	.6026	.6064	.6103	.6141	0.2
0.3	.6179	.6217	.6255	.6293	.6331	.6368	.6406	.6443	.6481	.6517	0.3
0.4	.6554	.6591	.6628	.6664	.6700	.6736	.6772	.6808	.6844	.6879	0.4
0.5	.6915	.6950	.6985	.7019	.7054	.7088	.7123	.7157	.7190	.7224	0.5
0.6	.7257	.7291	.7324	.7357	.7389	.7422	.7454	.7486	.7517	.7549	0.6
0.7	.7580	.7611	.7642	.7673	.7703	.7734	.7764	.7794	.7823	.7852	0.7
0.8	.7881	.7910	.7939	.7967	.7995	.8023	.8051	.8078	.8106	.8133	0.8
0.9	.8159	.8186	.8212	.8238	.8264	.8289	.8315	.8340	.8365	.8389	0.9
1.0	.8413	.8438	.8461	.8485	.8508	.8531	.8554	.8577	.8599	.8621	1.0
1.1	.8643	.8665	.8686	.8708	.8729	.8749	.8771	.8790	.8810	.8830	1.1
1.2	.8849	.8869	.8888	.8907	.8925	.8944	.8962	.8980	.8997	.90147	1.2
1.3	.90320	.90490	.90658	.90824	.90988	.91149	.91309	.91466	.91621	.91774	1.3
1.4	.91924	.92073	.92220	.92364	.92507	.92647	.92785	.92922	.93056	.93189	1.4
1.5	.93319	.93448	.93574	.93699	.93822	.93943	.94062	.94179	.94295	.94408	1.5
1.6	.94520	.94630	.94738	.94845	.94950	.95053	.95154	.95254	.95352	.95449	1.6
1.7	.95543	.95637	.95728	.95818	.95907	.95994	.96080	.96164	.96246	.96327	1.7
1.8	.96407	.96485	.96562	.96638	.96712	.96784	.96856	.96926	.96995	.97062	1.8
1.9	.97128	.97193	.97257	.97320	.97381	.97441	.97500	.97558	.97615	.97670	1.9
2.0	.97725	.97778	.97831	.97882	.97932	.97982	.98030	.98077	.98124	.98169	2.0
2.1	.98214	.98257	.98300	.98311	.98382	.98422	.98461	.98500	.98537	.98574	2.1
2.2	.98610	.98645	.98679	.98716	.98745	.98778	.98809	.98840	.98870	.98899	2.2
2.3	.98928	.98956	.98983	$.9^20097$	$.9^20358$	$.9^20613$	$.9^20863$	$.9^21109$	$.9^21344$	$.9^21576$	2.3
2.4	$.9^21802$	$.9^22024$	$.9^22240$	$.9^22451$	$.9^22656$	$.9^22857$	$.9^23053$	$.9^23244$	$.9^23431$	$.9^23613$	2.4
2.5	$.9^23790$	$.9^23963$	$.9^24132$	$.9^24297$	$.9^24467$	$.9^24614$	$.9^24766$	$.9^24915$	$.9^25060$	$.9^25201$	2.5
2.6	$.9^25339$	$.9^25473$	$.9^25604$	$.9^25731$	$.9^25855$	$.9^25975$	$.9^26093$	$.9^26207$	$.9^26319$	$.9^26427$	2.6
2.7	$.9^26533$	$.9^26636$	$.9^26736$	$.9^26833$	$.9^26928$	$.9^27020$	$.9^27110$	$.9^27197$	$.9^27282$	$.9^27365$	2.7
2.8	$.9^27445$	$.9^27523$	$.9^27599$	$.9^27673$	$.9^27741$	$.9^27814$	$.9^27882$	$.9^27948$	$.9^28012$	$.9^28074$	2.8
2.9	$.9^28134$	$.9^28193$	$.9^28250$	$.9^28305$	$.9^28359$	$.8^28411$	$.9^28462$	$.9^28511$	$.9^28559$	$.9^28605$	2.9
3.0	$.9^28650$	$.9^28694$	$.9^28736$	$.9^28777$	$.9^28817$	$.9^28856$	$.9^28893$	$.9^28930$	$.9^28965$	$.9^28999$	3.0
3.1	$.9^30324$	$.9^30646$	$.9^30957$	$.9^31260$	$.9^31553$	$.9^31836$	$.9^32112$	$.9^32378$	$.9^32636$	$.9^32886$	3.1
3.2	$.9^33129$	$.9^33363$	$.9^33590$	$.9^33810$	$.9^34024$	$.9^34230$	$.9^34429$	$.9^34623$	$.9^34810$	$.9^34991$	3.2
3.3	$.9^35166$	$.9^35335$	$.9^35499$	$.9^35658$	$.9^35811$	$.9^35939$	$.9^36103$	$.9^36242$	$.9^36376$	$.9^36505$	3.3
3.4	$.9^36631$	$.9^36752$	$.9^35869$	$.9^36982$	$.9^37091$	$.9^37197$	$.9^37299$	$.9^37398$	$.9^37493$	$.9^37585$	3.4
3.5	$.9^37674$	$.9^37759$	$.9^37842$	$.9^37922$	$.9^37999$	$.9^38074$	$.9^38146$	$.9^38215$	$.9^38282$	$.9^38347$	3.5
3.6	$.9^38409$	$.9^38469$	$.9^38527$	$.9^38583$	$.9^38637$	$.9^38689$	$.9^38739$	$.9^38787$	$.9^38834$	$.9^38879$	3.6
3.7	$.9^38922$	$.9^38964$	$.9^40039$	$.9^40426$	$.9^40789$	$.9^41158$	$.9^41504$	$.9^41838$	$.9^42159$	$.9^42468$	3.7
3.8	$.9^42765$	$.9^43052$	$.9^43327$	$.9^43503$	$.9^43848$	$.9^44094$	$.9^44331$	$.9^44558$	$.9^44777$	$.9^44988$	3.8
3.9	$.9^45190$	$.9^45385$	$.9^45573$	$.9^45753$	$.9^45926$	$.9^45092$	$.9^46253$	$.9^46406$	$.9^46554$	$.9^46696$	3.9

K_p	0.00	0.01	0.02	0.03	0.04	0.05	0.06	0.07	0.08	0.09	K_p
4.0	$.9^46833$	$.9^46964$	$.9^47090$	$.9^47211$	$.9^47327$	$.9^47439$	$.9^47546$	$.9^47649$	$.9^47748$	$.9^47843$	4.0
4.1	$.9^47934$	$.9^48022$	$.9^48106$	$.9^48186$	$.9^48263$	$.9^48338$	$.9^48409$	$.9^48477$	$.9^48542$	$.9^48605$	4.1
4.2	$.9^48665$	$.9^48723$	$.9^48778$	$.9^48832$	$.9^48882$	$.9^48931$	$.9^48978$	$.9^50226$	$.9^50655$	$.9^51066$	4.2
4.3	$.9^51460$	$.9^51837$	$.9^52199$	$.9^52545$	$.9^52876$	$.9^53193$	$.9^53497$	$.9^53788$	$.9^54066$	$.9^54332$	4.3
4.4	$.9^54587$	$.9^54831$	$.9^55065$	$.9^55288$	$.9^55502$	$.9^55706$	$.9^55902$	$.9^56089$	$.9^56268$	$.9^56439$	4.4
4.5	$.9^56602$	$.9^56759$	$.9^56908$	$.9^57051$	$.9^57187$	$.9^57318$	$.9^57442$	$.9^57561$	$.9^57675$	$.9^57784$	4.5
4.6	$.9^57888$	$.9^57987$	$.9^58081$	$.9^58172$	$.9^58258$	$.9^58340$	$.9^58419$	$.9^58494$	$.9^58566$	$.9^58634$	4.6
4.7	$.9^58699$	$.9^58761$	$.9^58821$	$.9^58877$	$.9^58931$	$.9^58983$	$.9^60320$	$.9^60789$	$.9^61235$	$.9^61661$	4.7
4.8	$.9^62067$	$.9^62453$	$.9^62822$	$.9^63173$	$.9^63508$	$.9^63827$	$.9^64131$	$.9^64420$	$.9^64696$	$.9^64958$	4.8
4.9	$.9^65208$	$.9^65446$	$.9^65673$	$.9^65889$	$.9^66094$	$.9^66289$	$.9^66475$	$.9^66652$	$.9^66821$	$.9^66981$	4.9

Threshold Limit Values for Indoor Air Pollutants

Substance [CAS No.]	TWA	STEL	Notations	MW	TLV® Basis-Critical Effect(s)
Acetic acid [64-19-7]	10 ppm	15 ppm	—	60.00	Irritation
Acetic anhydride [108-24-7]	5 ppm	—	—	102.09	Irritation
Acetone [67-64-1]	500 ppm	750 ppm	A4; BEI	58.05	Irritation
Acetonitrile [75-05-8]	20 ppm	—	Skin; A4	41.05	Lung
Acrylamide [79-06-01]	0.03 mg/m³	—	Skin; A3	71.08	CNS; dermatitis
Acrylic acid [79-10-7]	2 ppm	—	Skin; A4	72.06	Irritation; reproductive
Acrylonitrile [107-13-1]	2 ppm	—	Skin; A3	53.05	Cancer
Adipic acid [124-04-9]	5 mg/m³	—	—	146.14	Neurotoxicity; GI; irritation
Adiponitrile [111-69-3]	2 ppm	—	Skin	108.10	Lung
Aldrin [309-00-2]	0.25 mg/m³	—	Skin; A3	364.93	Liver
Allyl alcohol [107-18-6]	0.5 ppm	—	Skin; A4	58.08	Irritation
Allyl chloride [107-05-1]	1 ppm	2 ppm	A3	76.50	Liver
Allyl glycidyl ether (AGE) [106-92-3]	1 ppm	—	A4	114.14	Irritation; dermatitis; sensitization
‡ Allyl propyl disulfide [2179-59-1]	(2 ppm)	(3 ppm)	(-)	148.16	Irritation
Aluminum [7429-90-5] and compounds, as Al Metal dust	—	—	—		
Pyro powders	10 mg/m³	—	—	26.98	Irritation
‡ Welding fumes	5 mg/m³	—	(B2)	Varies	Lung
Soluble salts	5 mg/m³	—	—	N/A	Irritation
Alkyls (NOS)	2 mg/m³	—	—	Varies	Irritation
	2 mg/m³			Varies	Irritation
Aluminum oxide [1344-28-1]	10 mg/m³(E)	—	A4	101.96	Lung; irritation

Source: From American Conference of Governmental Industrial Hygienists (ACGIH®), *2003 TLVs® and BEIs®* Book. © 2003. With permission.

Substance [CAS No.]	TWA	STEL	Notations	MW	TLV® Basis-Critical Effect(s)
4-Aminodiphenyl [92-67-1]	—(L)	—	Skin; A1	169.23	Cancer (bladder)
2-Aminopyridine [504-29-0]	0.5 ppm	—	—	91.11	CNS
Amitrole [61-82-5]	0.2 mg/m³	—	A3	84.8	Reproductive; thyroid
Ammonia [7664-41-7]	25 ppm	35 ppm	—	17.03	Irritation
Ammonium chloride fume [12125-02-9]	10 mg/m³	20 mg/m³	—	53.50	Irritation
Ammonium perfluoroctanoate [3825-26-1]	0.01mg/m³		Skin; A3	431.00	Liver
Ammonium sulfamate [7773-06-0]	10 mg/m³	—	—	114.13	Irritation
tert-Amyl methyl ether (TAME) [994-05-8]	20 ppm	—	—	102.2	Neurologic; reproductive
Aniline [62-53-3]	2 ppm	—	Skin; A3; BEI	93.12	Anoxia
o-Anisidine [90-04-0]	0.5 mg/m³	—	Skin; A3; BEI₂	123.15	Anoxia
p-Anisidine [104-94-9]	0.5 mg/m³	—	Skin; A4; BEI₂	123.15	Anoxia
Antimony [7440-36-0] and compounds, as Sb	0.5 mg/m³	—	—	121.75	Irritation; lung; CVS
Antimony hydride [7803-52-3]	0.1 ppm	—	—	124.78	Irritation; blood
ANTU [86-88-4]	0.3 mg/m³	—	A4	202.27	Lung; irritation
Arsenic [7440-38-2] and inorganic compounds, as As	0.01mg/m³	—	A1; BEI	74.92 Varies	Cancer (lung, skin); lung
Arsine [7784-42-1]	0.05 ppm	—	—	77.95	Blood; kidney
Asbestos, all forms [1332-21-4]	0.1 f/cc (F)	—	A1	NA	Asbestosis; cancer
Asphalt (Bitumen) fume [8052-42-4] as benzne-soluble aerosol	0.5 mg/m³	—	A4	—	Irritation
Atrazine [1912-24-9]	5 mg/m³	—	A4	216.06	Irritation
Azinphos-methyl [86-50-0]	0.2 mg/m³ (I,V)	—	Skin; SEN: A4; BEI₁	317.34	Cholinergic
Barium [7440-39-3] and soluble compounds, as Ba	0.5 mg/m³	—	A4	137.30	Irritation; GI; muscles
Barium sulfate [7727-43-7]	10 mg/m³	—	—	233.43	Pneumoconiosis (baritosis)
Benomyl [17804-35-2]	10 mg/m³	—	A4	290.32	Dermititis; irritation; reproductive
Benz[a]anthracene [56-55-3]	—(L)	—	A2	228.30	Cancer
Benzene [71-43-2]	0.5 ppm	2.5 ppm	Skin; A1; BEI	78.11	Cancer
Benzidine [92-87-5]	—(L)	—	Skin; A1	184.23	Cancer (bladder)
Benzo[b]fluoranthene [205-99-2]	—(L)	—	A2	252.30	Cancer
Benzo[a]pyrene [50-32-8]	—(L)	—	A2	252.30	Cancer
Benzoyl chloride [98-88-4]	—	C 0.5 ppm	A4	140.57	Irritation
Benzoyl peroxide [94-36-0]	5 mg/m³	—	A4	242.22	Irritation
Benzyl acetate [140-11-4]	10 ppm	—	A4	150.18	Irritation
Benzyl chloride [100-44-7]	1 ppm	—	A3	126.58	Irritation; lung
‡ Beryllium [7440-41-7] and compounds, as Be	(0.002 mg/m³)	(0.01 mg/m³)	(-); A1	9.01	Cancer (lung); berylliosis

Substance [CAS No.]	TWA	STEL	Notations	MW	TLV® Basis-Critical Effect(s)
Biphenyl [92-52-4]	0.2 ppm	—	—	154.20	Lung
Bis (2-dimethyl-aminoethyl) ether (DMAEE) [3033-62-3]	0.05 ppm	0.15 ppm	Skin	160.26	Irritation; vision
Bismuth telluride				800.83	
Undoped [1304-82-1]	10 mg/m³	—	A4		Irritation
Se-doped, as Bi₂Te₃	5 mg/m³	—	A4		Irritation; lung
Borates, tetra, sodium salts					Irritation
Anhydrous [1330-43-4]	1 mg/m³	—	—	201.22	
Decahydrate [1303-96-4]	5 mg/m³	—	—	301.37	
Pentahydrate [12179-04-3]	1 mg/m³	—	—	291.30	
Boron oxide [13-3-86-2]	10 mg/m³	—	—	69.64	Irritation
Bromacil [314-4-9]	10 mg/m³	—	A3	261.11	Irritation
Bromine [7726-95-6]	0.1 ppm	0.2 ppm	—	159.81	Irritation
Bromine pentafluoride [7789-30-2]	0.1 ppm	—	—	174.92	Irritation
Bromoform [75-25-2]	0.5 ppm	—	Skin; A3	252.80	Irritation; liver
1,3-Butadiene [106-99-0]	2 ppm	—	A2	54.09	Cancer
‡ (Butane [106-99-0])	(800 ppm)	(-)	(-)	(58.12)	(Narcosis)
n-Butanol [71-36-3]	20 ppm	—	—	74.12	Irritation
sec-Butanol [78-92-2]	100 ppm	—	—	74.12	Irritation; narcosis
tert-Butanol [75-65-0]	100 ppm	—	A4	74.12	Narcosis; irritation
* 2-Butoxyethanol (EGBE) [111-76-2]	20 ppm	—	A3	118.17	Irritation; CNS
* 2-Butoxyethyl acetate (EGBEA) [112-07-2]	20 ppm	—	A3	118.17	Irritation; CNS
n-Butyl acetate [123-86-4]	150 ppm	200 ppm	—	116.16	Irritation
sec-Butyl acetate [105-46-4]	200 ppm	—	—	116.16	Irritation
tert-Butyl acetate [540-88-5]	200 ppm	—	—	116.16	Irritation
n-Butyl acrylate [141-32-2]	2 ppm	—	SEN; A4	128.17	Irritation
Butylated hydroxytoluene (BHT) [128-37-0]	2 mg/m³ (I, V)	—	A4	220.34	Irritation
n-Butyl glycidyl ether (BGE) [2426-08-6]	25 ppm	—	—	130.21	Irritation; sensitization
n-Butyl lactate [138-227-7]	5 ppm	—	—	146.19	Irritation; headache
n-Butyl mercaptan [109-79-5]	0.5 ppm	—	—	90.19	Irritation; CNS; reproductive
o-sec-Butylphenol [89-72-5]	5 ppm	—	Skin	150.22	Irritation
p-tert-Butyl toluene [98-51-1]	1 ppm	—	—	148.18	Irritation; CNS; CVS
Cadmium [7440-43-9] and compounds, as Cd	0.01 mg/m³ 0.002 mg/m³ (R)	— —	A2; BEI A2; BEI	112.40 Varies	Kidney
Calcium carbonate [471-34-1]	10 mg/m³ (E)	—	—	100.09	Irritation
Calcium chromate [13765-19-0], as Cr	0.001 mg/m³	—	A2	156.09	Cancer
Calcium cyanamide [156-62-7]	0.5 mg/m³	—	A4	80.11	Irritation; dermatitis
Calcium hydroxide [1305-62-0]	5 mg/m³	—	—	74.10	Irritation
Calcium oxide [1305-78-8]	2 mg/m³	—	—	56.08	Irritation

Substance [CAS No.]	TWA	STEL	Notations	MW	TLV® Basis-Critical Effect(s)
Calcium silicate, synthetic nonfibrous [1344-95-2]	10 mg/m³ (E)	—	A4	—	Irritation
Calcium sulfate [7778-18-9]	10 mg/m³ (E)	—	—	136.14	Irritation
Camphor, synthetic [76-22-2]	2 ppm	3 ppm	A4	152.23	Irritation; anosmia
* Caprolactam [105-60-2]	5 mg/m³ (I, V)	—	A5	113.16	Irritation
Captafol [2425-06-1]	0.1 mg/m³	—	Skin; A4	349.06	Dermatitis; sensitization
Captan [133-06-2]	5 mg/m³	—	SEN; A3	300.60	Irritation
Carbaryl [63-25-2]	5 mg/m³ (I)	—	A4	201.20	Cholinergic; reproductive
‡ Carbofuran [1563-66-2]	(0.1 mg/m³)	—	A4; BEI₁	221.30	Cholinergic
Carbon black [1333-86-4]	3.5 mg/m³	—	A4	—	Lung
Carbon dioxide [124-38-9]	5000 ppm	30,000 ppm	—	44.01	Asphyxiation
Carbon disulfide [75-15-0]	10 ppm	—	Skin; BEI	76.14	CVS; CNS
Carbon monoxide [630-08-0]	25 ppm	—	BEI	28.01	Anoxia; CVS; CNS; reproductive
Carbon tetrabromide [558-13-4]	0.1 ppm	0.3 ppm	—	331.65	Irritation; liver
Carbon tetrachloride [56-23-5]	5 ppm	10 ppm	Skin; A2	153.84	Liver; cancer
Carbonyl fluoride	2 ppm	5 ppm	—	66.01	Irritation; bone; fluorosis
Catechol [120-80-9]	5 ppm	—	Skin; A3	110.11	Irritation; CNS; lung
Cellulose [9004-34-6]	10 mg/m³	—	—	NA	Irritation
Cesium hydroxide [21351-79-1]	2 mg/m³	—	—	149.92	Irritation
Chlordane [57-74-9]	0.5 mg/m³	—	Skin; A3	409.80	Seizures; liver
Chlorinated camphene [8001-35-2]	0.5 mg/m³	1 mg/m³	Skin; A3	414.00	Seizures; liver
o-Chlorinated diphenyl oxide [31242-93-0]	0.5 mg/m³	—	—	377.00	Chloracne; liver
Chlorine [7782-50-5]	0.5 ppm	1 ppm	A4	70.91	Irritation
Chlorine dioxide [10049-04-4]	0.1 ppm	0.3 ppm	—	67.46	Irritation; bronchitis
2-Chloroacetophenone [532-27-4]	0.05 ppm	—	A4	154.59	Irritation; sensitization
Chloroacetyl chloride [79-04-9]	0.05 ppm	0.15 ppm	Skin	112.95	Irritation; lung
Chlorobenzene [108-90-7]	10 ppm	—	A3; BEI	112.56	Liver
Chlorobromomethane [74-97-5]	200 ppm	—	—	129.39	CNS; liver
Chlorodifluoromethane [75-45-6]	1000 ppm	—	A4	86.47	CVS
Chlorodiphenyl (42% chlorine) [53469-21-9]	1 mg/m³	—	Skin	266.50	Irriation; chloracne; liver
Chlorodiphenyl (54% chlorine) [11097-69-1]	0.5 mg/m³	—	Skin; A3	328.40	Irriation; chloracne; liver
Chloroform [67-66-3]	10 ppm	—	A3	119.38	Liver; reproductive
bis(Chloromethyl) ether [542-88-1]	0.001 ppm	—	A1	114.96	Cancer (lung)

Substance [CAS No.]	TWA	STEL	Notations	MW	TLV® Basis-Critical Effect(s)
Chloropentafluoro-ethane [76-15-3]	1000 ppm	—	—	154.47	CVS
Chloropicrin [76-06-2]	0.1 ppm	—	A4	164.39	Irritation; lung
1-Chloro-2-propanol [127-00-4] and 2-Chloro-1-propanol [78-89-7]	1 ppm	—	Skin; A4	94.54	Reproductive; genotoxic
-Chloroprene [126-99-8]	10 ppm	—	Skin	88.54	Irritation; liver; reproductive
2-Chloropropionic acid [598-78-7]	0.1 ppm	—	Skin	108.53	Irritation; reproductive
o-Chlorostyrene [2039-87-4]	50 ppm	75 ppm	—	138.60	Kidney; CNS; neurotoxic; liver
o-Chlorotoluene [95-49-8]	50 ppm	—	—	126.59	Irritation
* Chlorpyrifos [2921-88-2]	0.1mg/m^3 (I, V)	—	Skin; A4; BEI$_1$	350.57	Cholinergic
Chromite ore processing (Chromate), as Cr	0.05 mg/m^3	—	A1	—	Cancer (lung)
Chromium, [7440-47-3] and inorganic compounds, as Cr					
Metal and Cr II compounds	0.5 mg/m^3	—	A4	Varies	Irritation; dermatitis
Water-soluble Cr VI compounds	0.05 mg/m^3	—	A1; BEI	Varies	Liver; kidney; respiratory
Insoluble Cr VI compounds	0.01 mg/m^3	—	A1	Varies	Cancer; irritation
Chromyl chloride [14977-61-8]	0.025 ppm	—	—	154.92	Kidney; liver; respiratory
Chrysene [218-01-9]	—(L)	—	A3	228.30	Skin
Clopidol [2971-90-6]	10 mg/m^3	—	A4	192.06	Irritation
Coal dust					
Anthracite	0.4 mg/m^3 ®	—	A4	—	Lung fibrosis; lung function
Bituminous	0.9 mg/m^3 ®	—	A4	—	Lung fibrosis; lung function
Coal tar pitch volatiles [65996-93-2], as benzene-soluble aerosol	0.2 mg/m^3	—	A1	—	Cancer
Cobalt [7440-48-4] and inorganic compounds, as Co	0.02 mg/m^3	—	A3; BEI	58.93 Varies	Asthma; lung; CVS
Cobalt carbonyl [10210-68-1], as Co	0.1 mg/m^3	—	—	341.94	Lung edema
Cobalt hydrocarbonyl [16842-03-8], as Co	0.1 mg/m^3	—	—	171.98	Lung edema
Copper [7440-50-8]				63.55	Irritation; GI; metal fume fever
Fume	0.2 mg/m^3	—	—		
Dust and mists, as Cu	1 mg/m^3	—	—		
Cotton dust, raw	0.2 mg/m^3 (G)	—	—	—	Lung; byssinosis
Cresol, all isomers [1319-77-3; 95-48-7; 108-39-4; 106-44-5]	5 ppm	—	Skin	108.14	Dermatitis; irritation; CNS
Crufomate [299-86-5]	5 mg/m^3	—	A4; BEI$_1$	291.71	Cholinergic
Cumene [98-82-8]	50 ppm	—	—	120.19	Irritation; CNS
Cyanamide [420-04-2]	2 mg/m^3	—	—	42.04	Irritation
Cyanogen [460-19-5]	10 ppm	—	—	52.04	Irritation
Cyclohexane [110-82-7]	100 ppm	—	—	84.16	CNS
Cyclohexanol [108-94-1]	50 ppm	—	Skin	100.16	Irritation; CNS

Substance [CAS No.]	TWA	STEL	Notations	MW	TLV® Basis-Critical Effect(s)
* Cyclohexanone [108-94-1]	20 ppm	50 ppm	Skin; A3	98.14	Irritation; CNS; liver; kidney
Cyclohexene [110-83-8]	300 ppm	—	—	82.14	Irritation
Cyclohexylamine [108-91-8]	10 ppm	—	A4	99.17	Irritation
Cyclonite [121-82-4]	0.5 mg/m³	—	Skin; A4	222.26	Irritation; CNS; liver; blood
Cyclopentadiene [542-92-7]	75 ppm	—	—	66.10	Irritation
Cyclopentane [287-92-3]	600 ppm	—	—	70.13	Irritation; narcosis
Cyhexatin [13121-70-5]	5 mg/m³	—	A4	385.16	Irritation
DDT [50-29-3]	1 mg/m³	—	A3	354.50	Seizures; liver
Demeton [8065-48-3]	0.05 mg/m³ (I, V)	—	Skin; BEI₁	258.34	Cholinergic
Demeton-S-methyl [919-86-8]	0.05 mg/m³ (I, V)	—	Skin; SEN; A4; BEI₁	230.3	Cholinergic
Diacetone alcohol [123-42-2]	50 ppm	—	—	116.16	Irritation
* Diazinon [333-41-5]	0.01mg/m³ (I, V)	—	Skin; A4; BEI₁	304.36	Cholinergic
Diazomethane [334-88-3]	0.2 ppm	—	A2	42.04	Irritation; cancer (lung)
Diborane [19287-45-7]	0.1 ppm	—	—	27.69	CNS; lung function
2-N-Dibutylamino-ethanol [102-81-8]	0.5 ppm	—	Skin; BEI₁	173.29	Irritation; cholinergic
Dibutyl phenyl phosphate [2528-36-1]	0.3 ppm	—	Skin; BEI₁	286.26	Irritation; cholinergic
Dibutyl phosphate [107-66-4]	1 ppm	2 ppm	—	210.21	Irritation
Dibutyl phthalate [84-74-2]	5 mg/m³	—	—	278.34	Reproductive; irritation
o-Dichlorobenzene [95-50-1]	25 ppm	50 ppm	A4	147.01	Irritation; liver
p-Dichlorobenzene [106-46-7]	10 ppm	—	A3	147.01	Irritation; kidney
3,3'-Dichlorobenzidine [91-94-1]	—(L)	—	Skin; A3	253.13	Irritation; dermatitis
1,4-Dichloro-2-butene [764-41-0]	0.005 ppm	—	Skin; A2	124.99	Cancer; irritation
Dichlorodifluoro-methane [75-71-8]	1000 ppm	—	A4	120.91	CVS
1,3-Dichloro-5,5-dimethyl hydantoin [118-52-5]	0.2 mg/m³	0.4 mg/m³	—	197.03	Irritation
1,1-Dichloroethane [75-34-3]	100 ppm	—	A4	98.97	Liver; kidney; irritation
1,2-Dichloroethylene, all isomers [540-59-0; 156-59-2; 156-60-5]	200 ppm	—	—	96.95	Liver
Dichloroethyl ether [111-44-4]	5 ppm	10 ppm	Skin; A4	143.02	Irritation; lung
Dichlorofluoromethane [75-43-4]	10 ppm	—	—	102.92	Liver
Dichloromethane [75-43-4]	50 ppm	—	A3; BEI	84.93	CNS; anoxia
1,1-Dichloro-1-nitroethane [594-72-9]	2 ppm	—	—	143.96	Irritation
1,3-Dichloropropene [542-75-6]	1 ppm	—	Skin; A3	110.98	Irritation

Substance [CAS No.]	TWA	STEL	Notations	MW	TLV® Basis-Critical Effect(s)
2,2-Dichloropropionic acid [75-99-0]	5 mg/m³ (I)	—	A4	142.97	Irritation
Dichlorotetrafluoro-ethane [76-14-2]	1000 ppm	—	A4	170.93	CVS; narcosis; asphyxiation
Dichlorvos (DDVP) [62-73-7]	0.1 mg/m³ (I,V)	—	Skin; SEN; A4; BEI₁	220.98	Cholinergic
Dicrotophos [141-66-2]	0.05 mg/m³ (I,V)	—	Skin; A4; BEI₁	237.21	Cholinergic
Dicyclopentadiene [77-73-6]	5 ppm	—	—	132.21	Irritation
Dicyclopentadienyl iron [102-54-5]	10 mg/m³	—	—	186.03	Blood; liver
Dieldrin [60-57-1]	0.25 mg/m³	—	Skin; A4	380.93	Liver; CNS
Diesel fuel [68334-30-5; 68476-30-2; 68476-31-3; 68476-34-6; 77650-28-3], as total hydrocarbons	100 mg/m³ (V)	—	Skin; A3	Varies	Skin; irritation
Diethanolamine [111-42-2]	2 mg/m³	—	Skin	105.14	Liver; kidney; blood
Diethylamine [109-89-7]	5 ppm	15 ppm	Skin; A4	73.14	Irritation
2-Diethylaminoethanol [100-37-8]	2 ppm	—	Skin	117.19	Irritation; CNS
Diethylene triamine [111-40-0]	1 ppm	—	Skin	103.17	Irritation; sensitization
Di(2-ethylhexyl)-phthalate (DEHP) [117-81-7]	5 mg/m³	—	A3	390.54	Irritation
Diethyl ketone [96-22-0]	200 ppm	300 ppm	—	86.13	Irritation; narcosis
Diethyl phthalate [84-66-2]	5 mg/m³	—	A4	222.23	Irritation
Difluorodibromo-methane [75-61-6]	100 ppm	—	—	209.83	Irritation; liver; CNS
Digylcidyl ether (DGE) [2238-07-5]	0.1 ppm	—	A4	130.14	Irritation; reproductive; blood
Diisobutyl ketone [108-83-8]	25 ppm	—	—	142.23	Irritation
Diisopropylamine [108-18-9]	5 ppm	—	Skin	101.19	Vision; irritation
N,N-Dimethylacet-amide [127-19-5]	10 ppm	—	Skin; A4; BEI	87.12	Reproductive; liver
Dimethylamine [124-40-3]	5 ppm	15 ppm	A4	45.08	Irritation
Diemthylaniline (N,N-Dimethylaniline) [121-69-7]	5 ppm	10 ppm	Skin; A4; BEI₂	121.18	Anoxia; neurotoxicity
Dimethyl carbamoyl chloride [79-44-7]	—(L)	—	A2	107.54	Cancer (lung)
Dimethylethoxysilane [14857-34-2]	0.5 ppm	1.5 ppm	—	104.20	Irritation; headache
Dimethylformamide [68-12-2]	10 ppm	—	Skin; A4; BEI	73.09	Liver
1,1-Dimethylhydrazine [57-14-7]	0.01 ppm	—	Skin; A3	60.12	Irritation; neoplasia
Dimethylphthalate [131-11-3]	5 mg/m³	—	—	194.19	Irritation
Dimethyl sulfate [77-78-1]	0.1 ppm	—	Skin; A3	126.10	Irritation
Dinitolmide [148-01-6]	5 mg/m³	—	A4	225.16	Irritation

Substance [CAS No.]	TWA	STEL	Notations	MW	TLV® Basis-Critical Effect(s)
Dinitrobenzene, all isomers [528-29-0; 99-65-0; 100-25-4; 25154-54-5]	0.15 ppm	—	Skin; BEI$_2$	168.11	Anoxia
Dinitro-o-cresol [534-52-1]	0.2 mg/m^3	—	Skin	198.13	Metabolic disorders
Dinitrotoluene [25321-14-6]	0.2 mg/m^3	—	Skin; A3; BEI$_2$	182.15	CVS; reproductive
1,4-Dioxane [123-91-1]	20 ppm	—	Skin; A3	88.10	Irritation; liver; kidney
Dioxathion [78-34-2]	0.1 mg/m$^{3\,(I, V)}$	—	Skin; A4; BEI$_1$	456.54	Cholinergic
1,3-Dioxolane [646-06-0]	20 ppm	—	—	74.08	Blood; reproductive
Diphenylamine [122-39-4]	10 mg/m^3	—	A4	169.24	Liver; kidney; blood
Dipropyl ketone [123-19-3]	50 ppm	—	—	114.80	Irritation; liver; kidney; neurotoxicity
Diquat [2764-72-9]	0.5 mg/m$^{3\,(I)}$ 0.1 mg/m$^{3\,®}$	—	Skin; A4 Skin; A4	344.07	Irritation; eye Irritation; eye
Disulfiram [97-77-8]	2 mg/m^3	—	A4	296.54	GI; CVS
Disulfoton [298-04-4]	0.05 mg/m^3 $^{(I,V)}$	—	Skin; A4; BEI$_1$	274.38	Cholinergic
Diuron [330-54-1]	10 mg/m^3	—	A4	233.10	Irritation; blood
Divinyl benzene [1321-74-0]	10 ppm	—	—	130.19	Irritation
Emery [1302-74-5]	10 mg/m$^{3\,(E)}$	—	—	—	Irritation
Endosulfan [115-29-7]	0.1 mg/m^3	—	Skin; A4	406.95	Liver; CNS
Endrin [72-20-8]	0.1 mg/m^3	—	Skin; A4	380.93	CNS; liver
Enflurane [13838-16-9]	75 ppm	—	A4	184.50	CNS; CVS
Epichlorohydrin [106-89-8]	0.5 ppm	—	Skin; A3	92.53	Irritation; liver; kidney
* EPN [2104-64-5]	0.1mg/m$^{3\,(I)}$	—	Skin; A4; BEI$_1$	323.31	Cholinergic
‡ (Ethane [74-84-0])		(Simple asphyxiant $^{(D)}$)		(30.08)	(Asphyxiation)
Ethanol [64-17-5]	1000 ppm	—	A4	46.07	Irritation
Ethanolamine [141-43-5]	3 ppm	6 ppm	—	61.08	Irritation
* Ethion[563-12-2]	0.05 mg/m^3 $^{(I,V)}$	—	Skin; A4, BEI$_1$	384.48	Cholinergic
2-Ethoxyethanol (EGEE) [110-80-5]	5 ppm	—	Skin; BEI	90.12	Reproductive
2-Ethoxyethyl acetate (EGEEA) [111-15-9]	5 ppm	—	Skin; BEI	132.16	Reproductive
Ethyl acetate [141-786]	400 ppm	—	—	88.10	Irritation
Ethyl acrylate [140-88-5]	5 ppm	15 ppm	A4	100.11	Irritation; sensitization
Ehtylamine [75-04-7]	5 ppm	15ppm	Skin	45.08	Irritation
Ethyl amyl ketone [541-85-5]	25 ppm	—	—	128.21	Irritation
Ethyl benzene [100-41-4]	100 ppm	125 ppm	A3; BEI	106.16	Irritation; CNS
Ethyl bromide [74-96-4]	5 ppm	—	Skin; A3	108.98	Liver; kidney; CVS
Ethyl tert-butyl ether (ETBE) [637-92-3]	5 ppm	—	—	102.18	Irritation; lung function; reproductive
Ethyl butyl ketone [106-35-4]	50 ppm	75 ppm	—	114.19	Irritation; necrosis
Ethyl chloride [75-00-3]	100 ppm	—	Skin; A3	64.52	Liver; CNS

Substance [CAS No.]	TWA	STEL	Notations	MW	TLV® Basis-Critical Effect(s)
Ethyl cyanoacrylate [7085-85-0]	0.2 ppm	—	—	125.12	Irritation; necrosis
‡ Ethylene [74-85-1]		(Simple asphyxiant (D))	A4	28.00	(Asphyxiation)
Ethylene chlorohydrin [107-07-3]	—	C 1 ppm	Skin; A4	80.52	Irritation; liver; kidney; GI; CVS; CNS
Ethylenediamine [107-15-3]	10 ppm	—	Skin; A4	60.10	Irritation; asthma; sensitization
Ethylene dibromide [106-93-4]	—	—	Skin; A3	187.88	Irritation; liver; kidney
Ethylene dichloride [107-06-2]	10 ppm	—	A4	98.96	CVS
Ethylene glycol [107-21-1]	—	C 100 mg/m³ (H)	A4	62.07	Cancer; reproductive
Ethylene glycol dinitrate (EGDN) [628-96-6]	0.05 ppm	—	Skin	152.06	CVS
Ethylene oxide [75-21-8]	1 ppm	—	A2	44.05	Irritation; narcosis
Ethylenimine	0.5 ppm	—	Skin; A3	43.08	Irritation; bronchitis
Ethyl ether [60-29-76]	400 ppm	500 ppm	—	74.12	Irritation; narcosis
Ethyl formate [109-94-4]	100 ppm	—	—	74.08	Irritation
Ethyl mercaptan [75-08-1]	0.5 ppm	—	—	62.13	Irritation
2-Ethylhexanoic acid [149-57-5]	5 mg/m³ (I,V)	—	—	144.24	Reproductive
N-Ethylmorpholine [10-74-3]	5 ppm	—	Skin	115.18	Irritation; ocular
Ethyl silicate [78-10-4]	10 ppm	—	—	208.30	Irritation; kidney
Fenamiphos [22224-92-6]	0.1 mg/m³	—	Skin; A4; BEI₁	303.40	Cholinergic
Fensulfothion [115-90-2]	0.1 mg/m³	—	A4; BEI₁	308.35	Cholinergic
Fenthion [55-38-9]	0.2 mg/m³	—	Skin; A4; BEI₁	278.34	Cholinergic
Ferbam [14484-64-1]	10 mg/m³	—	A4	416.50	Irritation
Ferrovanadium dust [12604-58-9]	1 mg/m³	3 mg/m³	—	—	Irritation
Flour dust	0.5 mg/m³	—	SEN	—	Asthma; lung function; bronchitis
Fluorides, as F	2.5 mg/m³	—	A4; BEI	Varies	Irritation; bone; fluorosis
Fluorine [7782-41-4]	1 ppm	2 ppm	—	38.00	Irritation
Fonofos [944-22-9]	0.1 mg/m³	—	Skin; A3; BEI₁	246.32	Cholinergic
Formaldehyde [50-00-0]	—	C 0.3 ppm	SEN; A2	30.03	Irritation; cancer
Formamide [75-12-7]	10 ppm	—	Skin	45.04	Irritation; cancer
Formic acid [64-18-6]	5 ppm	10 ppm	—	46.02	Irritation
Furfural [98-01-1]	2 ppm	—	Skin; A3; BEI	96.08	Irritation
Furfuryl alcohol [98-00-0]	10 ppm	15 ppm	Skin	98.10	Irritation
Gasoline [86290-81-5]	300 ppm	500 ppm	A3	—	Irritation; CNS
Germanium tetrahydride [7782-65-2]	0.2 ppm	—	—	76.63	Blood
Glutaraldehyde [111-30-8], activated and inactivated	—	C 0.05 ppm	SEN; A4	100.11	Irritation; sensitization
Glycerin mist [56-81-5]	10 mg/m³	—	—	92.09	Irritation

Substance [CAS No.]	TWA	STEL	Notations	MW	TLV® Basis-Critical Effect(s)
Glycidol [556-52-5]	2 ppm	—	A3	74.08	Irritation; neoplasia
Glyoxal [107-22-2]	0.1 mg/m³ (I, V)	—	SEN; A4	58.04	Irritation
Grain dust (oat, wheat, barley)	4 mg/m³ (E)	—	—	NA	Irritation; bronchitis; pulmonary function
Graphite (all forms except graphite fibers) [7782-42-5]	2 mg/m³ ®	—	—	—	Pneumoconiosis
Hafnium [7440-58-6] and compounds, as Hf	0.5 mg/m³	—	—	178.49	Liver; irritation
Halothane [151-67-7]	50 ppm	—	A4	197.39	CNS; CVS; liver; reproductive
Helium [7440-59-7]		Simple asphyxiant (D)		4.00	Asphyxiation
Heptachlor [76-44-8] and Heptachlor epoxide [1024-57-3]	0.05 mg/m³	—	Skin; A3	373.32 389.40	CNS; liver; blood
Heptane [7440-59-7] (n-Heptane)	400 ppm	500 ppm	—	100.20	Irritation; narcosis
Hexachlorobenzene [118-74-1]	0.002 mg/m³	—	Skin; A3	284.78	Liver; metabolic disorders
Hexachlorobutadiene [87-68-3]	0.02 ppm	—	Skin; A3	260.76	Irritation; kidney
Hexachlorocyclo-pentadiene [77-47-4]	0.01 ppm	—	A4	272.75	Irritation; pulmonary edema
Hexachloroethane [67-72-1]	1 ppm	—	Skin; A3	236.74	Irritation; liver; kidney
Hexachloronapthalene [1335-87-1]	0.2 mg/m³	—	Skin	334.74	Liver; chloracne
Hexafluoroacetone [684-16-2]	0.1 ppm	—	Skin	166.02	Reproductive; kidney
Hexamethylene diisocyanate [822-06-0]	0.005 ppm	—	—	168.22	Irritation; sensitization
n-Hexane [110-54-3]	50 ppm	—	Skin; BEI	86.18	Neuropathy; CNS; irritation
Hexane, other isomers	500 ppm	1000 ppm	—	86.18	CNS; irritation
1,6-Hexanediamine [124-09-4]	0.5 ppm	—	—	116.21	Irritation
1-Hexene [592-41-6]	50 ppm	—	—	84.16	CNS; reproductive
sec-Hexyl acetate [108-84-9]	50 ppm	—	—	144.21	Irritation
Hydrazine [302-01-2]	0.01 ppm	—	Skin; A3	32.05	Irritation; liver
Hydrogen [1333-74-0]	Simple asphyxiant (D)			1.01	Asphyxiation
Hydrogenated terphenyls (nonirradiated) [61788-32-7]	0.5 ppm	—	—	241.00	Irritation; liver
* Hydrogen chloride [7647-01-0]	—	C 2 ppm	A4	36.47	Irritation; corrosion
Hydrogen cyanide and cyanide salts, as CN Hydrogen cyanide [74-90-8]	—	C 4.7 ppm	Skin	27.03	CNS; irritation; anoxia; lung; thyroid
Cyanide salts [592-01-8; 151-50-8; 143-33-9]	—	C 5 mg/m³	Skin	Varies	

Substance [CAS No.]	TWA	STEL	Notations	MW	TLV® Basis-Critical Effect(s)
Hydrogen fluoride [7664-39-3], as F	—	C 3 ppm	BEI	20.01	Irritation; bone; teeth; fluorosis
Hydrogen peroxide [7722-84-1]	1 ppm	—	A3	34.02	Irritation; pulmonary edema; CNS
Hydrogen selenide [7783-06-4]	0.05 ppm	—	—	80.98	Irritation; GI
‡ Hydrogen sulfide [7783-06-4]	(10 ppm)	(15 ppm)		34.08	(Irritation; CNS)
Hydroquinone [123-31-9]	2 mg/m^3	—	A3	110.11	CNS; dermatitis; ocular
2-Hydroxypropyl acrylate [999-61-1]	0.5 ppm	—	Skin; SEN	130.14	Irritation
Indene [95-13-6]	10 ppm	—		116.15	Irritation; liver; kidney
Indium [7440-74-6] and compounds, as In	0.1 mg/m^3	—	—	49.00	Pulmonary edema; bone; GI
Iodine [7553-56-2]	—	C 0.1 ppm	—	253.81	Irritation
Iodoform [75-47-8]	0.6 ppm	—	—	393.78	CNS; liver; kidney; CVS
Iron oxide (Fe$_2$O$_3$) [1309-37-1] dust & fume, as Fe	5 mg/m^3	—	A4	159.70	Pneumoconiosis
Iron pentacarbonyl [13463-40-6]	0.1 ppm	0.2 ppm	—	195.90	Pulmonary edema; CNS
Iron salts, soluble, as Fe	1 mg/m^3	—	—	Varies	Irritation
Isoamyl alcohol [123-51-3]	100 ppm	125 ppm	—	88.15	Irritation
Isobutanol [78-83-1]	50 ppm	—	—	74.12	Irritation; ocular
Isobutyl acetate [110-19-0]	150 ppm	—	—	116.16	Irritation
* Isoubutyl nitrite [542-56-3]	—	C 1 ppm$^{(I,V)}$	A3; BEI$_2$	103.12	Anoxia; blood
Isooctyl alcohol [26952-21-6]	50 ppm	—	Skin	130.23	Irritation
Isophorone [78-59-1]	—	C 5 ppm	A3	138.21	Irritation; narcosis
Isophorone diisocyanate [4098-71-9]	0.005 ppm	—	—	222.30	Dermatitis; asthma; sensitization
* Isopropanol [67-63-0]	200 ppm	400 ppm	A4	60.09	Irritation; CNS
2-Isopropoxyethanol [109-59-1]	25 ppm	—	Skin	104.15	Blood
* Isopropyl acetate [108-21-4]	100 ppm	200 ppm	—	102.13	Irritation; eye
Isopropylamine [75-31-0]	5 ppm	10 ppm	—	59.08	Irritation
N-Isopropylaniline [768-52-5]	2 ppm	—	Skin; BEI$_2$	135.21	Blood
Isopropyl ether [108-20-3]	250 ppm	310 ppm	—	102.17	Irritation
Isopropyl glycidyl ether (IGE) [4016-14-2]	50 ppm	75 ppm	—	116.18	Irritation
Kaolin [1332-58-7]	2 mg/m$^{3(E,R)}$	—	A4	—	Pneumoconiosis
* Kerosene [8008-20-6]/jet fuels [64742-47-8], as total hydrocarbon vapor	200 mg/m$^{3(P)}$	—	Skin; A3	Varies	Irritation; CNS; Skin
Ketene [463-51-4]	0.5 ppm	1.5 ppm	—	42.05	Lung irritation; lung edema
Lead [7439-92-1] and inorganic compounds, as Pb	0.05 mg/m^3	—	A3; BEI	207.20 Varies	CNS; blood; kidney; reproductive

Substance [CAS No.]	TWA	STEL	Notations	MW	TLV® Basis-Critical Effect(s)
Lead arsenate [3687-31-8], as $Pb_3(AsO_4)_2$	0.15 mg/m³	—	A3; BEI	347.13	CNS; anemia; kidney; reproductive
Lead chromate [7758-97-6], as Pb as Cr	0.05 mg/m³ 0.012 mg/m³	— —	A2; BEI A2	323.22	Cancer; CVS; reproductive
Lindane [58-89-9]	0.5 mg/m³	—	Skin; A3	290.85	CNS; liver
Lithium hydride [7580-67-8]	0.025 mg/m³	—	—	7.95	Irritation
‡ (LPG (Liquefied petroleum gas) [68476-85-7])	(1000 ppm)	(-)	(-)	(42-58)	(Asphyxiation)
Magnesite [546-93-0]	10 mg/m³(E)	—	—	84.33	Irritation; pneumoconiosis
* Magnesium oxide [1309-48-4]	10 mg/m³(I)	—	A4	40.32	Irritation; metal fume fever
* Malathion [121-75-5]	1 mg/m³(I,V)	—	Skin; A4; BEI₁	330.36	Cholinergic
Maleic anhydride [108-31-6]	0.1 ppm	—	SEN; A4	98.06	Irritation; asthma
‡ Manganese [7439-96-5] and inorganic compounds, as Mn	(0.2 mg/m³)	—	—	54.94 Varies	CNS (manganism); lung; reproductive
Manganese cyclopentadienyl tricarbonyl [12079-65-1], as Mn	0.1 mg/m³	—	Skin	204.10	CNS; pulmonary edema
Mercury [7439-97-6], as Hg					
Alkyl compounds	0.01mg/m³	0.03 mg/m³	Skin	200.59	CNS
Aryl compounds	0.1mg/m³	—	Skin	Varies	CNS; neuropathy; vision; kidney
Elemental and inorganic forms	0.025 mg/m³	—	Skin; A4; BEI	Varies Varies	CNS; kidney; reproductive
Mesityl oxide [141-79-7]	15 ppm	25 ppm	—	98.14	Irritation; narcosis; liver; kidney
Methacrylic acid {79-41-4]	20 ppm	—	—	86.09	Irritation
‡ (Methane [74-82-8])		(Simple asphyxiant (D))		(16.04)	(Asphyxiation)
Methanol [67-56-1]	200 ppm	250 ppm	Skin; BEI	32.04	Neuropathy; vision; CNS
Methomyl [16752-77-5]	2.5 mg/m³	—	A4; BEI₁	162.20	Cholinergic
Methoxychlor [72-43-5]	10 mg/m³	—	A4	345.65	CNS; liver
2-Methoxyethanol (EGME) [109-86-4]	5 ppm	—	Skin; BEI	76.09	Blood; reproductive; CNS
2-Methoxyethyl acetate (EGMEA) [110-49-6]	5 ppm	—	Skin; BEI	118.13	Blood; reproductive; CNS
4-Methoxyphenol [150-76-5]	5 mg/m³	—	—	124.15	Ocular; depigmentation
1-Methoxy-2-propanol (PGME) [107-98-2]	100 ppm	150 ppm	—	90.12	Irritation; anesthesia
bis-(2-Methoxypropyl) ether (DPGME) [34590-94-8]	100 ppm	150 ppm	Skin	148.20	Irritation; CNS
Methyl acetate [79-20-9]	200 ppm	250 ppm	—	78.04	Irritation; narcosis
Methyl acetylene [74-99-7]	1000 ppm	—	—	40.07	Anesthesia

Substance [CAS No.]	TWA	STEL	Notations	MW	TLV® Basis-Critical Effect(s)
Methyl acetylene-propadiene mixture (MAPP) [59355-75-8]	1000 ppm	1250 ppm	—	40.07	Anesthesia
Methyl acrylate [96-33-3]	2 ppm	—	Skin; SEN; A4	86.09	Irritation
Methylacrylonitrile [126-98-7]	1 ppm	—	Skin	67.09	Irritation; CNS
Methylal [109-87-5]	1000 ppm	—	—	76.10	Irritation; CNS
Methylamine [74-89-5]	5 ppm	15 ppm	—	31.06	Irritation
Methyl n-amyl ketone [110-43-0]	50 ppm	—	—	114.18	Irritation
N-Methyl aniline [100-61-8]	0.5 ppm	—	Skin; BEI_2	107.15	Anoxia; blood
Methyl bromide [74-83-9]	1 ppm	—	Skin; A4	94.95	Irritation
Methyl tert-butyl ether (MTBE) [1634-04-4]	50 ppm	—	A3	88.17	Reproductive; kidney
Methyl n-butyl ketone [591-78-6]	5 ppm	10 ppm	Skin; BEI	100.16	Neuropathy
Methyl chloride [74-87-3]	50 ppm	100 ppm	Skin; A4	50.49	Kidney; CNS; reproductive
Methyl chloroform [71-55-6]	350 ppm	450 ppm	A4; BEI	133.42	Anesthesia; CNS
Methyl 2-cyanoacrylate [137-05-3]	0.2 ppm	—	—	111.10	Irritation; dermatitis
Methylcyclohexane [108-87-2]	400 ppm	—	—	98.19	Narcosis; irritation
Methylcyclohexanol [25639-42-3]	50 ppm	—	—	114.19	Irritation; narcosis; liver; kidney
o-Methylcyclohexa-none [583-60-8]	50 ppm	75 ppm	Skin	112.17	Irritation; narcosis
2-Methylcyclopenta-dienyl manganese tricarbonyl [12108-13-3], as Mn	0.2 mg/m³	—	Skin	218.10	CNS; liver; kidney
Methyl demeton [8022-00-2]	0.5 mg/m³	—	Skin; BEI_1	230.30	Irritation; cholinergic
Methylene bisphenyl isocyanate (MDI) [101-68-8]	0.005 ppm	—	—	250.26	Irritation; lung edema; sensitization
4,4'-Methylene bis(20chloroaniline) [MBOCA; MOCA®] [1-1-14-4]	0.01 ppm	—	Skin; A2; BEI	267.17	Anoxia; kidney; cancer (bladder)
Methylene bis(4-cyclohexliso-cyanate) [5124-30-1]	0.005 ppm	—	—	262.35	Irritation; sensitization
4,4'-Methylene dianiline [101-77-9]	0.1 ppm	—	Skin; A3	198.26	Liver
Methyl ethyl ketone (MEK) [78-93-3]	200 ppm	300 ppm	BEI	72.10	Irritation; CNS
Methyl formate [107-31-3]	100 ppm	150 ppm	—	65.05	Irritation; narcosis; lung edema
Methyl hydrazine [60-34-4]	0.01 ppm	—	Skin; A3	46.07	Irritation; liver
Methyl iodide [74-88-4]	2 ppm	—	Skin	141.95	CNS; irritation
Methyl isoamyl ketone [110-12-3]	50 ppm	—	—	114.95	Irritation; narcosis; liver; kidney
Methyl isobutyl carbinol [108-11-2]	25 ppm	40 ppm	Skin	102.18	Irritation; anesthesia
Methyl isobutyl ketone [108-10-1]	50 ppm	75 ppm	BEI	100.16	Irritation; kidney

Substance [CAS No.]	TWA	STEL	Notations	MW	TLV® Basis-Critical Effect(s)
Methyl isocyanate [624-83-9]	0.02 ppm	—	Skin	57.05	Irritation; lung edema; sensitization
Methyl isopropyl ketone [563-80-4]	200 ppm	—	—	86.14	Irritation
Methyl mercaptan [74-93-1]	0.5 ppm	—	—	48.11	Irritation; CNS
Methyl methacrylate [80-62-6]	50 ppm	100 ppm	SEN; A4	100.13	Irritation; dermatitis
Methyl parathion [298-00-0]	0.2 mg/m^3	—	Skin; A4; BEI$_1$	263.23	Cholinergic
Methyl propyl ketone [107-87-9]	200 ppm	250 ppm	—	86.17	Irritation; narcosis
Methyl silicate [681-84-5]	1 ppm	—	—	152.22	Ocular; lung
α-Methyl styrene [98-83-9]	50 ppm	100 ppm	—	118.18	Irritation; dermatitis; CNS
Metribuzin [21087-64-9]	5 mg/m^3	—	A4	214.28	Blood; liver
* Mevinphos [7786-34-7]	0.01 mg/m$^{3(I,V)}$	—	Skin; A4; BEI$_1$	224.16	Cholinergic
Mica [12001-26-2]	3 mg/m$^{3®}$	—	—	—	Pneumoconiosis
Molybdenum [7439-98-7], as Mo				95.95	
Soluble compounds	0.5 mg/m$^{3®}$	—	A3		Lung; irritation
Metal and insoluble compounds	10 mg/m^{3} (I) 3 mg/m$^{3®}$	— —	— —		Lung; CNS Lung; CNS
Monocrotophos [6923-22-4]	0.05 mg/m$^{3(I,V)}$	—	Skin; A4; BEI$_1$	223.16	Cholinergic
Morpholine [110-91-8]	20 ppm	—	Skin; A4	87.12	Irritation; vision
Naled [300-76-5]	0.1 mg/m$^{3(I,V)}$	—	Skin; SEN; A4; BEI$_1$	380.79	Cholinergic; dermatitis
Naphthalene [91-20-3]	10 ppm	15 ppm	Skin; A4	128.19	Irritation; ocular; blood
-Naphthylamine [91-59-8]	—$^{(L)}$	—	A1	143.18	Cancer (bladder)
Neon [7440-01-9]	Simple asphyxiant $^{(D)}$			20.18	Asphyxiation
Nickel, as Ni Elemental [7440-02-0]	1.5 mg/m^3 (I)	—	A5	58.71	Dermatitis; pneumoconiosis
Soluble inorganic compounds (NOS)	0.1 mg/m^3 (I)	—	A4	Varies	CNS; irritation; dermatitis
Insoluble inorganic compounds (NOS)	0.2 mg/m^3 (I)	—	A1	Varies	Cancer; lung; irritation; dermatitis
Nickel subsulfide [12035-72-2], as Ni	0.1 mg/m^3 (I)	—	A1	240.19	Cancer; lung; irritation; dermatitis
Nickel carbonyl [13463-39-3], as Ni	0.05 ppm	—	—	170.73	Irritation; CNS
Nicotine [54-11-5]	0.5 mg/m^3	—	Skin	162.23	CVS; GI; CNS
Nitrapyrin [1929-82-4]	10 mg/m^3	20 mg/m^3	A4	230.93	Liver
Nitric acid [7697-37-2]	2 ppm	4 ppm	—	63.02	Irritation; corrosion; pulmonary edema
Nitric oxide [10102-43-9]	25 ppm	—	BEI$_2$	30.01	Anoxia; irritation; cyanosis
p-Nitroaniline [100-01-6]	3 mg/m^3	—	Skin; A4; BEI$_2$	138.12	Anoxia; anemia; liver
Nitrobenzene [98-95-3]	1 ppm	—	Skin; A3; BEI	123.11	Anoxia
p-Nitrochlorobenzene [100-00-5]	0.1 ppm	—	Skin; A3; BEI$_2$	157.56	Anoxia; blood; liver

Substance [CAS No.]	TWA	STEL	Notations	MW	TLV® Basis-Critical Effect(s)
4-Nitrodiphenyl [92-93-3]	—(L)	—	Skin; A2	199.20	Cancer (bladder)
Nitroethane [79-24-3]	100 ppm	—	—	75.07	Irritation; narcosis; liver
Nitrogen [7727-37-9]	Simple asphyxiant (D)			14.01	Asphyxiation
Nitrogen dioxide [10102-44-0]	3 ppm	5 ppm	A4	46.01	Irritation; pulmonary edema
Nitrogen trifluoride [7783-54-2]	10 ppm	—	BEI$_2$	71.00	Anoxia; blood; liver; kidney
Nitroglycerin (NG) [55-63-0]	0.05 ppm	—	Skin	227.09	CVS
Nitromethane [75-52-5]	20 ppm	—	A3	61.04	Thyroid
1-Nitropropane [108-03-2]	25 ppm	—	A4	89.09	Irritation; liver
2-Nitropropane [79-46-9]	10 ppm	—	A3	89.09	Liver; cancer
N-Nitrosodimethyl-amine [62-75-9]	—(L)	—	Skin; A3	74.08	Liver
Nitrotoluene, all isomers [88-72-2; 99-08-1; 99-99-0]	2 ppm	—	Skin; BEI$_2$	137.13	Anoxia; cyanosis
Nitrous oxide [10024-97-2]	50 ppm	—	A4	44.02	Reproductive; blood; CNS
Nonane [111-84-2], all isomers	200 ppm	—	—	128.26	CNS; skin; irritation
Octachloronaphthalene [2234-13-1]	0.1 mg/m³	0.3 mg/m³	Skin	403.74	Liver; dermatitis
Octane, all isomers [111-65-9]	300 ppm	—	—	114.22	Irritation
‡ (Oil mist, mineral)	(5 mg/m³(O))	(10 mg/m³)	(-)	—	(Lung)
Osmium tetroxide [20816-12-0]	0.0002 ppm	0.0006 ppm	—	254.20	Irritation; vision
Oxalic acid [144-62-7]	1 mg/m³	2 mg/m³	-	90.04	Irritation; burns
p,p'-Oxybis(benzene-sulfonyl hydrazide) [80-51-3]	0.1 mg/m³ (I)	—		326.00	Irritation
Oxygen difluoride [7783-41-7]	—	C 0.05 ppm	—	54.00	Irritation; kidney
Ozone [10028-15-6]				48.00	Lung function; irritation
Heavy work	0.05 ppm	—	A4		
Moderate work	0.08 ppm	—	A4		
Light work	0.10 ppm	—	A4		
Heavy, moderate, or light workloads (2 hours)	0.20 ppm	—	A4		
Paraquat [4685-14-7]	0.5 mg/m³ / 0.1 mg/m³®	— / —	—	257.18	Lung; irritation
* Parathion [56-38-2]	0.05 mg/m³(I,V)		Skin; A4; BEI	291.27	Cholinergic
* Particles (Insoluble or Poorly Soluble) Not Otherwise Specified	see Appendix E				
Pentaborane [19624-22-7]	0.005 ppm	0.015 ppm	—	63.17	CNS
Pentachloronaphthal-ene [1321-64-8]	0.5 mg/m³	—	Skin	300.40	Chloracne; liver
Pentachloronitroben-zene [83-68-8]	0.5 mg/m³	—	A4	295.36	Liver
Pentachlorophenol [87-86-5]	0.5 mg/m³	—	Skin; A3; BEI	266.35	CVS; CNS
Pentaerythritol [115-77-5]	10 mg/m³	—	—	136.15	Irritation

Substance [CAS No.]	TWA	STEL	Notations	MW	TLV® Basis-Critical Effect(s)
Pentane, all isomers [78-78-4; 109-66-0; 463-82-1]	600 ppm	—	—	72.15	Irritation; narcosis
Pentyl acetate, all isomers [628-63-7; 626-38-0; 123-92-2; 625-16-1; 624-41-9; 620-11-1]	50 ppm	100 ppm	—	130.20	Irritation
Perchloromethyl mercaptan [594-42-3]	0.1 ppm	—	—	185.87	Irritation; pulmonary edema
Perchloryl fluoride [7616-94-6]	3 ppm	6 ppm	—	102.46	Irritation; blood
Perlite [93763-70-3]	10 mg/m$^{3(E)}$	—	A4	—	Irritation
Persulfates, as persulfate	0.1 mg/m^3	—	—	Varies	Irritation
Phenol [108-95-2]	5 ppm	—	Skin; A4; BEI	94.11	Irritation; CNS; blood
Phenothiazine [92-84-2]	5 mg/m^3	—	Skin	199.26	Irritation; ocular; liver; kidney
N-Pheyl-beta-naphthyl amine [135-88-6]	—	—	A4	219.29	Irritation
o-Phenylenediamine [95-54-5]	0.1 mg/m^3	—	A3	108.05	Irritation; liver; blood
m-Phenylenediamine [108-45-2]	0.1 mg/m^3	—	A4	108.05	Irritation; liver
p-Phenylenediamine [106-50-3]	0.1 mg/m^3	—	A4	108.05	Sensitization; skin; eye
Phenyl ether [101-84-8], vapor	1 ppm	2 ppm	—	170.20	Irritation; nausea
Phenyl glycidyl ether (PGE) [122-60-1]	0.1 ppm	—	Skin; SEN; A3	150.17	Irritation; dermatitis
Phenylhydrazine [100-63-0]	0.1 ppm	—	Skin; A3	108.14	Dermatitis; anemia
‡ Phenyl mercaptan [108-98-5]	(0.5 ppm)	—	(-)	110.18	Irritation; (dermatitis)
Phorate [298-02-2]	0.05 mg/m^3	0.2 mg/m^3	Skin; BEI$_1$	260.40	Cholinergic
Phosgene [75-44-5]	0.1 ppm	—	—	98.92	Irritation; anoxia; lung edema
Phosphine [7803-51-2]	0.3 ppm	1 ppm	—	34.00	Irritation; CNS; GI
Phosporic acid [7664-38-2]	1 mg/m^3	3 mg/m^3	—	98.00	Irritation
Phosphorus (yellow) [12185-10-3]	0.1 mg/m^3	—	—	123.92	Irritation; liver; kidney; CVS; GI
Phosphorus oxychloride [10025-87-3]	0.1 ppm	—	—	153.35	Irritation; kidney
Phosphorus pentachloride [10026-13-8]	0.1 ppm	—	—	208.24	Irritation
Phosphorus oxychloride [10025-87-3]	0.1 ppm	—	—	153.35	Irritation; kidney
Phosphorus pentachloride [10026-13-8]	0.1 ppm	—	—	208.24	Irritation
‡ Pyridine [110-86-1]	(5 ppm)	—	(-)	79.10	Irritation; CNS; liver; kidney; (blood)
Quinone [106-51-4]	0.1 ppm	—	—	108.09	Irritation; eyes
Resorcinol [108-46-3]	10 ppm	20 ppm	A4	110.11	Irritation; dermatitis; blood

Substance [CAS No.]	TWA	STEL	Notations	MW	TLV® Basis-Critical Effect(s)
Rhodium [7440-16-6], as Rh				102.91	
Metal and insoluble compounds	1 mg/m^3	—	A4	Varies	Irritation
Soluble compounds	0.01 mg/m^3	—	A4	Varies	Irritation
Ronnel [299-84-3]	10 mg/m^3	—	A4; BEI$_1$	321.57	Cholinergic
Rosin core solder thermal decomposition products (colophoney) [8050-09-7]	—$^{(L)}$	—	SEN	NA	Irritation; asthma; sensitization
Rotenone (commercial) [83-79-4]	5 mg/m^3	—	A4	391.41	Irritation; CNS
Rouge	10 mg/m$^{3(E)}$	—	A4	159.70	Lung; siderosis; irritation
Rubber solvent (Naphtha) [8030-30-6]	400 ppm	—	—	97 (mean)	Irritation; CNS
Selenium [7782-49-2] and compounds, as Se	0.2 mg/m^3	—	—	78.96	Irritation
Selenium hexafluoride [7783-79-1]	0.05 ppm	—	—	192.96	Pulmonary edema
Sesone [136-78-7]	10 mg/m^3	—	A4	309.13	Irritation
Silica, Amorphous- Diatomaceous earth (uncalcined) [61790-53-2]	10 mg/m$^{3(E,I)}$ 3 mg/m$^{3(E,R)}$	— —	— —	—	Irritation; pneumoconiosis
Precipitated silica and silica gel [112926-00-8]	10 mg/m^3	—	—	—	Irritation
Silica fume [69012-64-2]	2 mg/m$^{3®}$	—	—	—	Irritation; fever
Silica, fused [60676-86-0]	0.1 mg/m$^{3®}$	—	—	60.08	Lung fibrosis
Silica, crystalline- Cristobalite [14464-46-1]	0.05 mg/m$^{3®}$	—	—	60.08	Lung fibrosis; silicosis
Quartz [14808-60-7]	0.05 mg/m$^{3®}$	—	A2	60.08	Silicosis; lung function; lung fibrosis; cancer
Tridymite [15468-32-3]	0.05 mg/m$^{3®}$	—	—	60.08	Lung fibrosis; silicosis
Tripoli [1317-95-9], as quartz	0.1 mg/m$^{3®}$	—	—	—	Lung fibrosis
Silicon [7440-21-3]	10 mg/m^3	—	—	28.09	Lung
Silicon carbide [409-21-2]				40.10	
* Nonfibrous	10 mg/m$^{3(I,E)}$	—	—		Lung function
* Fibrous (including whiskers)	3 mg/m$^{3(R,E)}$ 0.1 f/cc $^{(F)}$	—	A2		Lung function Lung fibrosis;cancer
Silver [7440-22-4]					Argyria (skin,
Metal	0.1 mg/m^3	—	—	107.87	eyes, mucosa)
Soluble compounds, as Ag		—	—	Varies	
Soapstone	6 mg/m$^{3(E)}$ 3 mg/m$^{3(E,R)}$	— —	— —	—	Pneumoconiosis
Sodium azide [26628-22-8]				65.02	
as sodium azide	—	C 0.29 mg/m^3	A4		CNS; CVS; lung
as hydrazoic acid vapor	—	C 0.11 ppm	A4		CNS; CVS; lung
Sodium bisulfide [7631-90-5]	5 mg/m^3	—	A4	104.07	Irritation

Substance [CAS No.]	TWA	STEL	Notations	MW	TLV® Basis-Critical Effect(s)
Sodium fluoroacetate [62-74-8]	0.05 mg/m³	—	Skin	100.02	CNS; CVS
Sodium hydroxide [1310-73-2]	—	C 2 mg/m³	—	40.01	Irritation
Sodium metabisulfite [7681-57-4]	5 mg/m³	—	A4	190.13	Irritation
Starch [9005-25-8]	10 mg/m³	—	A4	—	Dermatitis; lung
Stearates(J)	10 mg/m³	—	—	Varies	Irritation
Stoddard solvent [8052-41-3]	100 ppm	—	—	140.00	Irritation; narcosis; kidney
Strontium chromate [7789-06-2], as Cr	0.0005 mg/m³	—	A2	203.61	Cancer (lung)
Strychnine [57-24-9]	0.15 mg/m³	—	—	334.40	CNS
Styrene, monomer [100-42-5]	20 ppm	40 ppm	A4; BEI	104.16	Neurotoxicity; irritation; CNS
Sucrose [57-50-1]	10 mg/m³	—	A4	342.30	Lung
Sulfometuron methyl [74222-97-2]	5 mg/m³	—	A4	364.38	Irritation; blood
Sulfotep (TEDP) [3689-24-5]	0.2 mg/m³	—	Skin; A4; BEI₁	322.30	Cholinergic
Sulfur dioxide [7446-09-5]	2 ppm	5 ppm	A4	64.07	Irritation
Sulfur hexafluoride [2551-62-4]	1000 ppm	—	—	146.07	Asphyxiation
‡ Sulfuric acid [7664-09-5]	(1 mg/m³)	(3 mg/m³)	A2 (M)	98.08	Irritation; cancer (larynx)
Sulfur monochloride [10025-67-9]	—	C 1ppm	—	135.03	Irritation
Sulfur pentafluoride [5714-22-7]	—	C 0.01 ppm	—	254.11	Irritation
Sulfur tetrafluoride [7783-60-0]	—	C 0.1 ppm	—	108.07	Irritation
Sulfuryl fluoride [2699-79-8]	5 ppm	10 ppm	—	102.07	Irritation; CNS
Sulprofos [35400-43-2]	1 mg/m³	—	A4; BEI₁	322.43	Cholinergic
Synthetic vitreous fibers	1 t/cc (F)	—	A4	—	Irritation
Continuous filament glass fibers	5 mg/m³(I)	—	A4	—	Irritation
Continuous filament glass fibers	1 t/cc (F)	—	A3	—	Irritation; lung
Continuous filament glass fibers	1 t/cc (F)	—	A3	—	Irritation; lung
Glass wool fibers	1 t/cc (F)	—	A3	—	Irritation; lung
Rock wool fibers	0.2 t/cc (F)	—	A2	—	Lung fibrosis; cancer
Slag wool fibers					
Special purpose glass fibers					
Refractory ceramic fibers					
2,4,5-T [93-76-5]	10 mg/m³	—	A4	255.49	Irritation
Talc [14807-96-6]					
Containing no asbestos fibers	2 mg/m³(E,R)	—	A4	—	Lung
Containing asbestos fibers	Use asbestos TLV(K)	—	A1	—	Asbestos; cancer
Tantalum [7440-25-7] and tantalum oxide [1314-61-0] dusts, as Ta	5 mg/m³	—	—	180.95 / 441.90	Irritation; lung / Irritation; lung
Tellurium [13494-80-9] and compounds (NOS), as Te, excluding hydrogen telluride	0.1 mg/m³	—	—	127.60	CNS; cyanosis; liver
Tellurium hexafluoride [7783-80-4]	0.02 ppm	—	—	241.61	Irritation
Temephos [3383-96-8]	10 mg/m³	—	BEI₁	466.46	Cholinergic

Substance [CAS No.]	TWA	STEL	Notations	MW	TLV® Basis-Critical Effect(s)
Terbufos [13071-79-9]	0.1 mg/m$^{3(I,V)}$	—	Skin; A4	288.45	Cholinergic
Terephthalic acid [100-21-0]	10 mg/m^3	—	—	166.13	Lung; urinary
1,1,1,2-Tetrachloro-2,2-difluoroethane [76-11-9]	500 ppm	—	—	203.83	Liver; blood
1,1,2,2-Tetrachloro-1,2-difluoroehtane [76-12-0]	500 ppm	—	—	203.93	CNS; pulmonary edema
1,1,2,2-Tetrachloro-ethane [79-34-5]	1 ppm	—	Skin; A3	167.86	Liver; CNS; GI
Tetrachloroethylene [127-18-4] (perchloroethylene)	25 ppm	100 ppm	A3; BEI	165.80	Irritation; CNS
Tetrachloronaphthalene [1335-88-2]	2 mg/m^3	—	—	265.96	Liver
Tetraethyl lead [78-00-2], as Pb	0.1 mg/m^3	—	Skin; A4	323.45	CNS
Tetraethyl pyrophosphate (TEPP) [107-49-3]	0.05 mg/m^3	—	Skin; BEI$_1$	290.20	Cholinergic
Tetrafluoroethylene [116-14-3	2 ppm	—	A3	100.29	Kidney; liver
Tetrahydrofuran [109-99-9]	200 ppm	250 ppm	BEI	72.10	Irritation; narcosis
Tetramethyl lead [75-74-1], as Pb	0.15 mg/m^3	—	Skin	267.33	CNS
Tetramethyl succinonitrile [3333-52-6]	0.5 ppm	—	Skin	136.20	CNS
Tetranitromethane [509-14-8]	0.005 ppm	—	A3	196.04	Irritation
Tetrasodium pyrophosphate [7722-88-5]	5 mg/m^3	—	—	265.94	Irritation
Tetryl [479-45-8]	1.5 mg/m^3	—	—	287.15	Liver; dermatitis; sensitization
Thallium [7440-28-0] and soluble compounds, as Tl	0.1 mg/m^3	—	Skin	204.37 Varies	Irritation; CNS; CVS
4,4'-Thiobis (6-*tert*-butyl-*m*-cresol) [96-69-5]	10 mg/m^3	—	A4	358.52	Liver; kidney
Thioglycolic acid [68-11-1]	1 ppm	—	Skin	92.12	Irritation
Thionyl chloride [7719-09-7]	—	C 1ppm	—	118.98	Irritation
Thiram [137-26-8]	1 mg/m^3	—	A4	240.44	Irritation
Tin [7440-31-5], as Sn Metal	2 mg/m^3	—	—	118.69	Stannosis
Oxide & inorganic compounds, except tin hydride	2 mg/m^3 0.1 mg/m^3	— 0.2 mg/m^3	— Skin; A4	Varies Varies	Stannosis CNS; immunotoxicity; irritation
Organic compounds					
Titanium dioxide [13463-67-7]	10 mg/m^3	—	A4	79.90	Lung
o-Tolidine [119-93-7]	—	—	Skin; A3	212.28	Liver; kidney; blood
Toluene [108-88-3]	50 ppm	—	Skin; A4; BEI	92.13	CNS
‡ Toluene-2,4-diiso-cyanate (TDI) [584-84-9]	0.005 ppm	0.02 ppm	(-); A4	174.15	Irritation; sensitization
o-Toluidine [95-53-4]	2 ppm	—	Skin; A3; BEI$_3$	107.15	Anoxia; kidney
m-Toluidine [108-44-1]	2 ppm	—	Skin; A4; BEI$_2$	107.15	Anoxia; kidney
p-Toluidine [106-49-0]	2 ppm	—	Skin; A3; BEI$_2$	107.15	Anoxia; kidney

Substance [CAS No.]	TWA	STEL	Notations	MW	TLV® Basis-Critical Effect(s)
Tributyl phosphate [126-73-8]	0.2 ppm	—	BEI$_1$	266.32	Irritation; cholinergic
Trichloroacetic acid [76-03-9]	1 ppm	—	A3	163.39	Irritation
1,1,2-Trichloroethane [79-00-5]	10 ppm	—	Skin; A3	133.41	CNS; liver
Trichloroethylene [79-01-6]	50 ppm	100 ppm	A5; BEI	131.40	CNS; headache; liver
Trichlorofluoromethane [75-69-4]	—	C 1000 ppm	A4	137.38	CVS; CNS
Trichloronapthalene [1321-65-9]	5 mg/m^3	—	Skin	231.51	Liver
1,2,3-Trichloropropane [96-18-4]	10 ppm	—	Skin; A3	147.43	Liver; kidney
1,1,2-Trichloro-1,2,2-tribfluoroethane [76-13-1]	1000 ppm	1250 ppm	A4	187.40	Narcosis; CVS; asphyxiation
* Trichlorphon [52-68-6]	1 mg/m$^{3\ (I)}$	—	A4; BEI$_1$	257.60	Cholinergic
Triethanolamine [102-71-6]	5 mg/m^3	—	—	149.22	Irritation; liver; kidney
Triethylamine [121-44-8]	1 ppm	3 ppm	Skin; A4	101.19	Irritation; vision
Trifluorobromoethane [75-63-8]	1000 ppm	—	—	148.92	CNS; CVS
1,3,5-Triglycidyl-s-triazinetrione [2451-62-9]	0.05 mg/m^3	—	—	297.25	Blood; reproductive; dermatitis; sensitization
Trimethylamine [75-50-3]	5 ppm	15 ppm	—	59.11	Irritation
Trimethyl benzene (mixed isomers) [25551-13-7]	25 ppm	—	—	120.19	Irritation; CNS; blood
Trimethyl phospite [121-45-9]	2 ppm	—	—	124.08	Irritation
2,4,6-Trinitrotoluene (TNT) [118-96-7]	0.1 mg/m^3	—	Skin; BEI$_2$	227.13	Irritation; liver; blood; ocular
Triorthocresyl phosphate [78-30-8]	0.1 mg/m^3	—	Skin; A4; BEI$_1$	368.37	CNS; cholinergic
Triphenyl amine [603-34-9]	5 mg/m^3	—	—	245.33	Irritation
Triphenyl phosphate [115-86-6]	3 mg/m^3	—	A4	326.28	Irritation; dermatitis
Tungsten [7440-33-7], as W Metal and insoluble compounds Soluble compounds	5 mg/m^3 1 mg/m^3	10 mg/m^3 3 mg/m^3	— —	183.85 Varies Varies	Irritation CNS; irritation
* Turpentine [8006-64-2] and selected monoterpenes [80-56-8[127-91-3; 13466-78-9]	20 ppm	—	SEN; A4	136.00 Varies	Irritation; lung
Uranium (natural) [7440-61-1] Soluble and insoluble compounds, as U	0.2 mg/m^3	0.6 mg/m^3	A1	238.03 Varies	Kidney; blood; cancer
n-Valeraldehyde [110-62-3]	50 ppm	—	—	86.13	Irritation
Vanadium pentoxide [1314-62-1], as V$_2$O$_5$ Dust or fume	0.05 mg/m$^{3®}$	—	A4; BEI	181.90	Irritation; lung
Vegetable oil mists $^{(N)}$	10 mg/m^3	—	—	—	Lung

Substance [CAS No.]	TWA	STEL	Notations	MW	TLV® Basis-Critical Effect(s)
Vinyl acetate [108-05-4]	10 ppm	15 ppm	A3	86.09	Irritation
Vinyl bromide [593-60-2]	0.5 ppm	—	A2	106.96	Liver; CNS; cancer
Vinyl chloride [75-01-4]	1 ppm	—	A1	62.50	Cancer (liver)
4-Vinyl cyclohexene [100-40-3]	0.1 ppm	—	A3	108.18	Irritation; CNS; reproductive
Vinyl cyclohexene dioxide [106-87-6]	0.1 ppm	—	Skin; A3	140.18	Irritation; dermatitis; reproductive
Vinyl fluoride [75-02-5]	1 ppm	—	A2	46.05	Liver; cancer
* N-Vinyl-2-pyrrolidone [88-12-0]	0.05 ppm	—	A3	111.16	Liver; ototoxicity
Vinylidene chloride [75-35-4]	5 ppm	—	A4	96.95	CNS: liver; kidney
Vinylidene fluoride [75-38-7]	500 ppm	—	A4	64.04	Liver
Vinyl toluene [25013-15-4]	50 ppm	100 ppm	A4	118.18	Irritation
VM & P naphtha [8032-32-4]	300 ppm	—	A3	114.00	Irritation; CNS
Warfarin [81-81-2]	0.1 mg/m³	—	—	308.32	Blood; bleeding
Welding fumes (NOS)	5 mg/m³	—	B2	—	Lung; metal fume fever; irritation
‡ Wood dust ‡ (Certain hard woods as beech & oak) ‡ (Soft wood)	(1 mg/m³) (5 mg/m³)	(10 mg/m³)	A1 (-)	— —	Cancer; irritation; mucostasis; dermatitis Irritation; dermatisis; lung
Xylene [1330-20-7] (o, m & p isomers) [95-47-6; 108-38-3; 106-42-3]	100 ppm	150 ppm	A4; BEI	106.16	Irritation
m-Xylene α, α'-diamine [1477-55-0]	—	C 0.1 mg/m³	Skin	136.20	Irritation; blood
Xylidine (mixed isomers) [1300-73-8]	0.5 ppm [I,V]	—	Skin; A3; BEI₂	121.18	Cancer; genotoxic
Yttrium [7440-65-5] and compounds, as Y	1 mg/m³	—	—	88.91	Fibrosis
Zinc chloride fume [7646-85-7]	1 mg/m³	2 mg/m³	—	136.29	Irritation; lung edema
Zinc chromates [13530-65-9; 11103-86-9; 37300-23-5], as Cr	0.01 mg/m³	—	A1	Varies	Cancer (lung)
* Zinc oxide [1314-13-2]	2 mg/m³ ®	10 mg/m³ ®	—	81.37	Metal fume fever
Zirconium [7440-67-7] and compounds, as Zr	5 mg/m³	10 mg/m³	A4	91.22	Lung

Endnotes and Abbreviations

*	2003 adoption.
‡	See Notice of Intended Changes (NIC).
()	Adopted values enclosed are those for which changes are proposed in the NIC.
†	2003 Revision or Addition to the Notice of Intended Changes.
A	Refers to Appendix A : "Carcinogens."
B	Refers to Appendix B: "Substances of Variable Composition."
C	Ceiling limit; see definition in "Introduction to the Chemical Substances."
(D)	See definition in "Introduction to the Chemical Substances."
(E)	The value is for particulate matter containing no asbestos and < 1% crystalline silica.
(F)	Respirable fibers: length > 5 µm; aspect ratio 3:1, as determined by the membrane filter method at 400 to 450X magnification (4-mm objective), using phase-contrast illumination.
(G)	As measured by the vertical elutriator, cotton-dust sampler; see the TLV® documentation.
(H)	Aerosol only.
(I)	Inhalable fraction; see Appendix D, paragraph A.
(J)	Does not include stearates of toxic metals.
(K)	Should not exceed 2 mg/m^3 respirable particulate.
(L)	Exposure by all routes should be carefully controlled to levels as low as possible.
(M)	Classification refers to sulfuric acid contained in strong inorganic acid mists.
(N)	Except castor, cashew nut, or similar irritant oils.
(O)	Sampled by method that does not collect vapor.
(P)	Application restricted to conditions in which there are negligible aerosol exposures.
®	Respirable fraction; see Appendix D, paragraph C.
(T)	Thoratic fraction; see Appendix D, paragraph B.
(V)	Vapor and aerosol.
BEI	Substances for which there is a Biological Exposure Index or Indices (see BEI® section). BEI$_1$: see BEI® for acetylcholinesterase inhibiting pesticides. BEI$_2$: see BEI® for methemoglobin inducers.
CNS	Central nervous system.
CVS	Cardiovascular system.
GI	Gastrointestinal.
MW	Molecular weight.
NOS	Not otherwise specified.
SEN	Sensitizer; see definition in "Introduction to the Chemical Substances."
Skin	Danger of cutaneous absorption; see discussion in "Introduction to the Chemical Substances."
STEL	Short-term exposure limit; see definition in "Introduction to the Chemical Substances."
TWA	8-hour, time-weighted average; see definition in "Introduction to the Chemical Substances."
ppm	Parts of vapor or gas per million parts of contaminated air by volume at NTP conditions (25˚C; 760 torr).
mg/m^3	Milligrams of substance per cubic meter of air.

Thermal Physical Properties of Gases

T (K)	ρ (kg/m^e)	c_p (kJ/kg $\cdot$ K)	$\eta \cdot 10^7$ (N $\cdot$ s/m^2)	$k \cdot 10^3$ (W/m $\cdot$ K)	$D \cdot 10^6$ (m^2/s)
			Air		
100	3.5562	1.032	71.1	9.34	2.54
150	2.3364	1.012	103.4	13.8	5.84
200	1.7458	1.007	132.5	18.1	10.3
250	1.3947	1.006	159.6	22.3	15.9
300	1.1614	1.007	184.6	26.3	22.5
350	0.9950	1.009	208.2	30.0	29.9
400	0.8711	1.014	230.1	33.8	38.3
450	0.7740	1.021	250.7	37.3	47.2
500	0.6964	1.030	270.1	40.7	56.7
550	0.6329	1.040	288.4	43.9	66.7
600	0.5804	1.051	305.8	46.9	76.9
650	0.5356	1.063	322.5	49.7	87.3
700	0.4975	1.075	338.8	52.4	98
750	0.4643	1.087	354.6	54.9	109
800	0.4354	1.099	369.8	57.3	120
850	0.4097	1.110	384.3	59.6	131
900	0.3868	1.121	398.1	62	143
950	0.3666	1.131	411.3	64.3	155
1000	0.3482	1.141	424.4	66.7	168
1100	0.3166	1.159	449	71.5	195
1200	0.2902	1.175	473	76.3	224
1300	0.2679	1.189	496	82	238
1400	0.2488	1.207	530	91	303
1500	0.2322	1.230	557	100	350
1600	0.2177	1.248	584	106	390
1700	0.2049	1.267	611	113	435
1800	0.1935	1.286	637	120	482
1900	0.1833	1.207	663	128	534
2000	0.1741	1.337	689	137	589

T (K)	ρ (kg/m^e)	c_p (kJ/kg · K)	$\eta \cdot 10^7$ (N · s/m^2)	$k \cdot 10^3$ (W/m · K)	$D \cdot 10^6$ (m^2/s)
2100	0.1658	1.372	715	147	646
2200	0.1582	1.417	740	160	714
2300	0.1513	1.478	766	175	783
2400	0.1448	1.558	792	196	869
2500	0.1389	1.665	818	222	960
3000	0.1135	2.726	955	486	1570

Ammonia (NH$_3$)

T (K)	ρ	c_p	$\eta \cdot 10^7$	$k \cdot 10^3$	$D \cdot 10^6$
300	0.6894	2.158	101.5	24.7	16.6
320	0.6448	2.170	109	27.2	19.4
340	0.6059	2.192	116.5	29.3	22.1
360	0.5716	2.221	124	31.6	24.9
380	0.5410	2.254	131	34	27.9
400	0.5136	2.287	138	37	31.5
420	0.4888	2.322	145	40.4	35.6
440	0.4664	2.357	152.5	43.5	39.6
460	0.4460	2.393	159	46.3	43.4
480	0.4273	2.430	166.5	49.2	47.4
500	0.4101	2.467	173	52.5	51.9
520	0.3942	2.504	180	54.5	55.2
540	0.3795	2.540	186.5	57.5	59.7
560	0.3708	2.577	193	60.6	63.4
580	0.3533	2.613	199.5	63.8	69.1

Carbon dioxide (CO$_2$)

T (K)	ρ	c_p	$\eta \cdot 10^7$	$k \cdot 10^3$	$D \cdot 10^6$
280	1.9022	.830	140	15.20	9.63
300	1.7730	.851	149	16.55	11
320	1.6609	.872	156	18.05	12.5
340	1.5618	.891	165	19.7	14.2
360	1.4743	.908	173	21.2	15.8
380	1.3961	.926	181	22.75	17.6
400	1.3257	.942	190	24.3	19.5
450	1.1782	.981	210	28.3	24.5
500	1.0594	1.02	231	32.5	30.1
550	0.9625	1.05	251	36.6	36.2
600	0.8826	1.08	270	30.6	40.7
650	0.8143	1.10	288	44.5	49.7
700	0.7564	1.13	305	48.4	56.3
750	0.7057	1.15	321	51.7	63.7
800	0.6614	1.17	337	55.1	71.2

Carbon monoxide (CO)

T (K)	ρ	c_p	$\eta \cdot 10^7$	$k \cdot 10^3$	$D \cdot 10^6$
200	1.6888	1.045	127	17	9.63
220	1.5341	1.044	137	19	11.9
240	1.4055	1.043	147	20.6	14.1
260	1.2967	1.043	157	22.1	16.3
280	1.2038	1.042	166	23.6	18.8
300	1.1233	1.043	175	25	21.3
320	1.0529	1.043	184	26.3	23.9
340	0.9909	1.044	193	27.8	26.9
360	0.9357	1.045	202	29.1	29.8

T (K)	ρ (kg/m³)	c_p (kJ/kg · K)	$\eta \cdot 10^7$ (N · s/m²)	$k \cdot 10^3$ (W/m · K)	$D \cdot 10^6$ (m²/s)
380	0.8864	1.047	210	30.5	32.9
400	0.8421	1.049	218	31.8	36
450	0.7483	1.055	237	35	44.3
500	0.67352	1.065	254	38	53.1
550	0.61226	1.076	271	41.1	62.4
600	0.56126	1.088	286	44	72.1
650	0.51806	1.101	301	47	82.4
700	0.48102	1.114	315	50	93.3
750	0.44899	1.127	329	52.8	104
800	0.42095	1.140	343	55.5	116

Helium (He)

T (K)	ρ (kg/m³)	c_p (kJ/kg · K)	$\eta \cdot 10^7$ (N · s/m²)	$k \cdot 10^3$ (W/m · K)	$D \cdot 10^6$ (m²/s)
100	0.4871	5.193	96.3	73	28.9
120	0.4060	5.193	107	81.9	38.8
140	0.3481	5.193	118	90.7	50.2
160		5.193	129	99.2	
180	0.2708	5.193	139	107.2	76.2
200		5.193	150	115.1	
220	0.2216	5.193	160	123.1	107
240		5.193	170	130	
260	0.1875	5.193	180	137	141
280		5.193	190	145	
300	0.1625	5.193	199	152	180
350		5.193	221	170	
400	0.1219	5.193	243	187	295
450		5.193	263	204	
500	0.09754	5.193	283	220	434
550		5.193			
600		5.193	320	252	
650		5.193	332	264	
700	0.06969	5.193	350	278	768
750		5.193	364	291	
800		5.193	382	304	
900		5.193	414	330	
1000	0.04879	5.193	446	354	1400

Hydrogen (H₂)

T (K)	ρ (kg/m³)	c_p (kJ/kg · K)	$\eta \cdot 10^7$ (N · s/m²)	$k \cdot 10^3$ (W/m · K)	$D \cdot 10^6$ (m²/s)
100	0.24255	11.23	42.1	67	24.6
150	0.16156	12.6	56	101	49.6
200	0.12115	13.54	68.1	131	79.9
250	0.09693	14.06	78.9	157	115
300	0.08078	14.31	89.6	183	158
350	0.06924	14.43	98.8	204	204
400	0.06059	14.48	108.2	226	258
450	0.05386	14.5	117.2	247	316
500	0.04848	14.52	126.4	266	378
550	0.04407	14.53	134.3	285	445
600	0.04040	14.55	142.4	305	519
700	0.03463	14.61	157.8	342	676
800	0.03030	14.7	172.4	378	849
900	0.02694	14.83	186.5	412	1030

T (K)	ρ (kg/m^e)	c_p (kJ/kg $\cdot$ K)	$\eta \cdot 10^7$ (N $\cdot$ s/m^2)	$k \cdot 10^3$ (W/m $\cdot$ K)	$D \cdot 10^6$ (m^2/s)
1000	0.02424	14.99	201.3	448	1230
1100	0.02204	15.17	213	488	1460
1200	0.02020	15.37	226.2	528	1700
1300	0.01865	15.59	238.5	568	1955
1400	0.01732	15.81	250.7	610	2230
1500	0.01616	16.02	262.7	655	2530
1600	0.0152	16.28	273.7	697	2815
1700	0.0143	16.58	284.9	742	3130
1800	0.0135	16.96	296.1	786	3435
1900	0.0128	17.49	307.2	835	3730
2000	0.0121	18.25	318.2	878	3975

Nitrogen (N$_2$)

T (K)	ρ (kg/m^e)	c_p (kJ/kg $\cdot$ K)	$\eta \cdot 10^7$ (N $\cdot$ s/m^2)	$k \cdot 10^3$ (W/m $\cdot$ K)	$D \cdot 10^6$ (m^2/s)
100	3.4388	1.070	68.8	9.58	2.60
150	2.2594	1.050	100.6	13.9	5.86
200	1.6883	1.043	129.2	18.3	10.4
250	1.3488	1.042	154.9	22.2	15.8
300	1.1233	1.041	178.2	25.9	22.1
350	0.9625	1.042	200	29.3	29.2
400	0.8425	1.045	220.4	32.7	37.1
450	0.7485	1.050	239.6	35.8	45.6
500	0.6739	1.056	257.7	38.9	54.7
550	0.6124	1.065	274.7	41.7	63.9
600	0.5615	1.075	290.8	44.6	73.9
700	0.4812	1.098	321	49.9	94.4
800	0.4211	1.22	349.1	54.8	116
900	0.3743	1.146	375.3	59.7	139
1000	0.3368	1.167	399.9	64.7	165
1100	0.3062	1.187	423.2	70	193
1200	0.2807	1.204	445.3	75.8	224
1300	0.2591	1.219	466.2	81	256

Oxygen (O$_2$)

T (K)	ρ (kg/m^e)	c_p (kJ/kg $\cdot$ K)	$\eta \cdot 10^7$ (N $\cdot$ s/m^2)	$k \cdot 10^3$ (W/m $\cdot$ K)	$D \cdot 10^6$ (m^2/s)
100	3.945	.962	76.4	9.25	2.44
150	2.585	.921	114.8	13.8	5.8
200	1.930	.915	147.5	18.3	10.4
250	1.542	.915	178.6	22.6	16
300	1.284	.920	207.2	26.8	22.7
350	1.100	.929	233.5	29.6	29
400	0.9620	.942	258.2	33	36.4
450	0.8554	.956	281.4	36.3	44.4
500	0.7698	.972	303.3	41.2	55.1
550	0.6998	.988	324	44.1	63.8
600	0.6414	1.003	343.7	47.3	73.5
700	0.54989	1.031	38.8	52.8	93.1
800	0.4810	1.054	415.2	58.9	116
900	0.4275	1.074	447.2	64.9	141
1000	0.3848	1.090	477	71	169
1100	0.3498	1.103	505.5	75.8	196
1200	0.3206	1.115	532.5	81.9	229
1300	0.2960	1.125	588.4	87.1	262

T (K)	ρ (kg/m^e)	c_p (kJ/kg $\cdot$ K)	$\eta \cdot 10^7$ (N $\cdot$ s/m^2)	$k \cdot 10^3$ (W/m $\cdot$ K)	$D \cdot 10^6$ (m^2/s)
		Water vapor (steam)			
380	0.5863	2.060	127.1	24.6	20.4
400	0.5542	2.014	134.4	26.1	23.4
450	0.4902	1.980	152.5	29.9	30.8
500	0.4405	1.985	170.4	33.9	38.8
550	0.4005	1.997	188.4	37.9	47.4
600	0.3652	2.026	206.7	42.2	57
650	0.3380	2.056	224.7	46.4	66.8
700	0.3140	2.085	242.6	50.5	77.1
750	0.2931	2.119	260.4	54.9	88.4
800	0.2739	2.152	278.6	59.2	100
850	0.2579	2.186	296.9	63.7	113

Source: From Incropra and De Witt, John Wiley & Sons, New York, NY, 1990.

Diffusion Coefficients for More Gases

System	Temperature k						
	200	273.15	293.15	373.15	473.15	573.15	673.15
Large excess of air							
Ar-air		0.167	0.148	0.289	0.437	0.612	0.810
CH_4-air			0.106	0.321	0.485	0.678	0.899
CO-air			0.208	0.315	0.475	0.662	0.875
CO_2-air			0.160	0.252	0.390	0.549	0.728
H_2-air		0.668	0.627	1.153	1.747	2.444	3.238
H_2O-air			0.242	0.399	0.638	0.873	1.135
He-air		0.617	0.580	1.057	1.594	2.221	2.933
SF_6-air				0.150	0.233	0.329	0.438
Equimolar mixture							
Ar-CH_4				0.306	0.467	0.657	0.876
Ar-CO		0.168	0.187	0.290	0.439	0.615	0.815
Ar-CO_2		0.129	0.078	0.235	0.365	0.517	0.689
Ar-H_2		0.698	0.794	1.228	1.876	2.634	3.496
Ar-He	0.381	0.645	0.726	1.088	1.617	2.226	2.911
Ar-Kr	0.064	0.117	0.134	0.210	0.323	0.456	0.605
Ar-N_2		0.168	0.190	0.290	0.439	0.615	0.815
Ar-Ne	0.160	0.277	0.313	0.475	0.710	0.979	1.283
Ar-O_2		0.166	0.189	0.285	0.430	0.600	0.793
Ar-SF_6				0.128	0.202	0.290	0.389
Ar-Xe	0.052	0.095	0.108	0.171	0.264	0.374	0.498
CH_4-H_2			0.782.202	1.084	1.648	2.311	3.070
CH_4-SF_6				0.167	0.257	0.363	0.482
CO-CO_2			0.162	0.250	0.384		
CO-H_2	0.408	0.686	0.772	1.162	1.743	2.423	3.196
CO-He	0.365	0.619	0.698	1.052	1.577	2.188	2.882
CO-Kr		0.131	0.581	0.227	0.346	0.485	0.645
CO-N_2	0.133	0.208	0.231	0.336	0.491	0.673	0.878
CO-O_2			0.202	0.307	0.462	0.643	0.849
CO-SF_6				0.167	0.257	0.363	0.482
CO_2-C_3H_8			0.084	0.133	0.209		
CO_2-H_2	0.315	0.552	0.412	0.964	1.470	2.066	2.745
CO_2-H_2O			0.162	0.292	0.496	0.741	1.021
CO_2He	0.300	0.513	0.400	0.878	1.321		
CO_2-N_2			0.160	0.253	0.392	0.553	0.733
CO_2-N_2O	0.055	0.099	0.113	0.177	0.276		
CO_2-Ne	0.131	0.227	0.199	0.395	0.603	0.847	
CO_2-O_2			0.159	0.248	0.380	0.535	0.710
CO_2-SF_6				0.099	0.155		
D_2-H_2	0.631	1.079	1.219	1.846	2.778	3.866	5.103
H_2-He	0.775	1.320	1.490	2.255	3.394	4.726	6.242
H_2-Kr	0.340	0.601	0.682	1.053	1.607	2.258	2.999
H_2-N_2	0.408	0.686	0.772	1.162	1.743	2.423	3.196
H_2-Ne	0.572	0.982	0.317	1.684	2.541	3.541	4.677
H_2-SF_6		0.513	0.122	0.890	1.349	1.885	2.493

System	Temperature k						
	200	273.15	293.15	373.15	473.15	573.15	673.15
H_2-Xe		0.513	0.122	0.890	1.349	1.885	2.493
H_2O-O_2			0.244	0.403	0.645	0.882	1.147
He-Kr	0.330	0.559	0.629	0.942	1.404	1.942	2.550
He-N_2	0.365	0.619	0.698	1.052	1.577	2.188	2.882
He-Ne	0.563	0.948	1.066	1.592	2.362	3.254	4.262
He-O_2		0.641	0.697	1.092	1.640	2.276	2.996
He-SF_6			1.109	0.592	0.871	1.190	1.545
He-Xe	0.282	0.478	0.538	0.807	1.201	1.655	2.168
Kr-N_2		0.131	0.149	0.227	0.346	0.485	0.645
Kr-Ne	0.131	0.228	0.258	0.392	0.587	0.812	1.063
Kr-Xe	0.035	0.064	0.073	0.116	0.181	0.257	0.344
N_2-Ne			0.258	0.483	0.731	1.021	1.351
N_2-O_2			0.202	0.307	0.462	0.643	0.849
N_2-SF_6				0.148	0.231	0.328	0.436
N_2-Xe		0.107	0.123	0.188	0.287	0.404	0.539
Ne-Xe	0.111	0.193	0.219	0.332	0.498	0.688	0.901
O_2-SF_6			0.097	0.154	0.238	0.334	0.441

Source: Lide, *Handbook of Chemistry and Physics,* 82nd ed., CRC Press, Boca Raton, 2001–2002.

Permittivity (Dielectric Constant) of Gases

This table gives the relative permittivity ε (often called the *dielectric constant*) of some common gases at a temperature of 20°C and pressure of one atmosphere (101.325 kPa). Values of the permanent dipole moment μ in Debye units (1 D = 3.33564×10^{-30} C m) are also included.

The density dependence of the permittivity is given by the equation

$$\frac{\varepsilon - 1}{\varepsilon + 2} = \rho_m \left(\frac{4\pi N\alpha}{3} + \frac{4\pi N\mu^2}{9kT} \right)$$

where ρ_m is the molar density, N is Avogadro's number, k is Boltzmann's constant, T is the temperature, and α is the molecular polarizability. Therefore, in regions where the gas can be considered ideal, $\varepsilon - 1$ is approximately proportional to the pressure at constant temperature. For nonpolar gases ($\mu = 0$), $\varepsilon - 1$ is inversely proportional to temperature at constant pressure.

The number of significant figures indicates the accuracy of the values given. The values of for air, Ar, H_2, He, N_2, O_2, and CO_2 are recommended as reference values; these are accurate to 1 ppm or better.

The second part of the table, derived from Birnbaum and Chatterjee, gives the permittivity of water vapor in equilibrium with liquid water as a function of temperature.[4]

Molecular Formula	Name	ε	μ/D
	Compounds Not Containing Carbon		
	Air	(dry, CO_2-free)	1.0005364
Ar	Argon	1.0005172	0
BF_3	Boron trifluoride	1.0011	0
BrH	Hydrogen bromide	1.00279	0.827
ClH	Hydrogen chloride	1.00390	1.109
F_3N	Nitrogen trifluoride	1.0013	0.235
F_6S	Sulfur hexafluoride	1.00200	0
HI	Hydrogen iodide	1.00214	0.448
H_2	Hydrogen	1.0002538	0
H_2S	Hydrogen sulfide	1.00344	0.97
H_3N	Ammonia	1.00622	1.471
He	Helium	1.0000650	0
Kr	Krypton	1.00078	0
NO	Nitric oxide	1.00060	0.159
N_2	Nitrogen	1.0005480	0
N_2O	Nitrous oxide	1.00104	0.161
Ne	Neon	1.00013	0
O_2	Oxygen	1.0004947	0
O_2S	Sulfur dioxide	1.00825	1.633
O_3	Ozone	1.0017	0.534
Xe	Xenon	1.00126	0
	Compounds Containing Carbon		
CF_4	Tetrafluoromethane	1.00121	0
CO	Carbon monoxide	1.00065	0.110
CO_2	Carbon dioxide	1.000922	0
CH_3Br	Bromomethane	1.01028	1.822
CH_3Cl	Chloromethane	1.01080	1.892
CH_3F	Fluromethane	1.00973	1.858
CH_3I	Iodomethane	1.00914	1.62
CH_4	Methane	1.00081	0
C_2H_2	Acetylene	1.00124	0
C_2H_3Cl	Chloroethylene	1.0075	1.45
C_2H_4	Ethylene	1.00134	0
C_2H_5Cl	Chloroethane	1.01325	2.05
C_2H_6	Ethane	1.00140	0
C_2H_6O	Dimethyl ether	1.0062	1.30
C_3H_6	Propene	1.00228	0.366
C_3H_6	Cyclopropane	1.00178	0
C_3H_8	Propane	1.00200	0.084
C_4H_{10}	Butane	1.00258	0
C_4H_{10}	Isobutane	1.00260	0.132

Permittivity of Saturated Water Vapor

$t/C°$	ε	$t/C°$	ε
0	1.00007	60	1.00144
10	1.00012	70	1.00213
20	1.00022	80	1.00305
30	1.00037	90	1.00428
40	1.00060	100	1.00587
50	1.00095		

Source: From Lide, *Handbook of Chemistry and Physics*, 82nd ed., CRC Press, Boca Raton, 2001–2002.

REFERENCES

1. Maryott, A.A. and Buckley, F., *Table of Dielectric Constants and Electric Dipole Moments in the Gaseous State*, National Bureau of Standards Circular 537, 1953.
2. Younglove, B.A., *J. Phys. Chem. Ref. Data*, 11, Suppl. 1, 1982; 16, 577, 1987 (for data on N_2, H_2, O_2, and hydrocarbons over a range of pressure and temperature).
3. Landolt-Börnstein, *Numerical Data and Functional Relationships in Science and Technology*, New Series, Group IV, Vol. 4, Springer-Verlag, Heidelberg, 1980 (for data at high pressures).
4. Birnbaum, G., and Chatterjee, S.K., *J. Appl. Phys.*, 23, 220, 1952 (for data on water vapor).

Psychrometric Chart at Sea Level

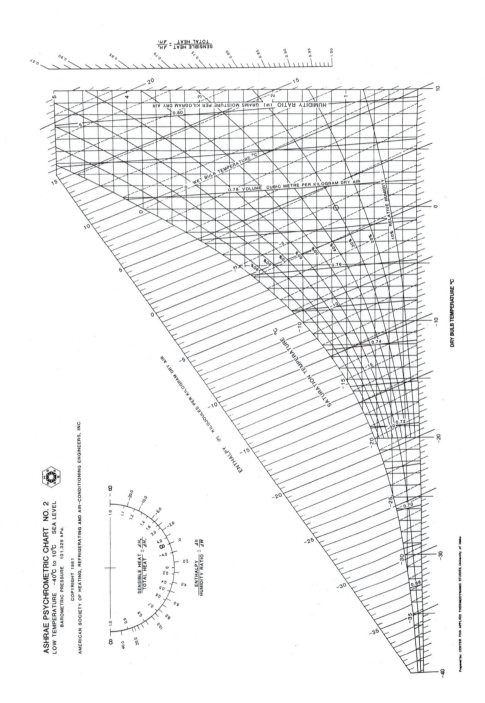

ASHRAE PSYCHROMETRIC CHART NO. 2
LOW TEMPERATURE   -40°C to 10°C   SEA LEVEL
BAROMETRIC PRESSURE 101.325 kPa.

COPYRIGHT 1981
AMERICAN SOCIETY OF HEATING, REFRIGERATING AND AIR-CONDITIONING ENGINEERS, INC.

Prepared by: CENTER FOR APPLIED THERMODYNAMIC STUDIES, University of Idaho

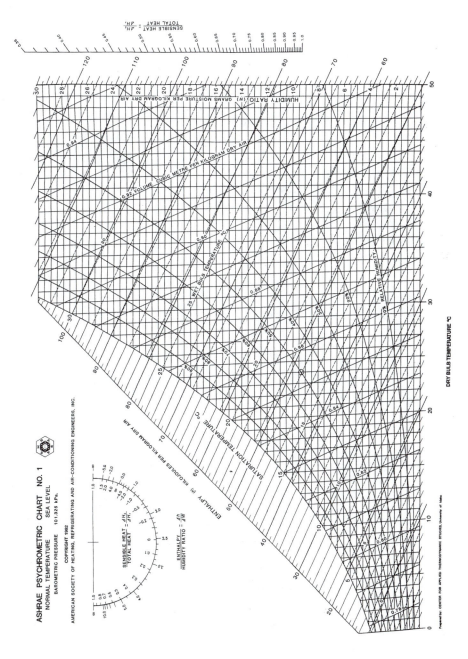

ASHRAE PSYCHROMETRIC CHART NO. 1
NORMAL TEMPERATURE   SEA LEVEL
BAROMETRIC PRESSURE   101.325 kPa.
COPYRIGHT 1992
AMERICAN SOCIETY OF HEATING, REFRIGERATING AND AIR-CONDITIONING ENGINEERS, INC.

DRY BULB TEMPERATURE °C

Prepared by: CENTER FOR APPLIED THERMODYNAMIC STUDIES, University of Idaho

Correlation Constants of Activated Carbon for VOC Adsorption*

---

* *Source:* Yaws et al., Determining VOC adsorption capacity, *Pollution Engineering,* February, 1995.

Formula	Name	$\log_{10} M_a = a_1 + a_2 \log_{10} C + a_3 [\log_{10} C]^2$			Min	Max	(M_a - g of compound/100 g of carbon, C - ppmv in air at 25 C and 1 atm)		
		a_1	a_2	a_3			M_a @ 10 ppmv	M_a @ 100 ppmv	M_a @ 1000 ppmv
$CBrCl_3$	Bromotrichloromethane	1.39842	0.23228	-0.02184	10	10000	40.63	59.65	79.20
$CBrF_3$	Bromotrichfluoromethane	-1.46247	0.58361	-0.01044	10	10000	0.13	0.46	1.56
CBr_2F_2	Dibromodifluoromethane	0.82076	0.30701	-0.01384	10	10000	13.00	23.96	41.42
CBr_3F	Tribromofluoromethane	-1.43748	0.55503	-0.00450	10	10000	0.13	0.45	1.54
CCl_2F_2	Dichlorodifluoromethane	-0.07350	0.40145	-0.01404	10	10000	2.06	4.71	10.10
CCl_2O	Phosgene	-0.64469	0.60428	-0.02986	10	10000	0.85	2.78	7.93
CCl_3F	Trichlorofluoromethane	0.17307	0.40715	-0.01915	10	10000	3.64	8.14	16.68
CCl_3NO_2	Chloropicrin	1.26745	0.20841	-0.01288	10	10000	29.04	42.93	59.81
CCl_4	Carbon tetrachloride	1.07481	0.28186	-0.02273	10	10000	21.57	35.29	51.98
$CHBr_3$	Tribromomethane	1.73184	0.19948	-0.02246	10	7238	81.07	109.89	134.33
$CHCl_3$	Chloroform	0.67102	0.36148	-0.02288	10	10000	10.22	20.07	35.45
CHN	Hydrogen cyanide	-4.39245	1.08948	-0.00740	10	10000	4.9E-04	5.7E-03	6.4 E-02
CH_2BrCl	Bromochloromethane	0.61399	0.41353	-0.02531	10	10000	10.05	21.87	42.35
CH_2BrF	Bromofluoromethane	0.45483	0.36332	-0.01606	10	10000	6.34	13.10	25.13
CH_2Br_2	Dibromomethane	1.08376	0.37211	-0.03238	10	10000	26.52	49.94	81.04
CH_2Cl_2	Dichloromethane	-0.07043	0.49210	-0.02276	10	10000	2.51	6.65	15.89
CH_2I_2	Diiodomethane	1.94756	0.14984	-0.01947	10	1583	119.66	147.70	166.67
CH_2O	Formaldehyde	-2.48524	0.69123	-0.00375	10	10000	1.6E-02	7.6E-02	0.36
CH_2O_2	Formic acid	-1.77731	1.09503	-0.06354	10	10000	0.18	1.44	8.63
CH_3Br	Methyl bromide	-1.23835	0.78564	-0.05521	10	10000	0.31	1.29	4.19
CH_3Cl	Methyl chloride	-1.91871	0.62053	-0.00549	10	10000	5.0E-02	0.20	0.78
CH_3Cl_3Si	Methyl trichlorosilane	1.07198	0.24275	-0.01911	10	10000	19.75	30.27	42.48
CH_3I	Methyl iodide	0.73997	0.32985	-0.01330	10	10000	11.39	22.21	40.72
CH_3NO	Formamide	1.30981	0.25274	—	10	80	36.52	—	—
CH_3NO_2	Nitromethane	-0.32847	0.70602	-0.05111	10	10000	2.12	7.57	21.36
CH_4	Methane	-4.31008	0.77883	-0.00628	10	10000	2.9E-04	1.7E-03	9.3E-03
CH_4Cl_2Si	Methyl dichlorosilane	0.73271	0.29305	-0.01822	10	10000	10.17	17.62	28.04
CH_4O	Methanol	-1.96739	0.82107	-0.01393	10	10000	6.9E-02	0.42	2.35

Formula	Name								
CH_4S	Methyl mercaptan	-1.12288	0.60573	-0.02094	10	10000	0.29	1.01	3.21
CH_5N	Methylamine	-1.93548	0.64710	-0.01057	10	10000	5.0E-02	0.21	0.81
CN_4O_8	Tetranitromethane	1.49047	0.18181	-0.01894	10	10000	45.01	60.02	73.35
CO	Carbon monoxide	-5.18782	0.90121	-0.01358	10	10000	5.0E-05	3.6 E-04	2.5 E-03
COS	Carbonyl sulfide	-1.42882	0.51061	0.00028	10	10000	0.12	0.39	1.27
CO_2	Carbon dioxide	-3.65224	0.80180	-0.00328	10	10000	1.4 E-03	8.7 E-03	5.3 E-02
CS_2	Carbon disulfide	-0.18899	0.47093	-0.01481	10	10000	1.85	4.94	12.32
$C_2Br_2F_4$	1,2-Dibromotetrafluoroethane	0.90388	0.25693	-0.00974	10	10000	14.16	23.92	38.64
C_2ClF_5	Chloropentafluoroethane	0.08264	0.34756	-0.01343	10	10000	2.61	5.30	10.10
$C_2Cl_3F_3$	1,1,2-Trichlorodifluoroethane	1.27368	0.18656	-0.01231	10	10000	28.05	39.59	52.79
C_2Cl_4	Tetrachloroethylene	1.40596	0.20802	-0.02097	10	10000	39.17	54.72	69.39
$C_2Cl_4F_2$	1,1,2,2-Tetrachlorodifluoroethane	1.37307	0.17625	-0.01465	10	10000	34.25	46.45	58.88
$C_2HBrClF_3$	Halothane	0.92405	0.31204	-0.02004	10	10000	16.45	29.38	47.84
C_2HCl_3	Trichloroethylene	1.02411	0.29929	-0.02539	10	10000	19.86	33.20	49.38
C_2HCl_3O	Dichloroacetyl chloride	1.23647	0.26219	-0.02596	10	10000	29.70	45.39	61.57
C_2HCl_3O	Trichloroacetaldehyde	1.17362	0.26971	-0.02513	10	10000	26.19	40.98	57.09
C_2HCl_5	Pentachloroethane	1.64566	0.13515	-0.01572	10	4829	58.22	71.30	81.22
$C_2HF_3O_2$	Trifluoroacetic acid	-0.12577	0.59373	-0.03445	10	10000	2.71	8.39	22.15
C_2H_2	Acetylene	-2.24177	0.82454	-0.03390	10	10000	3.5 E-02	0.19	0.84
$C_2H_2Br_4$	1,1,2,2-Tetrabromoethane	—	—	—	—	—	146.05	—	—
$C_2H_2Cl_2$	1,1-Dichloroethylene	0.48740	0.33282	-0.01622	10	10000	6.37	12.25	21.87
$C_2H_2Cl_2$	cis-1,2- Dichloroethylene	0.47567	0.39061	-0.02554	10	10000	6.93	14.28	26.16
$C_2H_2Cl_2$	trans-1,2-Dichloroethylene	0.47567	0.39061	-0.02554	10	10000	6.93	14.28	26.16
$C_2H_2Cl_2O_2$	Dichloroacetic acid	1.69237	0.09630	—	10	235	61.47	76.73	—
$C_2H_2Cl_4$	1,1,1,2-Tetrachloroethane	1.44097	0.19166	-0.01995	10	10000	40.99	55.52	68.61
$C_2H_2Cl_4$	1,1,2,2-Tetrachloroethane	1.52322	0.17848	-0.02019	10	6073	48.03	63.01	75.33
C_2H_3Cl	Vinyl chloride	-0.98889	0.66564	-0.04320	10	10000	0.43	1.48	4.16
C_2H_3ClO	Acetyl chloride	0.03627	0.45526	-0.02093	10	10000	2.96	7.30	16.35
$C_2H_3ClO_2$	Methyl chloroformate	0.41186	0.42776	-0.02776	10	10000	6.48	14.33	27.88
$C_2H_3Cl_3$	1,1,1-Trichloroethane	0.97331	0.28737	-0.02277	10	10000	17.29	28.64	42.70
$C_2H_3Cl_3$	1,1,2-Trichloroethane	1.17163	0.27791	-0.02746	10	10000	26.43	41.46	57.31
C_2H_3N	Acetonitrile	-0.79666	0.63512	-0.02598	10	10000	0.65	2.34	7.50
C_2H_3NO	Methyl isocyanate	-1.07579	0.85881	-0.06876	10	10000	0.52	2.33	7.62

$$\text{Log}_{10} M_a = a_1 + a_2 \log_{10} C + a_3 [\log_{10} C]^2$$

Formula	Name	a_1	a_2	a_3	Min	Max	M_a @ 10 ppmv	M_a @ 100 ppmv	M_a @ 1000 ppmv
C_2H_4	Ethylene	-2.27102	0.61731	-0.01467	10	10000	2.1 E-02	8.0 E-02	0.28
$C_2H_4Br_2$	1,1-Dibromoethane	1.37260	0.25671	-0.02516	10	10000	40.19	61.01	82.48
$C_2H_4Br_2$	1,2-Dibromoethane	1.44231	0.25500	-0.02666	10	10000	46.84	70.09	92.77
$C_2H_4Cl_2$	1,1-Dichloroethane	0.54485	0.36091	-0.02192	10	10000	7.65	15.10	26.93
$C_2H_4Cl_2$	1,2-Dichloroethane	0.55343	0.37072	-0.02161	10	10000	7.99	16.16	29.59
$C_2H_4Cl_2O$	Bis(chloromethyl)ether	0.95599	0.33784	-0.03200	10	10000	18.27	31.89	48.04
$C_2H_4F_2$	1,2-Difluoroethane	-3.97902	2.51862	-0.31617	10	10000	0.02	0.62	5.39
C_2H_4O	Acetaldehyde	-1.17047	0.62766	-0.02475	10	10000	0.27	0.97	3.09
C_2H_4O	Ethylene oxide	-2.42379	0.94878	-0.04062	10	10000	0.03	0.20	1.14
$C_2H_4O_2$	Acetic acid	-0.05553	0.68410	-0.06071	10	10000	3.70	11.74	28.21
$C_2H_4O_2$	Methyl formate	-0.99586	0.61693	-0.01847	10	10000	0.40	1.46	4.88
C_2H_4S	Thiacyclopropane	0.02258	0.45520	-0.02154	10	10000	2.86	7.03	15.64
C_2H_5Br	Bromoethane	0.31783	0.43549	-0.03072	10	10000	5.28	11.64	22.27
C_2H_5Cl	Ethyl chloride	-0.50828	0.50364	-0.02179	10	10000	0.94	2.58	6.40
C_2H_5ClO	2-Chloroethanol	0.74164	0.46933	-0.05158	10	9446	14.43	29.78	48.46
C_2H_5I	Ethyl iodide	1.00356	0.32123	-0.02405	10	10000	19.99	35.47	56.33
C_2H_5N	Ethyleneimine	-1.16912	0.91238	-0.07400	10	10000	0.47	2.29	7.98
C_2H_5NO	N-Methylformamide	1.23333	0.21723	—	10	333	28.22	46.54	—
$C_2H_5NO_2$	Nitroethane	0.44968	0.49708	-0.04612	10	10000	7.96	18.17	33.56
C_2H_6	Ethane	-2.40393	0.68107	-0.01925	10	10000	1.8E-02	7.6E-02	0.29
C_2H_6O	Ethanol	-0.51153	0.67525	-0.04473	10	10000	1.32	4.57	12.93
C_2H_6OS	Dimethyl sulfoxide	1.24042	0.31302	-0.04768	10	802	32.05	47.40	—
$C_2H_6O_2$	Ethylene glycol	1.40474	0.18738	-0.02663	10	121	36.77	47.09	—
$C_2H_6O_4S$	Dimethyl sulfate	1.34617	0.21539	-0.02336	10	890	34.53	48.25	—
C_2H_6S	Dimethyl sulfide	0.48472	0.37358	-0.02770	10	10000	6.77	13.22	22.71
C_2H_6S	Ethyl mercaptan	0.00552	0.40506	-0.01802	10	10000	2.47	5.54	11.44
$C_2H_6S_2$	Dimethyl disulfide	0.75878	0.35928	-0.02953	10	10000	12.26	22.87	37.23
C_2H_7N	Dimethylamine	-1.22492	0.63962	-0.03266	10	10000	0.24	0.84	2.51

Formula	Name								
C₂H₇NO	Monoethanolamine	1.21569	0.21994	—	10	485	27.27	45.24	—
C₂H₈N₂	Ethylenediamine	0.56504	0.46307	-0.04789	10	10000	9.55	19.94	33.36
C₃H₃Cl	Propargyl chloride	0.27135	0.40480	-0.02135	10	10000	4.52	9.90	19.66
C₃H₃N	Acrylonitrile	0.07669	0.49986	-0.03500	10	10000	3.48	8.64	18.25
C₃H₃NO	Oxazole	0.63350	0.30620	-0.02350	10	10000	8.25	14.19	21.91
C₃H₄	Methylacetylene	-2.52865	1.74715	-0.21635	10	10000	0.10	1.26	5.83
C₃H₄Cl₂	2,3-Dichloropropene	0.95417	0.30034	-0.02614	10	10000	16.92	28.20	41.68
C₃H₄O	Acrolein	-0.29632	0.49437	-0.02471	10	10000	1.49	3.92	9.21
C₃H₄O	Propargyl alcohol	0.22971	0.57711	-0.05441	10	10000	5.65	14.67	29.61
C₃H₄O₂	Acrylic acid	0.75549	0.47108	-0.05615	10	5221	14.81	29.72	46.06
C₃H₄O₃	Pyruvic acid	1.07410	0.41414	-0.05768	10	1679	26.95	46.95	62.72
C₃H₅Br	3-Bromo-1-propene	0.84815	0.32392	-0.02398	10	10000	14.06	25.12	40.19
C₃H₅Cl	3-Chloropropene	0.32792	0.36553	-0.01853	10	10000	4.73	9.66	18.10
C₃H₅ClO	alpha-Epichlorohydrin	0.83203	0.38983	-0.03932	10	10000	15.22	28.47	44.42
C₃H₅ClO₂	Methyl chloroacetate	1.07657	0.32514	-0.03617	10	9959	23.20	38.21	53.27
C₃H₅ClO₂	Ethyl chloroformate	0.94901	0.32529	-0.03201	10	10000	17.47	29.62	43.33
C₃H₅Cl₃	1,2,3-Trichloropropane	1.47241	0.18136	-0.02165	10	4843	42.87	56.04	66.32
C₃H₅I	3-Iodo-1-propene	1.33634	0.24222	-0.02271	10	10000	35.96	53.70	72.21
C₃H₅N	Propionitrile	0.05925	0.51747	-0.03781	10	10000	3.46	8.77	18.68
C₃H₅NO	Hydracrylonitrile	1.50994	0.11037	—	10	105	41.72	53.79	—
C₃H₅NO	Lactonitrile	1.44156	0.12689	—	10	157	37.02	49.58	—
C₃H₆	Propylene	-0.93674	0.57775	-0.03853	10	10000	0.40	1.16	2.82
C₃H₆Cl₂	1,1-Dichloropropane	0.95379	0.28791	-0.02487	10	10000	16.48	26.92	39.24
C₃H₆Cl₂	1,2-Dichloropropane	0.98872	0.28700	-0.02571	10	10000	17.78	28.83	41.43
C₃H₆Cl₂	1,3-Dichloropropane	1.10340	0.27837	-0.02824	10	10000	22.57	35.25	48.35
C₃H₆Cl₂	2,2-Dichloropropane	0.85314	0.29432	-0.02255	10	10000	13.33	22.47	34.13
C₃H₆O	Acetone	-0.14546	0.47497	-0.02286	10	10000	2.03	5.16	11.85
C₃H₆O	Allyl alcohol	0.32390	0.49368	-0.04370	10	10000	5.94	13.69	25.80
C₃H₆O	n-Propionaldehyde	0.05519	0.49738	-0.04331	10	10000	3.23	7.53	14.37
C₃H₆O	1,2-Propylene oxide	-0.42829	0.53858	-0.02757	10	10000	1.21	3.46	8.70
C₃H₆O	1,3-Propylene oxide	-0.50421	0.51872	-0.02296	10	10000	0.98	2.76	7.00
C₃H₆O₂	Ethyl formate	0.12618	0.42260	-0.02090	10	10000	3.37	7.72	16.06
C₃H₆O₂	Methyl acetate	0.13314	0.42849	-0.02188	10	10000	3.47	7.99	16.66

| Formula | Name | $\log_{10} M_a = a_1 + a_2 \log_{10} C + a_3 [\log_{10} C]^2$ | | | | | M_a - g of compound/100 g of carbon, C - ppmv in air at 25 C and 1 atm | | |
		a_1	a_2	a_3	Min	Max	M_a @ 10 ppmv	M_a @ 100 ppmv	M_a @ 1000 ppmv
$C_3H_6O_2$	Propionic acid	0.77846	0.44570	-0.05209	10	4872	14.86	28.94	44.34
$C_3H_6O_2S$	3-Mercaptopropionic acid	1.68823	0.05916	—	10	66	55.90	—	—
$C_3H_6O_3$	Lactic acid	1.60722	0.09225	—	10	107	50.06	61.90	—
$C_3H_6O_3$	Methoxyacetic acid	1.61885	0.08873	—	10	71	51.00	—	—
C_3H_6S	Thiacyclobutane	0.67420	0.37225	-0.03151	10	10000	10.35	19.62	32.16
C_3H_7Br	1-Bromopropane	0.83601	0.32406	-0.02407	10	10000	13.68	24.43	39.04
C_3H_7Br	2-Bromopropane	0.81137	0.31043	-0.02155	10	10000	12.60	22.18	35.37
C_3H_7Cl	Isopropyl chloride	0.31428	0.34779	-0.01661	10	10000	4.42	8.78	16.15
C_3H_7Cl	n-Propyl chloride	0.40133	0.34678	-0.01931	10	10000	5.36	10.41	18.53
C_3H_7I	Isopropyl iodide	1.26456	0.24157	-0.02122	10	10000	30.54	46.01	62.85
C_3H_7I	n-Propyl iodide	1.30623	0.24227	-0.02250	10	10000	33.57	50.21	67.69
C_3H_7N	Allylamine	0.16250	0.39815	-0.02105	10	10000	3.46	7.49	14.71
C_3H_7N	Propyleneimine	0.06919	0.43529	-0.02293	10	10000	3.03	7.05	14.75
C_3H_7NO	N,N-Dimethylformamide	0.90253	0.37875	-0.04523	10	5220	17.22	30.14	42.83
$C_3H_7NO_2$	1-Nitropropane	0.91328	0.34648	-0.03730	10	10000	16.69	28.64	41.40
$C_3H_7NO_2$	2-Nitropropane	0.83248	0.35732	-0.03608	10	10000	14.25	25.28	37.99
C_3H_8	Propane	-0.79460	0.49029	-0.02398	10	10000	0.47	1.23	2.89
C_3H_8O	Isopropanol	0.27183	0.46419	-0.03682	10	10000	5.00	11.30	21.53
C_3H_8O	n-Propanol	0.38644	0.48033	-0.04505	10	10000	6.63	14.69	26.42
$C_3H_8O_2$	2-Methoxyethanol	0.74339	0.41792	-0.04536	10	10000	13.06	24.99	38.81
$C_3H_8O_2$	Methylal	0.19079	0.38167	-0.01775	10	10000	3.59	7.64	15.00
$C_3H_8O_2$	1,2-Propylene glycol	1.48275	0.11594	—	10	170	39.69	51.84	—
$C_3H_8O_2$	1,3-Propylene glycol	1.58563	0.08395	—	10	58	46.73	—	—
C_3H_8S	n-Propylmercaptan	0.59031	0.31407	-0.02190	10	10000	7.63	13.52	21.65
C_3H_8S	Isopropyl mercaptan	0.55779	0.31539	-0.02051	10	10000	7.12	12.78	20.86
C_3H_8S	Ethyl-methyl-sulfide	0.62830	0.31889	-0.02320	10	10000	8.39	14.90	23.78
C_3H_9N	n-Propylamine	0.05768	0.34918	-0.01241	10	10000	2.48	5.09	9.85
C_3H_9N	Isopropylamine	0.07464	0.37106	-0.01568	10	10000	2.69	5.68	11.14

Formula	Name								
C_3H_9N	Trimethylamine	-0.09422	0.32583	-0.00337	10	10000	1.69	3.50	7.13
C_3H_9NO	1-Amino-2-propanol	1.25496	0.27456	-0.04254	10	617	30.69	43.04	—
C_3H_9NO	3-Amino-1-propanol	1.53733	0.08156	—	10	101	41.58	50.17	—
C_3H_9NO	Methylethanolamine	1.14745	0.30208	-0.04243	10	1422	25.53	38.18	46.97
$C_3H_9O_3P$	Trimethyl-phosphite	1.00568	0.21001	-0.01402	10	10000	15.91	23.42	32.32
$C_3H_9O_4P$	Trimethyl phosphate	1.48463	0.16933	-0.02290	10	1196	42.76	53.91	61.17
$C_3H_{10}N_2$	1,2-Propanediamine	0.90237	0.31904	-0.03400	10	10000	15.40	25.38	35.77
C_4H_4O	Furan	0.04084	0.40613	-0.01620	10	10000	2.70	6.14	12.99
$C_4H_4O_2$	Diketene	0.87430	0.37094	-0.03962	10	10000	16.06	28.69	42.71
C_4H_4S	Thiophene	0.80753	0.32166	-0.02654	10	10000	12.67	22.12	34.17
C_4H_5Cl	Chloroprene	0.72957	0.29786	-0.02111	10	10000	10.15	17.41	27.11
C_4H_5N	trans-Chrotonitrile	0.70791	0.37284	-0.03756	10	10000	11.05	20.11	30.79
C_4H_5N	cis-Chrotonitrile	0.58131	0.39427	-0.03590	10	10000	8.70	16.84	27.61
C_4H_5N	Methacrylonitrile	0.46655	0.38890	-0.03042	10	10000	6.68	13.26	22.88
C_4H_5N	Pyrrole	0.83128	0.38413	-0.04217	10	10000	14.90	26.97	40.19
C_4H_5N	Vinylacetonitrile	0.61844	0.40844	-0.04032	10	10000	9.70	18.80	30.26
$C_4H_5NO_2$	Methyl cyanoacetate	1.56587	0.09143	—	10	180	45.43	56.07	—
C_4H_6	1,3-Butadiene	-0.03359	0.34764	-0.01297	10	10000	2.00	4.07	7.81
C_4H_6	Dimethylacetylene	-0.06673	0.39387	-0.01524	10	10000	2.05	4.57	9.50
C_4H_6	Ethylacetylene	-0.02918	0.33636	-0.01056	10	10000	1.98	3.99	7.67
$C_4H_6Cl_2$	1,3-Dichloro-trans-2-butene	1.30208	0.19939	-0.02091	10	10000	30.24	41.42	51.53
$C_4H_6Cl_2$	1,4-Dichloro-cis-2-butene	1.40119	0.17876	-0.02041	10	5383	36.27	47.54	56.73
$C_4H_6Cl_2$	1,4-Dichloro-trans-2-butene	1.40904	0.18120	-0.02179	10	4503	37.02	48.34	57.09
$C_4H_6Cl_2$	3,4-Dichloro-1-butene	1.23394	0.21476	-0.02135	10	10000	26.75	37.85	48.54
C_4H_6O	trans-Crotonaldehyde	0.68353	0.36560	-0.03400	10	10000	10.35	19.00	29.81
C_4H_6O	2,5-Dihydrofuran	0.33990	0.40041	-0.02451	10	10000	5.20	11.03	20.92
C_4H_6O	Methacrolein	0.43461	0.37019	-0.02474	10	10000	6.03	11.91	21.01
$C_4H_6O_2$	gamma-Butyrolactone	1.29434	0.29719	-0.04658	10	592	35.07	50.39	—
$C_4H_6O_2$	cis-Crotonic acid	1.30871	0.25008	-0.03752	10	773	33.21	45.58	—
$C_4H_6O_2$	Methacrylic acid	1.23099	0.27648	-0.03903	10	1282	29.41	42.44	51.19
$C_4H_6O_2$	Methyl acrylate	0.45869	0.32104	-0.02001	10	10000	5.75	10.49	17.45
$C_4H_6O_2$	Vinyl acetate	0.61067	0.34797	-0.02595	10	10000	8.56	15.95	26.36
$C_4H_6O_3$	Acetic anhydride	1.07388	0.31083	-0.03575	10	7274	22.33	35.69	48.37

| Formula | Name | $\log_{10} M_a =$ $a_1 + a_2 \log_{10} C + a_3 [\log_{10} C]^2$ | | | | | M_a - g of compound/100 g of carbon, C - ppmv in air at 25 C and 1 atm) | | |
		a_1	a_2	a_3	Min	Max	M_a @ 10 ppmv	M_a @ 100 ppmv	M_a @ 1000 ppmv
C_4H_7N	n-Butyronitrile	0.64311	0.38787	-0.03822	10	10000	9.83	18.45	29.02
C_4H_7N	Isobutyronitrile	0.56697	0.38807	-0.03531	10	10000	8.31	15.92	25.90
C_4H_7NO	3-Methoxypropionitrile	1.13283	0.28534	-0.03732	10	2598	24.04	35.83	44.98
C_4H_8	1-Butene	0.07313	0.32701	-0.01452	10	10000	2.43	4.67	8.38
$C_4H_8Br_2$	1,2-Dibromobutane	1.69234	0.12766	-0.01497	10	4094	63.83	77.23	87.22
$C_4H_8Br_2$	2,3-Dibromobutane	1.68176	0.12916	-0.01492	10	4934	62.52	75.93	86.10
$C_4H_8Cl_2$	1,4-Dichlorobutane	1.38278	0.17796	-0.02030	10	5427	34.71	45.45	54.20
C_4H_8O	n-Butyraldehyde	0.45056	0.37372	-0.02689	10	10000	6.27	12.31	21.36
C_4H_8O	Isobutyraldehyde	0.39315	0.36715	-0.02379	10	10000	5.45	10.77	19.08
C_4H_8O	1,2-Epoxybutane	0.36719	0.37654	-0.02360	10	10000	5.25	10.61	19.25
C_4H_8O	Methyl ethyl ketone	0.46525	0.37688	-0.02801	10	10000	6.52	12.79	22.07
C_4H_8O	Ethyl vinyl ether	0.33311	0.33471	-0.01711	10	10000	4.47	8.59	15.25
C_4H_8O	Tetrahydrofuran	0.29856	0.35648	-0.01550	10	10000	4.36	8.90	16.93
$C_4H_8O_2$	Isobutyric acid	1.14021	0.29004	-0.03833	10	2388	24.66	36.89	46.27
$C_4H_8O_2$	n-Butyric acid	1.22589	0.26481	-0.03737	10	1244	28.40	40.37	48.30
$C_4H_8O_2$	1,4-Dioxane	0.66781	0.36208	-0.03034	10	10000	9.99	18.65	30.27
$C_4H_8O_2$	Ethyl acetate	0.63612	0.34441	-0.02691	10	10000	8.99	16.49	26.74
$C_4H_8O_2$	Methyl propionate	0.64273	0.34862	-0.02767	10	10000	9.20	16.96	27.52
$C_4H_8O_2$	n-Propyl formate	0.65855	0.34340	-0.02750	10	10000	9.43	17.19	27.62
$C_4H_8O_2S$	Sulfolane	1.77762	0.00118	0.00005	10	10000	60.10	60.28	60.48
C_4H_8S	Tetrahydrothiophene	0.93777	0.33197	-0.03877	10	10000	17.02	27.97	38.44
C_4H_9Br	1-Bromobutane	1.16698	0.24380	-0.02270	10	10000	24.44	36.62	49.44
C_4H_9Br	2-Bromobutane	1.13872	0.24481	-0.02203	10	10000	22.99	34.69	47.30
C_4H_9C	n-Butyl chloride	0.80024	0.29114	-0.02370	10	10000	11.69	19.40	28.87
C_4H_9Cl	sec-Butyl chloride	0.75046	0.29132	-0.02212	10	10000	10.46	17.56	26.63
C_4H_9Cl	tert-Butyl chloride	0.68673	0.28529	-0.01944	10	10000	8.97	15.12	23.32
C_4H_9N	Pyrrolidine	0.60693	0.36363	-0.03004	10	10000	8.72	16.37	26.76
C_4H_9NO	N,N-Dimethylacetamide	1.20026	0.25124	-0.03263	10	2631	26.23	37.34	45.74

Formula	Compound								
C$_4$H$_9$NO	Morpholine	1.00673	0.30572	-0.03294	10	10000	19.03	30.65	42.41
C$_4$H$_{10}$	n-Butane	0.03071	0.34304	-0.01596	10	10000	2.28	4.50	8.25
C$_4$H$_{10}$	Isobutane	-0.01676	0.33495	-0.01274	10	10000	2.02	4.00	7.47
C$_4$H$_{10}$O	n-Butanol	0.89881	0.32534	-0.03648	10	9276	15.41	25.33	35.20
C$_4$H$_{10}$O	sec-Butanol	0.76814	0.34611	-0.03478	10	10000	12.01	20.95	31.15
C$_4$H$_{10}$O	Diethyl ether	0.23477	0.36044	-0.02236	10	10000	3.74	7.35	13.03
C$_4$H$_{10}$O	Methyl propyl ether	0.36764	0.32893	-0.01787	10	10000	4.77	9.00	15.62
C$_4$H$_{10}$O	Methyl isopropyl ether	0.36373	0.31940	-0.01647	10	10000	4.64	8.64	14.92
C$_4$H$_{10}$O	Isobutanol	0.84818	0.33155	-0.03559	10	10000	13.94	23.38	33.30
C$_4$H$_{10}$O$_2$	1,3-Butanediol	—	—	—	—	—	48.89	—	—
C$_4$H$_{10}$O$_2$	1,4-Butanediol	—	—	—	—	—	51.19	—	—
C$_4$H$_{10}$O$_2$	2,3-Butanediol	1.50642	0.09239	—	10	239	39.70	49.11	—
C$_4$H$_{10}$O$_2$	1,2-Dimethoxyethane	0.74981	0.31330	-0.02616	10	10000	10.89	18.70	28.46
C$_4$H$_{10}$O$_2$	2-Ethoxyethanol	1.07911	0.27792	-0.03199	10	6983	21.14	32.14	42.16
C$_4$H$_{10}$O$_4$S	Diethyl sulfate	1.64797	0.05805	—	10	278	50.82	58.08	—
C$_4$H$_{10}$S	n-Butyl mercaptan	0.98086	0.24388	-0.02251	10	10000	15.93	23.91	32.35
C$_4$H$_{10}$S	Isobutyl mercaptan	0.93709	0.24802	-0.02179	10	10000	14.57	22.18	30.55
C$_4$H$_{10}$S	sec-Butyl mercapatan	0.92287	0.24856	-0.02146	10	10000	14.12	21.59	29.88
C$_4$H$_{10}$S	tert-Butyl mercaptan	0.84380	0.24937	-0.01939	10	10000	11.85	18.41	26.15
C$_4$H$_{10}$S	Diethyl sulfide	0.95993	0.24465	-0.02195	10	10000	15.23	22.98	31.35
C$_4$H$_{10}$S	Isopropyl methyl sulfide	0.92769	0.24689	-0.02132	10	10000	14.23	21.69	29.95
C$_4$H$_{10}$S	Methyl propyl sulfide	0.97217	0.24453	-0.02229	10	10000	15.65	23.55	32.00
C$_4$H$_{10}$S$_2$	Diethyl disulfide	1.42594	0.12564	-0.01373	10	5627	34.50	41.91	47.79
C$_4$H$_{11}$N	n-Butylamine	0.64570	0.31857	-0.02570	10	10000	8.68	15.14	23.45
C$_4$H$_{11}$N	Isobutylamine	0.60137	0.31538	-0.02369	10	10000	7.82	13.72	21.59
C$_4$H$_{11}$N	sec-Butylamine	0.57706	0.31254	-0.02250	10	10000	7.36	12.95	20.52
C$_4$H$_{11}$N	tert-Butylamine	0.50036	0.30334	-0.01897	10	10000	6.09	10.74	17.37
C$_4$H$_{11}$N	Diethylamine	0.54770	0.30799	-0.02107	10	10000	6.83	12.01	19.14
C$_4$H$_{11}$NO	Dimethylethanolamine	1.18381	0.23249	-0.02857	10	4183	24.42	34.24	42.09
C$_4$H$_{12}$Si	Tetramethylsilane	0.70867	0.23089	-0.01505	10	10000	8.40	12.89	18.45
C$_4$H$_{13}$N$_3$	Diethylene triamine	1.54270	0.06479	—	10	308	40.50	47.02	—

Index*

A

Absolute roughness, 331
Absorption, 336, 399–400, 414–415
ACH, *see* Air changes per hour (ACH)
Acid deposition, 6
Acid precipitation, 3
Acid rain, 2–3, 7
Activated carbon, 401–402, 405–406
Activation energy, 433
Active techniques, tracer gas methods, 453
Adhesion, 268–271, *269–271*
Adiabatic condition, 234
Adsorption, 398–403, 436
Aerodynamic air cleaners
    basics, 311–312
    collection efficiency, *326,* 326–329, *328–329*
    collector performance criteria, 345–346
    laminar flow, *321,* 321–326, *323–324*
    particle separation efficiency, 312–317, 321–326
    pressure drop, 317–319, *319,* 329–333, *330, 332*
    return cyclones, 312–319, *313*
    spray chamber scrubbers, 336–342, *337, 342*
    uniflow cyclones, 319–333
    venturi scrubber, 342–345, *343*
    wet scrubbers, 333–342
Aerodynamic diameter, 21, 91–94, *93*
Aerosols, 7
Aerospace, 299
Air age, *452,* 452–453, 460–465, *469*
Airborne pollutant, 1
Air changes per hour (ACH), 453, 489
Air cleaning efficiency, 449–452, *451*
Air cleaning impact, *537,* 537–538
Air distribution, 498–516
Airflow-measurement hood, 478–479, *479*
Air jets, 185–188, *186–187*

Air leakage, 441, 485, 527–533
Air quality
    basics, 1, 6–7
    chemical composition, 1, *2*
    indoor, 4–6
    outdoor, 2–4
    units, 7
    volumetric content, 1, *2*
Air samplers, 207–209, *208–209*
Air turnover rate, 453
Allen and Raabe studies, 84
Allergens, *25, see also* Pet allergies
Alternaria, 28, 32, *see also* Molds
Alvarez, Novick and, studies, 203
Ammonia-water system, 418, *418*
Analytical method, gaseous pollutants control, 425–430, *426*
Animal facilities
    allergies, 32–33
    basics, 12, *12, 25, 27,* 32–33
    incomplete mixing airspaces, 259, *259,* 495
    leakage of air, 530
    particle production rate, 260–261, *261*
    particulate matter, 5–6
    wet uniflow cyclone, 345, *346*
    zonal ventilation conditions, 497
Animal protein, *25*
Anisokinetic sampling, 211–220
Annular tubes, 161–163, *162*
Antisneakage, 391
Arendt and Kallmann studies, 366–367
Arithmetic mean speed, 110
Arrestance, 292–293
Arrhenius equation, 438
Asbestos, *25,* 25–26
Aspergillus, 28, *see also* Molds
Aspiration, 18, 207–208
Aureobasidium, 28, *see also* Molds
Averages, 43

* *Page numbers in italics reference figures.*

603

Y

Z